Algorithms for Intelligent Systems

This book series publishes research on the analysis and development of algorithms for intelligent systems with their applications to various real world problems. It covers research related to autonomous agents, multi-agent systems, behavioral modeling, reinforcement learning, game theory, mechanism design, machine learning, meta-heuristic search, optimization, planning and scheduling, artificial neural networks, evolutionary computation, swarm intelligence and other algorithms for intelligent systems.

The book series includes recent advancements, modification and applications of the artificial neural networks, evolutionary computation, swarm intelligence, artificial immune systems, fuzzy system, autonomous and multi agent systems, machine learning and other intelligent systems related areas. The material will be beneficial for the graduate students, post-graduate students as well as the researchers who want a broader view of advances in algorithms for intelligent systems. The contents will also be useful to the researchers from other fields who have no knowledge of the power of intelligent systems, e.g. the researchers in the field of bioinformatics, biochemists, mechanical and chemical engineers, economists, musicians and medical practitioners.

The series publishes monographs, edited volumes, advanced textbooks and selected proceedings.

More information about this series at http://www.springer.com/series/16171

Amit Sheth · Amit Sinhal · Abhinav Shrivastava ·
Amit Kumar Pandey
Editors

Intelligent Systems

Proceedings of SCIS 2021

Springer

Editors
Amit Sheth
Artificial Intelligence Institute
University of South Carolina
Columbia, SC, USA

Abhinav Shrivastava
University of Maryland
College Park, MD, USA

Amit Sinhal
JK Lakshmipat University
Jaipur, India

Amit Kumar Pandey
BeingAI Limited/Socients AI and Robotics
Paris, France

ISSN 2524-7565 ISSN 2524-7573 (electronic)
Algorithms for Intelligent Systems
ISBN 978-981-16-2247-2 ISBN 978-981-16-2248-9 (eBook)
https://doi.org/10.1007/978-981-16-2248-9

This Springer imprint is published by the registered company Springer Nature Singapore Pte Ltd.
The registered company address is: 152 Beach Road, #21-01/04 Gateway East, Singapore 189721, Singapore

Preface

This book is a collection of selected papers presented in International Conference on Sustainable Computing and Intelligent Systems (SCIS 2021), held at JK Lakshmipat University, Jaipur, India, during 5–6 February 2021. It includes novel and innovative work from experts, practitioners, scientists, and decision-makers from academia and industry. It covers chapters in the areas of intelligent systems, sustainable computing, artificial intelligence, and related topics. Sustainability includes energy efficiency, natural resources preservation, and use of multiple energy sources as needed in computing devices and infrastructure. It also includes a spectrum of related research issues such as applications of computing that can have ecological and societal impacts.

The objective of the conference was to bring together researchers from academia, government agencies, research laboratories, and the corporate sector from all over the world to present their research works carried out in intelligent systems and sustainable computing domain. Following are the four major tracks in which scholarly contributions were invited from the participants:

- Intelligent systems,
- Robotics,
- Computing and modelling,
- Computing for sustainability, health, and social good.

The conference could get very good response from all over the globe. As a result, 196 full-length papers were received from various participants, and 46 papers have been finally selected for the presentation in the conference, in four tracks under eight technical sessions, to be chaired by 16 senior academics and industry practitioners. In addition, many seasoned academics and industry professionals from India and abroad contributed to the academic deliberations of the conference.

We are thankful to Soft Computing Research Society (SCRS) for providing technical sponsorship and Springer for publication support. We extend our sincere thanks to all sponsors, speakers, reviewers, committee members, and participants for the success of the conference.

We hope that this conference proceedings will prove to be useful.

Columbia, USA
Jaipur, India
College Park, USA
Paris, France

Amit Sheth
Amit Sinhal
Abhinav Shrivastava
Amit Kumar Pandey

Contents

About the Editors

Prof. Amit Sheth is Educator, Researcher, and Entrepreneur. Prior to his joining the University of South Carolina as the founding director of the university-wide AI Institute, he was the LexisNexis Ohio Eminent Scholar and executive director of Ohio Center of Excellence in Knowledge-enabled Computing. He is Fellow of IEEE, AAAI, and AAAS. He is among the highly cited computer scientists worldwide. He has (co-)founded four companies, three of them by licensing his university research outcomes, including the first Semantic Web Company in 1999 that pioneered technology similar to what is found today in Google Semantic Search and Knowledge Graph. He is particularly proud of his students' exceptional success in academia and industry research laboratories and as entrepreneurs.

Prof. Amit Sinhal is working as Professor with the Department of Computer Science and Engineering at Institute of Engineering and Technology, JK Lakshmipat University, Jaipur. He has received his Bachelor of Engineering (B.E.) in the domain of Computer Engineering from Sardar Vallabhbhai National Institute of Technology (SVNIT), Surat. He did his Masters (CSE) and Doctorate (CSE) from Rajeev Gandhi Technical University, Bhopal (M.P.). Dr. Sinhal has more than two decades of experience including research, teaching, and IT industry. He worked in Atlanta (USA) for on-site project of CoreCard Inc. His research interest includes software engineering, soft computing, AI, ML, and NLP. He filed two patents and serving as Editor-in-Chief of *International Journal IJETAE*. He has to his credit four book chapters and one book with international publisher. He organized many international conferences & FDPs and published many research papers in international journals. Dr. Amit Sinhal is also the recipient of Best Academician of the Year 2018 at Mumbai, Best Faculty, and Best Head of the Department in the previous organizations. He is Life Member of ISTE, CSI, IAENG, AMLE, and CSTA.

Dr. Abhinav Shrivastava is Assistant Professor of Computer Science at University of Maryland with a joint appointment in the Institute of Advanced Computer Studies (UMIACS). Before that, he was a visiting research scientist at Google AI. He completed his Ph.D. in robotics from Carnegie Mellon University in 2017, where he was Microsoft Research Fellow. He serves as an area chair for CVPR 2018–2019/2021, ECCV 2018, WACV 2021, and AAAI 2021. His research is supported by DARPA (MediFor, SemaFor, GARD, SAIL-ON), IARPA (DIVA), Air Force (2x STTR), and gifts from Honda Research, Adobe, and Facebook Research. His research focuses on a wide variety of artificial intelligence topics, including computer vision, machine learning, and robotics. His research has been widely covered by international press, such as CNN, BBC, Forbes, and the Associated Press; and one of his projects, NEIL, was awarded the top-10 ideas in 2013 by CNN.

Dr. Amit Kumar Pandey (Ph.D., Robotics and AI) is robotics and AI scientist. He is Co-Founder of Being AI limited and serving as Chief AI Officer. He has served as President, Chief Science Officer (CSO), and CTO of Hanson Robotics, the creator of expressive humanoid robot, Sophia. He was Chief Scientist at SoftBank Robotics Europe, Paris, France, the creator of the mass produced sociable humanoid robots, Pepper and Nao. He worked as Researcher in Robotics and AI at LAAS-CNRS (French National Center for Scientific Research), Toulouse, France. His research interest includes Robotics, AI, and Society, addressing societal needs to achieve innovation through scientific advancements, new technologies, and ecosystem creation. He was appointed as General Chair of 28th IEEE International Ro-Man 2019 conference. He is also Founding Coordinator of Socially Intelligent Robots and Societal Applications (SIRo-SA) Topic Group (TG) of euRobotics (the European Union Robotics Community), contributing in the Multi-Annual Roadmap for robotics in Europe. He has served as PI of several European Union Horizon 2020 collaborative projects in Robotics and AI for Healthcare, Education, and Services, published 60+ research papers, and delivered talks in 100+ international venues. He is serving as Robot Design Competition Chair of International Conference on Social Robotics. His other recognitions include Best three Ph.D. theses in Robotics in Europe for the prestigious Georges Giralt Award by euRobotics; Innovation Leadership Award; Global Excellence Award; Achievers Award; Pravashi Bihari Samman Puruskar.

Chapter 1
Ontological Simulation of Quality Test Concepts in Seafood Domain

Vinu Sherimon, Alaa Ismaeel, Puliprathu Cherian Sherimon, Winny Anna Varkey, and B. Naveen

1 Introduction

In our society, lifestyle and food habits have changed considerably due to the increasing economic and technological developments. The level of nutrients, proteins, and dietary fiber are at times low in the typical day-to-day diet in some nations. Additionally, there is a decrease in the number of proactive tasks as well. All these have driven an increase in diseases such as obesity and hypertension [1]. The single most vital food one can consume for good health is seafood [2]. In this way, it is important to examine and comprehend the quality and tests that the fish goes through. The worldwide utilization of seafood has been growing a seemingly endless amount of time. Henceforth, research studies concerning their quality affirmation and dietary advantages are evolving day by day. The ecological dangers of fish include contamination due to heavy metals and bacteria. So, it is crucial to safeguard the quality of seafood. While accepting the raw material and during the processing stages, the seafood quality tests are conducted. This research paper presents the ontology model of seafood.

V. Sherimon (✉)
Department of IT, University of Technology and Applied Sciences, 74 Muscat, Sultanate of Oman

A. Ismaeel · P. C. Sherimon
Faculty of Computer Studies, Arab Open University, 1596 Muscat, Sultanate of Oman
e-mail: alaa.ismaeel@aou.edu.om

P. C. Sherimon
e-mail: sherimon@aou.edu.om

A. Ismaeel
Faculty of Science, Minia University, Minya 61519, Egypt

W. A. Varkey · B. Naveen
Saintgits College of Engineering, Kottayam 686532, Kerala, India

A. Sheth et al. (eds.), *Intelligent Systems*, Algorithms for Intelligent Systems,
https://doi.org/10.1007/978-981-16-2248-9_1

Ontology presents a standard platform of the thesaurus for the researchers to share knowledge in a domain. The fundamental concepts in the field and the associations between concepts are part of the ontology. Machines can interpret the ontology [3]. It is possible to overcome the limitations of keyword-based search using ontologies. They represent a critical part in the processing and sharing of knowledge. Ontology remains a vital backbone to semantic search. They provide a common and collaborative understanding of a domain between people and systems. Many information retrieval systems use ontologies to boost their performance [4].

The ontology describes several concepts in a domain and the relationships between those concepts. The ontology includes elements including individuals (objects instances), classes, attributes, relationships, restrictions, rules, etc. It is possible to represent knowledge using formal specifications such as vocabularies, taxonomies, and relational database schemas. But, since ontology facilitates the users to link multiple concepts to other concepts in numerous ways, it remains at the top of the list [5]. Automated data reasoning is also available in ontology since the relationships between the classes and concepts are expressed in the ontology. That is why ontology gained the fame and attention of many researchers all over the world [6].

The application of ontology for quality assurance has more success than taking surveys, integrated reviews, or cross-sectional studies. As ontology keeps on growing, it is possible to further refine in the future. More and more modifications can be made in ontology depending on need. An ontology defines relationships and semantics in meaningful ways [7]. Protégé is chosen as a tool because it is one of the best methods for implementing ontology.

The backbone of our system is ontology built in Protégé software which will assure the seafood quality. Even though many research and related works have been carried out for the quality assurance of seafood, this paper focuses its effectiveness not only in checking the quality of seafood but also on the importance and advantage in building ontology in Protégé software using OWL language. Laboratory technologists check several parameters in the quality tests conducted in seafood during different phases of processing (either in-company or government-approved laboratories). If the test values do not match with the quality criteria of the country of export (either EU or non-EU countries), it may lead to the rejection of seafood [8]. Here comes the significance of this seafood ontology, where the system shows the technologist if there is any deviation in the test results. It also confirms the correctness of the data. This paper illustrates the implementation of ontology.

The objective of this article is to introduce the seafood ontology for quality assurance and to enhance the information on various tests available for different seafood categories. The article illustrates the different steps in the development process of the seafood ontology as well as the various classes along with its properties.

The remainder of the paper is ordered as follows: Sect. 2 presents the ontology modeling. This section describes the ontology classes, subclasses, relationships, object properties, and data properties. Section 3 presents the results, and the Conclusion and Future is given in Sect. 4.

2 Ontology Modeling of Quality Tests in Seafood Domain

Numerous tests are done at different stages in a seafood processing lifecycle, to ensure its quality. The three primary tests performed on the samples are organoleptic, microbiological, and chemical tests. These tests provide the details of heavy metal content, pesticide content, texture, presence of microorganisms, etc. Assessing the microbiological quality of seafood has a crucial role in the seafood processing chain. Microbiological quality assessment of the seafood depends on the processing conditions within the seafood chain [4]. As pointed out earlier, it is quite easy to formally describe knowledge through an ontology using various concepts. Ontologies ensure a common understanding of information, and with it, they make explicit domain assumptions.

2.1 Ontology Classes and Subclasses

We have used Protégé 5.5 to model the Ontology.

Protégé is a free, open-source software that acts as an ontology editor and a knowledge management system. This software helps in the easy creation of ontologies. Taxonomies assist to organize the domain knowledge. The implementation process has many steps and can be represented in many ways by different tools in Protégé. Figure 1 shows the taxonomy of the seafood ontology using Protégé. We use classes, the building blocks of ontology to explain the domain concepts.

Fish and *Test* are two of the main classes. The subclasses of the *Test* class are *chemical, microbiological*, and *organoleptic*. Usually, we conduct quality tests on raw material and frozen products. Blanched frozen, cooked frozen, and uncooked frozen are the subcategories of frozen products. The specifications/standards are different for internal and external tests. Figure 2 depicts the subclass of the microbiological class. We add all the significant microbiological tests under this class. We include five main fish families in this study. Under each such fish family, we include the different fishes that belong to the corresponding family.

2.2 Object Properties, Data Properties and Instances

The binary relationship between two instances has been described as object properties. The object characteristics of seafood ontology are shown in Fig. 3. As shown in Fig. 4, different tests of the fish have been included as a data property for this ontology (including bony, cephalopods, crustaceans, molluscus and scombridae).

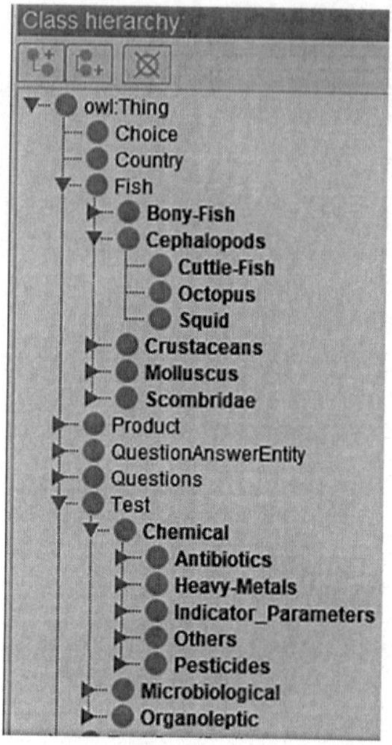

Fig. 1 Taxonomy of seafood ontology

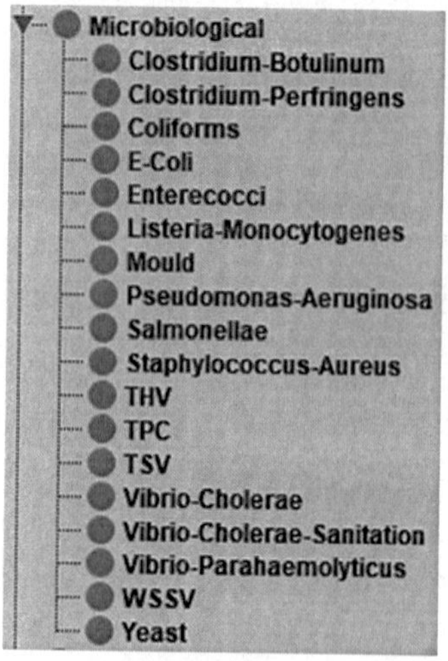

Fig. 2 Subclasses of microbiological class

2.3 *Representation of Ontology Relationships in OWLVIZ*

Protégé developed in OWL language renders many helpful visualization tools to specify the classes, their properties, and instances. It covers the pictorial representation using OWLViz and OntoGraf.

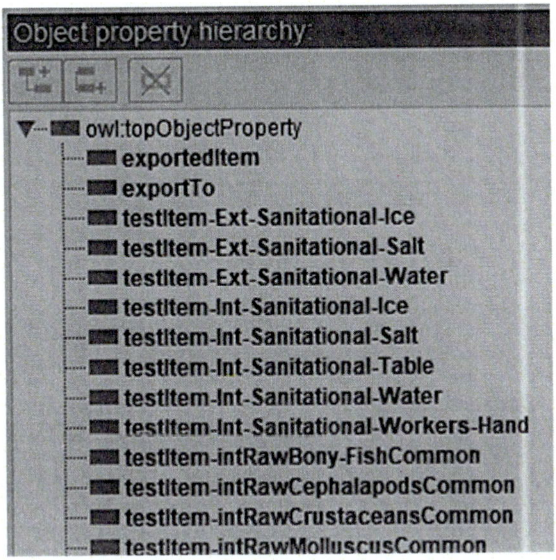

Fig. 3 Object property

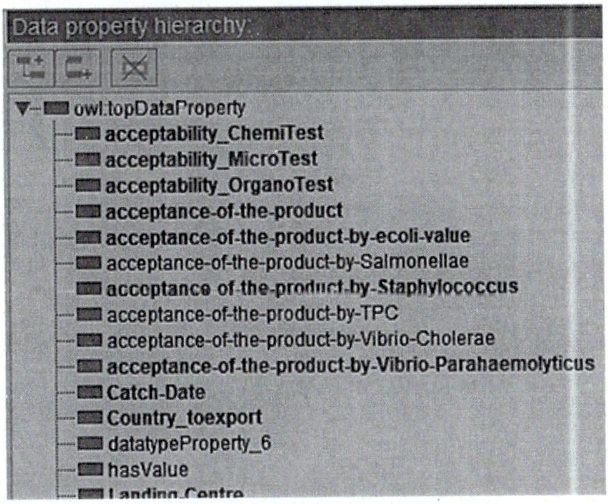

Fig. 4 Data property

Figure 5 portrays the Level 1 class hierarchy of seafood ontology in OWLViz which permits the viewing and incremental navigation of ontology.

The Level 2 class hierarchy representation and subclass representation of organoleptic test in OWLViz are shown in Figs. 6, 7. It allows not only the comparison of the asserted and inferred class hierarchy but also in differentiating the primitive and defined classes. The OWLViz tool of Protégé 5.5 aids to represent the class hierarchy and their relationships in a visualized manner [9]. The OWL Viz tool in Protégé 5.5 requires an open-source graph visualization software named Graphviz for the representation of structural information as the diagrams of a graph.

Fig. 5 Level 1 representation of the seafood ontology using OWLViz

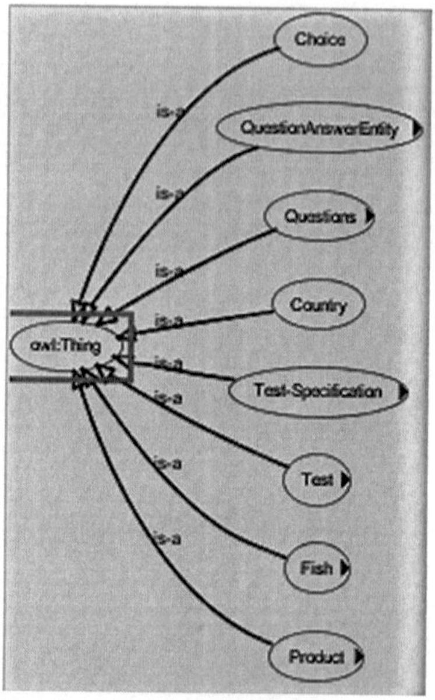

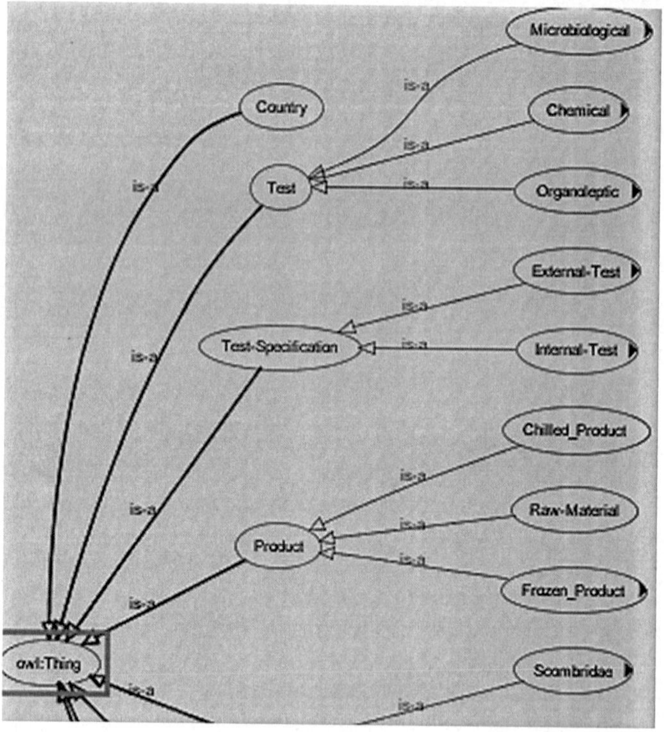

Fig. 6 Level 2 class hierarchy representation in OWLViz

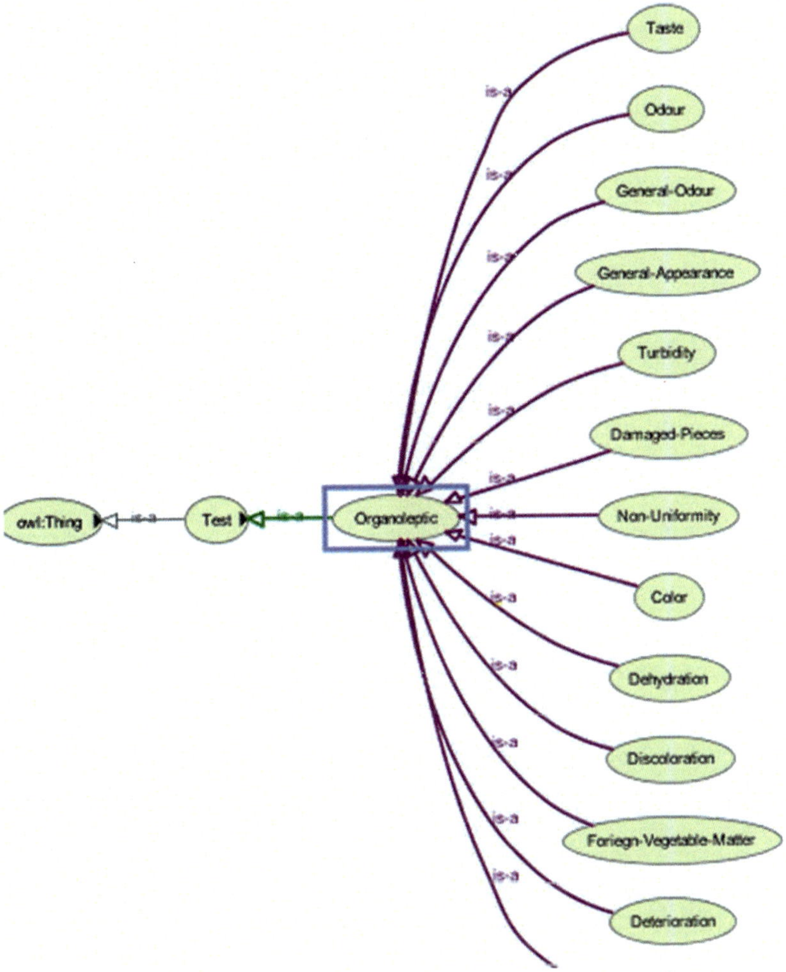

Fig. 7 OWLViz representation of subclasses of Organoleptic class

2.4 *Representation of Ontology Relationship in OntoGraf*

OntoGraf is a tool that requires different layouts to unify the construction of the Protége OWL application [10]. OntoGraf supports various relationships, including entity, equivalence, subclass, and properties of objects. It also shows the relation between classes and subclasses, but only for smaller ontologies OntoGraf may apply and is restricted with the expressiveness of the RDFS [11]. It is also possible for OWLViz.

OntoGraf enables a definition to be selected and interpreted in connection with a neighboring class from the defined class hierarchy. OntoGraf supports many formats

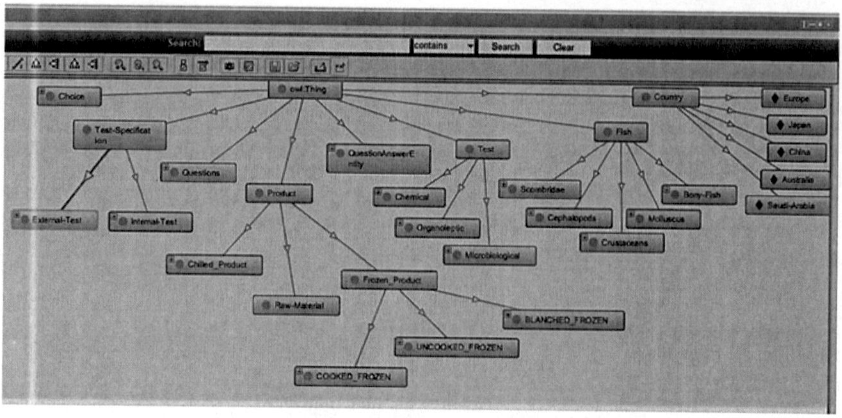

Fig. 8 Class hierarchy representation in OntoGraf

to organize the ontological structure automatically. Figure 8 shows the OntoGraf class hierarchy which supports interactive navigation of the seafood ontology relationship.

3 Results and Discussion

Protégé tool, SPARQL Query [12], facilitates the easy extraction of information from a large dataset with high performance. Figure 9 displays the query to list the different types of tests required for the Crab.

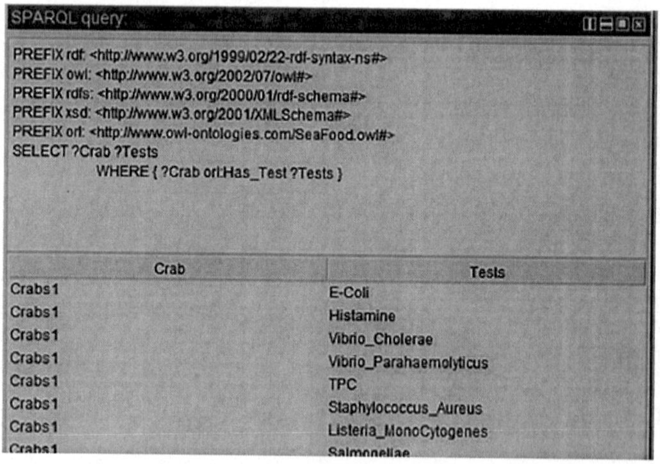

Fig. 9 SPARQL query which lists the tests of crab

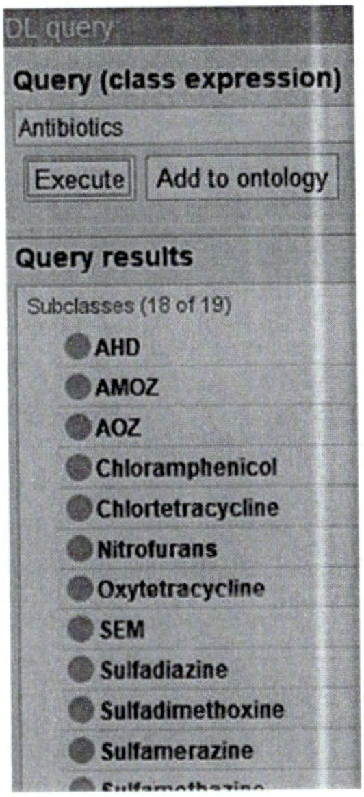

Fig. 10 DL query to display the instances of microbiological tests

The DL query tab, which is a powerful tool, made searching for a classified ontology effective. Figure 10 shows a basic query to show the instances of a class such as microbiological. This allows us to check for any class expression and returns its result with its subclasses and instances (if any). The Protégé 5.5 DL Query tab offers an interface to query and evaluate any class expression in a fraction of time. In the figure, the output was obtained as subclasses for each class expression. This indeed is really a quick and efficient method for searching a classified ontology.

4 Conclusion and Future

The marine and its quality prediction will have an everlasting scope in the future. Safety, quality, and nutritional benefits should be preserved, and proper management of food products should be assured. Seafarers must be encouraged to export because they have been fulfilling human beings' food security needs for centuries. Many studies have shown that outstanding fats, proteins, minerals, and vitamins can best be obtained through the consumption of seafood. The role of seafood in the maintenance and improvement of people's health should thus be strengthened to ensure the quality of the seafood.

The creation of the ontology of seafood is presented in this paper. The classes and subclasses are implemented in Protégé. Besides, the class hierarchy representations are shown in OWLViz and OntoGraf, their data characteristics, object characteristics, instances, and results obtained using SPARQL and DL Query are also shown. The potential future scope of this paper includes the ontology regulations used to enforce several fish family test descriptions and to implement the full quality assurance framework of seafood. Further, with the definite ontology, we can update and correct whenever future needs arise such as future development in technologies.

Funding The research leading to these results has received funding from the Research Council (TRC) of the Sultanate of Oman under the Block Funding Program BFP/RGP/ICT/18/113.

References

1. Jensen IJ (2014) Health benefits of seafood consumption-with special focus on household preparations and bioactivity in animal models
2. Seafood. https://en.wikipedia.org/wiki/Seafood. Accessed 03 Aug 2020
3. Noy NF, McGuinness DL (2001) Ontology development 101: a guide to creating your first ontology
4. Vinu PV, Sherimon PC, Krishnan R (2015) Modeling of test specifications of raw materials in seafood ontology using semantic web rule language (SWRL). In: Proceedings of the 2015 international conference on advanced research in computer science engineering & technology
5. What are ontologies? https://www.ontotext.com/knowledgehub/fundamentals/what-are-ontologies. Accessed 03 Aug 2020
6. Vinu PV, Sherimon PC, Reshmy K (2012) Development of seafood ontology for semantically enhanced information retrieval. Int J Comput Eng Technol
7. Vinu PV, Sherimon PC, Krishnan R (2014) Development of ontology for seafood quality assurance system. J Convergence Inf Technol 9(1):25
8. Vinu PV, Sherimon PC, Reshmy K (2013) Knowledge-base driven framework for assuring the quality of marine seafood export. Int J Artif Intell Knowl Discov 2(3):6–10
9. Cai Z, Shi K, Yang H (2015) A novel visualization for ontologies of semantic web representation. In: 2015 international conference on computational intelligence and communication networks (CICN). IEEE
10. Ramakrishnan S, Vijayan A (2014) A study on development of cognitive support features in recent ontology visualization tools. Artif Intell Rev 41(4):595–623
11. Dudáš M, Zamazal O, Svátek V (2014) Roadmapping and navigating in the ontology visualization landscape. In: International conference on knowledge engineering and knowledge management. Springer, Cham
12. Kremen P, Sirin E (2008) SPARQL-DL implementation experience. OWLED (Spring)

Chapter 2
Moroccan Stock Market Prediction Using LSTM Model on a Daily Data

Abdelhadi Ifleh and Mounime El Kabbouri

1 Introduction

Financial markets (FM) are one of the best methods to grow investors' money [1]. Toward to make bigger returns, investors have to predict accurately the upcoming price of assets. But FM is complex and noisy, and no one can detect when to buy or sell a stock [2–4].

Generally, there are two common approaches to forecast future price of SM [5, 6], the fundamental analysis and the technical analysis (TA). The first approach, the fundamental one, is founded on studying the features that impact the firm activities and determine the intrinsic value.

The decision of buying or shorting an asset is due to the comparing of the actual value of the asset to its fundamental value. Investors proceed to buy the stock if fundamental value is greater than the actual one and vice versa.

The second approach is the TA, which is an empirical technique that can predict asset prices. It is founded on the examination of the historical price to forecast the upcoming. Furthermore, it provides investors information before becoming public and helps them in investigating the behavior of the investors on the market.

There are a variety of technical indicators, and all of them are calculated using the high, open, low, close, and volume of the asset.

Due to the complexity and instability of FM, traditional forecasting methods cannot lead investors to estimate the future prices anymore. Hence, new methods of artificial intelligence (AI) are proposed to extract the useful information from the

A. Ifleh (✉) · M. El Kabbouri
Finance, Audit and Organizational, Governance Research Laboratory, National School of Commerce and Management, Settat, Hassan First University of Settat, Settat, Morocco
e-mail: a.ifleh@uhp.ac.ma

M. El Kabbouri
e-mail: mounime.elkabbouri@uhp.ac.ma

A. Sheth et al. (eds.), *Intelligent Systems*, Algorithms for Intelligent Systems,
https://doi.org/10.1007/978-981-16-2248-9_2

FM. There are multiple algorithms that might be mobilized in FM, and each model has its merits and restrictions [1].

In this article, we are going to use one of the most effective strength models in the AI, LSTM model, to estimate the close price of AttijariWafa Bank.

Our work will be presented as follows. Previous studies related to LSTM, AI, and FM are going to be presented in Sect. 2. The suggested LSTM methodology and predicting performance measures explained in Sect. 3. In Sect. 4, we would interpret the results of the model. Finally, a conclusion and the future works are described in Sect. 5.

2 Literature Review

In literature, researchers use various methods of AI to estimate the price of assets, and they are looking for the algorithm that has the highest accuracy. To decide when buying and shorting the stocks in Sao Paulo SM, Brasileiro et al. [7] used K-nearest neighbor. Chang et al. [5] proposed a model capable of recognizing buying and selling signals by combining piecewise linear representation and artificial neural network (ANN). Agrawal et al. [8] employed optimal LSTM to predict the evolution of SM and compared this model with other models like event LSTM, support vector machine (SVM), and linear regression (LR). Moreover, they reached that optimal LSTM has an accuracy upper than other models.

Qiu and Song [9] combined ANN with genetic algorithm (GA) to predict the Nikkie 225 index. Sezer et al. [10] employed neural networks (NN) to develop decision support systems capable of identifying buying and selling signals. Zhang and Wu [11] also combined backpropagation neural network and improved bacterial chemo taxis optimization to predict S&P 500, and the model provides a good accuracy.

Pang et al. [12] used LSTM and ELSTM to predict the Shanghai A-shares index, and the accuracies are 57.2% and 56.9%, respectively. Sethia and Raut [13] proposed LSTM, gated recurrent unit (GRU), ANN, and support vector regression (SVR) to predict the close price of S&P 500. They concluded that LSTM and GRU had an accuracy bigger than other models.

Almost a majority of previous studies discussed the developed markets, so the implementation of the LSTM on the Moroccan case will help Moroccan researchers and traders to make decisions in FM and to predict emergent markets.

3 Methodology

The data in this work consist of the daily closing prices of AttijariWafa Bank extracted from investing.com, and our data series cover the period going from 18/09/2008 to 02/03/2020. To build our model, we are going to use the LSTM model, and our model splits the data into 80% for training and the other 20% for testing. For training, we

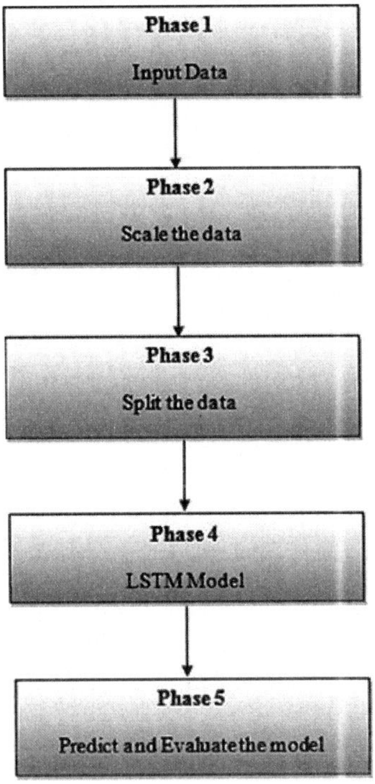

Fig. 1 Proposed method

use Adam optimizer and mean squared error as a loss function. Also, we used 40 epochs for training data. Our proposed method is divided into five main steps as shown in Fig. 1. We aim to predict the close prices with high accuracy in order to make more profit. In phase 2, we scale the data called in the phase 1; because in the time series of stock markets, the oldest observations are lower than the newest which will impact the calculating, and the higher values will dominate. In phase 3, we split the data into 80% for training and 20% for testing. In phase 4, we build our LSTM model and train it using 40 epochs, because they minimize the loss function. After the model was fitted in the previous phase, in the new one, we test the performance of our model on the train and test datasets using a wide range of accuracy metrics.

3.1 LSTM Model

The LSTM model was developed in 1997 by Hochreiter and Schmidhuber, which eliminates the vanishing gradient problem in RNN. This problem signifies that the information is not stored for a long time, and gradient becomes without any effect in the deepest layers.

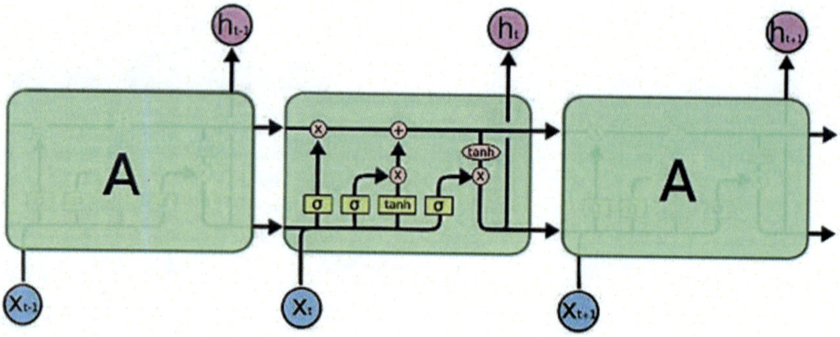

Fig. 2 LSTM internal architecture [14]

To solve this problem, the LSTM model includes a memory cell "C_t" that can store the information for a long sequence of time. Each memory cell has three gates, input gate "I_t," forget gate "f_t," and output gate "O_t" [13].

The input gate defines if the input needs to change the content of the cell, the forget gate decides whether to reset the cell content to zero, and the output gate decides if the cell content should offer the neuron's output.

Those gates are sigmoid functions and have a binary value 0 and 1, which 0 means let nothing pass and 1 let everything pass (see Fig. 2).

3.2 *Forecasting Performance Measures*

There is a wide range of performance measures to judge the precision of the prediction model; in our work, we use those measures:

MSLE:

$$\text{MSLE} = \frac{1}{n}\sum_{t=1}^{n}(\log(f_t + 1) - \log(y_t + 1))^2. \tag{1}$$

MAE:

$$\text{MAE} = \frac{1}{n}\sum_{t=1}^{n}|y_t - f_t|. \tag{2}$$

MSE:

$$\text{MSE} = \frac{1}{n}\sum_{t=1}^{n}(y_t - f_t)^2. \tag{3}$$

RMSE:

$$\text{RMSE} = \sqrt{\frac{1}{n}\sum_{t=1}^{n}(y_t - f_t)^2}. \tag{4}$$

RMSLE:

$$\text{RMSLE} = \sqrt{\frac{1}{n}\sum_{t=1}^{n}(\log(f_t + 1) - \log(y_t + 1))^2}. \tag{5}$$

where:

y_t Actual value in time period t;
f_t Forecast value in time period t;
n Number of periods forecasted.

They should be nearer to zero to offer the better predictions results [15].

4 Results and Discussion

The graphic below shows the forecasting of our model, the blue line is the trained data; the reddish orange color line is the rest of the data, and the orangish yellow color line is the predicted values. We see that the forecasted values are really closer to the real values (Fig. 3).

In addition, in our work, we used five performance measures to examine the precision of our predicting model, and they vary between 0.0001 and 23.76 for Attijariwafa Bank, as shown in Table 1.

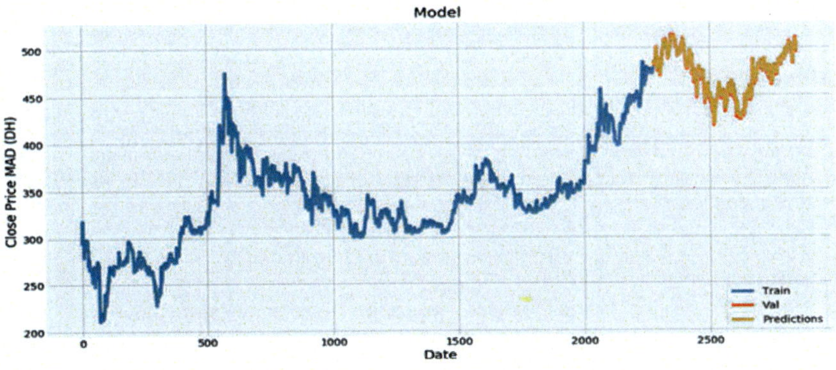

Fig. 3 Attijari stock close price predictions

Table 1 LSTM model accuracy metrics

MSE	RMSE	MAE	RMSLE	MSLE
23.7683	4.875274	3.77348	0.01042	0.000108

Table 1 shows that the LSTM model makes a respectable performance. MSLE and RMSLE are closer to zero; RMSE and MAE are lower than 10 and MSE equal to 23.7683

Compared to Ayşe work [1], our study gives predictions closer to actual values and gives accuracy metrics lower which mean more accurate.

The LSTM model can forecast the price of AttijariWafa Bank, and its fluctuation excellently agreed with the real values. The possible reason is that the LSTM be able to extract useful information from features by memorizing information using memory cell "C_t" and this conducts to superior predictive ability.

5 Conclusion

In this paper, we proposed the LSTM model to anticipate the CSE employing the dataset of daily closing prices of AttijariWafa Bank. The results depict that the forecasted values are closer to the real values which means that our model has good accuracy and is capable of detecting the direction of the price. The metrics used to examine the predictions in this work confirm the robustness of our model. Therefore, our model might be very helpful for traders to make right decisions in FM and can give a clear vision to the direction of the prices.

In the future works, we aim to utilize technical indicators as input features, and we will use our model on different market datasets and for different time windows. In addition, we will intend to compare the LSTM model with other models like SVM, ANN, and GA for the Moroccan case.

References

1. Dosdoğru AT (2019) Comparative study of hybrid artificial neural network methods under stationary and nonstationary data in stock market. Manage Decis Econ 1–12. http://doi.org/10.1002/mde.3016
2. Chopra S, Yadav D, Chopra AN (2019) Artificial neural networks based indian stock market price prediction: before and after demonetization. Int J Swarm Intell Evol Comput. http://doi.org/10.4172/2090-4908.1000174
3. Sahoo S, Mohanty MN (2020) Stock market price prediction employing artificial neural network optimized by gray wolf optimization. In: Advances in intelligent systems and computing, vol 1030. http://doi.org/10.1007/978-981-13-9330-3_8
4. Zhu C, Yin J, Li Q (2014) A stock decision support system based on DBNs. J Comput Inf Syst 10(2):883–893. https://doi.org/10.12733/jcis9653

5. Chang P-C, Warren Liao T, Lin J-J, Fan C-Y (2011) A dynamic threshold decision system for stock trading signal detection. Appl Soft Comput. http://doi.org/10.1016/j.asoc.2011.02.029
6. Yodele AA, Ayo CK, Adebiyi MO, Otokiti SO (2012) Stock Price prediction using a neural network with hybridized market indicators. J Emerg Trends Comput Inf Sci 3(1):1–9
7. Brasileiro RC, Souza VLF, Femandes BJT, Oliveira ALI (2013) Automatic method for stock trading combining technical analysis and the artificial bee colony algorithm. In: IEEE congress on evolutionary computation, June 20–23, Cancun, Mexico. http://doi.org/10.1109/CEC.2013.6557780
8. Agrawal M, Khan AU, Shukla PK (2019) Stock price prediction using technical indicators: a predictive model using optimal deep learning. Int J Recent Technol Eng (IJRTE) 8(2). ISSN: 2277-3878. http://doi.org/10.35940/ijrteB3048.078219
9. Qiu M, Song Y (2016) Predicting the direction of stock market index movement using an optimized artificial neural network model. PLoS ONE 11(5):e0155133. http://doi.org/10.1371/journal.pone.0155133
10. Sezer OB, Ozbayoglu M, Dogdu E (2017) A deep neural-network based stock trading system based on evolutionary optimized technical analysis parameters. In: Complex adaptive systems conference with theme: engineering cyber physical systems, CAS, Oct 30–Nov 1, Chicago, Illinois, USA. http://doi.org/10.1016/j.procs.2017.09.031
11. Zhang Y, Wu L (2009) Stock market prediction of S&P 500 via combination of improved BCO approach and BP neural network. Expert Syst Appl 36:8849–8854. https://doi.org/10.1016/j.eswa.2008.11.028
12. Pang X, Zhou Y, Wang P, Lin W, Chang V (2018) An innovative neural network approach for stock market prediction. J Supercomput. https://doi.org/10.1007/s11227-017-2228-y
13. Sethia A, Raut P (2019) Application of LSTM, GRU & ICA for stock price prediction. In: Proceedings of ICTIS 2018, vol 2. http://doi.org/10.1007/978-981-13-1747-7_46
14. Olah C (2015) Understanding lstm networks–colah's blog. colah.github.io
15. Klimberg RK, Sillup GP, Boyle K, Tavva V (2010) Forecasting performance measures—what are their practical meanings? http://doi.org/10.1108/S1477-4070(2010)0000007012

Chapter 3
A Fog–Cloud Computing-Inspired Image Processing-Based Framework for Lung Cancer Diagnosis Using Deep Learning

Aditya Gupta, Vibha Jain, and Wasaaf Hussain

1 Introduction

Cancer is a critical disorder that results in the death of a million people across the globe. Cancer is a genetic or chronic disorder that occurs when there is an uncontrolled growth of abnormal cells that may grow in any part of a body. Presently, there exist more than 200 types of cancers and the diagnosis of such a fatal disease is very difficult. However, there exist certain forms of cancers such as tumors that show visible growths and are highly life-threatening [1, 2]. Therefore, it is essential to develop advanced healthcare systems so that the diagnosis and treatment of such sorts of disorders can be carried out at the nascent stages without any loss to the patient's life.

1.1 Lung Cancer

Among several types of cancer diseases, the most prevalent and frequent type of cancer is lung cancer with around 2 million global cases per year. Many factors that may cause cancer include a poor diet, lack of physical activity, obesity, smoking, usual alcohol consumption, etc. The signs and symptoms of the cancer are not specific, and hence depend on the specific type and the stage of cancer, the patient is suffering from. However, certain symptoms that may be found in cancer patients include hair loss, fatigue, weight loss, persistent fever, cough, and many more [3]. Since the

A. Gupta (✉) · W. Hussain
Baba Ghulam Shah Badshah University, Rajouri, Jammu and Kashmir 185234, India
e-mail: aditya.gupta@bgsbu.ac.in

V. Jain
Netaji Subhas University of Technology, New Delhi 110078, India
e-mail: vibha.jain.cs19@nsut.ac.in

A. Sheth et al. (eds.), *Intelligent Systems*, Algorithms for Intelligent Systems,
https://doi.org/10.1007/978-981-16-2248-9_3

treatment of such a disorder is cumbersome and heavily costly, most of the people under the poverty line lose their lives due to such a fatal disorder.

Continuous supervision by concerned doctors and healthcare practitioners is important for patients with a significant risk of lung cancer. With the limited number of specialists and inadequate staff, it is a very difficult job for physicians to monitor each patient individually. Therefore, it is essential to have some intelligent system that may help diagnose the patients in the earlier stages of their lung cancer cases.

1.2 The Literature on Existing Work in Lung Cancer

In [4], the authors proposed an intelligent framework for early lung cancer detection. A glowworm swarm optimization technique, based on a high ordering recurrent neural network (RNN), was adopted to support the framework. The results obtained were found promising when compared to existing frameworks. In [5], the authors designed a fuzzy-based expert model for lung cancer prediction. The predictive model worked using four steps. The first step involved transforming collected values into fuzzy values. These fuzzy values were transformed into fuzzy soft-sets in step 2. A method namely parameter reduction was used in step 3 for obtaining new values from the values obtained in step 2. Finally, step 4 resulted in an output. In [6], the authors provided lung cancer epidemiology and several approaches for its early prediction. The data relevant to lung cancer was collected from multiple countries which include the USA, India, Egypt, etc. Several statistical analyses were made in terms of mortality and survival rates. It was observed that in India and Egypt, lung cancer cases have increased in the period from 1990 to 2014. In [7], the authors developed a calculator that gave the outcomes for estimating the mortality risk of lung cancer after a few weeks, months, and even years, once the user was diagnosed with lung cancer earlier in the diagnosed tests. In [8], the authors proposed a model for the prediction of lung cancer by utilizing the Gene Expression programming technique and using the microarray lung dataset. Lung cancer-relevant genes were selected using the gene selection techniques. The model was compared against the pre-existing machine learning techniques such as multilayer perceptron, support vector machine (SVM), and radial basis function neural networks. The GEP-based model performed well in terms of reliability as compared to the existing techniques. In [9], the authors employed an artificial neural network (ANN) model to detect the initial stages of lung cancer. The model was designed, trained, and validated using a survey of lung cancer datasets. This model showed an accuracy of 96.67%. In [10], the authors employed an innovative image processing approach for automatic lung cancer detection. Lung cancer was diagnosed by employing the optimized deep neural network trained using the modified gravitational search algorithm. Moreover, a technique namely linear discriminant analysis (LDA) was employed as a preprocessing stage for feature extraction.

Although several systems for solving lung cancer disease have existed in the literature, no one introduced the system that employs fog computing as a main component in the past. Keeping in view all such facts, we proposed a novel fog–cloud-based framework for lung cancer diagnosis and its two stages namely benign and malignant by employing a deep learning approach.

1.3 Motivation and Contributions

The increasing development in technologies such as fog–cloud computing, advanced image processing techniques, and machine learning techniques has made it feasible to design intelligent healthcare systems for the early diagnosis and prediction of fatal disorders. Moreover, with the generation of a huge amount of medical data and advancement in artificial intelligence, there is a vast scope for integration of the medical domain with artificial intelligence to solve real-time medical problems.

The major contributions of our proposed scheme are as follows:

- Designing a fog–cloud-based model for early lung cancer detection using image processing techniques.
- Formation of fog layer by using advanced image processing schemes for feature extraction.
- Artificial neural network-based stage classification and storage of results at the cloud layer for communication with patients.

2 Proposed Model

A fog–cloud-centric framework for the diagnosis and detection of lung cancer at the initial stages is depicted in Fig. 1. The framework consists of mainly three stages namely the device layer, fog layer, and cloud layer. The description of each of the layers is given in detail in the subsequent sections.

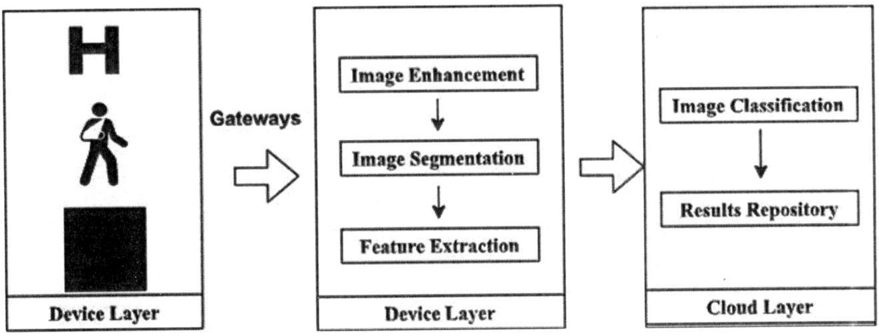

Fig. 1 Three-layered diagnosis architecture

The general description of the lung detection system consists of the following steps. The first step starts by collecting the 3D CT scan images of both the normal and abnormal behavior for further evaluation. The second stage uses different techniques for preprocessing collected images to improve the quality and clarity. Preprocessed images are better in quality and different segmentation and extraction algorithms can be applied to them using automated tools. These preprocessed images go through with different segmentation algorithms in the third image. Different abnormalities are extracted from these segmented images in stage four. Each type of abnormality indicates a different stage of lung cancer cells in images. Backpropagation artificial neural network (BP-ANN) is used to classify the abnormalities indicated by the prior stage. The complete chronological workflow of our proposed system is listed in Algorithm 1.

Algorithm 1. The Workflow of the Proposed System

Step 1. CT-scanned images of both normal and abnormal behavior are collected
Step 2. Different preprocessing image processing techniques are applied to the collected images to improve their quality and clarity at the fog layer
Step 3. Apply segmentation algorithms to the preprocessed images at the fog layer
Step 4. Fog layer applies feature extraction techniques to the segmented images for further investigations at the cloud layer
Step 5. Extracted features are trained using backpropagation artificial neural network to check whether the abnormality is cancerous or not

2.1 *Device Layer*

A device layer mainly forms the data collection component of the proposed model and may consist of several types of stakeholders such as hospitals, doctors, patients, and paramedic staff which are responsible for communicating the CT scanned images to the fog layer for investigation of possible lung cancer cases. The collected CT scanned images are communicated to the fog layer using mobile or Web-based health-care applications. In our proposed model, we have used the Lung Image Database Consortium image collection (LIDC-IDRI) [11] consisting of 1018 cases, each of which contains CT scanned images. Each image belongs to one of the three categories shown in Fig. 2.

2.2 *Fog Layer*

A fog layer uses image processing techniques as a preliminary step for further investigations. The image processing technique used at the fog layer consists of three steps

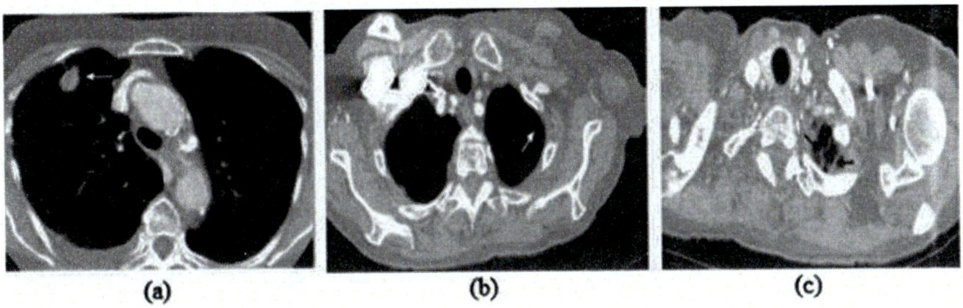

Fig. 2 **a** Stage 1: nodule_3 mm, **b** Stage 2: nodule < 3 mm, and **c** Stage 3: non-nodule_3 mm

namely image enhancement, image segmentation, and feature extraction. A detailed description of each stage is given ahead. Image preprocessing used different algorithms to convert a discrete image to a better-quality digital image by suppressing distortion and noise.

In our proposed model, we defined image preprocessing by the following steps. First, each collected CT scanned image is converted into the grayscale image using the predefined filters available in MATLAB. Although almost all the images are already in grayscale, however, this step is used to ensure the image format. These grayscale images are passed through different filters for image enhancement in the next step. At the last stage of preprocessing, images are segmented using different morphological operations to detect different spots and irregular cells in images. Besides, image segmentation is also used to find out the percentage coverage of and size of cancer affected cells and normal cells.

Image enhancement improves the quality of the collected input CT scanned images. There are majorly two broad classifications of image enhancement, viz. frequency domain method and spatial domain method. However, there is no evaluated scheme to check which method to use when. In our proposed scheme, we employed the following filters to improve the quality of the collected images. The complete procedure of image enhancement is explained in Fig. 3.

1. High-pass filters are used to pass the frequencies higher than the cut-off frequency while block all the frequencies which are below the cut-off frequency. This filter is used to sharpen the image.
2. Denoising using salt and pepper filtering removes salt-and-pepper noise from an image without reducing any further quality of the image. Salt-and-pepper noise is defined as the effect that is similar to the sparkling black and white dots on images.

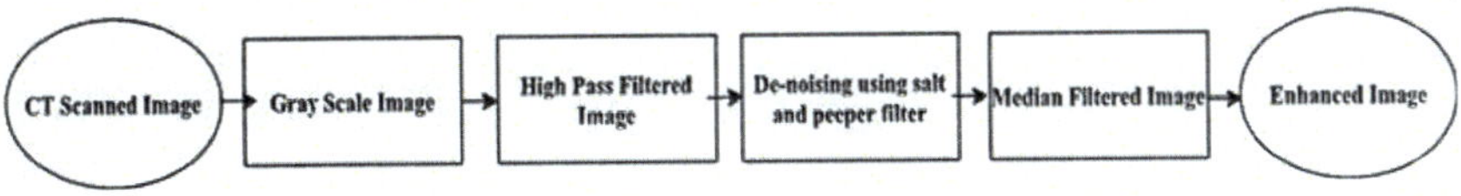

Fig. 3 Procedure of image enhancement

3. A nonlinear median filter is a smoothing function used for preserving edges in an image. To replace the centre pixel, a median value of the corresponding window is used by the median filter.

Image segmentation is the process of partitioning collected images into segments. Segmentation is done to make the representation of an image in a more simplified way or something more meaningful and easier to analyze.

1. Thresholding is a method to reduce a grayscale image into a binary image, involving only two colors. Image thresholding is used for portioning the image into foreground and background partitions.
2. Water shedding filters use the low and high elevation to capture light and dark pixels in the image. Different extracted abnormalities help us to identify the spread of cancerous cells in the lungs. These features can also be to identify the different stages of cancer by analyzing and designing different nodules analysis rules. In our proposed work, we use these extracted features as training sets in learning new rules in neural networks. After successful learning of neural networks, test images are used to check the validation of our network. Artificial neural networks (ANN) are used to classify the images into different classes of cancer, and it will distinguish whether the provided lung image is with malignant growth or not. Therefore, encouraging early identification and detection will improve the patient's survival rate. Complete working of the neural network is described in the next section.

2.3 *Cloud Layer*

A cloud layer forms an important component in the proposed model, where the actual processing takes place. A cloud layer provides huge processing and storage capabilities for offering innumerable healthcare services. A cloud layer consists of a classification component and the storage component. Classification component—the classification component is responsible for classifying the images with extracted abnormality into one of the two abnormal stages of lung cancer namely benign and malignant. The proposed model uses backpropagation artificial neural network (ANN) for the image classification, using different features extracted at prior stages. A neural network is used to emulate the human brain.

3 Results and Discussions

The proposed fog–cloud-assisted architecture for early diagnosis and classification of lung cancer is implemented using MATLAB. Different preprocessing and feature extraction tools are available in MATLAB for easy evaluation. Figure 4 shows the stagewise stage result in different operations. An artificial neural network (backpropagation ANN) is created using a Network/Data Manager Tool with an input layer, an

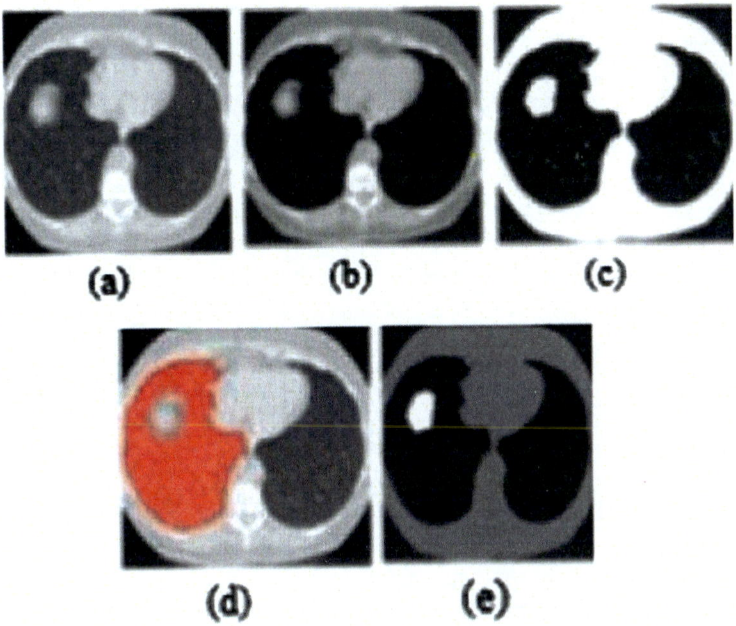

Fig. 4 **a** Original collected CT scan image. **b** Preprocessing image. **c** Image after thresholding. **d** Segmented lung region. **e** Cancerous nodule

output layer, and a hidden layer. Simulated results are presented in Table 1. To assess the performance measure of our proposed system, we have used three simulation parameters: sensitivity, specificity, and accuracy [12].

Accuracy effectively determines the correct observation ratio to the actual number of observations and is mathematically expressed as:

$$\text{Accuracy} = \frac{\sum \text{True Positive} + \sum \text{True Negative}}{\sum \text{True Positive} + \sum \text{True Negative} \sum \text{False Positive} + \sum \text{False Negative}} \tag{1}$$

Sensitivity also referred to as recall measures the ratio of correct positive observations to the total number of positive observations.

Table 1 Performance measure of ANN

Class/stage	Benign lung cancer	Malignant lung cancer	Normal
Sensitivity	100	80	100
Specificity	50	9.09	50
Accuracy	98.07	50	91.66

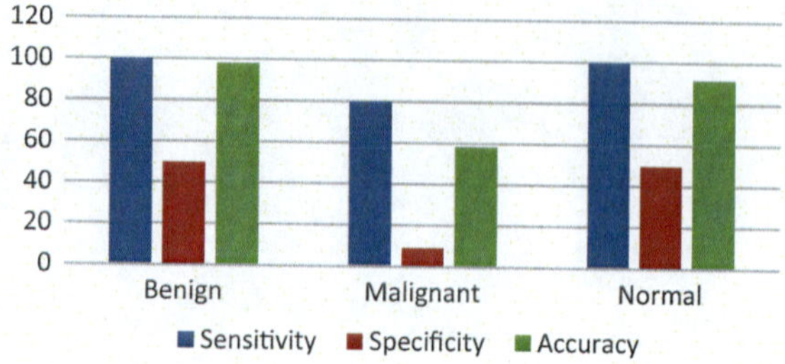

Fig. 5 Analysis of different performance metrics. **a** Sensitivity. **b** Specificity. **c** Accuracy

$$\text{Sensitivity} = \frac{\sum \text{True Positive}}{\sum \text{True Positive} + \sum \text{False Negative}} \tag{2}$$

Specificity is defined as the ability of a classifier to separate negative results.

$$\text{Specificity} = \frac{\sum \text{True Negative}}{\sum \text{True Negative} + \sum \text{False Positive}} \tag{3}$$

In the proposed method, the input stage consists of nine neurons, one for each extracted feature. Backpropagation neural networks use the principle of mean square error. Error is minimized at each stage after every epoch. Error at each stage is given by the difference between the output and respective input for each layer. For training and testing purposes, we use the Lung Image Database Consortium image collection (LIDC-IDRI) [11] consisting of 1018 cases with CT scanned images. The sensitivity of an artificial neural network assesses the impact of input parameters on generated output through the hidden layer. Results show that the proposed neural network is 80% sensitive while detecting malignant lung cancer (Stage 1). Sensitivity can also be termed as the total number of affirmative cases that are detected correctly whereas specificity is used to calculate the total of negative cases that are predicted as negative. Another term to represent sensitivity and specificity is truly positive and true negative, respectively. Graphical analysis of different performance metrics is shown in Fig. 5.

4 Conclusion

The proposed model helped in the early identification of cancerous diseases in the lung. Different image processing and feature extraction are used to detect different abnormalities from collected CT images. Using backpropagation ANN (BP-ANN), these collected features are successfully categorized into three classes. From our experimental results, we can conclude that by training the neural network using a large database, lung cancer can be efficiently detected at an early stage.

Acknowledgements The work was completed under the project titled "***Fog-Cloud Centric IoT assisted Technologies in Healthcare***" with sanction number BGSBU/TEQIP-III/RGS/004 supported under the TEQIP-III Research Grant Scheme (RGS) of the National Project Implementation Unit (NPIU), a unit of the Ministry of Education (MoE), Government of India and the World Bank. The authors are grateful to the funding agency for the financial and infrastructural support provided to carry out the research.

References

1. Cruz CSD, Tanoue LT, Matthay RA (2011) Lung cancer: epidemiology, etiology, and prevention. Clin Chest Med 32(4):605–644
2. Didkowska J, Wojciechowska U, Mańczuk M, Łobaszewski J (2011) Lung cancer epidemiology: contemporary and future challenges worldwide. Ann Transl Med 4(8)
3. Alberg AJ, Samet JM (2003) Epidemiology of lung cancer. Chest 123(1):21S–49S
4. Selvanambi R, Natarajan J, Karuppiah M, Islam SH, Hassan MM, Fortino G (2020) Lung cancer prediction using higher-order recurrent neural network based on glowworm swarm optimization. Neural Comput Appl 32(9):4373–4386
5. Khalil AM, Li SG, Lin Y, Li HX, Ma SG (2020) A new expert system in the prediction of lung cancer disease based on fuzzy soft sets. Soft Comput 1–29
6. Dubey AK, Gupta U, Jain S (2016) Epidemiology of lung cancer and approaches for its prediction: a systematic review and analysis. Chin J Cancer 35(1):71
7. Agrawal A, Misra S, Narayanan R, Polepeddi L, Choudhary A (2011) A lung cancer outcome calculator using ensemble data mining on seer data. In: Proceedings of the tenth international workshop on data mining in bioinformatics, pp 1–9
8. Azzawi H, Hou J, Xiang Y, Alanni R (2016) Lung cancer prediction from microarray data by gene expression programming. IET Syst Biol 10(5):168–178
9. Nasser IM, Abu-Naser SS (2019) Lung cancer detection using artificial neural network. Int J Eng Inf Syst (IJEAIS) 3(3):17–23
10. Lakshmanaprabu S, Mohanty SN, Shankar K, Arunkumar N, Ramirez G (2019) Optimal deep learning model for classification of lung cancer on ct images. Future Gener Comput Syst 92:374–382
11. Clark K, Vendt B, Smith K, Freymann J, Kirby J, Koppel P, Moore S, Phillips S, Maffitt D, Pringle M et al (2013) The cancer imaging archive (tcia): maintaining and operating a public information repository. J Digital Imaging 26(6):1045–1057
12. Hossin M, Sulaiman MN (2015) A review on evaluation metrics for data classification evaluations. Int J Data Min Knowl Manag Process 5(2):1

Chapter 4
Performance Analysis of Power System Stability of Four-Machine System by Using MBPSS and Static Compensator

Ajay Kumar Tiwari, Mahesh Singh, Shimpy Ralhan, and Nidhi Sahu

1 Introduction

The recent interconnected power systems are highly dynamic systems, and for improving the efficiency of modern power systems by taking care of transient stability, FACTs devices are very useful. Stability of the power system is very necessary, and for this purpose, rotor angle stability of synchronous machines is an important criterion. Transient stability is the type of rotor angle stability [1]. Rotor angle stability is basically generator stability because rotor angle is the angle between induced emf and terminal voltage of the generator. In the present scenario, competition between power plants is increasing day by day, so for this, efficiency of the power flow is the basic tool. Hence, the FACTs device is very useful for this purpose. In this paper, we are using STATCOM which provides reactive power as required and can deliver reactive power as per the requirement of the interconnected power system [2]. STATCOM is a very fast device which operates depending on the voltage of the system. And by observing the results, it is seen that the STATCOM is very fast in delivering the reactive power or absorbing the reactive power [3]. And rotor angle stability is also improved by connecting the STATCOM [4]. Cost of the static VAR compensator (SVC) is low as compared to STATCOM, but for large interconnected power systems, the cost of STATCOM does not affect the profit of the power plants as it improves the efficiency better than SVC [5]. Stability of the power system is divided into three types as shown in Fig. 1.

- Transient stability
- Steady-state stability
- Dynamic stability.

In this paper, we analyze the stability of the power system, and the stability of rotor angle is the type of transient stability [6]. So, by improving the rotor angle

A. K. Tiwari (✉) · M. Singh · S. Ralhan · N. Sahu
Shri Shankaracharya Technical Campus, Bhilai, Chhattisgarh 490020, India

A. Sheth et al. (eds.), *Intelligent Systems*, Algorithms for Intelligent Systems,
https://doi.org/10.1007/978-981-16-2248-9_4

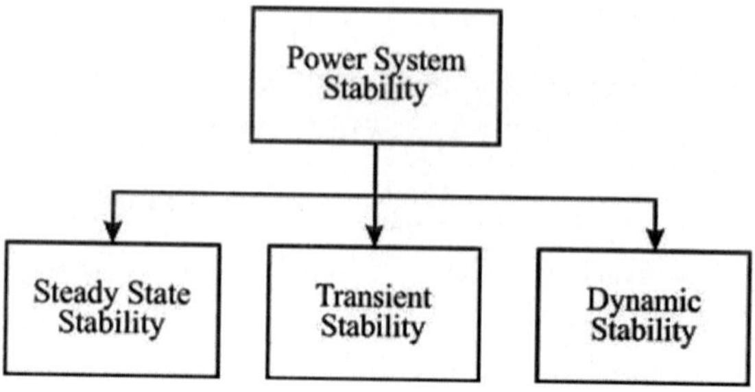

Fig. 1 Classification of power system stability

stability of the generator, transient stability of the system is automatically improved as shown in waveforms [7, 8]. Reactive power requirement by the system with the help of STATCOM is studied in the paper which is coordinated with the MBPSS. In this paper, stability of the system with and without STATCOM is discussed.

2 Transient Stability

Transient stability of the system mainly depends on the loading of the line and fault positioning of the line which affects the stability of the system performance [9, 10]. For stability improvement from overloading and due to fault is compensated by the FACTs devices used in the system which is helpful for improving the transient stability of the system as shown in Fig. 2 by compensating the overloading and voltage unbalancing of the line. Recently, FACTs devices are very useful for improving the system performance of the power system [11].

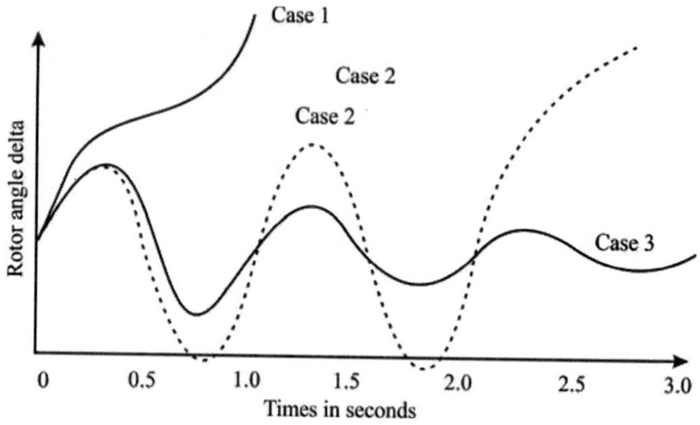

Fig. 2 Rotor angle response to a transient disturbance

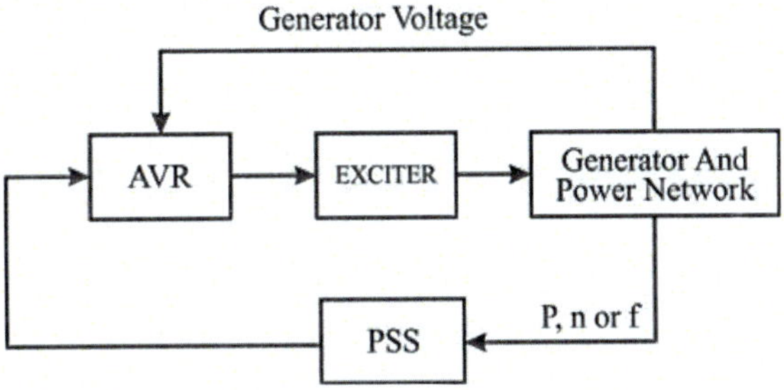

Fig. 3 Block diagram of PSS in feedback with generator

3 Power System Stabilizer

The PSS is an essential part of excitation whose function is to control the generator excitation by adding damping to the generator rotor oscillations. Purpose of PSS is to compensate for the oscillation of the rotor and help to regain the steady-state behavior. If automatic voltage regulator (AVR) is connected through input of the excitation, then the high gain of AVR will give a fine voltage control and it increases the opportunity of maintaining the synchronization of the generators during the large disturbances. Therefore, this conflict is almost resolved by limiting the output of PSS to 0.5% set point of the AVR. The input of the PSS may be acceleration $P_a = P_m - P_e$, i.e., differences of mechanical input power to electrical output power, or it may be frequency oscillations or rotor speed deviation [12]. PSS provides electrical torque outputs to the exciter as illustrated in Fig. 3.

4 Facts Controller (STATCOM and SVC)

STATCOM and SVC are power electronics-based controllers, and these controllers are very useful for improving efficiency of the power system. STATCOM and SVC are the shunt-connected devices which are used to provide the reactive power compensation of the power system and help to improve the response of the network. STATCOM operating speed is very fast as compared to SVC. STATCOM works as controllable voltage source, and SVC works as dynamically controllable reactance connected in parallel [5].

5 Static Synchronous Compensator (STATCOM)

Static compensator is a shunt FACTs device which operates by the variation in the voltages of the generators. The equivalent circuit of STATCOM is shown in Fig. 4. When the system voltage is low, static compensator generates the reactive power, and when the system voltage is high, it absorbs the reactive power and helps to maintain

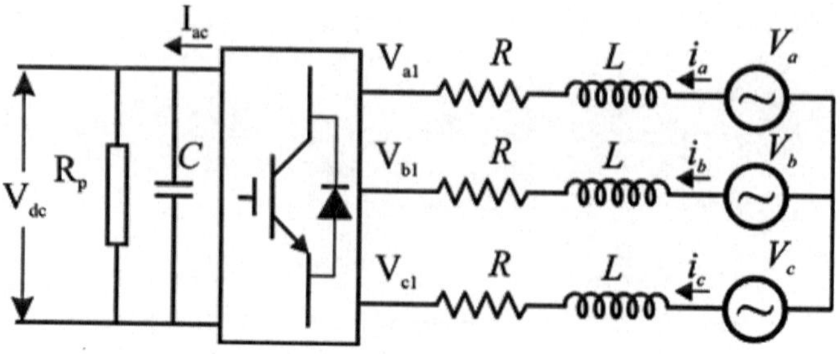

Fig. 4 Equivalent circuit of STATCOM

generator output and system performance [10]. It is helpful for large power system networks, and if installed for small power plants, then it is not much effective in terms of the initial cost of the small power plants as STATCOM is an expensive device. STATCOM can be applied with two basic controls:

1. Control of direct current voltage across capacitor.
2. AC voltage regulation of the power system at bus bar where static compensator is connected.

6 STATCOM V-I Characteristic

The V-I characteristics of static compensator is depicted in Fig. 5 which controls its output current independently over maximum capacitive or inductive range without

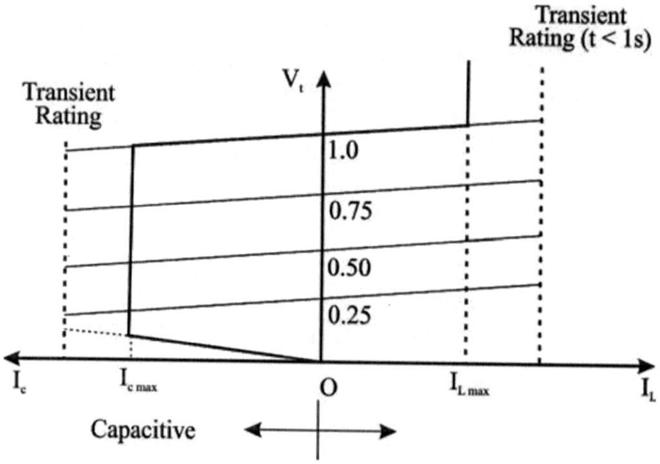

Fig. 5 V-I characteristic of an STATCOM

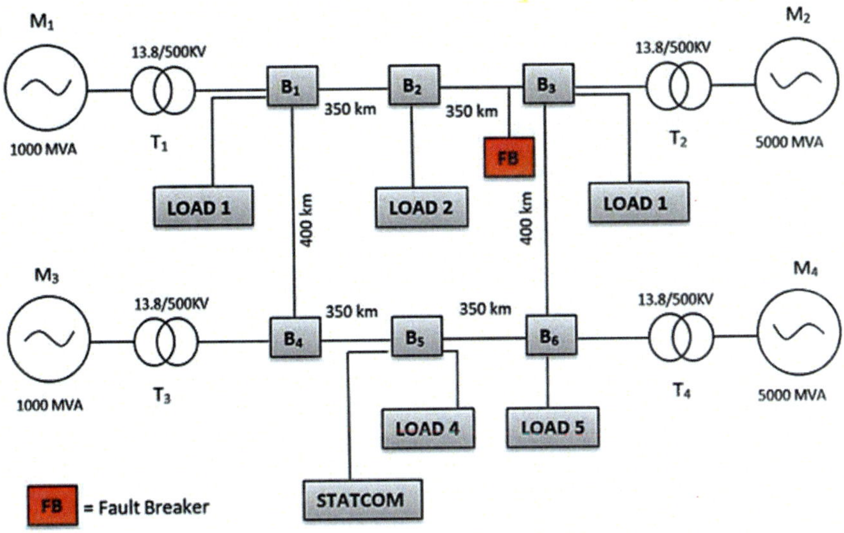

Fig. 6 One line diagram of MATLAB simulated of four machines and six-bus systems

depending on magnitude of alternating current (AC) system voltage. Strength of STATCOM is capable of generating the maximum output of capacitive generation; almost it is independent of the voltage of the system. The capability of the static compensator which is very useful for the conditions where static compensator is needed for supporting voltage regulation of the system during the disturbances; as we know that static compensator is capable of supply both inductive and capacitive compensation as and when required by the system [4].

7 System Description

The system model represents an interlinked transmission line network which consists of four generators with PSS interconnected in the excitation system. Complete system is interlinked through six buses B1 to B6. The test system has generators ratings as 5000 and 1000 MV A as shown in Fig. 6.

8 Result and Discussion

An outage is created between bus 2 and bus 3 which is considered a disturbance. PSS is a typical lead-lag controller which is used for improving damping. The new development of lead-lag power system stabilizer is multiband PSS, which contains three lead-lag power system stabilizer levels. The traditional PSS and MBPSS are coordinated and examined for outage analysis at the transmission line. We have here considered different cases with STATCOM and PSS and without STATCOM for outage. The system model is explained in the Appendix. The stability of the system which has four generators and six buses is examined for the following conditions:

Case 1: Outage with conventional PSS at generators.

Case 2: Outage with power system stabilizer connected at generator and STATCOM connected at transmission line.

Values of parameters of conventional power system stabilizer are given in Table 1. The results are shown in Fig. 7a, b which is the result of rotor speed of generators with static compensator and without static compensator.

Figure 7a shows the rotor speed of the generator without STATCOM which is unstable in nature. When an outage is created at 5.10–5.16 s, the rotor speed of the generators are observed and obtained. The deviation in the rotor speed shows a reduction in oscillations and settles fast with PSS and STATCOM as shown in Fig. 7b.

Analysis of the generator rotor angle due to the outage created at bus 2 and 3 is observed, and results are obtained. Simulation results of rotor angle for individual generators are shown in Fig. 8a–d with MBPSS, CPSS, and MBPSS and STATCOM, respectively.

Figure 8a–d is the results obtained from simulation with MBPSS, CPSS, and STATCOM which is shown in same graph for individual generators for analysis of the response of the generators during outage in transmission line which is created between bus 2 and bus 3 for 5.10–5.16 s; by observing the simulation results, we

Table 1 Generic lead-lag PSS parameters

Generators	Tw	K	T_1	T_2
G_1	0.71	2.00	0.0611	0.480
G_2	0.71	3.00	0.0595	0.540
G_3	0.71	2.00	0.0613	0.460
G_4	0.71	3.00	0.0588	0.520

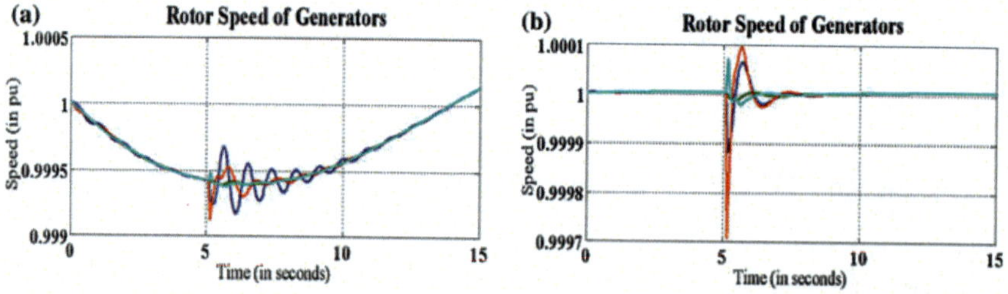

Fig. 7 **a** Output of all four generators rotor speed with PSS and no STATCOM. **b** Output of all four generators rotor speed with PSS and STATCOM

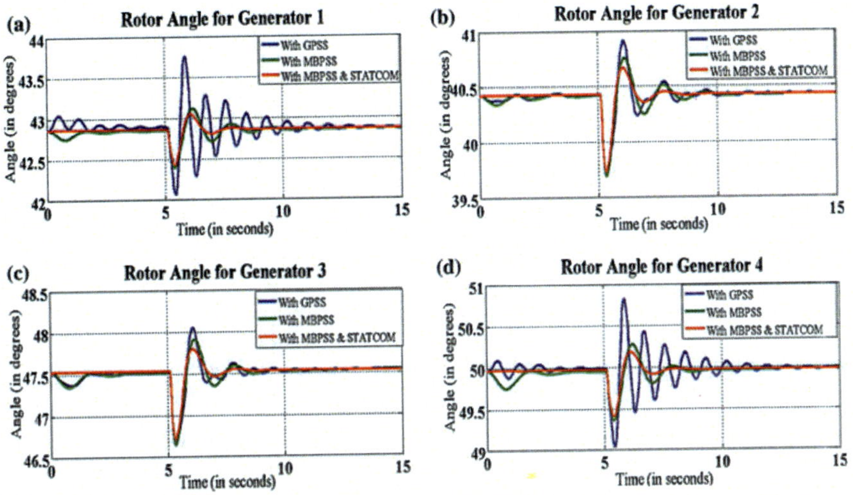

Fig. 8 **a** Rotor angle output of Generator 1 with different controllers. **b** Rotor angle output of Generator 2 with different controllers. **c** Rotor angle output of Generator 3 with different controllers. **d** Rotor angle output of Generator 4 with different controller

conclude that system behavior during outage is far better by using combination of STATCOM and MBPSS compare to conventional PSS.

Reactive power output of the system is shown in Fig. 9; this reactive power output result varies at 5.10–5.16 s which is outage period created by us for analysis of the system response; by seeing the figure, it shows variations between 5.10 and 5.16 s which is outage time of the system which means STATCOM helps system by absorbing or delivering the required reactive power during outage. Figure 10 shows the output voltage of static synchronous compensator.

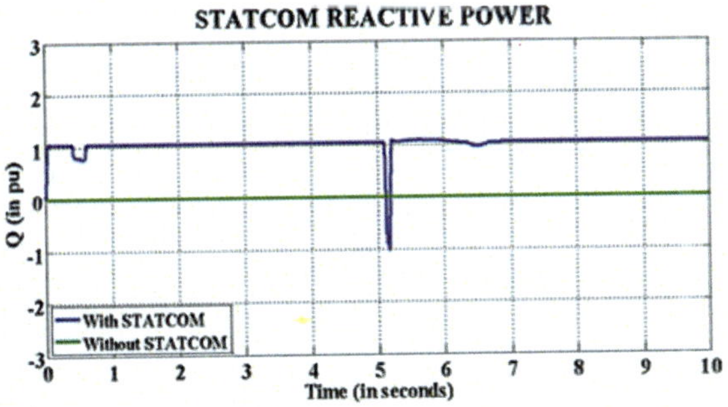

Fig. 9 Reactive power output of the STATCOM

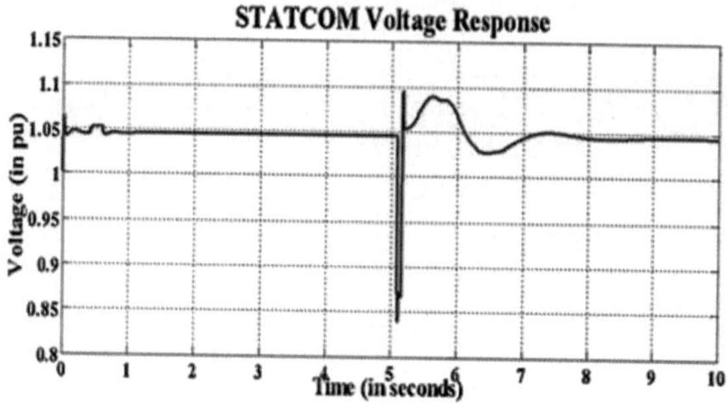

Fig. 10 Output voltage of STATCOM

9 Conclusion

The paper represents the four hydraulic generators and six buses transmission line in which outage is created between bus 2 and bus 3, and it is observed by simulation results and calculated results which are shown in the table. System is simulated and tested with different controllers like traditional power system stabilizer, multiband PSS (MBPSS), and MBPSS and STATCOM. Results obtained from multiband PSS and STATCOM are far better than MBPSS and generic PSS. MBPSS and STATCOM combination is connected with four-machine six-bus systems in MATLAB/Simulink. Three different outcomes are shown in the tables for individual generators.

We have obtained following conclusion from the tested system in MATLAB/Simulink.

- No PSS and no STATCOM controllers; the system is unstable and not able to maintain synchrony between the machines.
- Power system stabilizer is installed at the excitation system of each of the four generators, which helps to compensate the oscillations and gives the stable curves; but only power system stabilizer is not able for maintaining synchronism after outage in transmission line.
- The combination of MBPSS and STATCOM has given the smooth result for outage as compared to other controllers used in this model. So we conclude that the system transient stability is better and with the help of combination of MBPSS and STATCOM.

By seeing Tables 2, 3, 4, and 5, calculated results of peak overshoot and settling time are observed from the table; we can conclude that settling time of the generator

Table 2 Time response result of Generator 1

Controller	T_p	M_{po} (%)	T_s (s)	Peak (degree)
CPSS	5.40	2.64	14.98	43.80
MBPSS	5.90	2.08	8.24	43.10
MBPSS and STATCOM	6.10	0.44	3.34	42.90

Table 3 Time response result of Generator 2

Controller	T_p	M_{po} (%)	T_s (s)	Peak (degree)
CPSS	5.40	2.30	14.98	40.91
MBPSS	6.12	1.51	8.24	40.78
MBPSS and STATCOM	6.08	0.74	3.34	40.60

Table 4 Time response result of Generator 3

Controller	T_p	M_{po} (%)	T_s (s)	Peak (degree)
CPSS	5.40	1.98	14.98	48.36
MBPSS	6.10	1.18	8.24	47.95
MBPSS and STATCOM	6.18	0.64	3.34	47.70

Table 5 Time response result for Generator 4

Controller	T_p	M_{po} (%)	T_s (s)	Peak (degree)
CPSS	5.40	2.23	14.98	51.68
MBPSS	5.90	1.68	9.24	50.85
MBPSS and STATCOM	6.18	0.20	3.34	50.10

with MBPSS and STATCOM is far better than CPSS and only MBPSS connected systems.

References

1. Kundur P (1994) In: Balu NJ, Lauby MG (eds) Power system stability and control
2. Dhal PK, Christober C, Rajan A (2014) Design and analysis of STATCOM for reactive power compensation and transient stability improvement using intelligent controllers. In: International conference on electronics and communication systems. IEEE, pp 1–6
3. Singh M, Verma DK, Ralhan S, Patil M (2019) Performance analysis of SVC supplementary controller in multi-machine power system network. In: 2019 international conference on electrical, electronics and computer engineering. IEEE, pp 1–6
4. Rao P, Crow ML, Yang Z (2000) STATCOM control for power system voltage control applications. IEEE Trans Power Delivery 15(4):1311–1317
5. Pahade A, Saxena N (2013) Transient stability improvement by using shunt FACT device (STATCOM) with reference voltage compensation (RVC) control scheme. Int J Electr Electron Comput Eng 2(1):7–1
6. Mithulananthan N, Canizares CA, Reeve J, Rogers GJ (2003) Comparison of PSS, SVC, and STATCOM controllers for damping power system oscillations. IEEE Trans Power Syst 18(2):786–792
7. Hingorani NG, Gyugyi L, El-Hawary M (2000) Understanding FACTS: concepts and technology of flexible AC transmission systems, vol 1. IEEE Press, New York
8. Padiyar KR (2012) Analysis of subsynchronous resonance in power systems. Springer Science & Business Media, Berlin

9. Ahmadinia M, Ghazi R (2018) Coordinated control of STATCOM and ULTC to reduce capacity of STATCOM. In: Iranian conference on electrical engineering. IEEE, pp. 1062–1066
10. Mohanty AK, Barik AK (2011) Power system stability improvement using FACTS devices. Int J Mod Eng Res 2:666–672
11. Kale K, Meshram A, Sathawane R, Jethani S (2018) Power system stability improvement using STATCOM
12. Kamdar R, Kumar M, Agnihotri G (2014) Transient stability analysis and enhancement of IEEE-9 bus system. Electr Comput Eng Int J 2:41–51

Chapter 5
Performance Analysis of Sigmoid and Relu Activation Functions in Deep Neural Network

Akhilesh A. Waoo and Brijesh K. Soni

1 Introduction

Technologists around the world are continuously trying to uncover the root cause of having the capability for taking the right decision by a human being which means how a human mind takes appropriate decisions even in critical situations. The area of studying such types of causes and effects is broadly known as artificial intelligence. An artificial neural network is a tool for analyzing and expanding the existing capability of the human mind. An artificial neural network is a computational model parallel to a biological neural network. Similar to logical computing, neural computing also takes input, processes the input, and produces output. At the end of processing, a computing system reaches a state where it has to decide whether the system is in the condition of producing output or not. In an artificial neural network, there is a component that is wholly responsible for taking such a type of decision that is popularly known as activation function. Now in the depth of the artificial neural network domain, there are various types of activation functions used with their pros and cons. This paper describes two popular activation functions sigmoid and Relu, which are frequently used in various neural network models, and how both activation functions can be used with better performance.

2 Literature Review

The historical study of computational neuroscience started in 1890 when the great Spanish neuroscientist Santiago Ramón y Cajal discovered that neurons are interconnected as a network and can pass signals across the nervous system [1]. Professor

A. A. Waoo (✉) · B. K. Soni
AKS University, Satna, Madhya Pradesh 485001, India

A. Sheth et al. (eds.), *Intelligent Systems*, Algorithms for Intelligent Systems,
https://doi.org/10.1007/978-981-16-2248-9_5

Santiago Ramón y Cajal was awarded the Nobel Prize in 1906 for this great discovery in neuroscience and is now known as the father of modern neuroscience [2].

In the 1950s, English physiologists Sir Alan Lloyd Hodgkin and Sir Andrew Fielding Huxley conducted research on the giant neurons of the squid [3]. They were able to measure the voltage of electrical signals traveling through the axon, and the amount of voltage needed to fire a neuron cell is also known as an action potential [4].

In 1959, a research paper entitled "What the frog's eye tells the frog's brain" was published by the American cognitive scientist Jerome Ysroael Lettvin, and his two other teammates McCulloch and Pitts [5]. They discussed in his paper, how the visual cortex of a frog is structured and works. They also discussed the threshold value needed to fire neurons of the visual system [6].

In 1962, the first artificial perceptron was developed by the American psychologist Frank Rosenblatt [7]. He studied the theory of threshold value given by Walter Pitts and Warren McCulloch in 1959 [8]. Further, he considered the threshold function as an activation function, which is now a significant part of computational neuroscience.

In 1969 Professor Marvin Lee Minsky and Seymour Papert published a book entitled "Perceptrons: An Introduction to Computational Geometry," which overcome the major limitations of the perceptron. They committed that initially, the perceptron was even incapable of performing the XOR operation [9].

Nowadays, the activation function is a significant component of an artificial neural network in the domain of deep learning. Intentionally variations of the sigmoid and the Relu functions are most popular among deep learning researchers. These functions enabled researchers to achieve ever-expanding predictive power by stacking several layers of neuron units, assembled with activation functions, and trained using the back-propagation algorithm [10].

3 Concept

The activation function is a mathematical equation that determines the action potential of a neuron cell in an artificial neural network. Based on input value an activation function decides whether a neuron will fires or not. If the input value dominates the threshold limit neuron will fire, otherwise neurons will remain ideal.

3.1 Sigmoid Function

This is a nonlinear activation function which is also known as a logistic function. This is a very popular function since the early age of neural networks among researchers and engineers (Fig. 1). Conceptually, it works well only for the input values between 0 and 1, which means it provides probabilistic output. Even it also reduces the output value if it exceeds the limit of 0 and 1 [11].

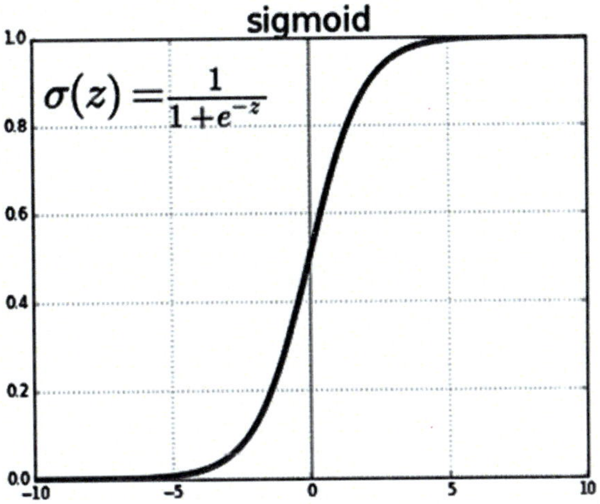

Fig. 1 Graphical representation of sigmoid activation function

3.2 Relu Function

This is a combination of a linear and nonlinear function, sometimes also known as a piecewise linear function or hinge function. The basic concept of the rectified linear unit activation function (Relu) is that it returns 0 for the input value 0 or less than 0 and returns the same value for other positive input values (Fig. 2) [12].

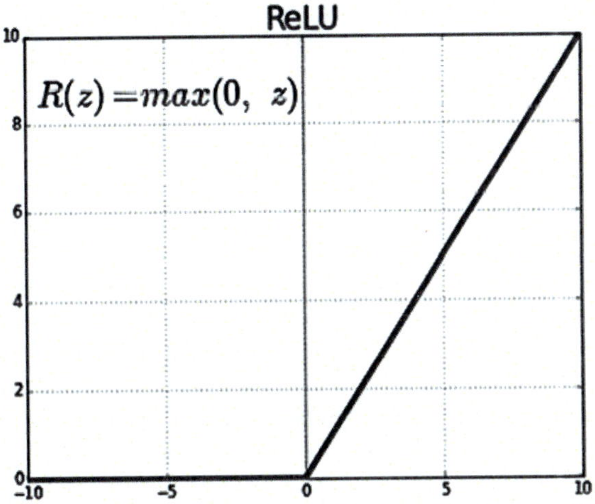

Fig. 2 Graphical representation of Relu activation function

4 Tools

There are some popular frameworks and tools for developing deep neural network models available such as Google Colab, TensorFlow, PyTorch, and Caffe. However, Google Colab is used for building and training models. There are some important repositories such as Keras, Scikit-learn, and NumPy used for using various methods for processing and manipulating data.

4.1 Keras

This is the most popular library for developing deep learning models. Two types of deep neural network models can be developed as recurrent neural network (RNN) and convolutional neural network (CNN), and other than this, a special variant of the convolutional neural network known as gradient adversarial neural network (GAN) can also be developed.

There are three categories of Keras's elements that are frequently used with the deep neural network. These are modules, layers, and models. Keras modules contain predefined classes, functions, and variables which are useful for developing deep learning models. Keras layers are the basic building block of Keras-based models. Each layer receives input data, processes them, and finally, generates output information. The output of one layer can flow into the next layer as its input, and so on. The Keras model represents the actual neural network model. Keras facilitates two modes to create the deep learning models, the first one is very simple and easy to use which is the sequential model, and the second one is a more flexible and advanced functional model [13].

4.2 Scikit-Learn

Scikit-learn, also known as Sklearn, is another useful and robust library for deep learning in Python which is frequently used for data manipulation and matrix processing. It is also a rich library for providing various built-in algorithms for deep learning, which can be easily used and customized according to appropriate datasets. Other than this, there are also available various statistical and probabilistic methods for classification, clustering, regression, dimensionality reduction, mode, and model selection. This library is written in Python and built upon the SciPy, NumPy, and Matplotlib supporting libraries [14].

4.3 *NumPy*

NumPy is another library for various types of matrix arrays processing in deep learning. It facilitates high-performance array objects and tools for working with these arrays. It also provides some mathematical functions to operate on these arrays and matrices. It is much faster for using the built-in Python methods. Some new data types can be defined using NumPy which allows NumPy to speedily integrate with various types of databases. Besides its scientific uses, NumPy can also be used as an efficient multi-dimensional container for generic datasets [15].

5 Method

Application programming interface (API) is also known as modules in the domain of deep learning programming. These modules are repositories having a wide range of methods, classes, and objects for handling various operations while building and training models. These are the names of modules used here in this model as datasets, models, layers, utils, and callbacks.

5.1 *Datasets*

The dataset module in Keras provides various datasets that can be used for training and testing deep learning models. There are various datasets available in the Keras library such as MNIST, Faison MNIST, CIFAR-10, CIFAR-100, and IMDB. However, in this experiment, a simple MNIST dataset is used [16].

5.2 *Models*

Keras facilitates defining the deep neural network models in fast and easy ways. Fundamentally, there are two broad ways to create neural network models such as *sequential* and *functional*. The sequential API is used to define the model layer-by-layer like a linear stack of multiple layers. It looks to be very easy to define and train a neural network model. However, sequential API has some disadvantages in that it does not allow to define neural network models that share layers or have multiple inputs or outputs. The functional API is another alternative way to define a deep neural network model. Unlike sequential API, it provides more flexibility to develop very complex networks with multiple inputs or outputs as well as a model that can also share layers [17].

5.3 Layers

This is a module in Keras having various classes, methods, and objects which provide support for defining layers and respective input data size. There are various types of layers available in the Keras such as core layers, convolution layers, pooling layers, recurrent layers, normalization layers, regularization layers, and preprocessing layers. However, in this model, the only dense layer is used which belongs to core layers [18].

5.4 Utils

This is a module in Keras having various classes, methods, and objects which provide support for data preprocessing. There are various types of utilities available in the Keras such as model plotting utilities, serialization utilities, backend utilities, and other NumPy and Keras utilities. However, in this model, the to_categorical utility is used [19].

5.5 Callbacks

This is a module in Keras having various classes, methods, and objects which provide support during the training at the start and end of the epochs. There are various callbacks available in Keras such as base callback class, model checkpoint, tensor board, early stopping, learning rate scheduler, reduce LR on the plateau, remote monitor, lambda callback, terminate on NaN, CSV logger, and prog bar logger [20].

6 Procedure

The defining and training process of a deep neural network for predicting target output is conceptualized as an algorithm having stepwise activities. Firstly, important libraries must be imported for availing various classes and methods for data processing. In the next step, the dataset should be loaded and preprocessed, and now, a new sequential model can be defined as having several layers and supporting activation functions. The defined model is now ready for compiling and evaluation accordingly.

6.1 Importing Libraries

In the first step, at which, necessary libraries or packages for supporting this experiment must be imported. There are three important libraries, namely NumPy, Sklearn, and Keras. NumPy is normally shortened and imported as the np alias name. In the next line, the train_test_split method was imported from the Scikit-learn library for splitting the overall dataset into two parts for training and testing, respectively. In the remaining lines of the MNIST dataset, sequential model structure, to_categorical method utility, and callback object are imported from Keras library.

```
Import NumPy as np
from sklearn.model_selection import train_test_split
from Keras.datasets import mnist
from Keras.models import Sequential
from Keras.layers import Dense
from Keras.utils import to_categorical
from Keras.callbacks import Callback
Loading Data Set
```

In this step, the MNIST dataset is loaded which will be further processed using the sigmoid- and Relu-based models. (X_train, y_train) and (X_val, y_val) are the subset of the training dataset, X_train indicates a set of training data, y_train indicates the set of labels applied on the training data, X_val indicates a set of validation data, and y_val indicates a set of labels applied on the validation data. load.data() is a respective method for loading a dataset.

```
(X_train, y_train), (X_val, y_val) = mnist.load_data()
```

6.2 Normalizing Dataset

This step loaded dataset is normalized into a single-precision floating-point format which is a computer number format, usually occupying 32 bits in computer memory. astype() method is used to cast the object to a specified data type. It is also capable of converting any suitable existing column to a categorical type.

```
X_train=X_train.astype('float32')/255
X_val=X_val.astype('float32')/255
```

6.3 Labeling Dataset

In this step, normalized data is labeled using the to_categorical method of Keras. This method converts data into a binary class matrix having ten classes and assigned appropriate labels. Here, both training and validation data subsets are labeled.

```
n_classes=10
y_train=to_categorical(y_train, n_classes)
y_val=to_categorical(y_val, n_classes)
```

6.4 *Flattening Dataset*

In this step, labeled data is flattened or reshaped using the np. reshape() method of the NumPy library. Flatten means to reshape a multidimensional array into a one-dimensional array. Here, both training and validation data subsets are flattened, 28X28 size two-dimensional array into 784 size one-dimensional array.

```
X_train=np.reshape(X_train, (60000, 784))
X_val=np.reshape(X_val, (10000, 784))
```

6.5 *Using the Sigmoid Activation Function*

This is the first task of defining a model having a sigmoid activation function for deciding whether a neuron fires or not. After preprocessing the data, now data is ready to get input into the suitable deep neural network model, providing the desired output.

6.5.1 Defining Model

In this step, a sequential model is defined using sigmoid activation function. Further, there are four dense layers having 600, 300, 100, and 10 nodes, respectively. In the first three hidden layers, sigmoid function is used, and in the output layer, softmax function is used for activation because this is a multiclass classification.

```
sigmoid_model=Sequential()
sigmoid_model.add(Dense(600,        input_dim=784,        activa-
tion='sigmoid'))
sigmoid_model.add(Dense(300, activation='sigmoid'))
sigmoid_model.add(Dense(100, activation='sigmoid'))
sigmoid_model.add(Dense(10, activation='softmax'))
```

6.5.2 Compiling Model

In this step, the created model is compiled using the compile method. Here, the categorical_crossentropy method is used for calculating loss, sgd is used for optimization, and accuracy for matrices.

```
sigmoid_model.compile(loss='categorical_crossentropy',
optimizer='sgd', metrics=['accuracy'])
```

6.5.3 Fitting Model

In this step, the compiled model is evaluated or fitted using the fit() method. Here, the number of epochs is 10, the batch size for data is 256, the validation subset is 20%, and the verbose value is 2.

```
sigmoid_model.fit(X_train, y_train, epochs=10,
batch_size=256,validation_split=0.2, verbose=2)
```

6.6 Using the Relu Activation Function

This is the second part of the defining model having the Relu activation function for deciding whether a neuron fire or not. It can be seen in the given code that how Relu activations are applied with hidden layers.

6.6.1 Defining Model

In this step, a sequential model is defined using the Relu activation function. Further, there are four dense layers having 600, 300, 100, and 10 nodes, respectively. In the first three hidden layers, sigmoid function is used, and in the output layer, softmax function is used for activation because this is a multiclass classification.

```
relu_model=Sequential()
relu_model.add(Dense(600,input_dim=784,activation='relu'))
relu_model.add(Dense(300,activation='relu'))
relu_model.add(Dense(100,activation='relu'))
relu_model.add(Dense(10,activation='softmax'))
```

6.6.2 Compiling Model

In this step, the created model is compiled using the compile () method. Here, the categorical_crossentropy method is used for calculating loss, sgd is used for optimization, and accuracy for matrices.

```
relu_model.compile(loss='categorical_crossentropy', opti-
mizer='sgd', metrics=['accuracy'])
```

6.6.3 Fitting Model

In this step, the compiled model is evaluated or fitted using the fit() method. Here, the number of epochs is 10, the batch size for data is 256, the validation subset is 20%, and the verbose value is 2.

```
relu_model.fit(X_train, y_train, epochs=10, batch_size=256, vali-
dation_split=0.2, verbose=2)
```

7 Result

This section depicts how the defined model is working with assigned training data and validation data. Here, the number of epochs is 10, and the batch size is 256. After running ten epochs, appropriate output is shown as 10 * 4 matrices having loss values and accuracy values.

7.1 Performance of Sigmoid Activation Function

This is the performance of the sigmoid-based model after training of ten epochs, shown in Fig. 3. There are five columns shown in the output image, the first column indicates the epoch numbers, the second and third columns show the loss values and accuracy values during the training, and the fourth and fifth columns show loss values and accuracy values during the validation, respectively.

```
Epoch 1/10
188/188 - 4s - loss: 2.3062 - accuracy: 0.1075 - val_loss: 2.3016 - val_accuracy: 0.1060
Epoch 2/10
188/188 - 4s - loss: 2.2987 - accuracy: 0.1140 - val_loss: 2.2983 - val_accuracy: 0.1060
Epoch 3/10
188/188 - 4s - loss: 2.2962 - accuracy: 0.1140 - val_loss: 2.2961 - val_accuracy: 0.1060
Epoch 4/10
188/188 - 4s - loss: 2.2934 - accuracy: 0.1141 - val_loss: 2.2927 - val_accuracy: 0.1060
Epoch 5/10
188/188 - 4s - loss: 2.2907 - accuracy: 0.1155 - val_loss: 2.2894 - val_accuracy: 0.1060
Epoch 6/10
188/188 - 4s - loss: 2.2879 - accuracy: 0.1170 - val_loss: 2.2866 - val_accuracy: 0.1060
Epoch 7/10
188/188 - 4s - loss: 2.2849 - accuracy: 0.1197 - val_loss: 2.2838 - val_accuracy: 0.1060
Epoch 8/10
188/188 - 4s - loss: 2.2818 - accuracy: 0.1162 - val_loss: 2.2810 - val_accuracy: 0.1060
Epoch 9/10
188/188 - 4s - loss: 2.2785 - accuracy: 0.1187 - val_loss: 2.2772 - val_accuracy: 0.1060
Epoch 10/10
188/188 - 4s - loss: 2.2751 - accuracy: 0.1194 - val_loss: 2.2733 - val_accuracy: 0.1060
<tensorflow.python.keras.callbacks.History at 0x7f47562825c0>
```

Fig. 3 Training and validation output of sigmoid function

```
Epoch 1/10
188/188 - 4s - loss: 1.7114 - accuracy: 0.5745 - val_loss: 1.0311 - val_accuracy: 0.7954
Epoch 2/10
188/188 - 4s - loss: 0.7582 - accuracy: 0.8283 - val_loss: 0.5397 - val_accuracy: 0.8734
Epoch 3/10
188/188 - 4s - loss: 0.4964 - accuracy: 0.8714 - val_loss: 0.4084 - val_accuracy: 0.8963
Epoch 4/10
188/188 - 4s - loss: 0.4067 - accuracy: 0.8895 - val_loss: 0.3523 - val_accuracy: 0.9043
Epoch 5/10
188/188 - 4s - loss: 0.3606 - accuracy: 0.8999 - val_loss: 0.3204 - val_accuracy: 0.9117
Epoch 6/10
188/188 - 4s - loss: 0.3310 - accuracy: 0.9071 - val_loss: 0.2986 - val_accuracy: 0.9160
Epoch 7/10
188/188 - 4s - loss: 0.3094 - accuracy: 0.9122 - val_loss: 0.2844 - val_accuracy: 0.9187
Epoch 8/10
188/188 - 4s - loss: 0.2922 - accuracy: 0.9174 - val_loss: 0.2685 - val_accuracy: 0.9239
Epoch 9/10
188/188 - 4s - loss: 0.2781 - accuracy: 0.9208 - val_loss: 0.2587 - val_accuracy: 0.9265
Epoch 10/10
188/188 - 4s - loss: 0.2660 - accuracy: 0.9237 - val_loss: 0.2478 - val_accuracy: 0.9285
<tensorflow.python.keras.callbacks.History at 0x7f4752b9d7f0>
```

Fig. 4 Training and validation output of Relu function

7.2 *Performance of Relu Activation Function*

This is the performance of the Relu-based model after training of ten epochs, shown in Fig. 4. There are five columns shown in the output image, the first column indicates the epoch numbers, the second and third columns show the loss values and accuracy values during the training, and the fourth and fifth columns show loss values and accuracy values during the validation, respectively.

8 Discussion

The results have shown that the Relu function performed better than the sigmoid function with the same dataset and the same training process applied.

8.1 *Comparative Training Performance*

In this section, a comparative discussion was presented between the Relu activation function performance and sigmoid activation function performance during the training. The given line graphs shown in Fig. 5 are a critical comparative analysis of loss values and accuracy values.

The above line graph shows that the first epoch starts with the initial loss values 2.3062 and 1.7114 of sigmoid function and Relu function which is a little bit closer to each other. But in the next epochs especially till third epoch, Relu gave a very

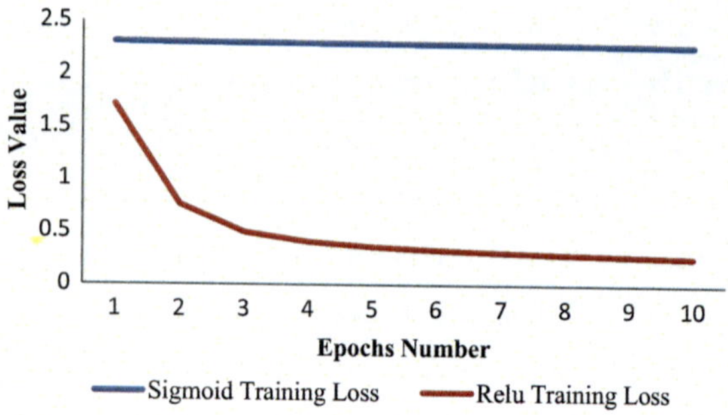

Fig. 5 Training loss performance of sigmoid and Relu function

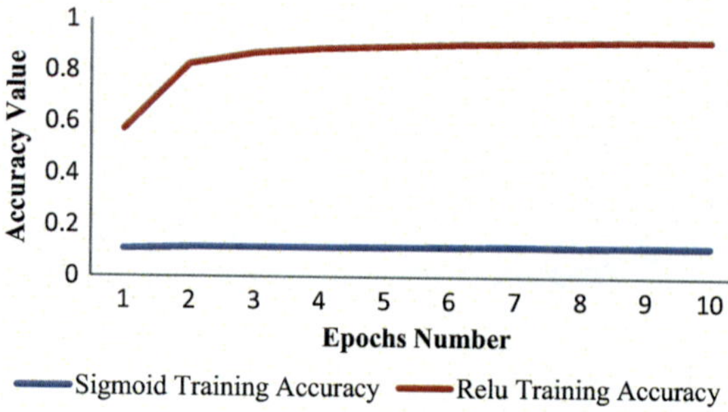

Fig. 6 Training accuracy performance of sigmoid and Relu function

good performance and decreased loss value and further continuously reduces the loss value up to 0.2660, and in the other hand, sigmoid function performed very poorly and stopped only at 2.7251, which means sigmoid was unable to decrease the loss value.

The above line graph (Fig. 6) shows that the first epoch starts with the initial accuracy values 0.1075 and 0.5745 of a sigmoid function and a Relu function which are a little bit closer to each other. But in the next epochs especially till the third epoch, Relu gave a very good performance and increased accuracy value and further continuously increases the accuracy value up to 0.9237, and on the other hand, the sigmoid function performed very poorly and stopped only at 0.1194 which means the sigmoid was unable to increase the accuracy value.

8.2 *Comparative Validation Performance*

In this section, a comparative discussion was presented between the Relu activation function performance and sigmoid activation function performance during the validation. The given line graphs are showing a critical comparative analysis of loss values and accuracy values.

The above line graph (Fig. 7) shows that the first epoch starts with the initial loss values 2.3016 and 1.0311 of a sigmoid function and a Relu function which are a little

Fig. 7 Validation loss performance of sigmoid and Relu function

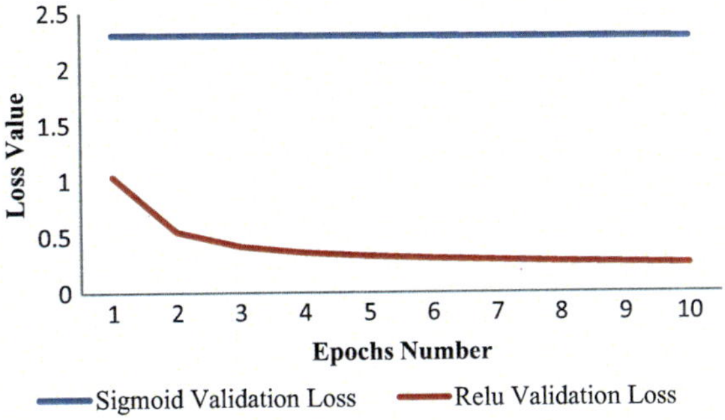

Fig. 8 Validation accuracy performance of sigmoid and Relu function

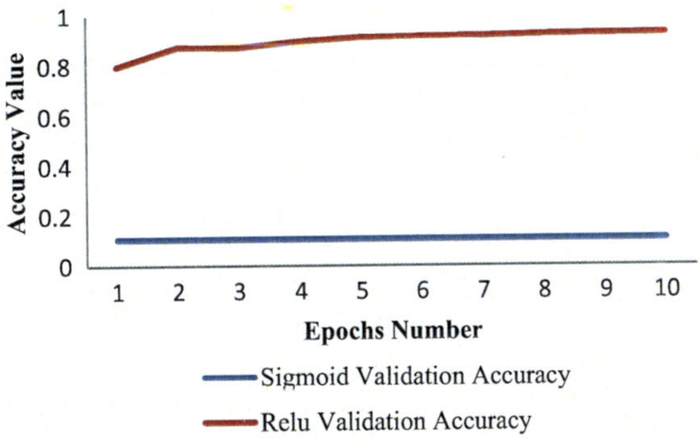

bit closer to each other. But in the next epochs especially till the third epoch, Relu gave a very good performance and decreased loss value and further continuously reduces the loss value up to 0.2478, and in the other hand, the sigmoid function performed very poorly and stopped only at 2.2733, which means sigmoid was unable to decrees the loss value.

The above line graph shows (Fig. 8) that the first epoch starts with the initial accuracy values 0.1060 and 0.7954 of sigmoid function and Relu function which is a little bit closer to each other. But in the next epochs especially till the third epoch, Relu gave a very good performance and increased accuracy value and further continuously increases the accuracy value up to 0.9285, and in the other hand, the sigmoid function performed very poorly and remain unchanged only at 0.1060 which means sigmoid was unable to increase the accuracy value.

References

1. De Carlos JA, Borrell J (2007) A historical reflection of the contributions of Cajal and Golgi to the foundations of neuroscience. Brain Res Rev 55:8–16. https://doi.org/10.1016/j.brainresrev.2007.03.010

2. Venkataramani PV (2010) Santiago Ramón y Cajal: father of neurosciences. Resonance 11(14):968–976
3. Schwiening CJ (2012) A brief historical perspective: Hodgkin and Huxley. J Physiol 590(11):2571–2575. https://doi.org/10.1113/jphysiol.2012.230458
4. Catterall WA, Raman IM, Robinson HPC, Sejnowski TJ, Paulsen O (2012) The Hodgkin-Huxley heritage: from channels to circuits. J Physiol 32(41):14064–14071. http://doi.org/10.1523/JNEUROSCI.3403-12.2012
5. Lettvin Y, Maturanat HR, McCulloch WS, Pitts WH (1959) What the frog's eye tells the frog's brain. In: Proceedings of the IRE 1940–1951
6. Myhrvold C (2013) In a frog's eye. MIT Technology Review
7. Olazaran M (1996) A sociological study of the official history of the perceptrons controversy. Soc Stud Sci 611–659
8. Seising R (2018) The emergence of fuzzy sets in the decade of the perceptron—Lotfi A Zadeh and Frank Rosenblatt's research work on pattern classification. Mathematics 6(7):110. https://doi.org/10.3390/math6070110
9. Minsky M, Papert S (1970) Perceptrons: an introduction to computational geometry. Inf Control 17(5):501–522. http://doi.org/10.1016/S0019-9958(70)90409-2
10. Szandała T (2020) Review and comparison of commonly used activation functions for deep neural networks. Bio-Inspired Neurocomput 203–224
11. Sun J, Binder A (2019) Generalized pattern attribution for neural networks with sigmoid activations. IJCNN 1–9. http://doi.org/10.1109/IJCNN.2019.8851761
12. Banerjee C, Mukherjee T, Pasiliao E (2019) An empirical study on generalizations of the ReLU activation function. In: ACM southeast conference, pp 164–167. http://doi.org/10.1145/3299815.3314450
13. Ketkar N (2017) Introduction to keras, deep learning with python, pp 95–109
14. Bisong E (2019) Introduction to scikit-learn. In: Building machine learning and deep learning models on Google cloud platform, pp 215–229. http://doi.org/10.1007/978-1-4842-4470-8
15. Harris CR, Millman KJ, van der Walt SJ, Gommers R (2020) Array programming with NumPy. Nature 585(7825):357–362. http://doi.org/10.1038%2Fs41586-020-2649-2
16. Datasets API. https://keras.io/api/datasets/. Accessed on Jan 2021
17. Models API. https://keras.io/api/models/. Accessed on Jan 2021
18. Keras layers API. https://keras.io/api/layers/. Accessed on Jan 2021
19. Utilities. https://keras.io/api/utils/. Accessed on Jan 2021
20. Callbacks API. https://keras.io/api/callbacks/. Accessed on Jan 2021

Chapter 6
CNC_Vision: Integrated Tool for Automated Creation of 2D Profiles for CNC Machine

Alarsh Tiwari, Ambuje Gupta, Harsh Kataria, Rushil Goomer, Sahaana Das, and Sonali Mehta

1 Introduction

The packaging is one of the most important sections in the shipping procedure of delicate instruments and goods; these are generally shipped enclosed in packages with protective measures such as package cushioning and padding. Expanded polyethylene (EPE) foam sheets are custom cut to encapsulate the article and shield it from its surroundings. The following paper [1] explains the procedure to manufacture and use EPE sheets. Apart from adding an extra layer of protection, this packaging approach also adds to the aesthetics of the packaged goods, as shown in Fig. 1.

To achieve aesthetically appealing and well-protected packaging, industries use computer numerical control (CNC) [2] fabrication devices such as plotters, routers, and cutters. A CNC router is a computer-controlled cutting machine that typically mounts a handheld router as a spindle used for cutting various materials.

Although CNC machines are very efficient, a specialist must operate and code the machine to move along a certain path to detect the boundary effectively. In the packaging industry, there are no standard objects to be packed. Hence, every new item they receive requires manual measurement of dimensions and an accurate outline drawing to be fed in the CNC machine, which are often tedious. The designer responsible for doing so needs to be skilled with the software and should have adequate expertise in using the tools to measure the object's dimensions to be packed accurately. This is a very time-consuming task and greatly increases the dependency of a packaging company on some individuals.

In this study, we intend to propose a new system that specifically aims to reduce the packaging industry's dependency on a few skilled individuals to efficiently produce designs or paths for the CNC machine. With our proposed solution, the whole multi-step process of switching between tools and software like AutoCAD [3] could potentially be replaced with a few clicks over the computer.

A. Tiwari (✉) · A. Gupta · H. Kataria · R. Goomer · S. Das · S. Mehta
Department of Computer Science Engineering, Bennett University, Greater Noida, India

A. Sheth et al. (eds.), *Intelligent Systems*, Algorithms for Intelligent Systems,
https://doi.org/10.1007/978-981-16-2248-9_6

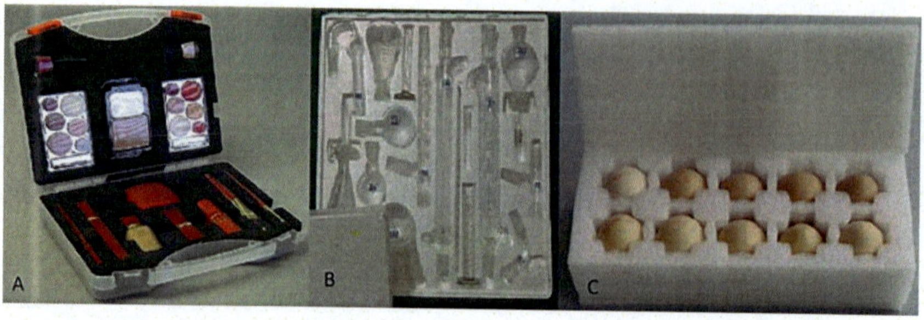

Fig. 1 Depicting CNC-cut products: tool kit packaging (**a**), glassware packaging (**b**), and packaged eggs (**c**)

The rest of the article is organized as follows: Related work is discussed in Sect. 2. The hardware setup in the proposed work is presented in Sect. 3. Section 4 presents the proposed methodology followed by experimental results and discussions in Sect. 5 and the conclusion in Sect. 6.

2 Related Work

There is plenty of research being currently done in the domains of digital fabrications and non-GUI interfacing. The flexible intelligent modelling system [4] is a significant instance of a computational-interface (CI) bridging between physical inputs and outputs, developed in the year 1980. The system is supposed to give support to the architecture of the design process. It consists of sensor-embedded blocks that can determine their configuration and provide output in the form of a 2D representation to a plotter.

Interactive fabrications [5] computerize a few plans and design steps to make an immediate connection between human input and machine yield. They present a progression of model gadgets that make use of real-time inputs to fabricate physical forms. CopyCAD [6] is a camera-projector framework on a CNC machine that aims to incorporate actual items into CAD; CopyCAD scans the canvas and requires the client to draw the 2D plan on that canvas. Similarly, VisiCut [7] attempts to attach a camera on a laser cutter that can be used to provide visual-positioning and preview. It was mainly developed to bring a consistent format for working with projects on a single device, saving precious time wasted in switching between lab computers and the target computer. PacCam [8] presents a definite study of how the client cooperates with a customary CNC machine and investigates the area of improvements. Nonetheless, their fundamental spotlight is on UI and design; they bring a ton of understanding about the innovation being embraced with the conventional techniques.

Our product's methodologies are mostly focused on either assisting users in making profiles for CNC machines or just improving the overall user interface (UI).

As per our best knowledge, we concluded that these works neither directly target the packaging industries nor remove the need for such industries to hire skilled professionals.

3 Hardware Setup

Capturing material shape and fitting a design is an essential component of the packaging task. Our proposed solution setup is simple, as illustrated in Fig. 2, and is quite easy to implement; it consists of a four-legged table with a glass top that serves as a base of dimensions 38.1 × 30.48 cm.

The base is illuminated from below to avoid shadow formation by the object, using a setup of multiple LED lights; the glass is covered with a special type of diffusing paper that helps evenly spread the light from the LED across the table. The overlooking camera in the displayed setup is a phone's camera having 3042 × 4032 pixels and is mounted on an adjustable stick. The camera could be easily replaced by a DSLR at a later stage, depending upon the use cases. Camera calibration, discussed later in the methodology section, needs to be carried out each time the camera's height is adjusted, or the setup is booted. After each system reboot, users would be required to capture two pictures: the first one of the empty tables and the second one of the materials to be packed on the table.

Fig. 2 First prototype of the proposed solution

4 Proposed Methodology

As shown in the hardware setup, there is a fixed 2 cm × 2 cm square drawn on the upper left corner of the base that is utilized by the overlooking camera for calibration purposes. A general overview of the process is shown in Fig. 3.

To begin the calibration using the corner square, the camera takes a photograph of the base and tries to find the square using OpenCV contour detection [9]. The base's upper left corner has the coordinates (0, 0) and contains the square for calibration. The algorithm then finds the Euclidean distance between the origin and the corners of the detected square described in Eq. 1.

$$\text{dist}((x, y), (a, b)) = \sqrt{(x - a) + (y - b)} \quad (1)$$

(x, y) represents the pixel location of the square's corner, and (a, b) represents the origin having the value (0, 0). With the Euclidean distance, the algorithm finds the pixels in a centimeter using the formula shown in Eq. 2.

$$\text{pixel per centimeter(ppc)} = \text{dip/dis.} \quad (2)$$

where ppc is the pixels per centimeter, the dip is the Euclidean distance calculated above, and dis is the distance of the edge of the square, which is pre-fed to the algorithm; in this case, 2 cm.

After calibration, the user is asked to place the object to be packed on the base. The overlooking camera captures the object's image as a top view. The image's edges are extracted using OpenCV's canny edge detection [10] and holistically nested edge detection [11] after applying a Gaussian Blur mask on the whole image. The output of both the edge detection methods is then displayed, and it is up to the user to select either one of the choices and proceed further.

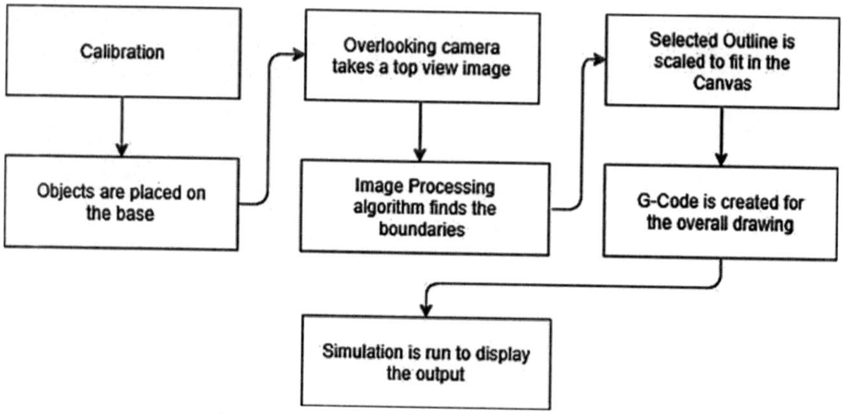

Fig. 3 Flowchart depicting the process to get the final G-code

The extracted edges using the above-mentioned process are then used to find the object's width and height, which is described in Eqs. 3 and 4, respectively.

$$\text{Height} = (Y(c) - Y(d))/\text{ppc}. \tag{3}$$

$$\text{Width} = (X(a) - X(b))/\text{ppc}. \tag{4}$$

where the $X(a) - X(b)$ denotes the difference of the x coordinates of the rightmost (a) and the leftmost (b) pixel of the object's boundary and $Y(c) - Y(d)$ denotes the difference of the y coordinates of the bottom-most (c) and topmost (d) pixel of the object's boundary. As shown in Eqs. 3 and 4, the difference between the coordinates is then divided by pixels per centimeter, which were stored in the calibration process to get the object's width and height in centimeters.

For a CNC [2] machine to operate, it requires a certain path to travel. In a few cases, like in the packaging industry, sharp edges and tight-fitting of the object in a base are not always required. Hence, we provide the user with two options to proceed further. The user could either continue with the boundaries detected by the algorithm or opt for a bounding shape approach. The bounding shape approach enables the user to select a shape that nearly fits the scanned object; the bounding shape could be any fixed polygon of size close to the scanned object's shape. If the user chooses to opt for bounding shapes and selects a preference as a rectangle, then the algorithm draws a rectangle considering the width and height extracted in the above steps.

To handle the noise in the environment, the hardware is custom designed to handle variability in the surroundings. The user must place the object on the system, and the hardware is under-lit to overcome the effects of shadows and external lighting conditions. Moreover, after clicking the photo, user has an option to select the region of interest, i.e., select the area just outside the object to get rid of the noise.

The detected boundary or the bounding shape is then resized to fit in the desired canvas. The canvas represents the physical sheet on which the CNC machine is to be operated on, whose shape and size are provided by the user. After resizing, the location of all the boundary pixels (X, Y) is found out relative to the origin using the pixels per centimeter methodology, as shown in Fig. 4.

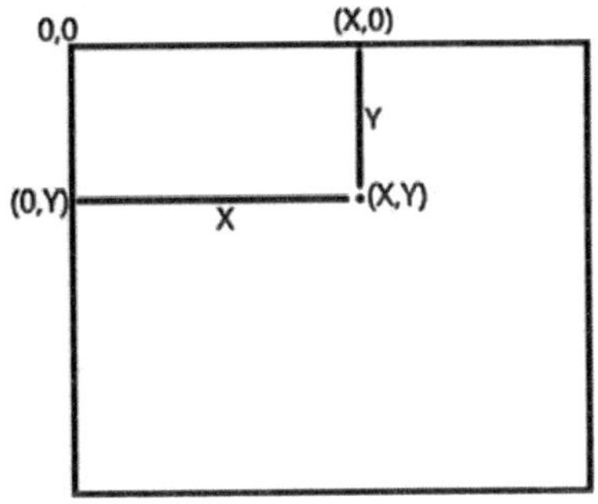

Fig. 4 Representing the canvas and distance of a point from the origin

This collection of the locations of boundary pixels in a vector format is then fed to the pygcode library to generate G-code [12] for the CNC machine to operate on. G-code is a set of instructions that guides the machine tool as to what type of action is to be performed at a certain point. The boundaries converted to G-code serve as a path, and the machine tool starts to cut through this path.

5 Experimental Results and Discussion

We conducted a few experiments with a small-scale packaging industry in Haryana, India, for validation. This section has discussed the results obtained on a scenario of a pen's packaging process. During the calibration phase, the corner square is being detected effectively as shown in Fig. 5 a, and the pixel per centimeter in our machine is calculated to be 47.5 pixels per centimeter. With this reference, the user places the pen with actual dimensions of 0.9 cm in width and 13.9 cm in height like Fig. 5 b. With our algorithm, we detected the pen's dimensions as 1 cm in width and 13.8 cm in height, and the output is shown in Fig. 5 c.

With our continuous experimentation, we found the algorithm to have an error of not more than 250 mm along both the axes. The pen's boundary detection with both the canny edges [10] and holistically nested edge detection is shown in Fig. 5d.

For our scenario, we opted to proceed with canny edge's output as they looked more precise and continuous. As shown in Fig. 6b, the detected edges are now resized and placed over a user-defined canvas, in our case of A4 standard size having a width and height of 210 cm and 297 cm, respectively. The user could also opt for a bounding shape as in Fig. 6a.

The location of the edges of the pen or the bounding shape is extracted to get the X–Y coordinates relative to the origin, which is then converted to G-code using the pygcode library in Python. The G-code simulation is done on NC Viewer [13], which is the actual path that the CNC machine will take and is shown in the figure for both bounding shape Fig. 6c and the object edge Fig. 6d selections. The G-code generated for the shapes is placed at the exact location as in the A4 canvas.

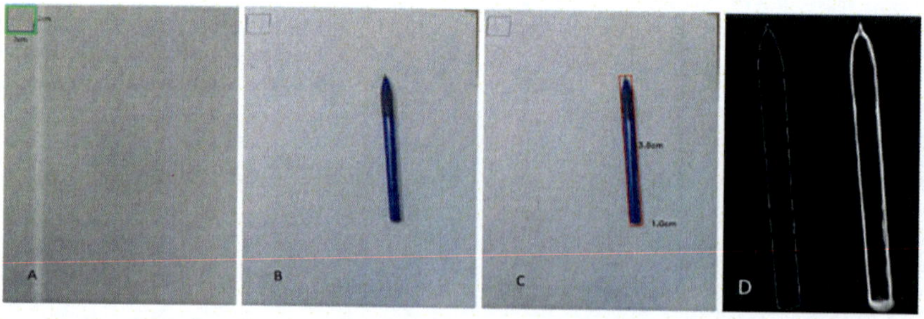

Fig. 5 a–d Represent the calibration, object placing, width detection steps, and output of canny edge detection (*D-Left*) and holistically nested edge detection (*D-Right*), respectively

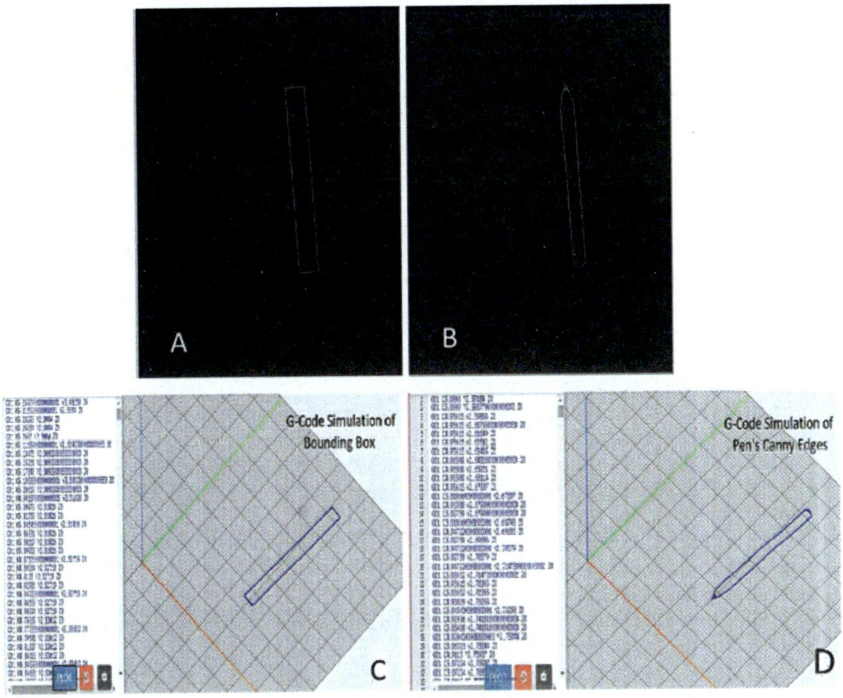

Fig. 6 **a, b** Represent the mapping, whereas **c, d** represent the simulation of the G-Code generated for bounding shapes and canny edges, respectively

6 Conclusion and Future Work

In this study, we demonstrated a stand-alone system that can help boost production in the packaging industry. It would offer more precision with a far less error rate compared to the traditional methods being followed as of now, owing to the reduced degree of human intervention in the process. This in turn could be a potential game changer in the industry as a cost-effective approach to packaging. The proposed idea has been currently implemented for a single object in the frame, which can be expanded to accommodate multiple objects at a time. Additionally, one can implement space optimization in the overall process to ensure minimum wastage of the material. The prototypes presented in this paper have only scratched the surface of a much larger area of exploration. Packaging is a complex process, and the authors hope their efforts would contribute toward automation in the packaging industry.

References

1. Carr CI (1963) Expanded polyethylene and method of making the same. US Patent 3098831
2. Ginting R, Hadiyoso S, Aulia S (2017) Implementation 3-axis cnc router for small scale industry. Int J Appl Eng Res 12(17):6553–6558
3. AutoDesk Autocad. https://www.autodesk.in/products/autocad/overview
4. Frazer J (1995) An evolutionary architecture, the architectural association. EG Bond Ltd., London, UK
5. Willis KD, Xu C, Wu K-J, Levin G, Gross MD (2010) Interactive fabrication: new interfaces for digital fabrication. In: Proceedings of the fifth international conference on Tangible, embedded, and embodied interaction, pp 69–72
6. Follmer S, Carr D, Lovell E, Ishii H (2010) Copycad: remixing physical objects with copy and paste from the real world. In: Adjunct proceedings of the 23nd annual ACM symposium on user interface software and technology, pp 381–382
7. Oster T (2011) Visicut: an application genre for laser cutting in persona fabrication. Bachelor's Thesis, RWTH Aachen University, Aachen
8. Saakes D, Cambazard T, Mitani J, Igarashi T (2013) Paccam: material capture and interactive 2d packing for efficient material usage on cnc cutting machines. In: Proceedings of the 26th annual ACM symposium on User interface software and technology, pp 441–446
9. Gurav RM, Kadbe PK (2015) Real time finger tracking and contour detection for gesture recognition using opencv. In: The 2015 international conference on industrial instrumentation and control (ICIC). IEEE, pp 974–977
10. Topal C, Akınlar C, Genc Y (2010) Edge drawing: a heuristic approach to robust real-time edge detection. In: 2010 20th international conference on pattern recognition. IEEE, pp 2424–2427
11. Xie S, Tu Z (2015) Holistically-nested edge detection. In: Proceedings of the IEEE international conference on computer vision, pp 1395–1403
12. Brown AC, De Beer D (2013) Development of a stereolithography (stl) slicing and g-code generation algorithm for an entry level 3-d printer. In: 2013 Africon. IEEE, pp 1–5
13. Viewer N Online G-code simulation tool. https://ncviewer.com/

Chapter 7
Survey of Real-Time Object Detection for Logo Detection System

Amarja Indapwar, Jaytrilok Choudhary, and Dhirendra Pratap Singh

1 Introduction

Object detection is one of the traditional issues where you work to perceive what and where—explicitly what articles are inside a given picture and furthermore where they are in the picture. As of late, profound learning has been effectively applied in different fields including computer vision, self-governing driving, informal community administrations, and logo detection frameworks. The issue of object discovery is more unpredictable than arrangement, which likewise can perceive protests however do not show where the article is situated in the picture. Moreover, characterization does not take a shot at pictures containing more than one object [1]. The layers of profound neural organization remove different highlights and subsequently give numerous degrees of deliberation. When contrasted with shallow organizations, this cannot concentrate or work on various highlights. The errand of discovering correspondences between two pictures of a similar scene or item is important for some computer vision applications. Camera alignment, 3D reproduction, picture enlistment, and article acknowledgment are only a couple [2, 3].

There are a few issues for preparing visual information from a picture which was debased by commotion or exposed to any change and furthermore its precision in coordinating. Logos are a portion of the developing exploration issues in this day and age. We see several logos consistently. We see them so much that we regularly do not consider them. Logo extraction from pictures is a major test issue with possibly wide business applications; numerous applications require continuous ordering. Ordering and extraction of logos pictures lead to flawlessness. Realistic logo is a phenomenal gathering of outline stuff very critical to charge the uniqueness acceptably. In industry, they have a fundamental function to review in the client desire related to a specific item. Logo location in live video transfer is certainly not a straightforward issue;

A. Indapwar (✉) · J. Choudhary · D. P. Singh
Computer Science Department, Maulana Azad National Institute of Technology, Madhya Pradesh, Bhopal 462003, India

A. Sheth et al. (eds.), *Intelligent Systems*, Algorithms for Intelligent Systems,
https://doi.org/10.1007/978-981-16-2248-9_7

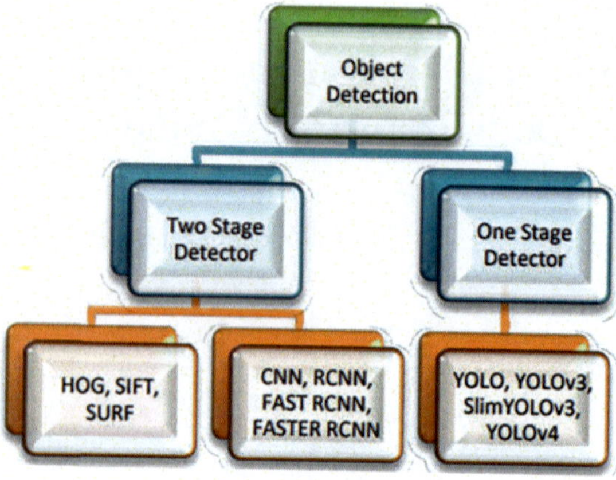

Fig. 1 Object detection flowchart

right off the bat logos can be of a few sorts. Logos can be made out of even lines, vertical lines, blend of both, text, designs or blend of every one of these elements. It is hard to recognize logos in video inclusion because of different sorts of change like scale, revolution, shearing, and interpretation [1, 4].

Picture enrollment is a significant strategy in numerous computer vision applications, for example, picture combination, picture recovery, object following, face acknowledgment, change recognition, etc. Neighborhoods include descriptors, i.e., how to identify highlights and how to depict them, play a key and significant part in picture enrollment measure, which straightforwardly impact the precision and power of picture enlistment [5]. Continuous item recognition models ought to have the option to detect the climate, parse the scene lastly, and respond in like manner. The model ought to have the option to distinguish what a wide range of articles are available in the scene. In this way, there are two capacities here. To start with, arranging the articles in the (picture order) and afterward finding the items with a bouncing box (object recognition). We can conceivably confront different difficulties when we are dealing with a continuous issue:

The varieties may be of contrast looking like articles, splendor level, deploying item location models. Finally, we should likewise keep a harmony between discovery execution and continuous prerequisites. For the most part, if the constant necessities are met, we see a drop in execution and the other way around. In this way, adjusting both these perspectives is additionally a test (Fig. 1).

2 Literature Survey

This section consists of a background and the literature survey of the paper.

2.1 Background

Harris Corner Detector (1998) gave us detection of interest points using the eigenvalues of the second-order moment matrix.

$$M = \sum_{(x,y)\in W} \begin{bmatrix} I_x^2 & I_x I_y \\ I_x I_y & I_x^2 \end{bmatrix}$$

This method was not scale invariant.

Laplaician of Gaussian-Lindeberg (1998) was based on automatic scale selection using Laplaician of Gaussian as well as the Hessian matrix

$$\nabla^2 G(x, y) = \frac{\delta^2}{\delta x^2} G(x, y) + \frac{\delta^2}{\delta y^2} G(x, y)$$

Harris-Laplace detector (2001) worked on the weak points of the above two methods and gave a better scale invariant detector. It used Harris Corner response or Harris determinant for interest point detection and Laplaician for scale detection. Scale invariant feature transform (SIFT) (2004) gave us a means of faster calculation of Laplaician of Gaussians by approximating it as a difference of Gaussian with a large 128-dimensional descriptor. Hessian-based detectors are supplementary unwavering in addition to repeatable than their Harris-based counterpart. Elevated dimensionality of the descriptor is a shortcoming of the SIFT algorithm and reduction of dimensionality accuracy.

2.2 Literature Review

In this section, various one-stage and two-stage real-time object detection methods are studied with their pitfalls.

2.2.1 Two-Phase Detectors

Two-stage detectors are used in these algorithms. The object detection accuracy is improved by each of the preceding algorithms.

SURF Algorithm

Herbert et al. [1] proposed speeded-up robust feature (SURF) algorithm. Speeded-up robust features which are another picture indicator and descriptors propelled by SIFT. SURF verges on or still beats an already extended plan by charmingly excited about

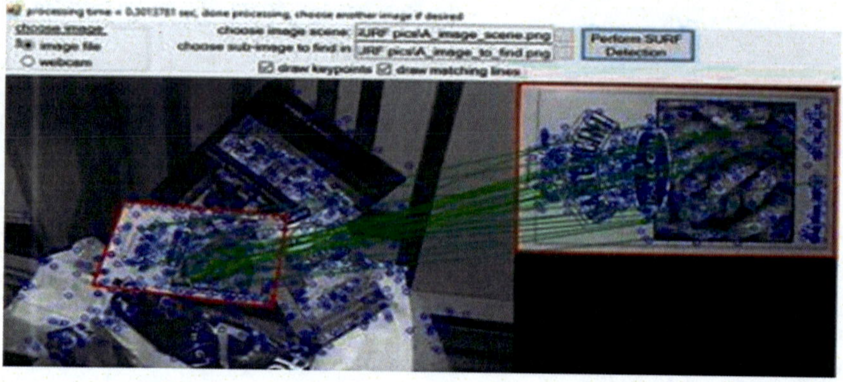

Fig. 2 SURF working [4]

idea repeatability, uniqueness, and strength, unmoving can be figured and analyzed amazingly quick. SURF calculation is valuable for extraction of the intrigue focus from reference logo and test picture instead of SIFT calculation. SURF is a lot quicker than SIFT. The fundamental distinction of SURF calculation is it essentially applied to the Determinant of Hessian (DoH) which is the extraction of the component focuses in the picture that is acknowledged through disentanglement and guess, contrasted with the SIFT calculation, and it diminishes the intricacy of highlight point extraction. It likewise has better constant execution. The main stepping stools of SURF are [4] Fast-Hessian detector, construct scale space, precise notification tip localization, attention end descriptor, compass understanding position, and descriptor strategy (Fig. 2).

The pitfalls of this method are large code size, feature locators cannot be prepared. Thus, it gives less exactness [6], feature building is more, abstract portrayal unrealistic, and other higher precision algorithms [7] are developed.

CNN Algorithm

Galvez et al. [7] have proposed this CNN model. Here, each object in the image is positioned as well as acknowledged by way of a convinced plane of exactitude. An image is passed enroot for the set of connections along with it is sent from beginning to end a mixture of convolutions furthermore pool layer [7]. At long last, the output is obtained from the purpose group.

Accomplishing new statues in object discovery and picture arrangement was made conceivable in view of convolution neural network (CNN). In any case, contrasted with picture order the item identification assignments are harder to break down, more energy devouring and calculation serious. To defeat these difficulties, a novel methodology is created for continuous item identification application on the road to progress the strictness along with vigor know-how of the innovation succession. This is consummate through incorporating the CNN in the midst of the SIFT calculation.

At this point, high exactness yields with little example information to prepare the model by coordinating the CNN and SIFT highlights which are achieved [8, 9].

Here, a webcam of eight super pixels, for the framework to be more powerful against the clamor, is utilized. Each picture prepared in the framework is intensified by utilizing various changes directly. These changes incorporate pivoted and scaled prepared pictures. In this model representation, a trademark CNN engineering is constructed commencing the starting point. The organization contains six layers, of which a convenient two max pooling layer follows through an absolutely allied layer. A piece moment in time there is a greatest pool sheet including the extent of channel search out multiply. The porthole amount of the channel considered here is 3 × 3 and afterward comes the maximum pool deposit by way of an extent of "2 × 2" is situated past each two CNN layer. Max pooling is acclimated with diminish the calculation used in favor of the more insightful layer in adding together to furthermore help in introducing interpretation invariance [10].

The drawbacks of this method are that objects in the picture may have various angles, proportions, and spatial areas, and some of the articles may be covering the vast majority of the picture while others are covering less, a large number of districts, more calculation time, and shapes might be unique [11].

RCNN Algorithm

Girshick et al. [12], instead of an immense quantity of region [12], proposed a horde of boxes in the illustration furthermore check qualification every of these boxes contain at all points. It uses discriminatory rummage around to dig out these boxes commencing reflection. Varying scales, colors, texture, and inclusion are the four regions that identify the object.

Discriminatory investigation to identify these patterns in a representation as well as region is proposed accordingly [13, 14]. It generates initial sub-segments to have multiple regions from one image. Then, considering color correspondence, consistency, amount along with silhouette compatibility, the similar regions are combined, and the region of interest is produced. It takes pre-trained CNN and retrains the model. Then, it trains the last layer of network based upon the quantity of classes to be detected. The region of interest for each image is taken into consideration, and all the regions are reshaped to match the CNN input size [14]. 1 binary SVM for each class is trained for object classification and background.

2000 areas are excerpted per picture. For each locale or part of the picture, we need to choose highlights utilizing CNN. For this, on the off chance that we have "I" number of pictures, at that point chosen districts determination be converted into $j \times 2000$. The entire strategy for intention distinguishing proof from beginning to end RCNN uses the accompanying three models:

Linear SVM classifier for the ID of article, CNN be utilizing designed for trademark extraction, and a relapse model is needed to fix the bouncing box. All these three cycles join to take a lot of time. It builds the running season of RCNN technique.

Consequently, RCNN requirements are just about 30–60 s to foresee the outcome on behalf of a few fresh pictures [15].

The pitfalls of this method are that training CNN models is slow and expensive, and it is time consuming as the number of models used are more [7].

Fast CNN Algorithm

Zhao et al. [15] proposed fast RCNN. In this method, instead of running a CNN 2000 times per image, it can be run just one per image to get all regular points of interest.

Instead of utilizing three distinct models of RCNN, fast RCNN [16] utilizes one model to extract qualities from the various areas. At that point, it disseminates the districts into a few classifications dependent on excerpted highlights, and the limit boxes of perceived divisions return together. Quick RCNN utilizes the technique for spatial pyramid pooling [6] to figure just a single CNN portrayal for the entire picture. It passes one locale for each image to a specific convolutional network model by substituting three particular models for excerption of qualities, appropriating them into divisions, and delivering jumping boxes.

The pitfalls are it still uses selective search and computation time still not up to the mark [15].

Faster RCNN Algorithm

Girshick et al. [16] proposed faster RCNN of enthusiasm surveying is the methodology that is increasing a lot of consideration during the countryside of article acknowledgment along with arrangement, a profound scholarship comes within reach of. An example possibly will be the location of items commencing a landscape of pictures containing numerous bits and pieces. The objective is to utilize a maximum pool resting on the sum total picture toward getting rid of included guides of fixed size. Region proposal network [15] and organization [12] are the two new concepts of this method for image localization.

The pitfalls are that it does not look at the entire image at once, it requires countless passes [6] from first to last a particular likeness to pull out every one of the bits and pieces, and it is highly system dependent at the same time as it uses channel [2].

2.2.2 One-Phase Detectors

Here, the algorithm uses only a single stage detector leading to faster computation. CNN is performed on the entire image leading to fewer mistakes.

YOLO Algorithm

Over feat Serman et al. [17] instruct a CN organization to act upon restriction in addition to fiddle with the intention of the localizer to complete discovery. Over feat in actual fact performs descending skylight place nevertheless, it is nevertheless a displaced skeleton Over feat advances for restraint by not naming implementations. Akin to DPM, the localizer perhaps observes near statistics when assembling an anticipation. Over feat insincerity basis about international surroundings in addition to subsequently require noteworthy post-handling to create cognizant discoveries multi-grasp. Our vocation is analogous within construction to arrangement by way of handling discovery [18]. The lattice line of attack to deal through jump box expectation depending on the multi-grasp skeleton intended for degeneration handling. Notwithstanding, handle location is a lot more straightforward assignment than object discovery.

Multi-grasp, in a minute foresee on the way to foresee alone intelligible area in favor of a visual rendering containing solitary articles. It does not necessitate weighing the bulk, vicinity, or confines of the objective or being hopeful of its course group, presently determining a borough apposite on behalf of accomplishment a knob resting on. YOLO predicts mutually full of beans boxes furthermore course group probability pro poles apart objects of plentiful curriculum in a depiction. It performs CNN on the entire image at once and hence leads to fewer mistakes.

By utilizing the methodology of one-phase identifier YOLO, the in-rank depiction is circulated toward an arrangement of $K \times K$ networks [15]. The locality of the article relies upon each framework of the info picture. Framework cells are utilized to foresee focus inside limit boxes. Five boundaries are anticipated for each limit box. These five components are A, b, c, d, and e. The focal point of focus inside the container is signified by "A" and "b" facilitates, "c," "d," and "e" speak to tallness, width, and score for certainty separately. "e" is estimated at the same time as the odds that contain the objective in the interior limit spar. YOLO utilizes the dark net system furthermore ImageNet1000 dataset to prepare the representation. It circulates an offered depiction to a network of $K \times K$ cells. In support of each cubicle in the organization, it registers certainty for "t" jumping boxes. The anticipated outcome is present keen on a tensor as $K \times K \times (t \times 5 + q)$ [15].

At this point, key depiction is unconnected to comprehensive $K \times K$ sub-depictions. In $t \times 5$, $5K \times K$ lattice speaks to the recognition of 5 credits meant for each bouncing spar, that is tallness, influence, certainty notch up, and focus coordinates (n, m) of identified articles. Here, "q" speaks to the likelihood of a category. YOLOv1 furthermore has numerous constraints. Along these lines, the deployment of the YOLO is confined fairly. A restriction on version1 depends upon the imminence of the article in the illustration. On top of the rancid opportunity to facilitate the articles let somebody notice out of bed since a cluster, they could not trace the modest bits and pieces. On the off chance that the components of the item are unique in relation to the picture utilized in preparing information, at that point, this design discovered trouble in the limitation and location of articles [19, 20]. The essential apprehension is enacting for finding bits and pieces in a given representation due to constraint mistake [21].

The pitfalls are services frozen spatial needs [18] and channel dependencies.

2.2.3 YOLOv3 Algorithm

Redmon et al. [22] proposed YOLOv3. Article discovery has been utilized in abundant domains. For illustration, wellbeing furthermore instruction, and furthermore, as all of these domains are embryonic hurriedly, subsequently to harmonize their rations, solitary segment models too need improvement. The subsequent progressed discrepancy is form (3) that utilizes intended setback to stature the largeness notches up. It gives the notch up on behalf of the entire objectives in apiece boundary spar. YOLOv3 canister awards the multi-label command seeing as it utilizes an intentional classifier designed for every one division as a like chalk and cheese of the softMax deposit utilized in YOLOv2. YOLOv3 considers darknet "53.." It has "53" layers of obscurity. These layers are supplementary surrounded along with contrast by way of darknet "19" considered in YOLOv2. Darknet- "53" contains for the most part "3 × 3" in addition to "1 × 1" channels at the sidestep join [18, 22, 23].

Pitfalls are high system dependency (channel), multi-label problem [6], and small object detection in image needs to be improved.

Slim YOLOv3 Algorithm

Xiao et al. [24], profounded Slim YOLOv3. YOLOv3, with the impair profundity of the convolutional footing, is another translation invoked by YOLOv3-Tiny. It was converse [22]. Subsequently, the continuous walk is essentially extended (severely about 440% quicker than the fore variations of YOLO), and however, identification nicety is reduced. Dark catch-53 designate of YOL v3 utilizes a few "1 × 1" volume seam abreast "3 × 3" volume footing for depart spotlight [23]. YOLOv3-Tiny utilizes pooling coping and lessens the digit for volume seams. It forecasts a three-dimensional tensor that includes an abject account, excessive spar, and brace foreseen at two singular scales. It compartments a likeness into $S \times S$ plexus cells. For unanswerable discoveries, we will neglect the excessive spar for which the abjectness record is not conception. For separating spotlight, volute seam and max pooling footing are useful in the fodder ardent series of demeanor of YOLOv3-Tiny [6, 24, 25]. Expectation of jumping boxes happens at two distinctive element map balances which are "13 × 13," as well as "26 × 26" converge with an up sample "13 × 13" section drawing (Figs. 3 and 4).

Readily available are various parts in the model. A scrap of the associations, a couple of cycles later, got repetitive, and subsequently, it may be eliminated from the associations on or after the representation. Eliminating the links is alluded to as pruning. Prune would not fundamentally sway the presentation of the model, and the calculation force would diminish altogether. Henceforth, during Slim YOLOv3,

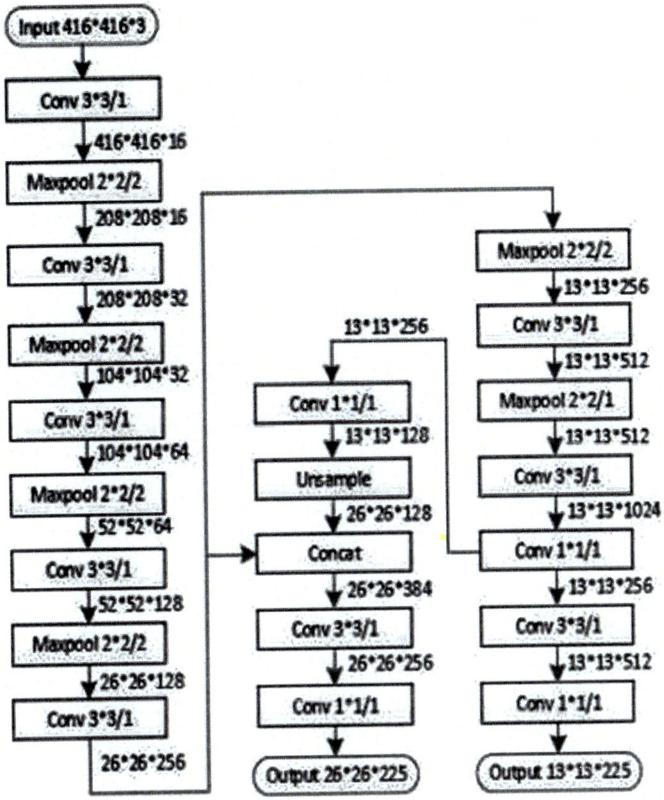

Fig. 3 Slim YOLOv3 architecture [6]

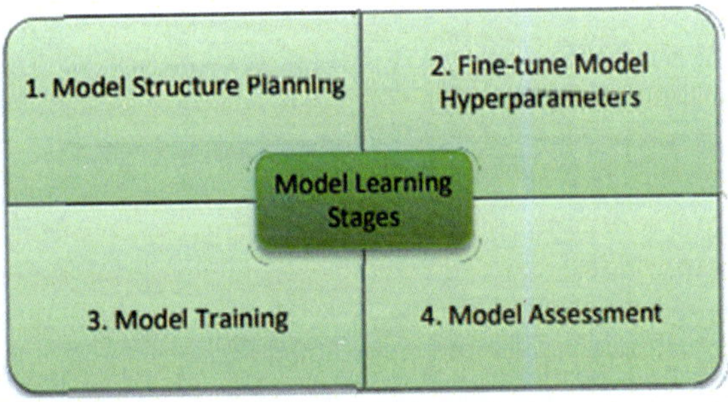

Fig. 4 Model learning stages

prune will perform a convolutional layer. We will get familiar with how this prune is through during the subsequent subdivision of this commentary.

Slim YOLOv3 is the change variant of YOLOv3. The CNN layer of YOLOv3 is piped in the direction of accomplishing a slender moreover more rapid form. Identifiers having a place with the YOLO arrangement plummet underneath on its own

Fig. 5 Slim YOLOv3 object detection [6]

juncture locators. It is an introverted juncture measure. The following representations bring into play the previously defined secures with the intention of spreading specific situations, weighing machines, and moreover angle proportions over a picture. Henceforth, we need not bother with an additional branch for separating area recommendations [22]. In view of the fact that all calculations are during a solitary institute, they are bound on the road to scamper more quickly compared to two-stage indicators. YOLO v3 be furthermore a solitary juncture indicator in addition to at in attendance the finest in class for object location.

Preparing with a bigger punishment factor of "$\alpha = 0.01$" prompts forceful rot of the scale aspect furthermore the representation begins to overfit. Hence, "$\alpha = 0.01$" is used here (Fig. 5).

The drawback of this method is that it does not have an efficient amount of average precision [22].

YOLOv4 Algorithm

Bochkovskiy et al. [26] proposed YOLOv4. The base of the research is to design a rapid in-service momentum based on the object detector in production system as well as to optimize parallel computation, slightly than the small calculation volume theoretical indicator (BFLOP) so that the intended item tin can subsist effortlessly skilled as well as be worn. Meant for illustration, a person who uses a conventional GPU to train and test containers accomplishes concurrent, sky-scraping eminence, furthermore compelling purpose recognition outcome, while the YOLOv4 consequences. The essence is a simple, and efficient model was developed which make each one container make use of a 1080 Ti or 2080 Ti GPU enroot for training a marvelous swift along with correct article detector, the confirmation of influence of state-of-the-art Bag-of-Freebies plus Bag-of-Specials method of entity detection throughout the detector teaching was done, and the alteration of state-of-the-art method in addition to construct them extra efficiently in addition to fitting meant for single GPU guidance, together with CBN, PAN, SAM, etc., was done [26].

The drawbacks are as follows: It is difficult to use, and it is not very preferable on custom data [27].

3 Conclusion

Here, the comparative study of one-stage and two-stage detector deep learning algorithms is done along with the challenges in each one of them. One-stage detector algorithms are far better than two-phase detector algorithms, and they overcome the various challenges faced by "two-stage detectors." The computational time of "one-stage detector" algorithms is very less compared to two-stage detector algorithms. The best one-stage detector algorithm is YOLOv4 as it is faster, easier, and good for production environments.

References

1. Bay H (2018) SURF: speeded up robust features. Home 7(4)
2. Redmon J (2016) You only look once: unified, real-time object detection. arXiv:1506.02640v5 [cs.CV]
3. Redmon J, Farhadi A (2018) YOLOv3: an incremental improvement. arXiv:1804.02767v1
4. Indapwar A, Dwivedi P, Choudhary J, Saritha K, Singh DP (2020) Logo detection system for live videos. IJST 29
5. Lengi C (2019) Local feature descriptor for image matching: a survey. Int J Comput Vis 60(2). https://doi.org/10.1109/ACCESS.2018.288885
6. Adarsh P (2020) YOLO v3-tiny: object detection and recognition using one stage improved model. In: 6th International Conference (ICACCS). 10.2010.1109/ICACCS48705.2020.9074315
7. Galvez R (2019) Object detection using convolutional neural networks. In: IEEE regulation on 10 annual international conference proceedings/TENCON, vol 2018, pp 2023–2027. https://doi.org/10.1109/TENCON.2018.8650517
8. Kim B-K (2016) Hierarchical committee of deep convolutional neural networks for robust facial expression recognition. J Multimodal User Interfaces 10(2):173–189
9. Sun B (2016) Facial expression recognition in the wild based on multimodal texture features. J Electron Imaging 25(6):061407–061407
10. Tripathi A (2018) Real time object detection using CNN. Home 7(4)
11. Ren S (2015) Faster RCNN: towards real-time object detection with region proposal networks. In: Advances in neural information processing systems, arXiv:1506.01497v3
12. Girshick R (2015) Fast RCNN. In: Proceedings of the IEEE international conference on computer vision, arXiv:1504.08083
13. Uijlings JRR (2013) Selective search for object recognition. Int J Comput Vis
14. Mazhar S, Singh S (2018) Region-based object detection and classification using faster RCNN. In: IEEE 4th international conference. 10.2010.1109/CIACT.2018.848041
15. Zhao ZQ, Zheng P, Xu ST, Wu X (2019) Object detection with deep learning: a review. IEEE Trans Neural Netw Learn Syst 30(11):3212–3232

16. Girshick R (2015) Fast RCNN. In: Proceedings of IEEE international conference on computer vision, ICCV 2015, pp 1440–1448
17. Redmon J (2016) You only look once: unified, real-time object detection. In: IEEE computer society conference
18. Redmon J, Divvala S, Girshick R, Farhadi A (2017) You only look once: unified, real-time object detection. In: IEEE computer society conference on computing vision pattern recognition. https://doi.org/10.1109/CVPR.2016.9
19. Jiao L (2019) A survey of deep learning-based object detection. IEEE Access 7:128837–128868
20. Wang D, Li C, Wen S, Chang X, Nepal S, Xiang Y (2019) Daedalus: breaking non-maximum suppression in object detection via adversarial examples. arXiv Prepr
21. Redmon J, Farhadi A (2017) YOLO9000: better, faster, stronger. In: 30th IEEE conference on computer vision pattern recognition, CVPR, pp 6517–6525
22. Redmon J, Farhadi A (2018) YOLOv3: an incremental improvement. arXiv Prepr
23. Fang W, Wang L, Ren P (2019) Tinier-YOLO: a real-time object detection method for constrained environments. IEEE Access 8:1935–1944
24. Xiao D, Shan F, Li Z, Le BT, Liu X, Li X (2019) A target detection model based on improved tiny—Yolov 3 under the environment of mining truck. IEEE Access 7:123757–123764. https://doi.org/10.1109/ACCESS.2019.2928603
25. Mao QC, Sun HM, Liu YB, Jia RS (2019) Mini-YOLOv3: real-time object detector for embedded applications. IEEE Access 7:133529–133538. https://doi.org/10.1109/ACCESS.2019.2941547
26. Bochkovskiy A (2020) YOLOv4: optimal speed and accuracy of object detection. Comput Vision Pattern Recogn (cs.CV), arXiv:2004.1093 [cs.CV]
27. Nelson J (2020) https://blog.roboflow.com/yolov5-is-here

Chapter 8
Identifying Tumor by Segmentation of Brain Images Using Combinatory Approach of Convolutional Neural Network and Biogeography Algorithm

Ashish Kumar Dehariya and Pragya Shukla

1 Introduction

Brain is the governor of the human body. It takes action for the activities like emotions, intelligence, speech, movements, senses, thoughts, physical activity, taste, creativity, etc. [1] therefore a small damage or defect in the brain can lead to total disturbance in the system of the human body. Its diagnosis is a must before it reaches a higher level so that in the right time proper treatment can be given to its patient. Higher stages of brain cancer can lead to death.

Brain tumors are diagnosed by almost 11,000 people per year [2]. Brain tumor is the anomalous growth of the flesh in the brain having uncontrolled growth and multiplication of cells [3]. There are two types of brain tumor, i.e., primary brain tumor and secondary brain tumor [4]. Primary brain tumor is the type that stems inside while secondary brain tumor has the origin in some other part of the body like lungs but the stems drift toward brains mostly because of the stream of blood [5].

MRI is very popular for scanning brain tumors and related images. It provides the digital representation of tissues and its characteristics in any tissue plane. MRI is the technique used to diagnose brain tumors. Digitally images can be clear and segmentation is used to gain the information from its complexity [6]. Different modalities of patients depending upon the handling of complexity, time and constraints are to take the computer-based image analysis method [7].

The segmentation process aims in identifying the region of the images and labeling the meaning indicated by that pixel, for each pixel rather than finding out the meaning of the entire image [8]. Imposing classification methods in segmented regions can improve the accuracy level of the diagnosis.

Biogeography-based genetic algorithm segments Brain MRI images. Incorporating Convolutional neural networks in segmented blocks of images, classify the

A. K. Dehariya (✉) · P. Shukla
Department of Computer Engineering, IET, DAVV, Indore 452017, Madhya Pradesh, India

A. Sheth et al. (eds.), *Intelligent Systems*, Algorithms for Intelligent Systems,
https://doi.org/10.1007/978-981-16-2248-9_8

patterns in different available categories. Using Image Segmentation by Biogeography and Convolutional Neural Network (ISBCNN), classification of brain tumors and the normal structure of the brain is possible with greater accuracy.

2 Literature Survey

In 2017, Bharath et al. [9] used blind source separation approaches to get tissue-specific profiles and their distribution from MRSI data. A window method is used to boost the peaks and scale down the length of the spectra then construct a 3-D MRSI tensor. For the decomposition of the tensor, they used nonnegative canonical polyadic decomposition (NCPD) with common factors in mode-1 and mode-2. Here tissue-specific spectral profiles were generated using NCPD factor matrices.

Zotin et al. [10] presented a model for brain tumor detection. Here input image was preprocessed using median filter and balance contrast enhancement technique. Segmentation was done using FCM clustering method then Canny edge detector constructed the edge map of brain tumor. More number of iterations are required here to get a justified result for detection.

Li et al. [11] experimented with a combination of the multi-modal information fusion and convolution neural network for detection of brain tumors. Slow network convergence and over fitting problem were observed here so they added a real normalization layer in convolution and a pooling layer to improve the convergence speed moreover weighted loss function added then developed the feature learning.

Vijayakumar [12] used a capsule neural network (CapsNet) to classify the brain cancer type. CapsNet succeeded in having improved accuracy while comparing with convolutional neural networks, NN and ResNet and when using less training data. But its keen observation concluded that it struggles while differentiating the objects which are very similar [12].

Huang et al. [13] used FCM clustering algorithm with rough set for the image segmentation. They constructed the attribute table by the values obtained from the FCM segmentation result and the image is divided into the small areas on the basis of attributes. Weighted values are obtained by value reduction and used for the calculation of difference between the region and similarity of region. Later, the model was realized through equivalence difference degrees. Final values of equivalence degree were used to evaluate segmentation of images and merge regions. The method has limitations up to only MRI images of brain and CT, generated artificially. The other observed limitations are: (1) Different masks are required to treat various sets of images. (2) Execution time increases due to calculation of the information table for each image that is required to find the segmentation class of sub-images. (3) Detailed description of noise removal or skull identification are not given in image preprocessing part.

Noreen et al. [14] used two models that are Inception V-3 and DensNet201 for creation of the two different modes for the identification and diagnosis of the brain tumor. Initially the features of the inception model are extracted from the pre-trained inception model V3 then concatenation is performed. Then values passed through the SoftMax classifier for the classification of the brain tumor. Similar process is done with DensNet201. Dens Netblock extracted the features then concatenation performed. Later, SoftMax classifiers were used for determining the tumor.

Dehariya et al. [15], Segmented Brain MRI images using Biogeography Optimization. Here each time a random segment center reduces its accuracy with the same image set.

3 Proposed Model

Model for identifying brain tumor by Biogeography and Convolutional Neural Network (ISBCNN) is sectioned into two parts. First part describes the segmentation model and the second part explains the classification model. In segmentation, a Biogeography genetic algorithm was used for finding the segmented region. Classification model trains neural networks by a set of images (segmented pixels of tumor and non-tumor region). Trained convolutional neural networks give classification of tumorous and non-tumorous brains. Some of the notations used in explanation are I = Input Image, B = Block of Image, C = Number of Segments, H = Habitat, h = Number of Habitat, λ = Habitat Immigration Rate, α = Habitat Emigration Rate, HIS = Rank of Habitat Fitness Value, F = Fitness Value, M_h = Mutation Rate, M_p = Mutation Probability, S = Stride of Filter in CNN, P = Padding in CNN, x = Neuron Input Value, W = Weight between Layers.

3.1 Image Preprocessing

For noise removal, a wiener filter was utilized. In order to remove unwanted spots this filter works on spots and other important information like edge, corner remains unaffected. The input image has a skull part as well so identification and removal of

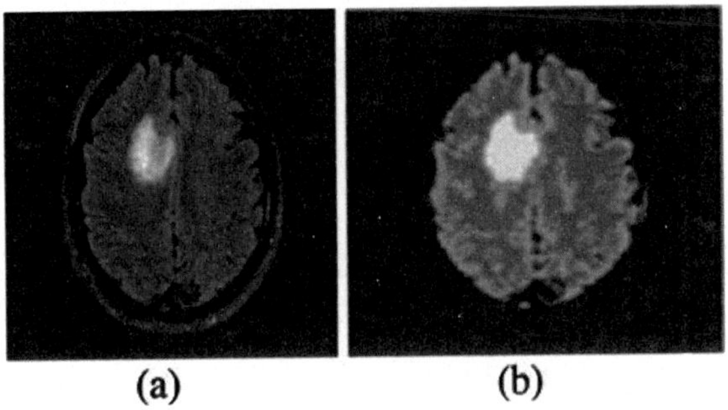

Fig. 1 **a** Input image, **b** image after preprocessing

this part was also performed in preprocessing. Figure 1a is showing the sample input image while Fig. 1b is showing the image after preprocessing.

3.2 Biogeography Genetic Algorithm

Image segmentation can be performed by Biogeography genetic algorithm [15]. Biogeography genetic algorithm used to find the cluster centers from the input preprocessed image by segmenting it into tumor and non-tumor regions.

Immigration and Emigration Rate: Biogeography genetic algorithm uses terms known as Immigration and Emigration represented by λ, α. Immigration is a rate of habitat h to accept new species and Emigration rate of a habitat is transfer of existing species.

$$\lambda_{\mathrm{h}} = 1 - \alpha_{\mathrm{h}}. \tag{1}$$

$$\alpha_{\mathrm{h}} = \frac{HSI_{\mathrm{h}}}{h}. \tag{2}$$

HSI is a rank index of h as per suitability estimation for a habitat.

Generate Population: Cluster sets were randomly generated by Gaussian distribution function ranged between 0 and 255. If h number of habitats were required to generate for segmenting image into c number of segments, then population obtained by Eq. (3).

$$H \leftarrow \mathrm{Population}(c, h). \tag{3}$$

Habitat Suitability Index: In order to understand the feasibility of the cluster center, habitat cluster center pixel values were used for clustering of images into c number of segments. So, the difference between the segment center pixel and image pixels $(m \times n)$ were summed as a fitness value by Eq. (4).

$$F_{\mathrm{h}} = \sum_{m=1}^{\mathrm{column}} \sum_{n=1}^{\mathrm{Row}} \mathrm{Min}\big(H_{\mathrm{h,c}} - I(m, n)\big)^{c}. \tag{4}$$

$$\mathrm{HSI} \leftarrow \mathrm{Rank}(F, h). \tag{5}$$

As per the distance sum, fitness values were arranged in increasing order to get the HSI of the current population matrix.

Crossover and Mutation of Habitat: This operation was applied on habitat sets which have low Emigration and Immigration rate value as compared to a constant. So crossover needs one habitat fulfilling a condition of emigration higher than emigration constant and immigration rate higher than immigration constant. During crossover

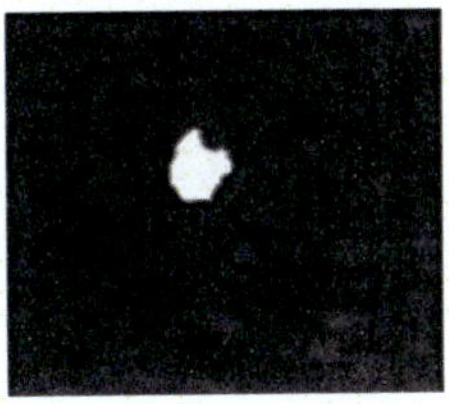

Fig. 2 Segmented image showing white tumor region in brain

segment center pixel value from emigration habitat replaces the pixel value of same segment in another habitat. This process generated a new set of habitats which improved the searching capability of work. To modify the individual habitat mutation operation involved mutation probability estimated by Eq. (5). As per HIS value mutation was performed in selected habitats. Mutation rank generated by Eq. (6) and mutation probability generated by Eq. (7).

$$M_{\mathrm{h}} = \frac{\mathrm{HSI}_h}{\mathrm{Sum}(h)}. \tag{6}$$

$$M_{\mathrm{p}} = \frac{M_{\mathrm{h}}}{\mathrm{Max}(h)}. \tag{7}$$

3.3 Image Segmentation

After a sufficient number of iterations of the Biogeography algorithm, images get segmented into segment representative pixels. Hence image gets segmented into two regions, i.e., tumor in white color and non-tumor in black color shown in Fig. 2.

3.4 Convolution Neural Network

Segmented image was further processed by a convolutional neural network. Each training image was divided into a fixed-size block of $b \times b$ [shown in Eq. (8)], which was passed in a convolutional filter to get the training input vector. Working block diagram of the convolution training model is shown in Fig. 3.

$$B \leftarrow \mathrm{Block}(\mathrm{SI}, b). \tag{8}$$

Convolution: In this step, a filter of window size F_{c} was prepared and moved on each block of segmented image from left to right and top to bottom. This movement reduces the size of block as per stride s and padding p values. Filter values multiply the block values at each movement. Filter matrix is a combination of 1 and 0.

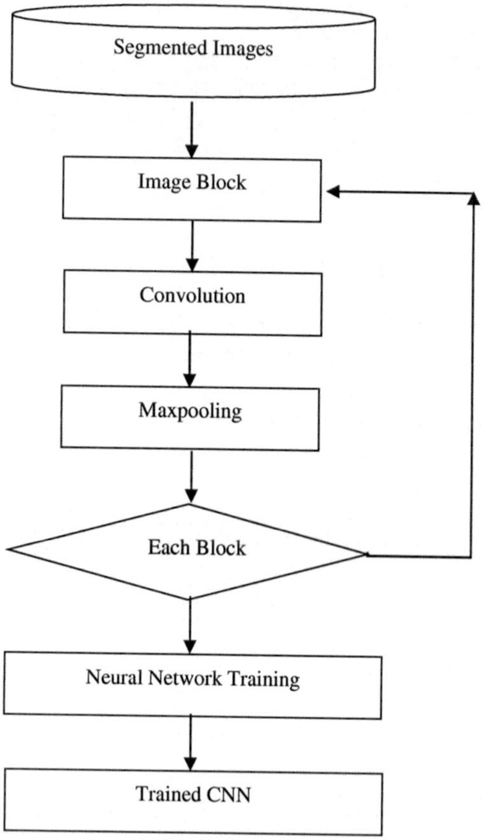

Fig. 3 Training of convolution neural network for image segmentation

$$C \leftarrow \text{Convolution}(B, s, p, F_c). \quad (9)$$

Stride is a movement speed control variable having integer values. Padding is a null row or column added in the block if required.

$$C \leftarrow \text{Maxpooling}(C, s, p, F_m). \quad (10)$$

Maxpooling: In this step, a filter of window size F_m was prepared and moved on each block of segmented image from left to right and top to bottom. This movement reduces the size of block as per stride s and padding p values [shown in Eq. (10)]. Maximum value of the filter area for the corresponding block replaces other values.

Training of Neural Network: Output from convolution and maxpooling was arranged as a training vector with segment class as output. Sigmoidal activation function [shown in Eq. (11)] was used in the Process of CNN. Input vector x and output vector o were used to adjust the weights of layers as per desired output segment class.

$$y = \frac{1}{e^{-xw}} \tag{11}$$

After a sufficient number of iterations trained CNN was obtained.

4 Experiment and Results

Brain tumor Segmentation dataset (code 105) obtained from the repository maintained on URL https://ijsret.com/2017/12/14/computer-science/. was used for experimental purposes. This dataset contains actual MRI images and related ground truth images. Combining these type of images considered as 1 set. Proposed model was implemented using MATLAB software. The comparison model is FCMRS [13].

Table 1 shows ISBCNN tumor detection is better as compared to FCMRS. Improvement was achieved by use of trained convolutional neural networks for identification of pixels. Testing image was preprocessed and passed in blocks of the bxb neural network. ISBCNN has improved the accuracy value by 2.27%. Precision values of image segmentation for different testing images are shown in Table 2. Average precision value of ISBCNN is 0.96916 and FCMRS is 0.957. In Table 3, recall values comparison shows that tumor detection by ISBCNN is better as compared to FCMRS [13]. Table 4 shows that ISBCNN f-measure values are better than FCMRS values.

Table 1 Accuracy value-based comparison

Images	FCMRS	ISBCNN
Set 1	90.0467	90.3137
Set 2	97.1351	97.8696
Set 3	91.1805	92.4157
Set 4	91.6686	97.8601
Set 5	92.805	95.7108

Table 2 Precision value-based comparison

Images	FCMRS	ISBCNN
Set 1	0.9163	0.9153
Set 2	0.9824	0.9826
Set 3	0.9709	0.9701
Set 4	0.9928	0.9929
Set 5	0.9226	0.9849

Table 3 Recall value-based comparison

Images	FCMRS	ISBCNN
Set 1	0.98203	0.9849
Set 2	0.988461	0.9959
Set 3	0.936821	0.9509
Set 4	0.922492	0.9858
Set 5	0.993147	0.9710

Table 4 F-measure value-based comparison

Images	FCMRS	ISBCNN
Set 1	0.947231	0.9488
Set 2	0.985446	0.9892
Set 3	0.953573	0.9604
Set 4	0.956377	0.9891
Set 5	0.956583	0.9779

5 Conclusion

This paper has proposed a brain tumor detection model which finds segmentation regions without any supervision or prior information. Paper has utilized a biogeography genetic algorithm for image segmentation which gives segmented images pixels set. Each segmented image was blocked for preparing the training vector of convolutional neural networks. In CNN, convolution and maxpooling operator increased the efficiency for learning. Experiments were performed on real MRI image dataset and the result revealed that the proposed model has increased the accuracy of tumor detection by 2.27% as compared to FCMRS algorithm. It was also shown that the proposed model has increased the F-measure value by 1.36%. In future scholars can adopt other deep neural networks for increasing the segmentation accuracy of work. As CNN capability of learning increases by increasing the number of inputs at the time of training hence one can achieve better accuracy by considering large datasets.

References

1. Chandra SK, Bajpai MK (2020) Fractional Crank-Nicolson finite difference method for benign brain tumor detection and segmentation. Biomed Signal Process Control 60. https://doi.org/10.1016/j.bspc.2020.102002
2. Filho PPR, da Silva Barros AC, Almeida JS, Rodrigues JPC, de Albuquerque VHC (2019) A new effective and powerful medical image segmentation algorithm based on optimum path snakes. Appl Soft Comput J 76:649–670
3. Brain tumor: introduction Cancer.Net. [online] https://www.cancer.net/cancer-types/brain-tumor/introduction. Accessed: 25/02/2019

4. Brain Tumors-Classifications, Symptoms, Diagnosis and Treatments. [Online]. Available: https://www.aans.org/Patients/Neurosurgical-Conditions-and-Treatments/Brain-Tumors
5. Deepa AR, Sam Emmanuel WR (2018) An efficient detection of brain tumor using fused feature adaptive firefly backpropagation neural network. Multim Tools Appl 1–16
6. Avate PS (2018) Classification of the brain tumor using PCA and PNN classifier. Int J Sci Res Eng Trends 4(3)
7. Kumar A, Singh A (2020) An implementation of brain tumor detection using convolutional neural network algorithm. Int J Sci Eng Technol 8(4)
8. Bindhu V (2019) Biomedical image analysis using semantic segmentation. J Innov Image Process 1(02):91–101. https://doi.org/10.36548/jiip.2019.2.004
9. Bharath HN, Sima DM, Sauwen N, Himmelreich U, De Lathauwer L, Van Huffel S (2017) Nonnegative canonical polyadic decomposition for tissue-type differentiation in gliomas. IEEE J Biomed Health Inf 21:1124–1132
10. Zotin A, Simonovb K, Kurakoc M, Hamadc Y, Kirillova S (2018) Edge detection in MRI brain tumor images based on fuzzy C-means clustering. In: International conference on knowledge-based and intelligent information and engineering systems. https://doi.org/10.1016/j.procs.2018.08.069
11. Li M, Kuang L, Xu S, Sha Z (2019) Brain tumor detection based on multimodal information fusion and convolutional neural network. IEEE Access 7:180134–180146. https://doi.org/10.1109/ACCESS.2019.2958370
12. Vijayakumar T (2019) Classification of brain cancer type using machine learning. J Artif Intel 1(02):105–113. https://doi.org/10.36548/jaicn.2019.02.006
13. Huang H, Meng F, Zhou S, Jiang F, Manogaran G (2019) Brain image segmentation based on FCM clustering algorithm and rough set. IEEE Access Special Section on New Trends In Brain Signal Processing And Analysis
14. Noreen N, Palaniappan S, Qayyum A, Ahmad I, Imran M, Shoaib M (2020) A deep learning model based on concatenation approach for the diagnosis of brain tumor. IEEE Access 8:55135–55144
15. Dehariya A, Shukla P (2020) Brain cancer prediction through segmentation of images using biogeography based optimization. Int J Adv Res Eng Technol (IJARET) 11(11)

Chapter 9
New Center Folding Strategy Encoding for Reversible Data Hiding in Dual Stego Images

C. Shaji and I. Shatheesh Sam

1 Introduction

Data hiding [1–3] on dual stegano images has advantages such as high payload and high security. Dual stegano image-based data hiding provides two stego images after hiding the data, while a single stego image-based data hiding provides a single stego image. Data hiding can be classified as reversible data hiding (RDH) and non-reversible data hiding. Non-reversible data hiding does not recover the original cover image, while the RDH exactly recovers the original host image and data. In a communication system, the transmitter uses the data embedding algorithm, while the receiver uses the data extraction algorithm. Recently, so many data hiding approaches have been proposed which include reversible and non-reversible data hiding algorithms, least significant bit (LSB) [4], exploiting modification direction (EMD) [5], and pixel-value differencing (PVD) [6]. LSB Substitution and other irreversible data hiding methods are widely used. Out of these, LSB methods provide high-quality stego image and high embedding rate. In [7], Zang and Wang introduced the EMD data hiding technique that provides low distortion, where Kieu [8] and increase the embedding rate on a pixel pair to several bits. Difference expansion (DE) [9–11] is one of the commonly used RDH methods. This scheme forecasts the highest pixel value by the second-highest pixel and forecasts the least pixel by the second-least pixel value. Higher embedding rate can be achieved by the methods [12–14] where these methods provide a lower PSNR. Several encoding techniques are [15, 16] C. Shaji et al. available that provides a high embedding rate and visual quality.

Register number of the organisation is 17213112162017.

C. Shaji (✉) · I. Shatheesh Sam
Department of Computer Science, Nesamony Memorial Christian College, Manonmaniam Sundaranar University, Abishekapatti, Tirunelveli, Tamil Nadu 627012, India

A. Sheth et al. (eds.), *Intelligent Systems*, Algorithms for Intelligent Systems,
https://doi.org/10.1007/978-981-16-2248-9_9

2 Proposed Method

The proposed new center folding strategy encoding approach has two important phases such as (i) embedding phase (ii) extraction phase.

2.1 *Embedding Phase*

Figure 1 displays pictorial representation of the proposed method. Initially, group the bits and transform it to decimal values. The decimal numbers are represented as m_i. The decimal intensity is applied to an intensity reduction process which calculates the index values $U_{i.}$ The intensity reduction process initially estimates a codebook. The codebook has a size of 2^D. The intensity reduction is applied such that the index is calculated corresponding to the decimal intensity m_i. After calculating the index for the first decimal intensity $m_{i,}$ the codebook is updated such that the previously calculated decimal intensity moves to the top. This process is repeated to obtain the complete index U_i. The index U_i is converted to encoded index EU_i using Eq. (1),

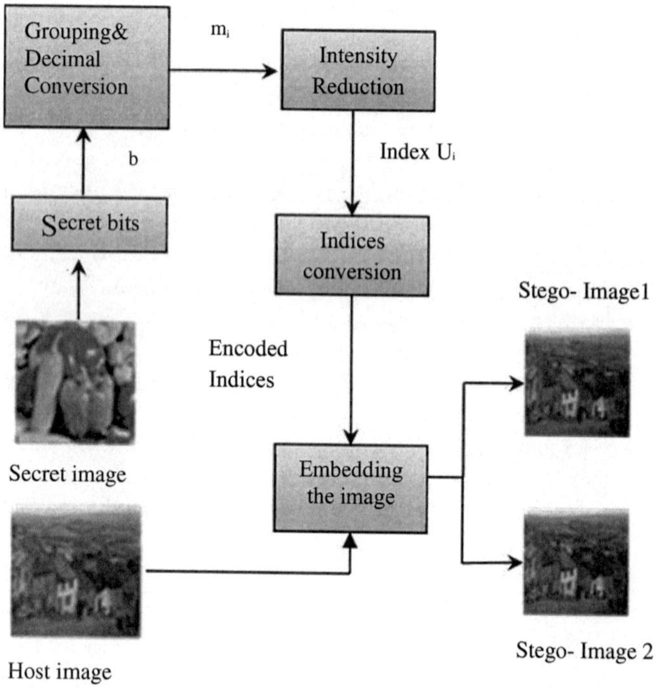

Fig. 1 Pictorial representation embedding the data

$$EU_i = \begin{cases} -\frac{U_i+1}{2} & \text{if } U_i \text{ is odd} \\ \frac{U_i}{2} & \text{if } U_i \text{ is even} \end{cases} \tag{1}$$

The encoded index is scrambled using a predetermined key K. The stego pixel pairs $T_1(c, d)$ and $T_2(c, d)$ can be calculated from the host pixel $T(c, d)$ as Eq. (2)

$$T_1(c, d) = T(c, d,) + \left\lfloor \frac{EU_i}{2} \right\rfloor \tag{2}$$

Consider the binary message as $b = \{1, 0, 1, 1, 0, 1, 0, 1, 0, 0, 1, 0\}$. Let $D = 3$; therefore, group the binary data such that 3 bits in a group, and converting to decimal, we get the decimal data as $m_i = \{5{,}5{,}2{,}2\}$. A codebook is generated having a size of 2^D. Here, $D = 3$; therefore, the size of codebook is 2^3. Initially, the code word and index are written sequentially from 0 to 2^{D-1}, i.e., 0–7. First the decimal data is 5; therefore, $U_1 = 5$. Move the previous code word 5 to the top and search the index for next data. Here, the index of next data is 0. Therefore, the second index is $U_2 = 0$; repeat the same procedure for all decimal data m_i to obtain the complete index $U_i = \{5, 0, 30\}$. From the indices U_i, the encoded intensity EU_i can be calculated using Eq. (1) as $EU_i = \{-3, 0, 1, 0\}$. Let the cover pixels represented be $T(c, d) = \{83, 42, 127, 22\}$. The stego pixels $T_1(c, d)$ and $T_2(c, d)$ are calculated using Eqs. (2) and (3) as $T_1(c, d) = \{81, 42, 127, 22\}$ and $T_2(c, d) = \{84, 42, 128, 22\}$.

$$T_2(c, d) = T(c, d) - \left\lfloor \frac{EU_i}{2} \right\rfloor \tag{3}$$

2.2 Extraction Phase

Figure 2 is a block diagram of the proposed data extraction process. The encoded index can be estimated from $T_1(c, d)$ and $T_2(c, d)$ using the relation in Eq. (4).

$$EU_i = T_1(c, d) - T_2(c, d) \tag{4}$$

From the encoded index EUi, estimate the index U_i as,

$$U_i = \begin{cases} 2 \times |EU_i| - 1 & EU_i < 0 \\ 2 \times EU_i & EU_i \geq 0 \end{cases} \tag{5}$$

$$T(c, d) = \left\lceil \frac{T_1(c, d) + T_2(c, d)}{2} \right\rceil \tag{6}$$

Fig. 2 Block diagram representation of proposed data

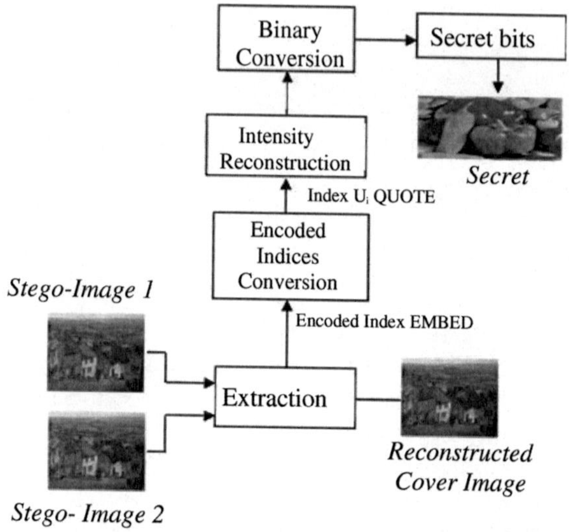

To demonstrate the use of an intensity reconstruction method to estimate decimal data m_i. The intensity reconstruction process initially estimates a codebook. The codebook has a size of 2^D. The intensity reconstruction is applied such that the codeword is calculated corresponding to the index $U_{i.}$ After calculating the codeword (decimal intensity m_i) for the first index U_i, the codebook is updated such that the previously calculated decimal intensity moves to the top. This process is repeated to obtain the decimal intensity m_i. From the decimal intensity m_i, the secret data is obtained by modifying intensity of m_i. The original cover pixels $T(c, d)$ can be recovered from $T_1(c, d)$ and $T_2(c, d)$ as Eq. (6). Consider the stego images as $T_1(c, d) = \{81, 42, 127, 22\}$ and $T_2(c, d) = \{84, 42, 126, 22\}$. Using Eq. (4), the encoded index can be calculated as $EU_i = \{-3, 0, 1, 0\}$. From the encoded index EU_i the index U_i can be calculated using Eq. (5) as $U_i = \{5, 0, 3, 0\}$. A codebook is generated having a size of 2^D. Here, $D = 3$; therefore, the size of codebook is 2^3. Initially the codeword and index are written sequentially from 0 to 2^{D-1}, i.e., 0–7. Initially, the index is 5, and the codeword for the index $U_1 = 5$ is 5; therefore, $m_1 = 5$. Move the previous codeword 5 to the top and search the codeword for the next index. Here, the codeword for next index is 5. Therefore, the second procedure for all index U_i to obtain the complete decimal intensity $m_i = \{5, 5, 2, 2\}$. The decimal intensity m_i is converted to 3-bit binary decimal intensity is $m_2 = 5$. Repeat the same numbers as $b = \{1, 0, 11, 0, 1, 01, 0, 0, 1, 0\}$ to obtain the secret bits.

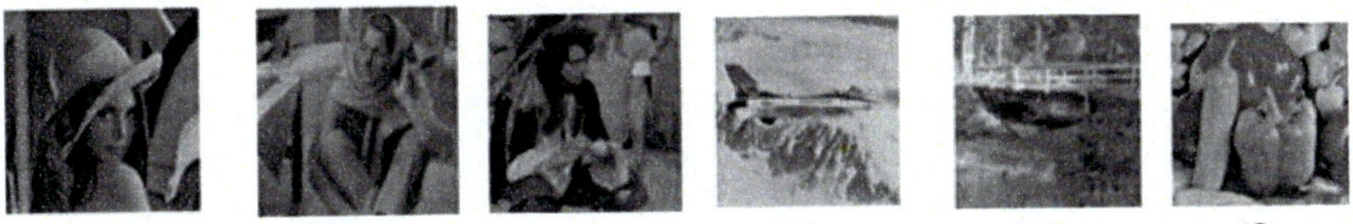

(a) (b) (c) (d) (e) (f)

Fig. 3 Test cover images and test secret image **a** Lena, **b** Barbara, **c** Man, **d** Bridge, **e** Jet plane, **f** test secret image: Pepper

3 Experimental Results

MATLAB was used to test the proposed dual stegano developed RDH algorithm. The test cover image and test secret images are denoted in Fig. 3. The test cover images and secret images are 8-bit gray scale images having capacity of 512 × 512 and 256 × 256, respectively. The proposed method was verified using parameters, namely s PSNR and embedding rate. The embedding rate (bpp) and PSNR are calculated using Eqs. (7) and (8).

$$\mathrm{PSNR}_n = 10\log_{10}\left[\frac{255^2}{\frac{1}{\mathrm{MN}}\sum_{j=1}^{M}\sum_{k=1}^{N}(I - I_n)^2}\right] \quad (7)$$

$$\mathrm{bpp} = \frac{C}{2 \times \mathrm{MN}} \quad (8)$$

where C is the embedding capacity, MN is the size of the image, and $n = 1, 2$. Here, $n = 1$ relate to first stego image and $n = 2$ relates to the second stego image. Let $\mathrm{PSNR}_{\mathrm{avg}}$ shows the average PSNR of the two stego images.

The comparison of the proposed technique is presented in Table 1, and it is compared with the traditional methods such as Ki-Hyun Jung [11] and Wang et al. The proposed scheme provides a high embedding capacity compared to the existing methods. The embedding capacity is around 943,718 bits, and the average PSNR is around 46.5 dB which is slightly lesser than Ki-Hyun Jung's method [10].

Figure 4 shows the change of PSNR for various embedding rates. As the value of D increases, the PSNR decreases, but the embedding capacity increases. The extreme embedding rate is around 1.8 bpp for $D = 4$ and 1.4 bpp for $D = 3$.

4 Conclusion

This paper recommended a method for data hiding on dual stego image-based reversible using new center folding strategy encoding. The outline aims to reduce the changes in intensities of pixels. The embedding process estimates the index from the decimal intensity using the intensity reduction process, and it is converted

Table 1 Performance comparison for PSNR and embedding capacity

Method	Metric	Lena	Barbara	Man	Jet plane	Bridge
Ki-Hyun Jung	$PSNR_1$	48.13	48.11	48.13	44.22	48.11
	$PSNR_2$	47.12	47.10	47.13	43.32	47.15
	$PSNR_{avg}$	47.625	47.605	47.63	43.77	47.63
	Capacity	519,180	519,180	519,180	519,180	519,180
Wang et al	$PSNR_1$	42.52	42.43	42.65	42.52	42.61
	$PSNR_2$	38.49	39.13	39.13	39.17	38.19
	$PSNR_{avg}$	40.505	40.78	40.89	40.845	40.4
	Capacity	786,432	786,432	786,432	786,432	786,432
Proposed ($D = 4$)	$PSNR_1$	46.1303	46.0924	46.0894	46.1212	46.0894
	$PSNR_2$	46.0231	46.9729	46.9699	46.0122	46.9699
	$PSNR_{avg}$	46.076	46.532	46.529	46.066	46.529
	Capacity	943,718	943,620	943,718	943,718	943,518

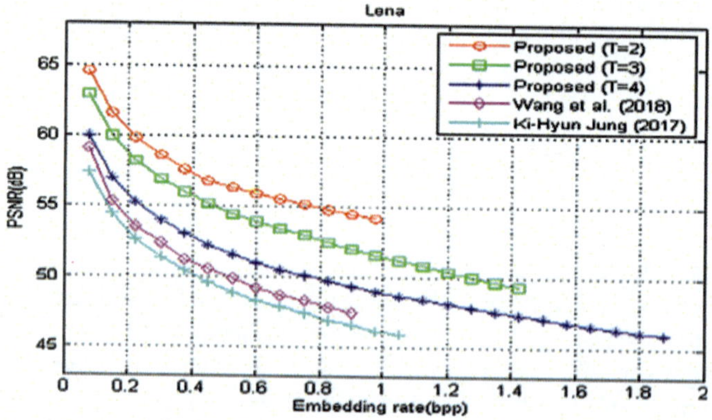

Fig. 4 Performance evaluation of proposed and traditional methods

into encoded intensity before embedding. Similarly, during the extraction method, the encoded index is extracted from two stego images and it is converted into an index. The index is converted into decimal intensity using an intensity reconstruction process, from which the secret image can be reconstructed. Experimental results reveal that the proposed method shows better results than the state of methods, in terms of embedding rate (bpp) and PSNR.

References

1. Ni Z, Shi YQ, Ansari N, Su W (2006) Reversible data hiding. IEEE Trans Circ Syst Video Technol 16:354–361. https://doi.org/10.1109/TCSVT.2006.869964
2. Li X, Li J, Li B, Yang B (2013) High-fidelity reversible data hiding scheme based on pixel-value-ordering and prediction-error expansion. Signal Process 93:198–205. https://doi.org/10.1016/j.sigpro.2012.07.025
3. Gui X, Li X, Yang B (2014) A high-capacity reversible data hiding scheme based on generalized prediction-error expansion and adaptive embedding. Signal Process 98:370–380. https://doi.org/10.1016/j.sigpro.2013.12.005
4. Mielikainen J (2006) LSB matching revisited. IEEE Signal Process Lett 13:285–287. https://doi.org/10.1109/LSP.2006.870357
5. Kieu TD, Chang CC (2011) A steganographic scheme by fully exploiting modification directions. Expert Syst Appl 38:10648–10657. https://doi.org/10.1016/j.eswa.2011.02.122
6. Wu DC, Tsai WH (2003) A steganographic method for images by pixel-value differencing. Pattern Recogn Lett 24:1613–1626. https://doi.org/10.1016/S0167-8655(02)00402-6
7. Zhang X, Wang S (2006) Efficient steganographic embedding by exploiting modification direction 10:781–783
8. Alattar AM (2004) Reversible watermark using the difference expansion of a generalized integer transform. IEEE Trans Image Process 13:1147–1156. https://doi.org/10.1109/TIP.2004.828418
9. Tian J (2003) Reversible data embedding using a difference expansion. IEEE Trans Circ Syst Video Technol 13:890–896. https://doi.org/10.1109/TCSVT.2003.815962
10. Chang C-C, Kieu TD, Chou Y-C (2007) Reversible data hiding scheme using two steganographic images. In: TENCON 2007—2007 IEEE Reg. 10 Conference, pp 1–4. https://doi.org/10.1109/TENCON.2007.4483783
11. Lu TC, Wu JH, Huang CC (2015) Dual-image-based reversible data hiding method using center folding strategy. Signal Process 115:195–213. https://doi.org/10.1016/j.sigpro.2015.03.017
12. Lee CF, Huang YL (2013) Reversible data hiding scheme based on dual stegano-images using orientation combinations. Telecommun Syst 52:2237–2247. https://doi.org/10.1007/s11235-011-9529-x
13. Jung K (2017) Authenticable reversible data hiding scheme with less distortion in dual stego-images. Multimedia Tools Appl 77:6225–6241. https://doi.org/10.1007/s11042-017-4533-0
14. Wang Y, Shen J, Hwang M (2018) A novel dual image-based high payload reversible hiding technique using LSB matching 20:801–804. https://doi.org/10.6633/IJNS.201807
15. Shaji C, Sam IS (2019) A new data encoding based on maximum to minimum histogram in reversible data hiding. Imaging Sci J 0:1–13. https://doi.org/10.1080/13682199.2019.1592892
16. Shaji C, Sam IS (2020) Two level data encoding approach for reversible data hiding in dual Stego images. Multimedia Tools Appl. https://doi.org/10.1007/s11042-020-09273-y

Chapter 10
Role of Privacy Concern and Control to Build Trust in Personalized Social Networking Sites

Darshana Desai

1 Introduction

Over the last decade, technology has influenced socializing in society in amazing and innovative ways with social networking sites (SNS) and used by all the strata of user groups with a different mindsets, backgrounds, and demographics. Personalization is used in such Web sites to address the need of the diverse user in the form of highly relevant content presented based on their search history, contacts, interactions with their friends, and content that is liked while interacting on these SNS. Social media Web sites are also considered as an effective medium for marketing and influence the users with personalized information. Users' interactions with such Web sites are tracked and the identification of users' preferences which have also raised privacy concerns among the users. Users can control the personalized information and offerings with privacy settings provided on the SNS to some extent. The hacking of social networking Web sites and users' personal information breaching have raised more privacy concerns and motivated users to have control over information sharing [1]. Though personalization is researched in recent years, it has a deficit in the study of the effect of personalization on users' information processing and behavioral intentions affecting control, privacy concerns, and trust on social networking Web sites. This research attempts to fill this gap in exploring various factors affecting users' behavioral intentions like satisfaction, trust through intrinsic behavior of privacy concerns, desire to control over the personalization of a Web site through privacy settings, and different features used in social networking site Facebook. The main objective is to explore the attitude of the user toward personalization in Facebook social networking site with privacy concerns, desire to control, and trust.

D. Desai (✉)
Department of MCA, Indira College of Engineering and Management, Parandwadi, Pune 410506, India

A. Sheth et al. (eds.), *Intelligent Systems*, Algorithms for Intelligent Systems,
https://doi.org/10.1007/978-981-16-2248-9_10

2 Related Work

Personalization is a continuous procedure to cater to users' needs with highly relevant information of users' interest identified through explicit source through customization choices or implicit need identified by analyzing users' browsing behavior and interaction with the Web sites [2–4]. Personalized content with high relevance to users' needs reduces cognitive efforts needed for information access and services leading to higher satisfaction [5]. The high relevance of personalized information builds more trust in the user which inspires users to revisit the social networking Web sites [6]. Users' personalized interaction helps in building strong customer relationships and increases satisfaction leading to desirable positive behavior and purchase intentions [6, 7]. Some researchers also suggest that more personalization stimulates privacy concerns in users and instigates negative behavior toward information sharing with the Web site. This study examines potential moderators of the negative effects of privacy concern on behavioral intentions in the context of personalized online interactions. Results show that users' positive behavioral intentions can be increased by providing more control to the users in the form of user-initiated customization and settings toward personalization.

3 Conceptual Framework and Hypotheses

3.1 Content Personalization and Privacy Concern

Chellappa [4] defined perceived privacy as "the individual probability with which consumers believe that the collection and subsequent access, use, and disclosure of their private and personal information is consistent with their expectations." Highly relevant personalized information reduces cognitive efforts and increases satisfaction in the user [8]. However, users experience higher privacy concerns when the personalization process uses users' information without consent and has negative feelings about personalization.

To provide personalized content, applications require users' personal information [9, 10] and understand users' implicit need by observing interaction with the Web site like social media behavior and likings which have the potential for an invasion of privacy. Individuals might have higher privacy concerns about online personalization if they are not aware of the intentions behind content personalization which reduces trust in social networking sites.

H1: User exhibits higher privacy concern with higher content personalization.

3.2 *Content Personalization and Control*

Users experience higher perceived control and a sense of involvement in the personalization process when offered customization choices lead to higher satisfaction and enjoy the interaction within personalized sites [8]. Users experience an intrinsic feeling of control when given choice of information sharing, viewing of commercial advertisement, setting of privacy, accessibility of information, and consent asked for cross-app communication. So, our work hypothesizes:

H2: User exhibits higher desire to control with higher content personalization.

3.3 *Privacy Concern and Control*

Users develop higher privacy concerns when perceiving loss of privacy when asked for more information to personalize or track their behavioral interaction using SNS. Privacy concern is the user's inbuilt desire for controlling the procurement and information usage upon sharing or acquired through Web site interaction and transactions. Privacy concerns of the user are related to the user's anonymity preservation and are strongly associated with the control of the user on personalization, explicit customization through settings and preferences [11, 12]. Earlier research has found that information control is a major factor in the privacy concerns of users in an online environment [13, 14]. Users have high privacy concerns when perceived with high risk and threats to privacy upon information sharing and less control over personalization settings like users will have less concern about privacy when provided with the option to control the visibility of posted content. In contrast, users experience lower privacy risks provided with higher control with exposure to privacy policies and sharing personal information [13]. Our research postulates:

H3: User exhibits higher desire to control with higher privacy concerns toward personalization.

3.4 *Privacy Concern and Trust*

Personalization of Web sites reduces cognitive efforts by the user to search information, and users feel higher satisfaction with more relevant personalized information [8] eventually developing trust toward Web sites. Users with higher privacy concerns need more control over personalization to develop trust over social networking Web sites. Privacy concerns of the user lead to trust building upon fulfillment [15]. Personalization has cognitive benefits and includes the cost of information disclosure. In the process to achieve personalization, Web sites need to collect more information about users' personal and behavioral characteristics like interests and implicit needs. This increases privacy concerns when privacy risks are associated particularly in wary of personalized content [10]. So, the research postulates:

H4: Users exhibit higher trust with addressed privacy concerns.

3.5 Control and Trust

Trust is users' intrinsic feeling and willingness to be vulnerable to the actions with expectation and addresses their expectation [5]. Users provided with control over information sharing, visibility of profile, post, and protection of information build more trust toward social networking Web sites [16]. Users having control over information flow, profile visibility, and protection of their personal information are more likely to develop trust in social networking Web sites [14]. Highly relevant personalized content offering on Web sites like a targeted advertisement, recommendation, location tracking, and offering suggestions based on geographic's increased users' satisfaction induces intrinsic feeling trust. The researcher proposes that

H5: Users experience more trust with higher control over the personalization.

3.6 Trust and Behavioral Intention

Research shows a positive relation in users' trust over digital interaction and willingness to use. Evidently, we can infer the significant role of trust in behavioral intention to revisit social networking sites offering online personalization. Trust in the Web site instigates users to share more personal information for personalization to feel more confident and motivated to revisit social networking Web sites. Users with less trust are less likely to perceive, and value benefits from the personalization of Web sites, on the contrary, are likely to frequently visit the trusted Web sites having personalization features [10]. So, author postulates

H6: Users experience positive behavioral intention with higher trust toward personalized social networking sites.

4 Research Method

4.1 Data Collection and Sampling

Research focuses on the behavioral intention of Facebook users with personalization, so we adopted a survey-based method for data collection and test proposed research hypotheses. The random sampling method is used to target a population of users who are using social networking sites Facebook for more than two years and experienced personalized service by the Web site in different forms like a recommendation of friends, post, and content based on users' likings and interactions. All the constructs identified from the prior studies are content personalization, trust, privacy concerns, control, and revisit intentions and are measured with a five-point Likert scale. All the responses were collected through online questionnaires from the respondents who are using Facebook's social networking Web site having personalization as a prime feature.

4.2 Measurement Model

The reliability of the constructs was checked through a pilot study of 50 responses to confirm the validity of variables and survey questions. The population of data collection is the respondents from India having Facebook accounts and has experienced personalization directly or indirectly. Three hundred and fifty-six valid responses were collected from a total of 400 responses after data preprocessing by removing noise and cleaning incomplete and inconsistent data. Responses with a standard deviation below 0.30 were removed to fetch final valid responses for further data analysis. Factors were identified and confirmed using exploratory and confirmatory factor analysis methods using SPSS 20.0. The structural equation modeling (SEM) technique is used to identify the model fit of the proposed model. The Cronbach's Alpha coefficient for assessing the reliability of construct is in the range of 0.70–0.90 which is above 0.70 reliability coefficient criterion for internal consistency of constructs, indicates higher consistency of scale of questionnaire items used within this survey, and construct items loaded are satisfactory with correct factors with loadings of more than 0.60 in confirmatory factor analysis showing construct validity and reliability. The validity of the model is checked using confirmatory factor analysis (CFA) and two-step validity measurement by Anderson and Gerbing [17] identifying convergent validity and discriminant validity of measurement model. Subsequently, testing of the hypothesis is performed, and the model is identified using the structural equation modeling technique.

4.3 Confirmatory Factor Analysis and Validity Test

Confirmatory factor analysis is used to test the measurement model using SPSS AMOS 21.0. Model fit is proved if the model fit indices reach accepted standards and recommended values, and the result shown in Table 1 exhibits adequate fit as the model fit index exceeds the recommended value.

Composite reliability measure shows internal consistency of construct item, and CR value is above 0.7 indicating good reliability and measurement items stability of

Table 1 Statistics of construct items

Construct	Items	Factor loadings	Composite reliability (CR)	The average variance extracted (AVE)	Cronbach's Alpha
Content personalization	SNIP1 SNIP2 SNIP4 SNIP3 SNIP5	0.781 0.717 0.732 0.776 0.623	0.864	0.516	0.863
Control	SNCON1 SNCON2	0.980 0.927	0.973	0.943	0.973
Privacy concern	SNPRIVACY1 SNPRIVACY2 SNPRIVACY3 SNPRIVACY4 SNPRIVACY5 SNPRIVACY6	0.742 0.756 0.652 0.695 0.787 0.720	0.85	0.462	0.871
Trust	SNTRUST1 SNTRUST2 SNTRUST3 SNTRUST4	0.748 0.718 0.780 0.659	0.828	0.547	0.828
Behavioral intention	SNINT1	0.969	0.967	0.907	0.967
	SNINT2	0.946			
	SNINT3	0.944			

each construct. Every CR scored above 0.8, which is above recommended value [18], indicating good reliability and stability for the measurement items of each construct.

Convergent validity of measurement model standards is recommended by Bagozzi and Yi [19] as follows: (1) factor loadings of the construct should exceed 0.5 [20]; (2) CR should be above 0.7; and (3) average variance extracted (AVE) score of each construct should exceed 0.5 [18]. The result shown in Table 2 depicts the factor

Table 2 Discriminant validity

Construct	Trust	Privacy concern	Content personalization	Behavioral intention	Control
Trust	0.740				
Privacy concerns	0.680	0.729			
Content personalization	0.457	0.461	0.718		
Behavioral intention	0.431	0.459	0.412	0.952	
Control	0.394	0.469	0.376	0.497	0.974

Table 3 Fit indices for the measurement models

Fit indices	Recommended value (Author)	Measurement model
v^2/df	< 5.0 (Bentler and Bonnett 1980) [21]	1.88
Goodness of fit index (GFI)	> 0.9 [22]	0.96
Adjusted for degrees of freedom (AGFI)	> 0.8 [22]	0.86
Normed fit index (NFI)	> 0.9 [21]	0.94
Comparative fit index (CFI)	> 0.9 [23]	0.945
Root mean square error of approximation (RMSEA)	< 0.06 [24]	0.043

loading score of each construct item exceeded 0.7, CR of the construct is above 0.8, and AVE score ranges from 0.46 to 0.95 which satisfies all the conditions for convergent validity showing measurement items correlates strongly with its theoretical constructs.

The research suggested that for the discriminant validity of the model, the AVE of the construct should exceed the correlation coefficients of the constructs [18]. Table 3 shows the correlation coefficient matrix of all the constructs in the research model. Also, diagonal element values are square roots of AVE score for the constructs. The correlation coefficient values for any two constructs are lesser than the square root of the AVE score for the constructs. Research shows good discriminant validity of the research constructs in model measurement as constructs are different from each other. So, research shows that the measurement model is having good construct reliability, discriminant validity, and convergent validity.

4.4 *Test Model Fit with Structural Equation Modeling*

Research model is tested with a structural equation model using AMOS 21.0. The model fit indices $v^2/df = 1.88$, GFI = 0.94, AGFI = 0.92, NFI = 0.95, CFI = 0.97, and RMSEA = 0.043 indicate a good model as shown in Table 3. All the indices for model fit in structural equation modeling indicate that the proposed model has good fit.

Figure 1 displays the standardized path coefficients, variance explained (R^2), and path significance values for the path as per the proposed hypothesis, and all hypotheses are supported except correlation of trust and control. As with variance explained (R^2), R^2 values of revisit intention and trust are 0.34 and 0.54 which is above 0.3 and shows a good research model.

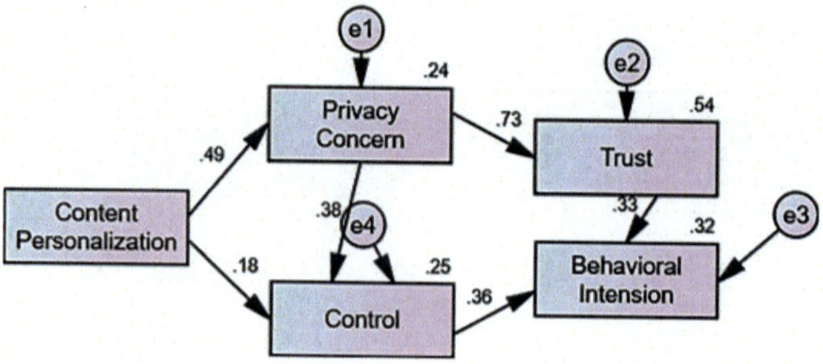

Fig. 1 Structural equation modeling result

Table 4 Hypotheses testing result

Hypotheses	Estimate	S.E	C.R	*P*-value
H1:Privacy Concern: Content Personalization	0.492	0.040	12.188	***
H2: Control: Content Personalization	0.196	0.049	4.028	***
H3:Trust: Privacy Concern	0.718	0.031	23.424	***
H4: Control: Privacy Concern	0.405	0.048	8.396	***
H5: Behavioral Intention: Trust	0.348	0.043	8.103	***
H6: Behavioral Intention: Control	0.357	0.040	8.954	***

*** indicates p-value < 0.001

5 Result and Discussion

SEM result in Table 4 shows the high correlation between constructs content personalization, privacy concerns, control, trust, and revisit intention. All the hypotheses proposed in the model satisfy users' trust dependency on control over the personalized Web site. Users' privacy concern and desire to control over the personalization of Web sites are highly dependent on content personalization. Users given with the choices for customization and settings for personalization and information sharing develop more privacy concerns and more desire to control the personalization of social networking Web sites. Research shows that users develop more trust toward the personalized Web site and privacy concerns. Users with high privacy concerns are less likely to develop trust in social networking Web sites having personalization.

6 Conclusions and Future Scope of Research

This research is a qualitative study on users' behavior toward personalization in social networking Web sites. Users' privacy concerns and control over the personalization process in the form of explicit customization play a significant role in developing trust toward social networking Web sites. Users with higher desire to control show high privacy concerns and are less likely to share personal information, also such users avoid more interaction with the Web site in the form of non-disclosure of information and restricting users to view their post on social media sites. Personalized content with high relevance is the key factor in developing trust mediated by users' privacy concerns and desire to have control over Web sites subsequently develop trust and motivate users to repeat visits to the Web site. Research is more useful in designing the personalization of the Web site and understanding users' attitudes and key factors affecting decision making to revisit the Web site. The survey-based research methodology adopted for the study can be further researched with users' behavioral intention in a controlled laboratory environment with live interaction with the personalization of social networking Web sites.

References

1. Al Qudah DA, Al-Shboul B, Al-Zoubi A, Al-Sayyed R, Cristea AI (2020) Investigating users' experience on social media ads: perceptions of young users. Heliyon 6(7)
2. Appel G, Grewal L, Hadi R, Stephen AT (2020) The future of social media in marketing. J Acad Mark Sci 48(1):79–95
3. Chellappa RK, Shivendu S (2006) A model of advertiser—portal contracts: personalization strategies under privacy concerns. Inf Technol Manage 7(1):7–19
4. Chellappa RK, Sin RG (2005) Personalization versus privacy: an empirical examination of the online consumer's dilemma. Inf Technol Manage 6:181–202
5. Dwyer C, Hiltz SR, Passerini K (2007) Trust and privacy concerns within social networking sites: a comparison of Facebook and MySpace. In: Americas conference on information systems, proceedings of the thirteenth Americas conference on information systems, Keystone, 9–12 Aug, Colorado, USA, p 339
6. Desai D (2019) Personalization aspects affecting users' intention to revisit social networking site. Int J Trend Sci Res Dev (IJTSRD) 4(1):612–621. ISSN: 2456-6470. https://doi.org/10.31142/ijtsrd29631
7. Taylor DG, Davis D, Jillapalli R (2009) Privacy concern and online personalization: the moderating effects of information control and compensation. Electron Commerce Res 9:203–223. https://doi.org/10.1007/s10660-009-9036-2
8. Desai D (2019) An empirical study of website personalization effect on users intention to revisit E-commerce website through cognitive and hedonic experience. In: Balas V, Sharma N, Chakrabarti A (eds) Data management, analytics and innovation. Advances in intelligent systems and computing, vol 839. Springer
9. Gupta A, Dhami A (2015) Measuring the impact of security, trust and privacy in information sharing: a study on social networking sites. J Direct, Data Digital Market Pract 17(1):43–53
10. Stevenson D, Pasek (2015) Privacy concern, trust, and desire for content personalization (March 30, 2015). In: The 43rd research conference on communication, information and internet policy
11. Mohamed N, Ahmad IH (2012) Information privacy concerns, antecedents and privacy measure use in social networking sites: evidence from Malaysia. Comput Human Behavior 28(6):2366–2375
12. Shin D (2010) The effects of trust, security and privacy in social networking: a security-based approach to understand the pattern of adoption. Interact Comput 22(5):428–438. https://doi.org/10.1016/j.intcom.2010.05.001
13. Senthil Kumar N, Saravanakumar K, Deepa K (2016) On privacy and security in social media—a comprehensive study. Phys Procedia 78:114–119

14. Tucker CE (2014) Social networks, personalized advertising, and privacy controls. J Mark Res 51(5):546–562
15. Komiak S, Benbasat I (2006) The effects of personalization and familiarity on trust and adoption of recommendation agents. MIS Q 30(4):941–960. https://doi.org/10.2307/25148760
16. Aldhafferi N, Watson C, Sajeev ASM (2013) Personal information privacy settings of online social networks and their suitability for mobile internet devices. Int J Secur Privacy Trust Manag 2(2):1–17
17. Anderson JC, Gerbing DW (1988) Structural equation modeling in practice: a review and recommended two-step approach. Psychol Bull 103(3):411–423. https://doi.org/10.1037/0033-2909.103.3.411
18. Fornell C, Larcker D (1981) Evaluating structural equation models with unobservable variables and measurement error. J Mark Res 18(1):39–50. https://doi.org/10.2307/3151312
19. Bagozzi RP, Yi Y (1988) On the evaluation of structural equation models. JAMS 16:74–94
20. Hair Jr, J. F. et al. (1998) Multivariate Data Analysis with Readings. Englewood Cliffs, NJ: Prentice-Hall.
21. Bentler PM, Bonett Douglas G (1980) Significance tests and goodness of fit in the analysis of covariance structures. Psychological Bulletin 88(3):588–606.
22. Jöreskog KG, Sörbom D (1981) LISREL V user's guide. Chicago: Int Educational Services.
23. Hu LT, Bentler PM (1999) Cutoff criteria for fit indexes in covariance structure analysis: conventional criteria versus new alternatives. Struct Eq Model 6:1–55
24. Browne MW, Cudeck R (1993) Alternative Ways of Assessing Model Fit. Sage Publications, Newbury Park, CA.

Chapter 11
Hybrid Restricted Boltzmann Algorithm for Audio Genre Classification

Dhruvika Taunk and Mayank Patel

1 Introduction

Due to nonlinear representation of hypothesis function, the neural networks that are artificial have the capability of grasping high level complex knowledge from the given inputs [1]. The most challenging task is to train a neural network as well as generalize it with various hidden layers using traditional techniques. As stated in some similar research literature [2], the neural networks get bewildered in local minima when trained by the conventional algorithms of back propagation. No doubt, better results are achieved with the enhancement in restricted Boltzmann machine algorithms and pre-training algorithms like auto-encoders [3]. Nowadays, the interest in deep learning algorithms is increasing in every field, either it be academic or non-academic. This is due to recent enhancement in computational techniques that now deep learning algorithms can be easily trained. Support vector machines and random forest are the traditional classification algorithms that beat the deep neural network in a few instances according to recent findings [4]. Automatic speech recognition is the major concentration of this exercise. The major motivation of this project is to classify music genres and analyze their performance by implementing different deep learning algorithms [5].

2 Setup and Theory

The common setup of the whole procedure is described here. The basic concept behind the working of neural networks depends on the simulation to the brain that means how the brain studies a particular concept. And this completely depends upon the neural network [6]. The input to the brain is the signal received by each neuron, and

D. Taunk (✉) · M. Patel
Geetanjali Institute of Technical Studies, Udaipur 313001, Rajasthan, India

A. Sheth et al. (eds.), *Intelligent Systems*, Algorithms for Intelligent Systems,
https://doi.org/10.1007/978-981-16-2248-9_11

then, the activation signal is formed by combining and transforming these signals as an output. For the other neurons, this activation signal acts as an input. In this scenario, the perceptron unit is referred to an individual neuron. As given in Eq. (1), the difference between the perceptron unit and the common perceptron is just that the logistic function is used for computation of activation:

$$h(\overrightarrow{x}) = \text{logistic}(\overrightarrow{w} \cdot \overrightarrow{x} + \theta) \tag{1}$$

The choice of the logistic function used for the computation of activation signal depends upon the specific problem [7]. In this case, a sigmoid function is used, $f(z) = \frac{1}{1+e^{-z}}$. The diagrammatic representation of feed-forward neural network is given in Fig. 1 [8]. In this diagram, different layers of feed-forward network have been shown, which include input, hidden, and output layers. The nodes present in the input layer are basically an element of the data points feature vector. To be more precise, Sect. 3 talks about the feature vectors and these feature vectors are the vectors of the transformed amplitudes. As shown in Eq. (2), the previous hidden layer's activation value is accepted by an output layer as a logistic function in order to perform multiclass classification tasks by using softmax function. The expectation is that it will be capable of predicting the data's label.

$$\begin{aligned} \Pr \Pr(x, w, \theta) &= \text{softmax}_i(w \cdot x + \theta) \\ &= \frac{e^{w \cdot x_i + \theta_i}}{\sum_j e^{w \cdot x_j + \theta_j}} y_{\text{pred}} = \arg\max_i \Pr(x, w, \theta) \end{aligned} \tag{2}$$

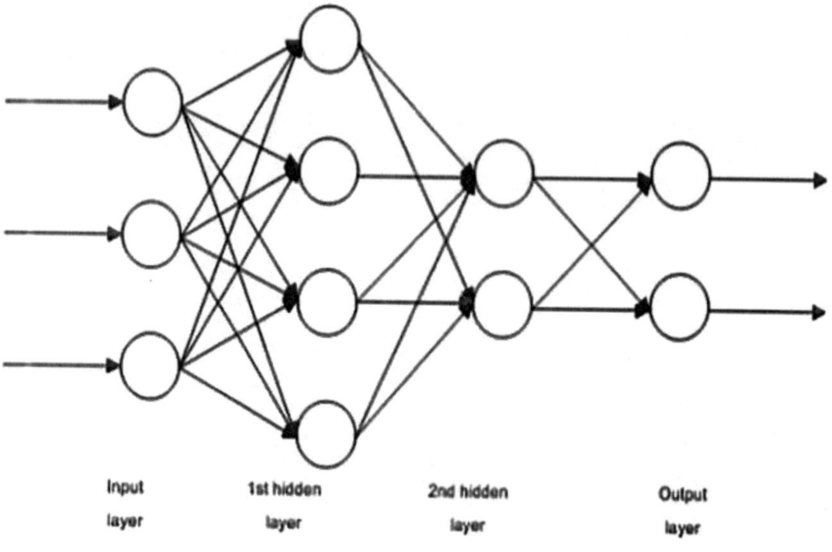

Fig. 1 Neural network representation

How dense a network is, depends on the number of hidden layers present in a network. As shown in Fig. 1 with a greater number of hidden layers, the network becomes complex and becomes capable of representing complicated concepts with degree of correlation being very high like in the case of audio data. In this paper, the aim is to train the neural network in such a manner that it is capable of predicting the accurate genre.

2.1 Algorithms

As discussed above, generally bad results are found whenever the number of hidden layers is more than 2 using conventional training methods (that is, backward propagation with randomized initial weights). For the requirement of efficient training of neural networks, two algorithms are discovered.

2.1.1 An Auto-encoder

Here is the good learning of hidden layers for good initial representation [9]. An easy solution to this problem has been given by Vincent and collaborators. Unsupervised learning algorithms can be considered for solving this purpose wherein the input data representing the hidden sparse is used. This can be achieved by designing the neural network in such a way that the hidden layers present in the network can efficiently learn the input. Through the algorithm of the neural network, one can clearly conclude that

$$\begin{aligned} y &= s(\vec{w} \cdot \vec{x} + \theta) \\ \text{And } z &= s\left(\vec{w}' \cdot \vec{x} + \theta\right) \end{aligned} \tag{3}$$

The aim is to find the weights and reduce the error loss in reconstruction, $\sum_i (z_i - x_i)^2$. This imposed hidden representation is useful only if in the case of audio data there is observed a correlation of high degree. In Eq. 3, here the sparse constraint plays a vital role. If there is no constraint of sparse, then the auto-encoder will just be able to reach identity mapping. In order to create a complex network, the auto-encoders can place in a form a stack [10]. When done in this form, the hidden representation learned of the kth-1 layer will be passed as an input to the kth layer that will also further encode it and the process goes on like this. The logistic unit performs the classification on the top most layer of the auto-encoders. Initially, the hidden layers are pre-trained, and later through the back propagation, it is finely tuned. In layer sections, there is a study about getting the constraint problem resolved and efficient working after the implementation. This can be studied through Fig. 2 [11]. Till now, the use of auto-encoder is not a success that is why the results so far

Fig. 2 Neural network mapping input feature vector on itself [13]

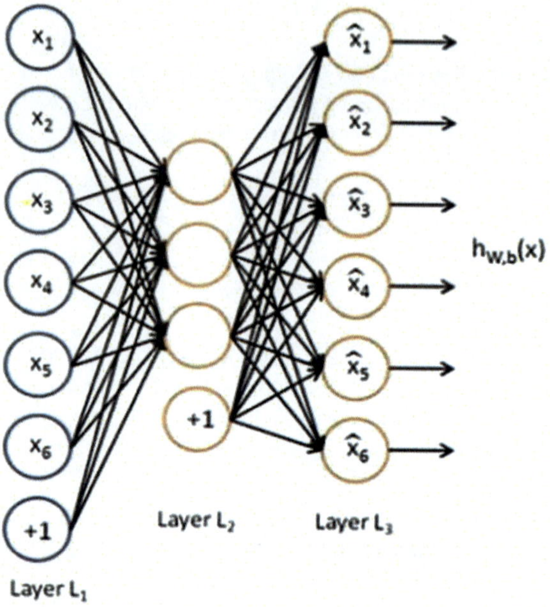

are not displayed. Although the implementation of restricted Boltzmann machine provides results which are given in later sections.

2.1.2 Restricted Boltzmann Machine

There's a different method of pre-training the deep neural networks parameters. A model based on energy was presented in the year 2006 by Hinton and Salakhutdinov. The probability function can be addressed as a function of energy present in the system.

$$p^{(x)} = \frac{e^{-E(x)}}{Z} \tag{4}$$

$$Z = \sum_{\text{all config}} e^{-E(x)} \tag{5}$$

From the above-mentioned Eqs. (4) and (5), one thing is clear that this scenario corresponds to identifying parameters that reduces the amount of energy used by parameters configuration. However, shown in Eq. (6) this can be achieved through reduction in the model's log-likelihood that is negative.

$$\frac{1}{N}\sum_{\text{all config}} \log\log p(x) \tag{6}$$

In statistical mechanics, this is a popular approach of physics known as canonical ensemble approach [12]. Reduction in the negative log-likelihood is similar to reduction in energy of the model through which equilibrium state of the system is identified, since the model was not watched in a restricted Boltzmann machine (RBM) that is why hidden layers are found there. The special energy function used is Hamiltonian in Eq. (7).

$$E(v, h) = -b'v - c'h - h'Wv \tag{7}$$

Hamiltonian function is similar to the free energy of the below form as shown in Eq. (8).

$$F(v) = -b'v - \sum_i \log \sum_{h_i} e^{h_i(e_i + w_i v)} \tag{8}$$

Since both the units, visible and hidden, are independent, one shall understand,

$$\begin{aligned} \Pr(v) &= \prod_i \Pr(v) \\ \Pr(h) &= \prod_j \Pr(h) \end{aligned} \tag{9}$$

If the meaning of free energy is understood, then free energy can be easily minimized through using stochastic gradient descent shown in Eq. (9). Generally, Eq. (2) cannot be used alone for computing free energy, so Monte Carlo algorithm is deployed to calculate the expected value of stochastic gradient. For using Monte Carlo, RBM is trained, which is discussed in Sect. 5.1.

3 Dataset

In this section, we briefly explain the different datasets used for evaluating our proposed audio genre classification.

3.1 Data Accumulation

After comparing many open audio datasets with the metadata associated and after the selection of genre collection GTZAN. This consists of 1 k audio tracks each of which are of duration 30 s. Basically, representation is of ten genres which are collections of 100 tracks each [14]. Each of these tracks are in .au format. These tracks are 16-bit audio files of 22,050 Hz. This includes ten audio genres, namely pop, jazz, metal, classical, country, hiphop, disco, blues, rock, and reggae.

3.2 *Feature Selection Through Mel Frequency Cepstral Coefficient (MFCC)*

In the previous section, the representation of snippets is done as 22,050 sample/second $\times$ 30 s $=$ 661,500 length of vectors. And a convention machine cannot take so much of a heavy load. The previous studies from different researchers helps us to draw a conclusion that MFCC is one of the most important features for representation of the long-time domain waveform and also helps in minimizing the dimensions [15]. At the same time, MFCC is also capable of capturing various information. Firstly, we use a hamming window, which is of 25 with 10 ms timing of overlapping. This is used for the generation of smooth frames in one sequence. After that Fourier Transform is applied in order to obtain the frequency component. Later, frequency is mapped to Mel scale frequency. This helps to model human consciousness of pitch variation that is approximately 1 kHz below in linear and 1 kHz above in logarithm. This leads to a kind of group mapping of frequencies, taking its product with the frequencies and computing logarithmic value. Then, the discrete cosine transform (DCT) is computed in order to reduce autocorrelation of the frequency component. Lastly, the 13 out of these 20 frequencies are kept because higher frequencies make less difference to the perception of humans and have lesser information about the song. In this setup, we have moreover divided the MFCC features into four different sections of equal size and then determined the initial 40 from each of them. Basically, we have created MFCC features of length 2080, that is 13 $\times$ 160. This is done for the representation of an audio file of 30 s, which shall be used later. In Fig. 3a, the scattering of the plot for distribution of data is represented. It depicts that the complete dataset can be segregated into two class cases, but the increase in the number classes is directly proportional to the increase in mixing of data points. In the case of ten classes, the data becomes very complicated, and the classification of the audio genre becomes quite challenging.

4 Proposed Algorithm

As mentioned in the above section, in order to train RBM, approximate Ptrain $(v) \approx P(v)$ is assumed (the underlying data distribution) with the condition distribution derived Eqs. (10) and (11).

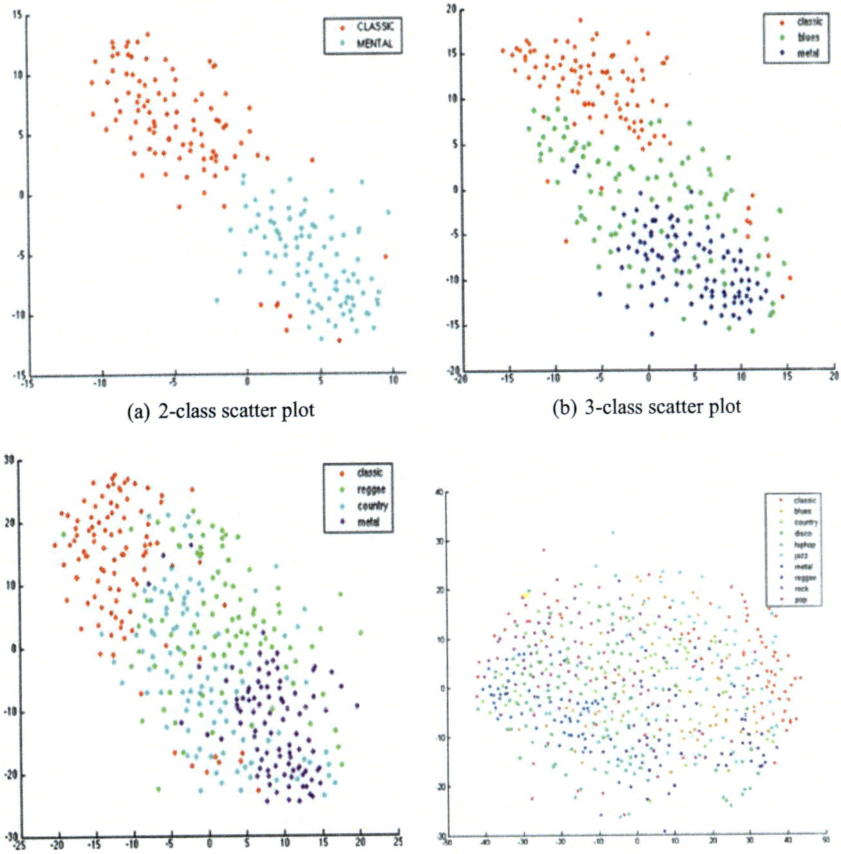

(a) 2-class scatter plot

(b) 3-class scatter plot

(c) 4 class scatter plot

(d) 10-class scatter plot

Fig. 3 Package t-SNE (Lvan der Maaten & Hinton)

$$\Pr(v) = g\left(b_i + \sum_j w_{i,j} v_j\right) \tag{10}$$

$$\Pr(h) = g\left(a_i + \sum_j w_{i,j} h_j\right) \tag{11}$$

For the purpose of updating of parameters, the contrastive divergence learning is applied which is a similar version of Gibbs sampling:

Algorithm 1: Contrastive Divergence

Computation of weight combination $W \times vi$, where $v1$ is the training sample for the hidden layer.

Hidden layer activation vector h1 is generated by: $h1 : h1 = f(W \times v1)$
v2 is generated by making through $h1$ to input layer by: $v2 = g(W \times h1)$
$h2$ is generated by $v2 : h2 : h2 = f(W \times v2)$ here $h1$, $v2$ are considered as binary values
Assume $m1 = h1 \times v1, m2 = h2 \times v2$
Updating the following

$$b = (v1 - v2)$$

$$c = (h1 - h2)$$

$$W = (m1 - m2)$$

As specified in (2), contrastive divergence's one iteration yields good results. Although in the experiment, it is found that number iterations lead to more efficient performance; therefore, here in this experiment, there are five experiments.

5 Experimental Setup

In this section, experimental setup configuration is explained, and how RBM is connected with a multilayer network.

5.1 *Connection of RBM with Multilayer Network*

A neural network of five layer is built having three hidden layers as the basic structure. Then after that, the RBMs are trained for every layer in the neural network as the weight initialization. Although the output layer remains an exception. The main concern is about training RBMs in a loop of layers and then arranging them in a form of stack in the end of a multilayer architecture. The training steps of RBM are mentioned below:

6 Pre-training Steps

- RBM is trained for the initial layer with untrained input data.
- Next, RBM is trained for another layer.

7 Training Network

- RBMs are stacked to the equivalent layers as the network's initial weight.
- To train the multilayer network, forward and backward propagation is implemented either, conjugate gradient method is used.

7.1 Experimental Setup

Since the number of iterations is fixed that is 5, so RBMs training is also fixed. The number of hidden nodes and iteration sides according to the experiments is shown in Fig. 4, so we state that it depends upon the parameters, and optional results arc achieved. Firstly, the experiment is performed on two-class classifications, then three-class classifications, and later on ten-class classification.

8 Results

This section described the results obtained when the different classification is applied to the data sets. Results obtained show that the proposed classification model gives better results. Results according to different class classification are as follows.

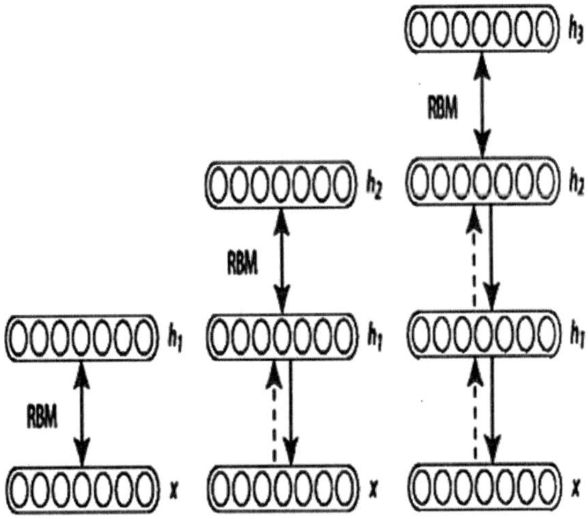

Fig. 4 RBMs getting trained iteratively

Table 1 Two-class classification

Classic versus metal	NN	DBN
	CG 3-line search	CG with 3-line search
Train set accuracy	100% (120 out of 120)	100% (120 out of 120)
Test set accuracy	97.5% (78 out of 80)	98.75% (79 out of 80)
More stats: H hidden nodes T running time	H ~ [500 500 1000] T ~ 18 s	H ~ [1000 1500 2000] T ~ 7 min

Table 2 Three-class classification

Classic, metal, and blues	NN	DBN
	CG 3-line search	CG with 3-line search
Train set accuracy	90% (162 out of 180)	91% (164 out of 180)
Test set accuracy	70.8% (85 out of 120)	69.16% (83 out of 120)
More stats: H hidden nodes T running time	H ~ [500 500 1000] T ~ 26 s	H ~ [1000 1000 2500] T ~ 20 + min

8.1 First Experiment Results

The result obtained from the two-class classification is that the neural network whether it be the neural or deep belief neural network yields the same classification for two genres. Moreover, both the networks provide 100% accuracy and efficiency on the training set. However, 97.5% for neural network and 98.75% for deep belief neural network for testing set. This result is given in Table 1. T'he training and testing of deep belief neural networks are more intensive than normal because it requires a greater number of hidden layers and that makes the computation complex. In the case of three-class classification, both the networks are more competitive shown in Table 2. Here in the case of deep belief neural network, the training set achieves high accuracy but the testing set gets less accuracy. In the case of four-class classification, the neural network provides better results as compared to deep belief neural networks given in Table 3.

8.2 Second Experiment Results

This result of the second experiment makes us clear that we were expecting deep belief neural network to perform well, but it is not performing that good as expected. The results are given in Tables 4 and 5. We observed that maybe it was because the small datasets used in this experiment. By implementing the same feature selection technique mentioned in Sect. 3.2, in order to generate more samples, we reduce the

Table 3 Four-class classification

Classic, metal, blues, and disco	NN	DBN
	CG 3-line search	CG with 3-line search
Train set accuracy	90.83% (218 out of 240)	69.17% (166 out of 240)
Test set accuracy	63.75% (102 out of 160)	51.88% (83 out of 160)
More stats: H hidden nodes T running time	H ~ [500 500 1000] T ~ 26 s	H ~ [1000 1000 2500] T ~ 20 + min

Table 4 Three-genre classification with larger dataset

Classic, metal, and blues	NN	DBN
	CG 3-line search	CG with 3-line search
Train set accuracy	92.89% (2508 out of 2700)	94.5% (2521 out of 2700)
Test set accuracy	77.17% (1389 out of 1800)	77.94% (1403 out of 1800)
More stats: H hidden nodes T running time	H ~ [500 500 1000] T ~ 50 + min	H ~ [1000 1500 2500] T ~ 120 + min

large amount of 13 × 2600 features into small subsets of 15 160 × 13 of MFCCS. This means that now we are generating 15 samples for each track, this provides us efficient results because now in spite of 100 we have 1500 samples. And after this modification, run this experiment again and receive excellent results:

- After this modification, we have got 15 × more datasets, which has enhanced the accuracy of both the networks.
- Moreover, this leads to the outperformance of deep belief neural networks both in terms of accuracy of train set and test set.
- In three-genre class cases, the deep belief neural network shows improvement of 3.5% in training set and improvement of 8.78% in testing set. This leads to overcoming the problem of over-fitting in small datasets.
- Similarly, in case of four-genre classification, the deep belief neural network shows improvement of 6.13% in training set and improvement of 9.27% in testing set.

Table 5 Four-genre classification with larger dataset

Classic, metal, blues, and disco	NN	DBN
	CG 3-line search	CG with 3-line search
Train set accuracy	94.69% (3406 out of 3600)	75.3% (2712 out of 3600)
Test set accuracy	60.46% (1451 out of 2400)	61.15% (1467 out of 2400)
More stats: H hidden nodes T running time	H ~ [500 500 1000] T ~ 60 min	H ~ [1000 1500 2500] T ~ 160 + min

So generally, if we observe the trend, belief neural networks perform better than the neural network, and with some modifications like large dataset, its performance stands out. And we promise if this could be achieved with an even larger dataset.

9 Conclusions

Improvement has been seen between the two experiments. Second experiment turned out to be a great success. However, the main goal is to provide proper audio classification four 10 × 100 (1000) tracks instead of, for 10 × 1500 sub-samples. The solution for this problem is that we can reduce the track by generating 15 samples having labels of the same class, and then after that, the majority vote of these samples can be used as the final foretold of the original audio. This turns out to be a good solution for ensembles of deep neural networks. Further, we would like to work on the training process wherein the project runs in parallelepiped format.

References

1. Bengio Y (2009) Learning deep architectures for AI. Found Trend Mach Learn 2(1):1–127
2. Lee H, Largman Y, Pham P, Ng AY (2009) Unsupervised feature learning for audio classification using convolutional deep belief networks. In: NIPS 2009
3. Le QV et al. Building high-level features using large scale unsupervised learning
4. Vincent P, Larochelle H, Bengio Y, Manzagol PA (2008) Extracting and composing robust features with denoising autoencoders. In: Proceedings of the twenty-fifth international conference on machine learning (ICML'08). ACM, pp 1096–1103
5. Hinton GE, Salakhutdinov RR (2006) Reducing the dimensionality of data with neural networks. Science 313:504–507
6. Bastien F, Lamblin P, Pascanu R, Bergstra J, Good-Fellow I, Bergeron A, Bouchard N, Warde-Farley D, Bengio Y (2012) Theano: new features and speed improvements? NIPS 2012 deep learning workshop
7. Giri KC, Patel M, Sinhal A, Gautam D (2019) A novel paradigm of melanoma diagnosis using machine learning and information theory. In: 2019 international conference on advances in computing and communication engineering (ICACCE), Sathyamangalam, Tamil Nadu, India, pp 1–7. https://doi.org/10.1109/ICACCE46606.2019.9079975
8. Patel M, Sheikh R (2019) Handwritten digit recognition using different dimensionality reduction techniques. Int J Recent Technol Eng 8(2):999–1002. ISSN: 2278-3075
9. Patel M, Badi N, Sinhal A (2019) The role of fuzzy logic in improving accuracy of phishing detection system. Int J Innov Technol Explor Eng 8(8):3162–3164. ISSN: 2278-3075
10. Shekhawat VS, Tiwari M, Patel M (2021) A secured steganography algorithm for hiding an image and data in an image using LSB technique. In: Singh V., Asari VK, Kumar S, Patel RB (eds) Computational methods and data engineering. Advances in Intelligent Systems and Computing, vol 1257. Springer, Singapore. https://doi.org/10.1007/978-981-15-7907-3-35
11. Menaria HK, Nagar P, Patel M (2020) Tweet sentiment classification by semantic and frequency base features using hybrid classifier. In: Luhach A, Kosa J, Poonia R, Gao XZ, Singh D (eds) First international conference on sustainable technologies for computational intelligence. Advances in Intelligent Systems and Computing, vol 1045. Springer, Singapore. https://doi.org/10.1007/978-981-15-0029-9-9

12. Joshi S, Patel M, Natural language processing for classifying text using Naïve Bayes model. Paideuma J 13(10):72–77. ISSN No: 0090-5674. 11.3991/Pjr.V13I10.85307
13. Sungheetha A, Sharma R (2020) Trans capsule model for sentiment classification. J Artif Intell 2(03):163–169
14. Suma V (2019) Computer vision for human-machine interaction—review. J Trend Comput Sci Smart Technol (TCSST) 1(02):131–139
15. Alam MR, Bennamoun M, Togneri R, Sohel F (2016) Deep Boltzmann machines for i-vector based audio-visual person identification. In: Bräunl T, McCane B, Rivera M, Yu X (eds) Image and video technology. PSIVT 2015. Lecture Notes in Computer Science, vol 9431. Springer, Cham. https://doi.org/10.1007/978-3-319-29451-3-50

Chapter 12
A Design of Current Starved Inverter-Based Non-overlap Clock Generator for CMOS Image Sensor

Hima Bindu Katikala and G. Ramana Murthy

1 Introduction

As a part of implantable bio-signal acquisition, the sensor part has to be digitized by using an A/D converter and the signal generated should be stored for analysis [1]. In this view, memory architecture should be developed to store the data without any mismatches, and interference with less delay at this instant delay elements is considered to produce reduced delay output [2, 3]. To control the rising and falling edge of the clock or any other signal in an integrated circuit, a special variable element is used known as delay element (Variable Delay Element). It is a circuit that produces an output signal with a certain amount of delay. Delay elements (DE) are classified into three categories transmission gate-based, cascaded inverter-based, and voltage-controlled-based. Each classified DE performance varies with delay, power consumption, signal integrity and area. Mainly the delay elements are used in modern application-specific integrated circuit (ASIC), delay-locked loop (DLL), phase-locked loop (PLL), microcontrollers, memory architectures generate non-overlap clocks, etc., and its precision affects the performance of the circuit [4, 5]. As memory design is used for storing the data in terms of two basic read and write operations, for the large storage purpose, it is better to use DE to slow down the data consecutively to avert data corruption, over and above to condense skew (spatial variation in clock edges) between the propagation paths. In this paper, we discuss the DE functionality, especially in memory architectures used in complementary metal oxide semiconductor (CMOS) image sensor (CIS).

H. Bindu Katikala · G. Ramana Murthy (✉)
Vignan's Foundation for Science, Technology and Research, Vadlamudi, Guntur, AP 522213, India
e-mail: drgrm_ece@vignan.ac.in

A. Sheth et al. (eds.), *Intelligent Systems*, Algorithms for Intelligent Systems,
https://doi.org/10.1007/978-981-16-2248-9_12

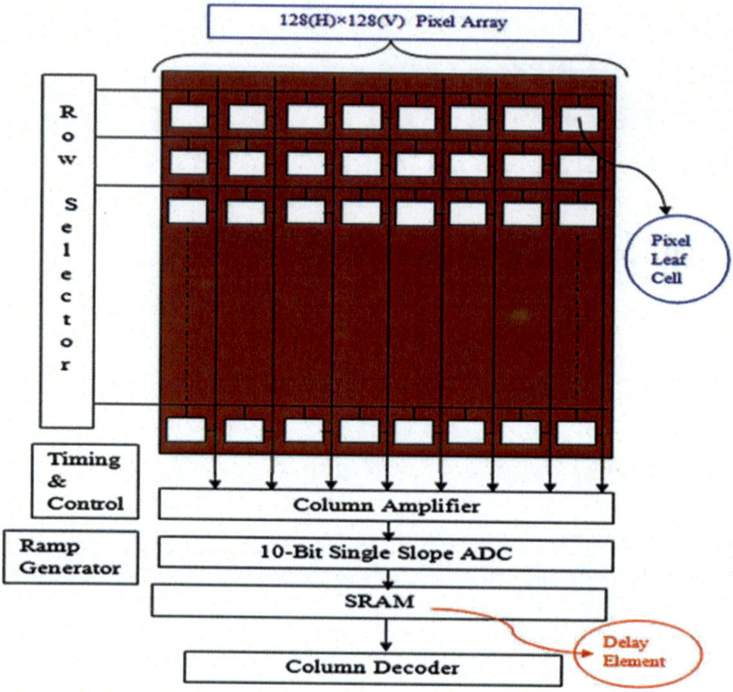

Fig. 1 CMOS image sensor (CIS) with SRAM-delay element [1]

As conventional CIS consists of pixel array architecture, column amplifiers, 10-bit single-slope analog-to-digital converter (ADC), static random-access memory (SRAM), and column decoder are in Fig. 1 [6]. In CIS, especially in SRAM architecture, a DE is used to generate non-overlap clocks for sense amplifier. The clocks generated are differing with certain delay to ensure precharge operation especially for Read operation. From Fig. 1, single-slope ADC uses 10-bit asynchronous up–down counter to generate output later to lock the counter's output; a 10-bit latch is used. Finally, the latched data is written and stored in the SRAM [7, 8].

In CIS memory, architecture contains SRAM cell and sense amplifier with non-overlap clock generator. SRAM cell as illustrated in Fig. 2a has control signals like data (Din), write signals (Øwr and Øwr-b) that are two opposite to each other, bit-line signals (BL, BLB), read signal (Ørd), and output signals (Q, Q'). For write operations, let us consider that, initially $Q = 0$, $Q' = 1$ and both the bit-lines act as input signals, then make Øwr to high value and apply BL and BLB as '0' and '1'. To write high value in a SRAM cell, make N2 transistor stronger than P2 transistor, it is possible by changing the aspect ratios of both the transistors. Now $Q = 1$ and $Q' = 0$ depends upon Din, and finally the write operation (stores the new value in SRAM cell) is completed successfully.

To have a read operation, sense amplifier is essential to sense and amplify low power signals to recognizable logic levels (as '0' or '1'). As from Fig. 2b, sense

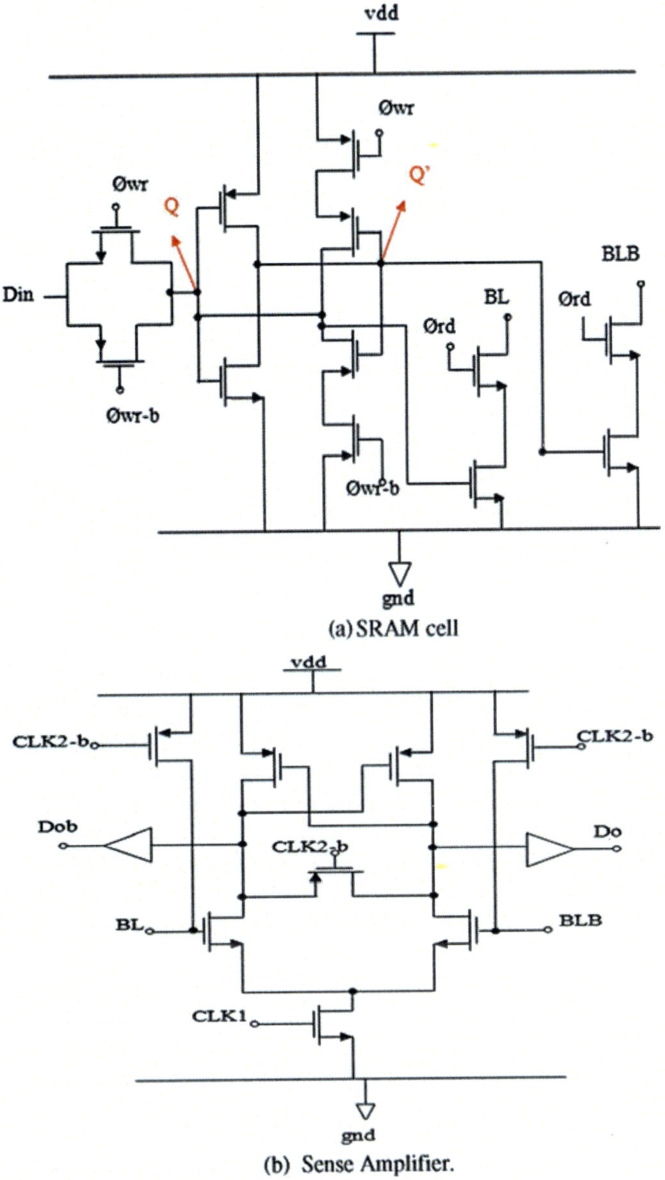

Fig. 2 **a** SRAM cell. **b** Sense amplifier

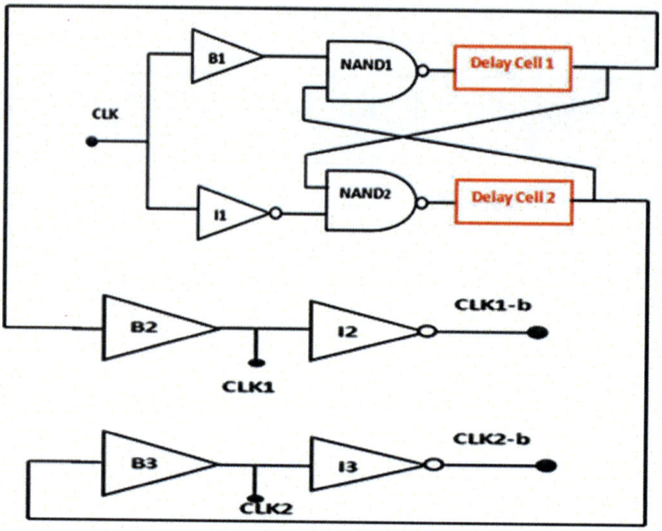

Fig. 3 Non-overlap clock generator for sense amplifier

amplifier has two clock signals which are generated by the non-overlap clock generator. Non-overlap clock generator as in Fig. 3 produces two clock outputs (CLK1 and CLK2-b) through its subcircuits like Buffers (B1, B2, B3), Inverters (I1, I2, I3) with two NAND gates and delay element cells. To perform read operation, activate the read signal (Ørd), CLK2-b signal should be '0' in line to precharge the both bit-lines to high (Vdd). By using Non-overlap clock generators, we can achieve CLK2-b as logic0 here BL and BLB act as output signals.

2 Operation of Non-Overlap Clock Generator

As shown in Fig. 3, the NAND gates and delay cells are in a cross-coupled manner.

Step-1: when CLK = 1, then output of B1 = 1 and I1 = 0.

Step-2: B1 and I1 act as input signals to NAND gates.

Step-3: As I1 = 0 then NAND2 gate output is 1, further it is processed as input to delay cell 2.

Step-4: As both the delay cells are designed by two current starved-based inverters, it generates the same output as '1' as in Fig. 6.

Step-5: Now O/P of delay cell 2 act as second, I/P of NAND1 gate, it generates output as '0'.

Step-6: Further the delay cells (1 and 2) O/P's are given to (B2, B3 and I2, I3). Therefore, the finalized O/P's is taken as CLK1 = 0, CLK1-b = 1 and CLK2 = 1, CLK2-b = 0.

Accordingly, BL and BLB are precharged to logic high (vdd). On hold state, i. e., if Øwr = 0, then the output signals Q, Q' have previously stored bits ($Q = 1$ and $Q' = 0$). Now the point is how the delay element is designed.

3 Delay Element Construction

Recent advances in CMOS technology improves the performance of the digital world and also scales down both analog and digital devices in terms of size. Coming to sensor devices, these advanced technologies improve the design performance at the level of speed, low power supply, and production. Especially in image sensors, the output processed from ADC is to be digitalized due to increased reliability, self-correcting capability, more resistant to environmental variations, and improved noise performance. Fully digital-to-analog converter uses DE to convert amplitude variants of input signal to delay variation and then the delay is digitized. As mentioned above, the pixel array output should be digitized and processed into DE to avoid data interferences.

According to the literature survey, Watanabe et al. proposed a new fully digitalized ADC with a chain of DE, where each inverter has a supply voltage of (Vin) and the input signal pulse is given to the chain of DE [9]. The measurement of delay is done by placing N input latch and connecting its inputs to the outputs of the DEs in the delay unit. Even if these techniques are simple, it requires an 'N' number of DE for the N-bit of ADC. Reduced chain of DE design is proposed with an addictive delay loop, but this architecture design suffers with nonlinearity and more delay variation as a decrease in supply voltage. To solve the above-mentioned reasons, a new current starved inverter-based DE was proposed with good linearity (linear relation between delay and input) and placed in a parallel manner, thus the DE delay is controlled by the delay of the two inverters and its operating conditions are tabulated in Table 1 [10–12].

In order to have better linearity and control over delay, in this paper, we used parallel connected inverters as DE as in Fig. 4 for non-overlap clock generator to have clock outputs (CLK1, CLK1-b, CLK2, CLK2-b) which are utilized by sense

Table 1 Conditions of delay element

In	Vin	Out	State
0	0	1	Normal inverter
0	1	1	Delay varied due to falling edge
1	0	0	Delay varied due to falling edge
1	1	0	Normal inverter

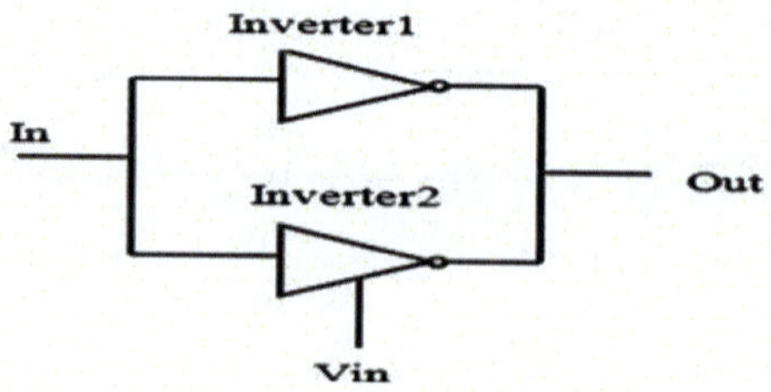

Fig. 4 Delay element

amplifier. Now both the BL and BLB are pre-charged to vdd. Here, $Q = 1$, so there is no discharge, but $Q' = 0$ and BLB = 1, there exist voltage difference in BLB voltage same is compared by the sense amplifier (act as a comparator), final process '1' as output depends on Din, (that indicates $Q =$ Do and $Q' =$ Dob) means read operation completed successfully. Before processing (ADC) signal to the SRAM cell, an intermediate latch of 10-bit operates based on level sensitivity is used to hold the data.

3.1 Simulation Results

The write and read operation of SRAM cell is plotted in Fig. 5a, b. Here for write operation: When Din = 0, Øwr = 1, BL = 0, BLB = 1, then $Q = 0$ and $Q' = 1$ as shown in Fig. 5a similarly When Din = 1, Øwr = 1, BL = 0, BLB = 1, then $Q = 1$ and $Q' = 0$, hence SRAM cell stores a new value based on Øwr.

Next for reading operation: After precharging the sense amplifier through clock signals generated from non-overlap circuit and also activating the read and write control signal Ørd, Øwr, which means when Øwr = 1, Din = 0 and CLK is high, then $Q = 0 =$ Do and $Q' = 1 =$ Dob as in Fig. 5b When Øwr = 1, Din = 1 and CLK is high, then $Q = 1 =$ Do and $Q' = 0 =$ Dob.

The implementation of the non-overlap Clock generator and its transient response are simulated in Cadence Virtuoso 180 nm technology as in Fig. 6a, b.

As in Fig. 6b, the delay element inputs are Del1in and Del2in and output responses are Del1out = clk1 and Del2out = Clk2. It is evident that the DE outputs are responsible for generating a clock signal.

4 Conclusion

Finally, the DE operates at 1.8v implemented in Cadence Virtuoso 180 nm technology with the delay of 0.0022 ns at 20 kHz frequency and has power dissipation as 18.43 nW can be used in SRAM memory architecture. And can also be used in CMOS image sensors to improve linearity.

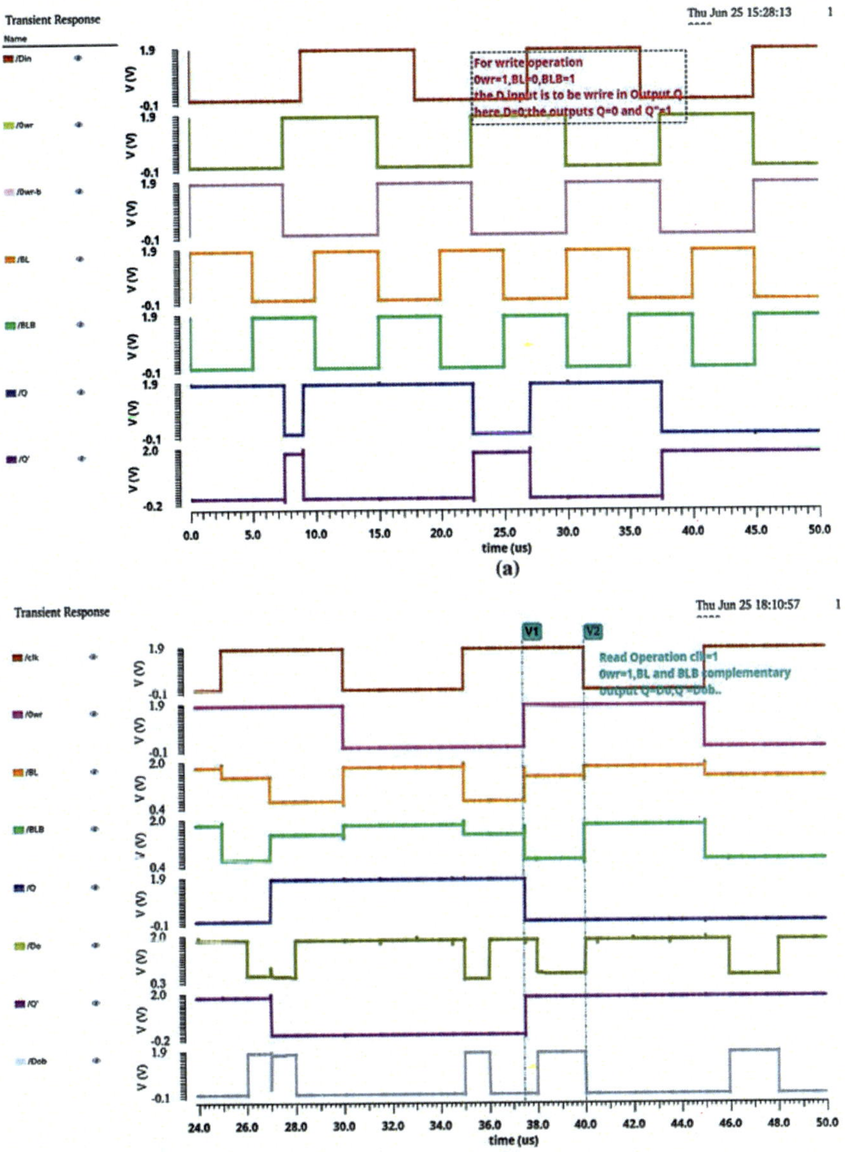

Fig. 5 Transient response of SRAM cell. **a** Write operation. **b** Read operation

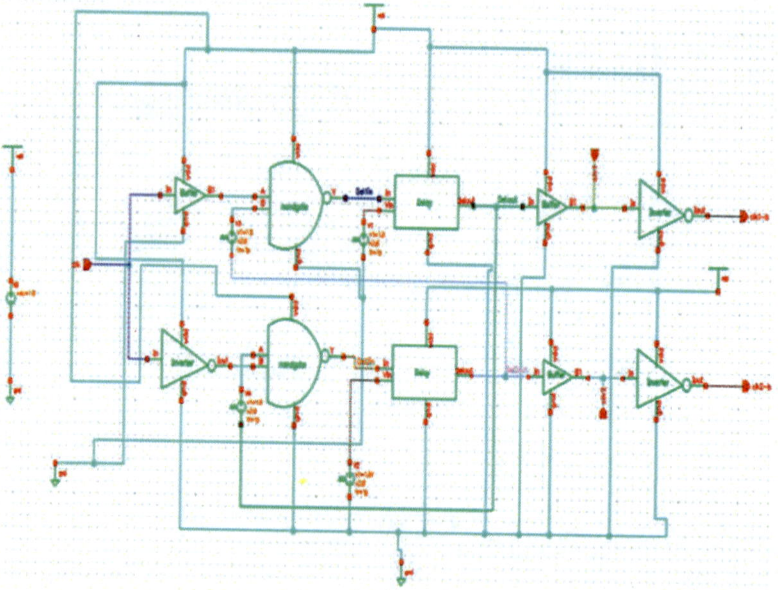

(a)

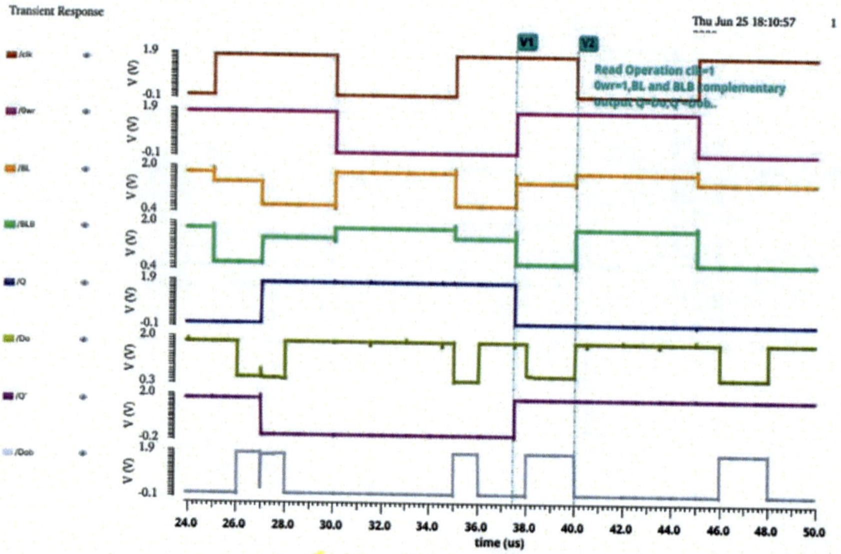

(b)

Fig. 6 a, b Schematic of non-overlap clock generator and its output response in 180 nm technology

References

1. Levski D, Wäny M, Choubey B (2018) A 1-μs ramp time 12-bit column-parallel flash TDC-interpolated single-slope ADC with digital delay-element calibration. IEEE Trans Circuit Syst Regul Pap 66(1):54–67. https://doi.org/10.1109/TCSI.2018.2846592
2. Mroszczyk P, Dudek P (2014) Tunable CMOS delay gate with improved matching properties. IEEE Trans Circuits Syst I Regul Pap 61(9):2586–2595. https://doi.org/10.1109/TCSI.2014.2312491
3. Bindu Katikala H, Shaik S (2018) A survey of the vision restoration eye. J Int Pharm Res 45(1):501–505. ISSN: 1674-0440
4. Kobenge SB, Yang H (2008) Digitally controllable delay element using switched-current mirror. WSEAS Trans Circuit Syst 8(7):599–608. ISSN: 1109-2734
5. Maymandi-Nejad M, Sachdev M (2003) A digitally programmable delay element: design and analysis. In: IEEE Trans actions on very large-scale integration (VLSI) systems, vol 11(5), pp 871–878. https://doi.org/10.1109/TVLSI.2003.810787
6. Wang F, Theuwissen AJP (2014) Linearity analysis of a CMOS image sensor. In: Proceedings of electronic imaging, San Francisco, pp 84–90. https://doi.org/10.2352/ISSN.2470-1173.2017.11.IMSE-191
7. Hemachandran K, Mary George P, Subhashree NS, Divya P (2020) Delay functionalities in SRAM: a technical review. Int J Grid Distrib Comput 13(1):433–442. ISSN: 2005-4262
8. Verma P, Halba R, Patel H, Baghini MS (2016) On-chip delay measurement circuit for reliability characterization of SRAM. In: 2016 IEEE computer society annual symposium on VLSI (ISVLSI), pp 331–336. https://doi.org/10.1109/ISVLSI.2016.24
9. Mahapatra NR, Tareen A, Garimella SV (2002) Comparison and analysis of delay elements. In: The 45th midwest symposium on circuits and systems MWSCAS. IEEE, vol 2, pp II-II. https://doi.org/10.1109/MWSCAS.2002.1186901
10. Farkhani H, Meymandi-Nejad M, Sachdev M (2008) A fully digital ADC using a new delay element with enhanced linearity. In: IEEE international symposium on circuits and systems, Seattle, pp 2406–2409. https://doi.org/10.1109/ISCAS.2008.4541940
11. Moazedi M, Abrishamifar A, Sodagar AM (2011) A highly-linear modified pseudo-differential current starved delay element with wide tuning range. In: 19th Iranian conference on electrical engineering. IEEE, pp 1–4. ISSN: 2164-7054
12. Jovanović GS, Stojčev M (2005) Linear current starved delay element. In: Proceedings of ICEST. https://doi.org/10.1080/00207210600560078

Chapter 13

A Comparison of the Best Fitness Functions for Software Defect Prediction in Object-Oriented Applications Using Particle Swarm Optimization

Hurditya, Ekta Rani, Mridul Gupta, and Ruchika Malhotra

1 Introduction

Any software faces defects or faults of some sort during its life cycle, and a lot of effort goes into minimizing the quantity and severity of these defects. A software defect prediction model is an approach aimed towards reducing such defects, and it does so by training on past data of fault-proneness and predicting the faults on a set of data that it has never seen before.

While several defect prediction models are in existence, only a few are solely focused on identifying and predicting defects in software written in object-oriented (OO) programming languages. We have focused our work on OO software and have chosen the dataset sources and analysis metrics accordingly.

For defect prediction, we have used a hybridized approach combining a search-based algorithm (SBA) along with a classification technique. The SBA we have chosen is particle swarm optimization (PSO). An SBA selects a global best solution from a variety of different possible solutions, and PSO works on that principle. PSO searches through the entire dataset, considering each data point as a particle, and finds its own best position and its global best position with respect to all other data points. The position and velocity of each particle are updated in each iteration, and these values are calculated and assessed using a performance metric in a function known

Hurditya (✉) · E. Rani · M. Gupta (✉) · R. Malhotra
Department of Software Engineering, Delhi Technological University, Bawana Road, Rohini, Delhi 110042, India
e-mail: hurditya_2k17se48@dtu.ac.in

M. Gupta
e-mail: mridulgupta_2k17se63@dtu.ac.in

E. Rani
e-mail: ektarani_2k17se35@dtu.ac.in

R. Malhotra
e-mail: ruchikamalhotra@dtu.ac.in

A. Sheth et al. (eds.), *Intelligent Systems*, Algorithms for Intelligent Systems,
https://doi.org/10.1007/978-981-16-2248-9_13

as the fitness or objective function. How the global best solution is selected depends on this fitness function. Thus, it is clear that the performance of PSO depends greatly on the fitness function underuse.

The fitness functions we have studied are G-mean, accuracy, specificity, sensitivity, and precision. Our study will give an insight into the effect these fitness functions have on the performance of a defect prediction model. Altering the fitness functions of PSO for improving its performance has been done in the past as well.

The PSO optimizer performs the task of dimensionality reduction. It reduces the number of metrics to be used by the ML technique for model building so that its performance measured using the G-mean metric is improved. The metrics which have been selected the greatest number of times across all datasets can be termed as the most significant.

The G-mean values of each ML technique paired with the different fitness function variants of the PSO optimizer are calculated across the datasets of the five applications and subjected to the Friedman test in which each technique's performance is compared by its mean rank. The classification is done on the optimized dataset using artificial neural networks (ANN). The G-mean values for each fitness function are calculated using the ROC analysis and compared with the G-mean values obtained from applying these techniques on their own. To show that, there is indeed a significant improvement offered by a particular fitness function, and a non-parametric statistical test (Friedman test) has been used which compares the variance by ranks of sets of numerical values and indicates whether a significant difference exists between them.

2 Related Work

Defect prediction has been an area of great interest in the predictive modelling domain, and there has been some valuable work done in the field using PSO in the past. Carvalho et al. [1] proposed a multi-objective PSO (MOPSO) algorithm for the identification of fault-prone classes and methods showing the capability of PSO for defect prediction tasks and how it compares to other ML models.

The recent study by Wang et al. [2] gives a comprehensive overview of the PSO algorithm and its various applications. The various methods of calculating particle fitness in the past have been discussed in their paper.

Can et al. [3] proposed a P-SVM model for defect prediction. Their study combined PSO with SVM to obtain results that were observed to be better than the results given by SVM classifier alone and GA-SVM, which is a combination of genetic algorithms (GA), another type of search-based optimization technique, with SVM. PSO and GA perform the task of optimizing the dataset by reducing their dimensionality, and SVM is the classifier that performs prediction on the optimized dataset.

In the past, some work has been done to observe the difference in the performance of optimization algorithms by changing the objective/fitness functions. Malhotra and

Khanna's study [4] gave a model to select an appropriate CPSO fitness function for each class in a dataset rather than the whole dataset, showing that the optimization of a dataset can be done to a very large extent, and thus, showing a better objective function is the key to building a better optimizer based on PSO. The study by Aslam [5] used different fitness functions of genetic programming and showed the difference in their performance.

The use of OO metrics has also increased in the past as seen in studies by Li and Henry [6] and Gyimothy et al. [7] who used OO metrics for building predictive models. This study has also been carried out using these OO metrics which is beneficial in the sense that the applications over which defects have been predicted are written mostly in an OO programming language, i.e. Java.

We have used an ANN classifier for the task of classification, whose compatibility with SBAs has been explored in the past in studies by John et al. [8] and Da and Xiurun [9]. This study aims at progressing these studies forward; it combines and builds on the previous work and uses the modifications to PSO's objective function to build a better defect prediction model and observe the changes.

3 Methodology and Experimental Design

This section provides the explanation regarding the empirical process followed throughout the course of this study.

3.1 Overview

In this study, four LOC statistics along with 23 OO metrics as defined by the understand tool have been used as the independent variables, and the dependent variable is named "Fault" which signifies whether a class is defect-prone or not. The data used in this study has been collected from the source codes of five Android applications—camera, contacts, MMS, framework base, and settings.

These datasets were then subjected to analysis through optimization and classification techniques as detailed further in this section.

3.2 Optimization Technique—Particle Swarm Optimization

PSO is a swarm population-based optimization technique that was developed by Eberhart and Kennedy. Each potential solution is termed a *particle*, and it traverses through the d-dimensional space. This study has been done using the PySwarms library's pyswarms.discrete.binary module which is the implementation of a binary PSO where a particle's velocity and position are updated using the following equations:

$$V_{ij}(t+1) = w \cdot V_{ij}(t) + c_1 r_{1j}\big(p_{ij}(t) - X_i(t)\big) + c_2 r_2\big(\hat{p}_j(t) - X_i(t)\big) \quad (1)$$

$$X_{ij}(t+1) = \begin{cases} 0, & \text{if rand}() \geq S\big(V_{ij}(t+1)\big) \\ 1, & \text{if rand}() \geq S\big(V_{ij}(t+1)\big) \end{cases} \quad (2)$$

where $S(x)$ is the sigmoid function defined as

$$S(x) = \frac{1}{1 + R^{-x}} \quad (3)$$

$p_{i,j}$ is the best previously visited position of the particle, and $\hat{p}_j$ is the global best value among all particle positions. The cost of each particle at iteration t is calculated using the objective function:

$$C(X_{ij}(t)) = \alpha(1 - p) + (1 - \alpha)\left(1 - \frac{N_f}{N_i}\right) \quad (4)$$

We create and train the ANN classifier for the PSO optimizer likewise to any classification task. For each particle, the ANN is trained on a training sample, and the G-mean is calculated over the test samples and is returned to the objective function for the calculation of a particle's cost. This cost determines the particles updated position in the next iteration according to the equation:

$$p_{ij}(t+1) = \begin{cases} X_{ij}(t), & \text{if } C(X_i(t)) < C(p_{ij}(t)) \\ p_{ij}(t), & \text{if } C(X_i(t)) \geq C\big(p_{ij}(t)\big) \end{cases} \quad (5)$$

The global best position is also updated as necessary at each iteration according to the equation (Table 1):

$$\hat{p}_j(t) = \begin{cases} p_{ij}(t), & \text{if } C\big(p_{ij}(t)\big) < C\big(\hat{p}_j(t-1)\big) \\ p_{ij}(t-1), & \text{if } C\big(p_{ij}(t)\big) \geq C\big(\hat{p}_j(t-1)\big) \end{cases} \quad (6)$$

The PSO algorithm has been given below:

Table 1 PSO parameters

Parameter	Value
Number of particles (Q)	50
Number of iterations (I_t)	60
Trade-off hyperparameter (α)	0.55
Inertia weight (w)	0.45
Cognitive parameter (c_1)	0.5
Social parameter (c_2)	0.5

```
PSO Algorithm
1. Initialize position and p_(ij )value of each particle randomly.
2. do
3. for each particle i=1 to Q, do
4.     for the set of features 1 to d, do
5.           r_1=rand(0,1)
6.           r_2=rand(0,1)
7.           Update the particle's velocity according to equation
(1)
8.     Update the particle's position according to equation (2)
9.     Send the particles' positions as a subset to the objective
       function for cost calculation
10.    Update best position of the particle according to equation
(5)
11.    Update the global best position according to equation (6)
12. while a termination criterion is not met
```

3.3 Classification Technique—Artificial Neural Networks

ANN classifiers work like an imitation of the human brain. In this study, a multi-layer perceptron with one hidden layer has been used. The multi-layer perceptron involves improving the accuracy of the classifier by forward and backward propagation in each epoch/iteration. The ANN algorithm begins with the assignment of random weights to the linkages or synapses between the nodes in each layer. The activation function calculates and records the activation rate of each node in each forward iteration, and the error rate is similarly recorded in each backward iteration. The weights of the synapses are correspondingly updated to improve the performance in the next iteration, and the process is repeated for a pre-defined number of epochs. In this study, the ANN classifier has been run for 100 epochs for each classification task, and we have used the rectified linear unit (ReLU) activation function.

We have chosen the ANN classifier due to its adaptability towards complex and varied data and its ability to implicitly detect complex nonlinear relationships between dependent and independent variables. Historically, there have been many studies discussing the success of neural networks for defect prediction like [10]. The effectiveness of neural networks along with the optimization by swarm intelligence for defect prediction in software is evident.

3.4 Analysis Process

(1) The dataset obtained from all five applications is pre-processed before any analysis task begins. This involves standardizing the data using the standard scaling function $Z = (x - u)/s$.

(2) The dataset is divided into training and testing samples.
(3) Dimensionality reduction is done using PSO with the first fitness function.
(4) Classification is done using an ANN classifier.
(5) The G-mean values are recorded.
(6) Steps 3–5 are repeated for all fitness function variants.
(7) The G-mean values obtained from all fitness function variants are compared, and it is determined if a significant performance improvement is observed or not by using a statistical test, the Friedman test in this case.

4 Results and Inferences

This section summarizes the results obtained from all optimization, classification, and statistical tests that have been performed in this study. The optimization results give us the feature subset of OO metrics as selected by the PSO optimizer. The classification is done after choosing each fitness function separately. The statistical tests help us determine whether the results obtained are significant or not.

4.1 Dimensionality Reduction

The PSO optimizer selects a feature subset from the given 23 OO metrics and four LOC metrics in each dataset. A summary of the results is given in Fig. 1.

It can be seen from Fig. 1 that RFC (number of methods, including inherited ones), LOC (number of lines containing source code), IFANIN (number of immediate base classes), number of children (NOC), lack of cohesion in methods (LCOM), and depth of inheritance tree (DIT) are the metrics which have been selected the greatest number of times. It must be noted that these results are in accordance with the results of previous well-reputed works like that of [7] where it was established that LOC, RFC, LCOM, CBO, and DIT are very significant metrics for defect prediction which serves as a validation of our results stating the most significant OO metrics.

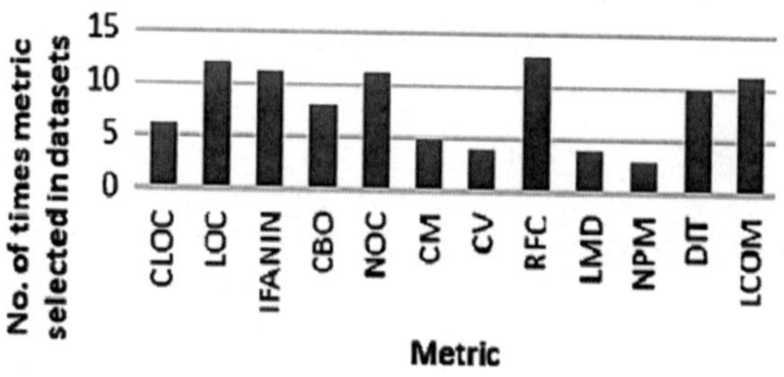

Fig. 1 Metrics selected by PSO

4.2 Model Prediction

The G-mean values calculated using each different fitness function were calculated and recorded by using tenfold cross-validation. Table 2 depicts the averages of the G-mean values calculated for each dataset when the cross-validation technique is applied. The highest values have been highlighted in bold.

It can be seen that the G-mean function is the most successful in providing the best AUC and G-mean scores when compared to the performance measured using other fitness functions.

According to the Friedman test, there was a statistically significant difference in AUC values depending on the fitness function variant used, $\chi^2(2) = 19.1889$, **p = 0.0072**. The null hypothesis of the Friedman test is rejected as the p-value is less than 0.05. Hence, the results can be said to be significant. The corresponding ranking based on the mean of ranks (higher is better) has been given below:

Our results are in accordance with the previous studies on the topic like [4] which showed how well the G-mean fitness function performs in a CPSO algorithm. The superior performance of functions like G-mean and precision indicates that when a model's defect prediction capabilities are tested over several datasets for a comprehensive study, the biases posed by simple functions like accuracy in unbalanced

Table 2 Predicted G-mean values for the datasets

Parameter		Accuracy	G-Mean	Sensitivity	Specificity	Precision
Settings	Kitkat	**0.767**	0.756	**0.767**	**0.767**	**0.767**
	Lollipop	0.767	**0.811**	0.74	0.764	0.773
	Marshmallow	0.74	**0.793**	0.733	0.726	0.757
	Nougat	0.786	**0.805**	0.753	0.789	0.773
	Oreo	0.706	**0.723**	0.682	0.698	0.703
MMS	Lollipop	0.354	**0.571**	0.546	0.546	0.599
	Marshmallow	**0.401**	**0.401**	0.378	0.37	0.4
	Nougat	**0.497**	**0.497**	0.395	0.388	0.388
	Oreo	0.482	0.491	0.375	0.433	**0.54**
Framework base	Oreo	0.538	0.546	0.525	**0.579**	0.577
Contacts	Kitkat	0.555	**0.571**	0.555	0.552	0.555
	Lollipop	0.447	**0.556**	0.415	0.415	0.45
	Marshmallow	0.325	0.456	0.318	**0.513**	0.51
	Nougat	0.65	**0.735**	0.679	0.606	0.643
	Oreo	0.45	**0.698**	0.593	0.59	0.588
Camera	Marshmallow	0.556	**0.611**	0.488	0.487	**0.611**
	Nougat	**0.611**	0.583	0.418	0.584	**0.611**
	Oreo	0.803	**0.842**	0.806	0.894	**0.842**
Mean	0.579722	**0.635889**	0.564778	0.5945	0.615944	

Table 3 Friedman test mean ranks

Rank	Fitness function	Mean ranks
1	G-mean	4.10
2	Precision	3.47
3	Accuracy	2.89
4	Specificity	2.50
5	Sensitivity	2.03

datasets can be overcome. It can be said that G-mean is one of the best functions to be paired with PSO for defect prediction.

5 Conclusion

The following conclusions have been reached during the course of the study:

i. RFC, LOC, IFANIN, NOC, LCOM, and DIT were the metrics that were selected most often by the PSO optimizer.
ii. G-mean and sensitivity were seen to be the best and worst PSO fitness functions for software defect prediction, respectively, and the significance of these results was confirmed by using the Friedman test. The results show how the selection of a particular fitness function might significantly affect the performance of an optimization and classification model and thus also helps determine which fitness functions are to be used for a particular purpose. Among the functions studied, G-mean stood out to be a notably superior fitness function to maximize the performance of the PSO model and its application, and further testing is advised for future researchers to maximize generalizability.

References

1. de Carvalho AB, Pozo A, Vergilio SR (2010) A symbolic fault-prediction model based on multiobjective particle swarm optimization. J Syst Softw 83(5):868–882
2. Wang D, Tan D, Liu L (2018) Particle swarm optimization algorithm: an overview. Soft Comput 22(2):387–408
3. Can H, Jianchun X, Ruide Z, Juelong L, Qiliang Y, Liqiang X (2013) A new model for software defect prediction using Particle Swarm Optimization and support vector machine. Chinese Control and Decision Conference (CCDC), Guiyang, China
4. Malhotra R, Khanna M (2019) Dynamic selection of fitness function for software change prediction using Particle Swarm Optimization. Inform Softw Technol 112:51–67
5. Aslam MW (2015) Selection of fitness function in genetic programming for binary classification. In: Science and information conference (SAI), London
6. Li W, Henry S (1993) Object-oriented metrics that predict maintainability. J Syst Softw 23:111–122

7. Gyimothy T, Ferenc R, Siket I (2005) Empirical validation of object-oriented metrics on open-source software for fault prediction. IEEE Trans Softw Eng 31:897–910
8. John S, Sikdar S, Swamy PK, Das S, Maity B (2008) Hybrid neural–GA model to predict and minimise flatness value of hot rolled strips. J Mater Proces Technol 195:314–320
9. Da Y, Xiurun G (2005) An improved PSO-based ANN with simulated annealing technique. Neurocomputing 63:527–533
10. Arar ÖF, Ayan K (2015) Software defect prediction using cost-sensitive neural networks. Appl Soft Comput 33:263–277

Chapter 14
Stable Zones of Fractional Based Controller for Two-Area System Having Communication Delay

Idamakanti Kasireddy, A. W. Nasir, Rahul, and R. V. D. Rama Rao

1 Introduction

The main parameters in the power system are frequency and voltage. These parameters fluctuate based on demand of electrical energy consumers. Hence, these fluctuations should maintain within a specified range to provide quality power to consumers [1, 2]. The frequency is the important parameter for the LFC problem. In the literature [3, 4], the linear models of the power system are considered to analyze the LFC problem. However, the communication delay is the most significant problem of the power system. It is nothing but the time consumed by a signal to reach the sink from the source. These delays may occur in two cases. One is while measuring the frequency as well as the power flow through RTU and control center. Another one occurs between control center output and various generating units [5–8]. The communication delay is given by a function e^{ST_d} which is exponential where T_d is the approximated communication delay of each area.

In the literature [9, 10], many LFC techniques are available for power system control. In this work, a fractional based controller is used due to its advantages such as more design flexibility, large stability region, and selectivity [11, 12]. Any system may be considered as a stable system if it has the ability to regain its normal operating condition after the clearance of fault or any disturbance. So, in order to prevent the system entering unstable regions during static or dynamic conditions, the parameters of the controller are to be selected carefully. Some literature [13–16] has been presented on the stable and unstable regions of fractional and simple PI/PID

I. Kasireddy (✉) · R. V. D. Rama Rao
EEE, Vishnu Institute of Technology, Bhimavaram 534201, India

A. W. Nasir
CMR Institute of Technology, Bangalore 560037, India
e-mail: 2014rsee003@nitjsr.ac.in

Rahul
IIT BHU, Varanasi 221005, India

A. Sheth et al. (eds.), *Intelligent Systems*, Algorithms for Intelligent Systems,
https://doi.org/10.1007/978-981-16-2248-9_14

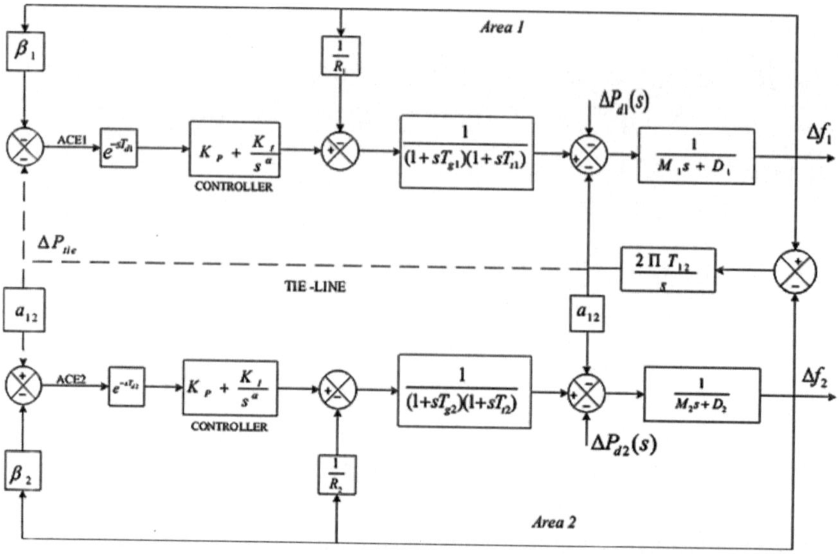

Fig. 1 Two-area power system with communication delay

controllers. In [17], the stable areas of fractional controllers for single-area LFC systems are discussed. This paper describes the stable and unstable zones of fractional PI controllers using graphical methods for different α values in $K_{p,}K_i$ plane. The graphical method can provide an easy way of finding stable and unstable zones of the controller. Using this method, the stability zones of the controller can be easily found by making imaginary and real parts of the characteristic equation of the system to zero. From Fig. 1., T_{d1} is total communication delay in area 1,T_{d2} is total communication delay in area 2, T_{g1}, T_{t1}, M_1, D_1, β_1, R_1 are the governor and turbine time constants in area 1, governor moment of inertia in area 1, governor damping factor in area 1, frequency bias factor in area 1, governor speed regulation in area 1, K_{ps1}, and T_{ps1} are the system gain and time constants in area 1. T_{g2}, T_{t2}, M_2, D_2, β_2, R_2 are the governor and turbine time constants in area 2, governor moment of inertia in area 2, governor damping factor in area 2, frequency bias factor in area 2, governor speed regulation in area 2, and K_{ps2} and T_{ps2} are the system gain and time constants in area 2.Δf_1,Δf_2, ΔP_{tie1} and ΔP_{tie1} are deviations in frequency deviations and tie-line power in area 1 and area 2. ACE 1 and ACE 2 are the system area control errors. The values of system parameters are taken from [8].

2 Estimation of Stable and Unstable Zones of Fractional PI Controller Parameters

To represent stable and unstable zones of fractional controller parameters (K_P, K_I & α), an equation is developed from Fig. 1 by considering following assumptions.

1. Tie-line power deviation among the areas is neglected.
2. Communication delay is assumed to be the same in both areas.
3. Two areas are identical.

That is, $R_1 = R_2 = R$, $T_{t1} = T_{t2} = T_t$, $T_{g1} = T_{g2} = T_g$, $M_1 = M_2 = M$, $\beta_1 = \beta_2 = \beta$, and $D_1 = D_2 = D$.

The characteristic equation of Fig. 1 is given by (1).

$$f(S, T_d, \alpha) = 1 + \frac{\{\beta\, e^{-S\,T_d}\left(K_P + \frac{K_I}{S^\alpha}\right) + \frac{1}{R}\}}{\left(1 + S\,T_g\right)(MS + D)(1 + S\,T_t)} = 0 \tag{1}$$

$$\begin{aligned} f(S, T_d, \alpha) = {} & RS^\alpha(1 + ST_t)\left(1 + ST_g\right)(Ms + D) \\ & + \beta R\, e^{-ST_d}(K_P\, S^\alpha + K_I) + S^\alpha = 0 \end{aligned} \tag{2}$$

$$f(S, T_d, \alpha) = \left\{ \begin{array}{l} \left(T_t\, T_g\, MR\right) S^{3+\alpha} + \left(R\, T_t\, T_g\, D + MR\, T_t + MRT_g\right) s^{2+\alpha} \\ +(DRT_g + DRT_t + MR)\, S^{1+\alpha} \\ + S^\alpha(RD + 1) + \beta R\, e^{-S\,T_d}(K_P\, S^\alpha + K_I) \end{array} \right\} = 0 \tag{3}$$

$$\begin{aligned} f(S, T_d, \alpha) = {} & K_3\, S^{3+\alpha} + K_2\, s^{2+\alpha} + K_1\, S^{1+\alpha} + K_0\, S^\alpha \\ & + \beta R\, e^{-S\,T_d}(K_P\, S^\alpha + K_I) = 0 \end{aligned} \tag{4}$$

where

$$K_3 = T_t\, T_g\, MR,\ K_2 = T_t\, T_g\, MR,\ K_1 = DR\, T_g + DR\, T_t + MR,\ K_0 = DR + 1$$

put $s = \omega e^{j\frac{\pi}{2}}$ and $e^{-ST_d} = \cos(\omega T_d) - j\sin(\omega T_d)$

$$f(j\omega, T_d, \alpha) = \left\{ \begin{array}{l} K_3\omega^{3+\alpha} e^{j\frac{\pi}{2}(3+\alpha)} + K_2\omega^{(2+\alpha)} e^{j\frac{\pi}{2}(2+\alpha)} + K_1\omega^{1+\alpha} e^{j\frac{\pi}{2}(1+\alpha)} + \\ K_0\omega^\alpha e^{j\frac{\pi\alpha}{2}} + \beta R(K_P\omega^\alpha e^{j\frac{\pi\alpha}{2}} + K_I)\{\cos(\omega T_d) - j\sin(\omega T_d)\} \end{array} \right\} = 0 \tag{5}$$

$$\left\{ \begin{array}{l} \beta R\,\omega^\alpha . \cos\left(\omega\, T_d - \frac{\pi\alpha}{2}\right) K_P + \beta R\cos(\omega\, T_d)\, K_I + K_0\,\omega^\alpha\cos\left(\frac{\pi\alpha}{2}\right) \\ - K_1\,\omega^{1+\alpha}\sin\left(\frac{\pi\alpha}{2}\right) - K_2\,\omega^{2+\alpha}\cos\left(\frac{\pi\alpha}{2}\right) + K_3\,\omega^{3+\alpha}\sin\left(\frac{\pi\alpha}{2}\right) \end{array} \right\} = 0 \tag{6}$$

$$\left\{\begin{array}{l} -\beta R\,\omega^{\alpha}\,.sin\left(\omega\, T_d - \frac{\pi\alpha}{2}\right) K_P - \beta R \sin(\omega\, T_d)\, K_I + K_0\,\omega^{\alpha} \sin\left(\frac{\pi\alpha}{2}\right) \\ + K_1\,\omega^{1+\alpha} \cos\left(\frac{\pi\alpha}{2}\right) - K_2\,\omega^{2+\alpha} \sin\left(\frac{\pi\alpha}{2}\right) - K_3\,\omega^{3+\alpha} \cos\left(\frac{\pi\alpha}{2}\right) \end{array}\right\} = 0 \quad (7)$$

The simple equations are

$$K_P * X_1(\omega, \alpha) + K_I * Y_1(\omega, \alpha) + M_1(\omega, \alpha) = 0 \quad (8)$$

$$X_2(\omega, \alpha) * K_P + Y_2(\omega, \alpha) * K_I + M_2(\omega, \alpha) = 0 \quad (9)$$

Solving Eqs. (8) and (9)

$$K_P = \frac{M_2(\omega, \alpha)\, Y_1(\omega, \alpha) - Y_2(\omega, \alpha)\, M_1(\omega, \alpha)}{X_1(\omega, \alpha)\, Y_2(\omega, \alpha) - X_2(\omega, \alpha)\, Y_1(\omega, \alpha)} \quad (10)$$

$$K_I = \frac{X_2(\omega, \alpha)\, M_1(\omega, \alpha) - X_1(\omega, \alpha)\, M_2(\omega, \alpha)}{X_1(\omega, \alpha)\, Y_2(\omega, \alpha) - X_2(\omega, \alpha)\, Y_1(\omega, \alpha)} \quad (11)$$

where
$X_1(\omega, \alpha) = \beta R\,\omega^{\alpha} \cos\left(\omega\, T_d - \frac{\pi\alpha}{2}\right)$, $Y_1(\omega, \alpha) = \beta R \cos(\omega\, T_d)$;
$M_1(\omega, \alpha) == K_0\,\omega^{\alpha} \cos\left(\frac{\pi\alpha}{2}\right) - K_1\,\omega^{1+\alpha} \sin\left(\frac{\pi\alpha}{2}\right) - K_2\,\omega^{2+\alpha} \cos\left(\frac{\pi\alpha}{2}\right) + K_3\,\omega^{3+\alpha} \sin\left(\frac{\pi\alpha}{2}\right)$;
$X_2(\omega, \alpha) = -\beta R\,\omega^{\alpha} \sin\left(\omega\, T_d - \frac{\pi\alpha}{2}\right)$, $Y_2(\omega, \alpha) = -\beta R \sin(\omega\, T_d)$;

$$\begin{aligned} M_2(\omega, \alpha) = {} & K_0\omega^{\alpha} \sin\left(\frac{\pi\alpha}{2}\right) + K_1\omega^{1+\alpha} \cos\left(\frac{\pi\alpha}{2}\right) \\ & - K_2\omega^{2+\alpha} \sin\left(\frac{\pi\alpha}{2}\right) - K_3\omega^{3+\alpha} \cos\left(\frac{\pi\alpha}{2}\right) \end{aligned}$$

The locus of K_P, K_I, α, ω is obtained by evaluating Eqs. (10) and (11) in the $K_P - K_I$ plane.

3 Results

Figure 2a depicts various stable and unstable zones of fractional PI controller for $\omega = \in [0, 3.5]$ at distinct α values. Figure 2b depicts stable and unstable zones of simple PI controller $\omega = \in [0, 3.5]$.

A point (K_P, K_I) in regions R_1, R_2, R_3, and R_4 is considered for dynamic analysis of the controlled system for distinct α values. Various simulation analysis has been conducted by considering controller parameters in different regions which is discussed below.

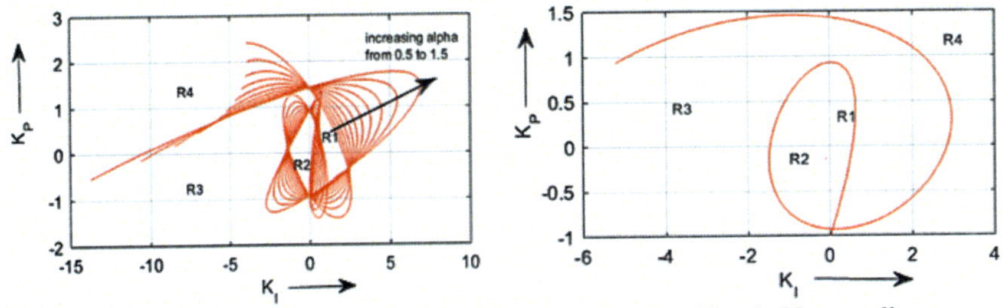

(a) Fractional controller for various α values (b) Simple PI controller

Fig. 2 Stable and unstable zones of controllers for $T_d = 2.28$ s

Case 1: *PI*

The PI controller parameters $K_I = 0.2$ and $K_P = 0.4$ (see Table 1) are considered from region $R1$ of Fig. 2b. Then, the controlled system is simulated with 10% load disturbance in the MATLAB environment. The obtained deviations are depicted in Fig. 3a. It can be noticed from Fig. 3a that AGC is absolutely stable in R1. Similarly,

Table 1 Parameters of controller for various regions

Controller	Region (R_1)		Region (R_2)	
	K_P	K_I	K_P	K_I
PI	0.4	0.2	0.6	−0.5
(Fractional PI) $\alpha = 1.3$	1	0.5	0.5	−0.5
(Fractional PI) $\alpha = 1.5$	1	0.9	0.5	−5

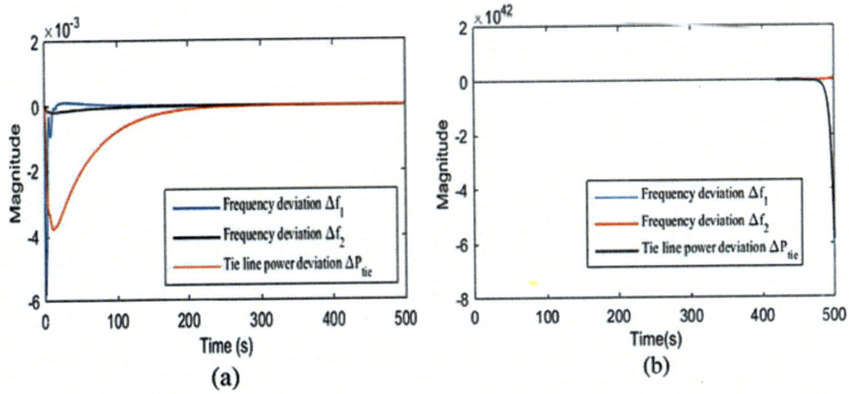

Fig. 3 System dynamic responses. **a** $K_I = 0.2$ and $K_P = 0.4$ in region $R1$. **b** $K_I = -0.5$ and $K_P = 0.6$ in region $R2$

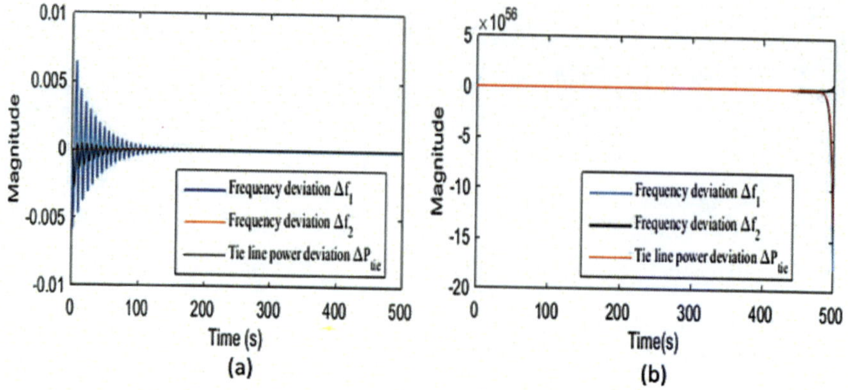

Fig. 4 System dynamic responses. **a** $K_I = 0.5$ and $K_P = 1$ in region $R1$. **b** $K_I = -0.5$ and $K_P = 0.5$ in region $R2$

the PI controller parameters $K_I = -0.5$ and $K_P = 0.6$ (see Table 1) are considered from region $R2$ of Fig. 2b. Then, the controlled system is tested with 10% load disturbance. The obtained frequency dynamic responses (for area 1 and area 2) and tie-line power dynamic response are depicted in Fig. 3b. It is clear from Fig. 3b that the power system is unstable in $R2$.

Case 2: $\alpha = 1.3$ (Fractional PI)

The fractional PI controller parameters $K_I = 0.5$ and $K_P = 1$ (see Table 1) are considered from region $R1$ of Fig. 2a. Then, the controlled system is simulated with 10% load disturbance in the MATLAB environment. The obtained frequency dynamic responses (for area 1 and area 2) and tie-line power dynamic response are depicted in Fig. 4a. It can be noticed from Fig. 4a that AGC is absolutely stable in $R1$. Similarly, the PI controller parameters $K_I = -0.5$ and $K_P = 0.5$ (see Table 1) are considered from region $R2$ of Fig. 2a. Then, the controlled system is simulated with 10% load disturbance in the MATLAB environment. The obtained frequency dynamic response (for area 1 and area 2) and tie-line power dynamic response are depicted in Fig. 4b. It is clear from Fig. 4b that power system is unstable in $R2$.

Case 3: $\alpha = 1.5$ (Fractional PI)

The fractional PI controller parameters $K_I = 0.5$ and $K_P = 1$ (see Table 1) are considered from region $R1$ of Fig. 2a. Then, the controlled system is simulated with 10% load disturbance in the MATLAB environment. The obtained frequency dynamic responses (for area 1 and area 2) and tie-line power dynamic response are depicted in Fig. 5a. It can be noticed from Fig. 5a that AGC is absolutely stable in region $R1$. Similarly, the PI controller parameters $K_I = -5$ and $K_P = 0.5$ (see Table 1) are considered from region $R2$ of Fig. 2a. Then, the controlled system is simulated with 10% load disturbance in the MATLAB environment. The obtained frequency dynamic responses (for area 1 and area 2) and tie-line power dynamic response are

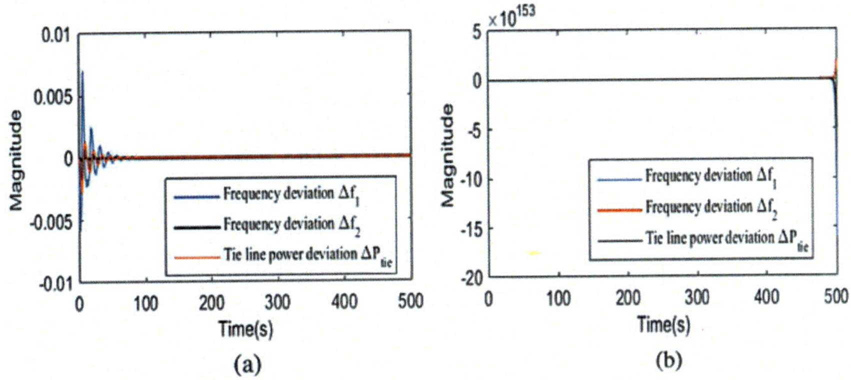

Fig. 5 System dynamic responses. **a** $K_I = 0.5$ and $K_P = 0.9$ in region $R1$. **b** $K_I = -5$ and $K_P = 0.5$ in region $R2$

depicted in Fig. 5b. It is clear from Fig. 5b that the power system is unstable in region $R2$.

Also, it is clear from Fig. 2a, b that the stability area of a fractional PI controller is more when compared to a simple PI controller. Therefore, one can conclude that the fractional PI has more choices of controller parameters for various α values in order to protect the power system with delay in communication.

4 Conclusion

In this work, a graphical based technique has been adopted to find the stable and unstable zones of fractional controller parameters. It has been observed that the fractional controller contains wide stable regions when compared to simple PI controllers.

Hence, the fractional based controller has more choices of K_P and K_I for various α values in order to protect the two-area system including delay of communication.

References

1. Elgered OI (1976) Electrical energy systems theory. Tata McGraw-Hill, New York
2. Kundur P (1994) power system stability and control. Tata McGraw Hill, New York
3. Sivaramakrishnan AY, Hariharan MV, Srisailam MC (1984) Design of variable structure load frequency controller using pole assignment technique. Int J Control 40(3):487–498
4. Reddy IK, Nasir AW, Singh AK (2019) Application of FOPID-FOF controller based on IMC theory for automatic generation control of power system. IETE J Res
5. Fu C, Wang C, Wang LY, Shi D (2020) An alternative method for mitigating impacts of communication delay on load frequency control. Int J Electric Power Energy Syst 119

6. Singh VP, Kishor N, Samuel P (2015) Communication time delay estimation for load frequency control in two-area power system. Elsevier, Amsterdam, pp 69–85
7. Tan N, Kaya I, Yeroglu C, Atherton DP (2006) Computation of stabilizing PI and PID controllers using the stability boundary locus. Energy Convers Manage 47:3045–3058
8. Sönmez S, Ayasun S (2016) Stability region in the parameter space of PI controller for a single-area load frequency control system with time delay. IEEE Trans Power Syst 31(1):829–830
9. Tan W (2010) Unified tuning of PID load frequency controllers for power systems via IMC. IEEE Trans Power Syst 25(1):341–350
10. Sahu RK, Panda S, Pradhan PC (2015) Design and analysis of hybrid firefly algorithm-pattern search based fuzzy PID controller for LFC of multi area power systems. Int J Electr Power Energy Syst 69:200–212
11. Muresan C, Dulf EH, Both R (2015) A novel tuning algorithm for fractional order IMC controllers for time delay processes. Int J Mech Eng Robot Res 4(3):218–221
12. Cxelik V, Özdemir MT, Bayrak G (2016) The effects on stability region of the fractional-order PI controller for one-area time-delayed load–frequency control systems. Trans Instit Measur Control 1–13
13. Ruszewski A (2008) Stability regions of closed loop system with time delay inertial plant of fractional order and fractional order PI controller. Bull Polish Acad Sci Tech Sci 56(4)
14. Bongulwar MR, Patre BM (2017) Stability regions of closed loop systems with one non-integer plus time delay plant by fractional order PID controller. Int J Dynam Control 5:159–167
15. Çelik V, Özdemir MT, Bayrak G (2017) The effects on the stability region of the fractional-order PI controller for one-area time-delayed load–frequency control systems. Trans Inst Meas Control 39(10):1509–1521
16. Chen S, Huang H (2019) Design of fractional order proportional integral controller using stability and robustness criteria in time delay system. Measur Control 52(9–10):1552–1566
17. Kumar R, Kasireddy I, Kumar A, Singh AK (2019) Estimation of stability regions of fractional PI controllers for LFC of power system. In: 2019 IEEE international conference on sustainable energy technologies and systems (ICSETS), Bhubaneswar, India, pp 313–318

Chapter 15
Lifting Scheme and Schur Decomposition based Robust Watermarking for Copyright Protection

K. Prabha and I. Shatheesh Sam

1 Introduction

Resources of nowadays like cloud computing and big data create an urgent need to develop a scheme for the copyright protection of multimedia big data. Securing the shared digital content is a major problem to be solved in this multimedia environment. Digital watermarking is an incredible strategy for copyright assurance of multimedia content. This area attracted many researchers attention and leads to the development of several schemes [1–3] for securing the copyright. The significant necessity of the watermarking approach includes robustness and imperceptibility. Mainly, the watermark embedding process is executed in two groups, namely spatial domain and frequency domain [4]. The watermark embedding process of the spatial domain technique is performed by changing the pixels. This technique is easy, computationally simple, and less robust to attacks. The embedding process of the frequency domain technique is implemented in the modified coefficients of the image using some transforms. Different transformation techniques like Discrete Wavelet Transform (DWT) [5], Discrete Cosine Transform (DCT) [6], Lifting Wavelet Transform (LWT) [7], etc., are utilized to obtain robust watermarking. The advantage of these methods is robustness against attacks. Several hybrid transform [8, 9] methods are recently proposed to obtain better performance.

The transform domain technique along with some decomposition techniques is a good choice for achieving imperceptible and robust watermarking schemes. The color cover image is preprocessed by Quaternion Hadamard Transform (QHT) [10], and the unitary matrix of Schur decomposition is altered to embed the watermark.

K. Prabha (✉)
Department of Computer Science, Nesamony Memorial Christian College Affiliated to Manonmaniam Sundaranar University, Abishekapatti, Tirunelveli 627012, Tamil Nadu, India

I. Shatheesh Sam
Department of PG Computer Science, Nesamony Memorial Christian College Affiliated To Manonmaniam Sundaranar University, Abishekapatti, Tirunelveli 627012, Tamil Nadu, India

A. Sheth et al. (eds.), *Intelligent Systems*, Algorithms for Intelligent Systems,
https://doi.org/10.1007/978-981-16-2248-9_15

In order to reduce visual distortion, the upper triangular matrix and unitary matrix of Schur decomposition [11] are chosen for embedding the watermark information. In [12], the watermark embedding is performed in the last two rows of the Walsh Hadamard Transform (WHT) coefficients because these elements do not influence the image quality and therefore improve its robustness.

Schur decomposition achieves less complexity in the frequency domain while comparing it with other matrix decomposition techniques. Therefore, the proposed scheme uses LWT in combination with Schur decomposition. The watermark robustness and security are improved by scrambling the watermark image using Arnold transform [13]. The contribution of the proposed work includes less visual distortion, watermark imperceptibility, watermark robustness, and watermark security. The objective of the proposed scheme is to achieve high robustness than the traditional methods.

The overall paper is arranged as follows. Section II discusses the presented blind watermarking scheme. Section III illustrates the performance analysis of the presented scheme. Section IV explains the effectiveness of the method as a conclusion.

2 Proposed Watermarking Scheme

The watermark insertion and retrieval are the two stages of the proposed scheme.

2.1 Watermark Insertion Procedure

Figure 1 shows the process of inserting watermark which is expressed by the below steps.

Step 1: Split the watermark image W into RGB channels. The Arnold transform with the secret key K_1 is applied to each channel to provide security and robustness of the watermark image. Generate binary information of 8-bit by converting each pixel value. Combine all the 8-bit binary information to get a binary watermark sequence.

Step 2: Partition the color cover image into RGB channels. Apply LWT [14] using Haar wavelet to B channel for obtaining the LL, HL, LH, and HH subbands as the blue channel is sensitive to human eyes and thus achieves robustness. LL subband of LWT is chosen since it consists of low frequency component, and it is split into 4×4 non-overlapping image blocks. The security and robustness are improved by choosing suitable embedding blocks [15] using pseudo random sequence with secret key K_2.

Step 3: Each embedding block $E_{i,j}$ of the LL subband of the cover image is processed with Schur decomposition to obtain the unitary matrix $Q_{i,j}$ and the upper triangular

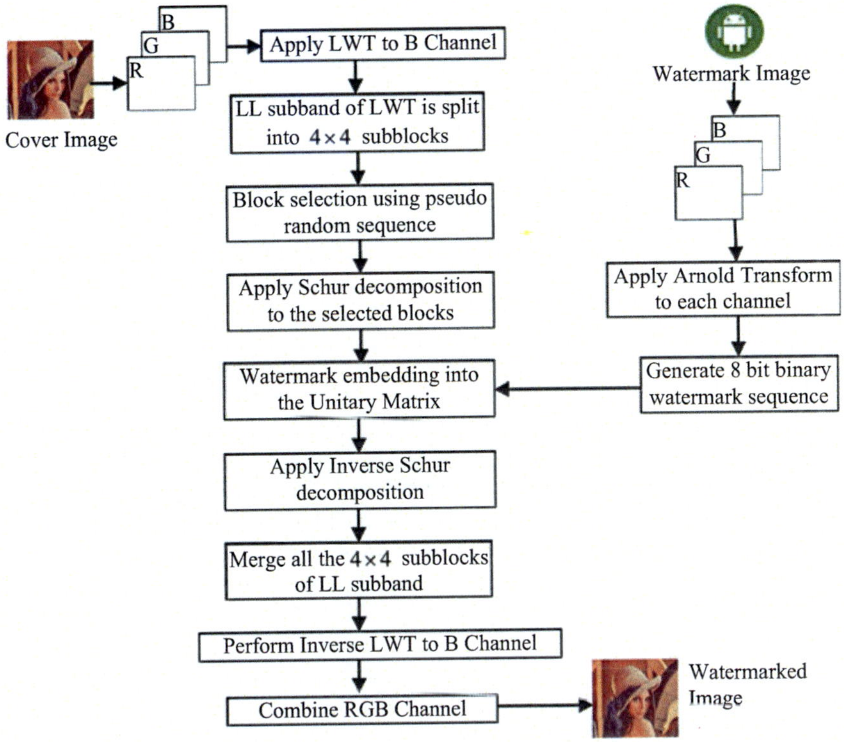

Fig. 1 Diagrammatic representation of the proposed watermark insertion procedure

matrix $U_{i,j}$ as follows.

$$[Q_{i,j}, U_{i,j}] = \text{schur}(E_{i,j}) \tag{1}$$

Here, the embedding block position is represented by i, j.

Step 4: The largest eigenvalue $U_{\max}$ of the upper triangular matrix $U_{i,j}$ and the column C of the largest eigenvalue are computed by Eqs. (2) and (3).

$$U_{\max} = \text{find}(\max(U_{i,j})) \tag{2}$$

$$C = \text{determine}(\max(u_{\max})) \tag{3}$$

where *find* function searches the element and the *max* function returns the largest eigenvalue; the column of the largest eigenvalue is determined by the *determine* function.

Step 5: The unitary matrix $Q_{i,j}$ is modified to embed the preprocessed watermark data *wm* by the following procedure,

- Choose the C th vector of $U_{i,j}$ as per Eq. (3). Perform the embedding process using Eqs. (4) and (5).

$$Q'_{2,C} = \begin{cases} \text{sign}(Q_{2,C}) * \left(Q_{\text{avg}} + {}^{T}\!/_{2}\right) & \text{if } wm = 1 \\ \text{sign}(Q_{2,C}) * \left(Q_{\text{avg}} - {}^{T}\!/_{2}\right) & \text{if } wm = 0 \end{cases} \tag{4}$$

$$Q'_{3,C} = \begin{cases} \text{sign}(Q_{3,C}) * \left(Q_{\text{avg}} - {}^{T}\!/_{2}\right) & \text{if } wm = 1 \\ \text{sign}(Q_{3,C}) * \left(Q_{\text{avg}} + {}^{T}\!/_{2}\right) & \text{if } wm = 0 \end{cases} \tag{5}$$

where $Q_{avg} = \left(abs\left(Q_{2,C}\right) + abs(Q_{3,C})\right)/2$; $Q'_{2,C}$ and $Q'_{3,C}$ represents the modified $Q_{2,C}$ and $Q_{3,C}$ after inserting the watermark *wm*. The sign of the corresponding coefficient is indicated by $sign(.)$. To balance the robustness and invisibility of the watermark, the embedding coefficient T is selected as 0.01.

- The watermarked image block $E'_{i,j}$ is obtained by applying inverse Schur decomposition using Eq. (6).

$$E'_{i,j} = Q'_{i,j} U_{i,j} {Q'_{i,j}}^{T} \tag{6}$$

Step 6: Repeat steps 3 to 5 to insert all the watermark data into the LL subband of the cover image. Merge all the 4 × 4 non-overlapping image blocks of LL subband and then perform inverse LWT to B channel. Finally, the watermarked image E' is generated by combining the RGB channels.

2.2 Watermark Retrieval Procedure

The watermark reconstruction is the converse of watermark insertion as depicted in Fig. 2.

Step 1: Perform step 2 of the watermark insertion process to choose the watermarked image blocks $E'_{i,j}$.

Step 2: Apply Schur decomposition to $E'_{i,j}$ for attaining $Q'_{i,j}$ and $U'_{i,j}$.

Step 3: Retrieve the watermark data wm' from the unitary matrix $Q'_{i,j}$ in accordance with the column of the largest eigenvalue using the below equation.

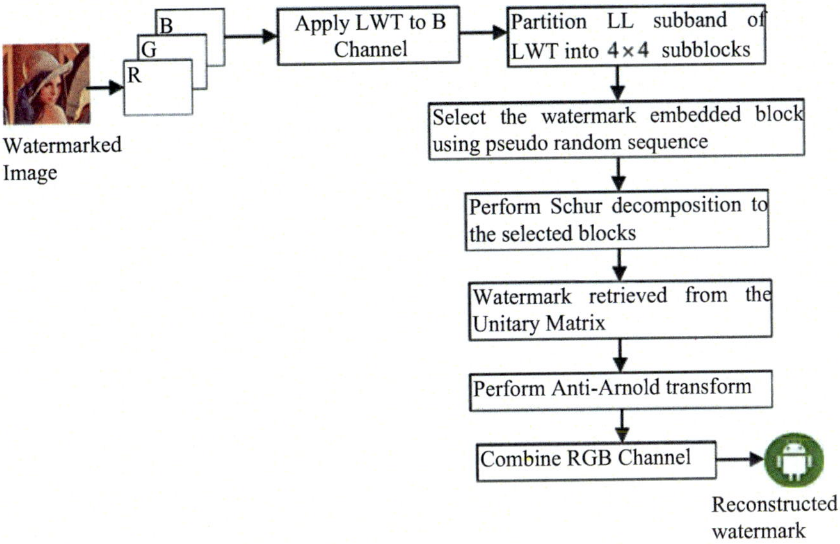

Fig. 2 Diagrammatic representation of the proposed watermark retrieval procedure

$$wm' = \begin{cases} 0 \text{ if } Q'_{2,C} > Q'_{3,C} \\ 1 \text{ if } Q'_{2,C} < Q'_{3,C} \end{cases} \tag{7}$$

Step 4: From all embedded image blocks, the watermark bits are extracted by iterating steps 2 and 3. Partition the retrieved watermark bits into 8-bit sequence and convert it to decimal values. Perform anti-Arnold transform with the secret key K_1 and then the RGB channels are combined to reconstruct the watermark image.

3 Experimental Analysis

The proposed method is validated for its imperceptibility, robustness, and security. Imperceptibility is assessed by metrics such as PSNR and structural similarity index measurement (SSIM). The higher PSNR and SSIM reveal that the watermarking scheme has better imperceptibility. NCC metric measures the watermark robustness, and the higher NCC implies more robustness. The Arnold transform and pseudo random sequence algorithm ensure the security of the proposed scheme.

For performance analysis, the color cover images of dimensions 512 × 512 are chosen from [16], and the three color images are used as the watermark images which is of dimension 32 × 32 as shown in Fig. 3.

The image quality is analyzed by PSNR, and it is calculated by

(i) Avion (ii) Baboon (iii) Peppers (iv) Lena (v) Android (vi) Apple (vii) Peugeot logo

Fig. 3 Images used for testing (i)–(iv) RGB cover images (v)–(vii) RGB watermark images

Table 1 Performance analysis of the proposed with several test images under 'NO ATTACK'

Cover image	Secret image					
	Peugeot logo		Apple		Android	
	PSNR (dB)/SSIM	NCC	PSNR (dB)/SSIM	NCC	PSNR (dB)/SSIM	NCC
Avion	38.0171/0.9771	0.9996	38.0560/0.9760	0.9993	38.0886/0.9768	0.9990
Baboon	39.1903/0.9824	0.9911	39.7834/0.9835	0.9952	39.8921/0.9867	0.9972
Peppers	38.9938/0.9841	0.9960	39.0275/0.9839	0.9914	39.0184/0.9842	0.9968
Lena	41.0447/0.9858	0.9957	41.1155/0.9858	0.9973	41.1241/0.9860	0.9967

$$\text{PSNR} = 10\log_{10}\left(255^2/\frac{1}{3\text{pq}}\sum_{F=1}^{3}\sum_{U=1}^{p}\sum_{V=1}^{q}\left(E(U,V,F) - E'(U,V,F)\right)^2\right) \tag{8}$$

$E(U, V, F)$ is the cover image, $E'(U, V, F)$ is the watermarked image, and (U, V) denotes the pixel coordinate.

The quality perception is more correlated with SSIM which is given by

$$\text{SSIM}(E, E') = l(E, E')c(E, E')s(E, E') \tag{9}$$

where $l(E, E')$ is the luminance distortion, $c(E, E')$ is the contrast distortion, and $s(E, E')$ is the loss of correlation.

Robustness analysis is done for the original watermark $wm(U, V, F)$ and the retrieved watermark $wm'(U, V, F)$ using the following relation.

$$\text{NCC} = \frac{\sum_{F=1}^{3}\sum_{U=1}^{p}\sum_{V=1}^{q} wm(U,V,F)\times wm'(U,V,F)}{\sqrt{\sum_{F=1}^{3}\sum_{U=1}^{p}\sum_{V=1}^{q} wm(U,V,F)^2}\sqrt{\sum_{F=1}^{3}\sum_{U=1}^{p}\sum_{V=1}^{q} wm'(U,V,F)^2}} \tag{10}$$

where the row and column size of the watermark image is denoted by p and q, respectively.

Table 1 describes the performance analysis without any attack. The high PSNR and SSIM attained indicate the imperceptibility of the presented approach. The NCC value attained is close to 1, and this indicates better robustness.

The presented method robustness is verified by undergoing several attacks, and the resultant NCC is stated in Table 2. The NCC value obtained for different attacks is nearer to 1 except cropping attack, and this shows better robustness. The proposed method (LWT + Schur decomposition) is compared with QHT with Schur decomposition [10] and Schur decomposition [11]-based method which is described in Fig. 4 with Lena as cover image and Apple as watermark. The proposed method has

Table 2 NCC under various attacks

Attacks	Lena and Apple	Avion and Peugeot logo	Attacks	Lena and Apple	Avion and Peugeot logo
Brighten (+ 90)	 0.9663	 0.9306	Cropping 20%	 0.8934	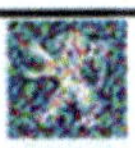 0.8842
Darken (- 80)	 0.9217	 0.9786	Scaling	 0.9700	 0.9554
Sharpening	 0.9826	 0.9645	Motion Blur	 0.9983	 0.9996
Gaussian Low pass filter (size '3')	 0.9770	 0.9566	Rotation 10°	 0.9497	 0.9561

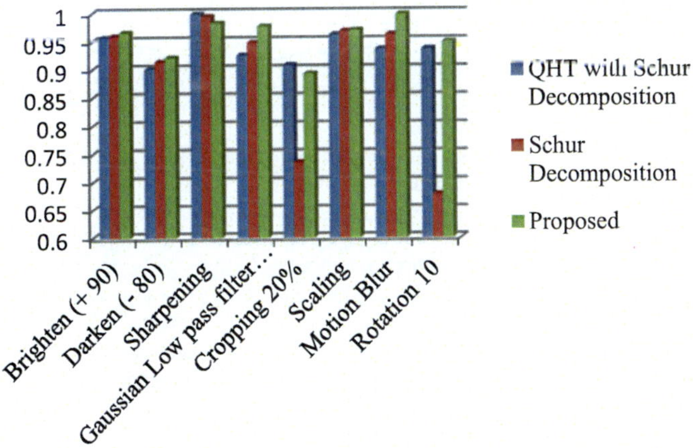

Fig. 4 NCC comparison of the proposed with traditional methods

high NCC than the traditional schemes. The proposed method shows an enhanced outcome in respect to imperceptibility and robustness.

4 Conclusion

The proposed watermarking scheme is dependent on LWT and Schur decomposition which is appropriate for ownership protection. The projected scheme inserts the watermark information into the selected image blocks by pseudo random sequence. The watermark invisibility is enhanced by the watermarked image blocks, and the robustness is improved by embedding coefficient and Arnold transform. This scheme exposes good performance for watermark invisibility and robustness. In the future, this scheme may be considered for video media in big data.

References

1. Kaur S, Sidhu RK (2016) Robust digital image watermarking for copyright protection with SVD-DWT-DCT and Kalman filtering. Int J Emerg Technol Eng Res 4:59–63
2. Liu XL, Lin CC, Yuan SM (2016) Blind Dual Watermarking for Color Images' Authentication and Copyright Protection. IEEE Trans Circuits Syst Video Technol 28:1047–1055
3. Zhang Y, Wang C, Wang X, Wang M (2017) Feature-based image watermarking algorithm using SVD and APBT for copyright protection. Futur Internet 9:1–15
4. Kumar C, Singh AK, Kumar P (2017) A recent survey on image watermarking techniques and its application in e-governance. Multimed Tools Appl 77:3597–3622
5. Kashyap N, Sinha GR (2012) Image watermarking using 3-level discrete wavelet transform (DWT). Int J Mod Educ Comput Sci 4:50–56
6. Ko HJ, Huang CT, Horng G, Wang SJ (2020) Robust and blind image watermarking in DCT domain using inter-block coefficient correlation. Inf Sci (Ny) 517:128–147
7. Kazemivash B, Moghaddam ME (2016) A robust digital image watermarking technique using lifting wavelet transform and firefly algorithm. Multimed Tools Appl 76:20499–20524
8. Prabha K, Vaishnavi MJ, Shatheesh Sam I (2019) Quaternion Hadamard transform and QR decomposition based robust color image watermarking. In: Proceedings of the international conference on trends in electronics and informatics, ICOEI 2019. pp 101–106
9. Prabha K, Sam IS (2020) A novel blind color image watermarking based on walsh hadamard transform. Multimed Tools Appl 79:6845–6869
10. Li J, Yu C, Gupta BB, Ren X (2018) Color image watermarking scheme based on quaternion Hadamard transform and Schur decomposition. Multimed Tools Appl 77:4545–4561
11. Su Q, Zhang X, Wang G (2020) An improved watermarking algorithm for color image using Schur decomposition. Soft Comput 24:445–460
12. Prabha K, Shatheesh Sam I (2020) An effective robust and imperceptible blind color image watermarking using WHT. J King Saud Univ Comput Inf Sci. https://doi.org/10.1016/j.jksuci.2020.04.003
13. Min L, Ting L, Yu-jie H (2013) Arnold transform based image scrambling method. In: 3rd international conference on multimedia technology, pp 1309–1316

14. Daubechies I, Sweldens W (1998) Factoring wavelet transforms into lifting steps. J Fourier Anal Appl 4:247–269
15. Su Q, Chen B (2017) An improved color image watermarking scheme based on Schur decomposition. Multimed Tools Appl 76:24221–24249
16. CVG–UGR—Image database. http://decsai.ugr.es/cvg/dbimagenes/c512.php

Chapter 16
Role and Significance of Internet of Things in Combating COVID-19: A Study

Kirti Vijayvargia

1 Introduction

Today is an era of Smart systems such as smart city, smart home, smart farming, smart parking and smart healthcare. The backbone of smart systems is IoT that provides the base for building smart systems. IoT applications are widely used in healthcare [1–3] to detect the disease and provide real-time monitoring of patients. During the year 2020 healthcare has gone through various challenges and opportunities due to the pandemic COVID-19. In addition to the research done in the field of medical sciences, use of technology such as IoT, AI, Block chain, etc., to prevent pandemic has been explored. IoT provides ways for early diagnosis of disease and enforcing social distancing as well as monitoring quarantined and isolated people and thus it is often helpful for fighting against COVID-19.

1.1 *Internet of Things (IoT)*

IoT is an acronym for Internet of Things provided by Kevin Asthon in 1999 [4]. "Internet of Things" is defined as "worldwide network of interconnected objects having unique identity and following standard communication protocols" [5]. It allows connecting anything to anyone, anywhere, anytime, without or very less human intervention. With the concept of smart systems, sensors and actuators have become part of our daily life.

K. Vijayvargia (✉)
International Institute of Professional Studies, Devi Ahilya University, Takshashila Campus Khandwa Road, Indore 452001, India

A. Sheth et al. (eds.), *Intelligent Systems*, Algorithms for Intelligent Systems,
https://doi.org/10.1007/978-981-16-2248-9_16

1.2 COVID-19

COVID-19 is an acronym for "Coronavirus Disease 2019." It is a respiratory and contagious illness that is caused by the severe acute respiratory syndrome coronavirus-2 (SARS-CoV-2) [6]. The first COVID-19 case affecting human life was reported in Wuhan City, in the Hubei province of China in December 2019. Since then, the COVID-19 has taken a giant form spreading like wildfire in almost the entire world. This rapid increase has enforced the need to invent and take various actions for it. Technology can play a vital role in combating this pandemic.

1.3 COVID-19 and IoT

Any pandemic like COVID-19 handling includes four steps as shown in Fig. 1. These steps are prevention, diagnosis, treatment, and monitoring. At every stage, technology plays a significant role. IoT is helpful in providing an environment where everything surrounding us such as our TV and washing machine. Microwave, lights, AC, and even our footwear knows what we want and what we need to do. This characteristic of IoT is very helpful to fight against COVID-19 as quarantine and isolation are two major stages of COVID-19 handling process. At both quarantine and isolation stages, human–human interaction is less because of its contagious nature, so machine–machine communication (M2M)—a Core element of IoT will be helpful.

IoT provides many ways to fight against COVID-19. *Wearables* like smart helmets, wrist bands, smart thermometers, and smart glasses help to monitor temperature, respiratory conditions, and other parameters of COVID-19 patients and other

Prevention
Maintain social distancing
Various alerts to take precautions like using mask, hand wash and avoid surface touching
Contact tracing, cluster identification and compliance of quarantine

Diagnosis
Using various wearables to measure the health conditions and give proper alerts
Using contact tracing app, give alert messages to all person in contact and also monitor their health

Treatment
Inform the concerned medical staff during any emergency
Automated treatment process
Telehealth consultation

Monitoring
Smart (remote) monitoring of health conditions of infected patients
Inform the concerned medical staff and family members during any emergency

Fig. 1 COVID-19 handling stages and use of IoT at each stage suspected persons. *Robots* can help in touchless sample collection as well as providing medicines, monitoring the health conditions of patients, and assisting them in their daily activities. *Drones* are helpful in finding infected people in crowd as well as at distant locations. *IoT buttons* are useful in making alerts to patient's family members about their health [7].

Healthcare professionals and medical companies are doing a lot of efforts for vaccines and medicines to fight with COVID-19; but at the same time, efforts using technology are also needed to combat this giant. Among various technologies, IoT can provide many promising solutions for monitoring quarantined and isolated people and spreading of the virus. This study focuses on surveying these efforts and gives an insight for future research.

2 Surveyed Contributions

Wearables are very helpful for monitoring temperature, heart rate, respiration rate, and oxygen saturation. Different wearable monitoring devices and respiratory support systems such as CPAP devices, ventilators, and oxygen therapy are discussed in [7]. Devices are explored based on their working, offering services, cost, merits, and demerits. This comparative discussion is helpful in selecting the best wearable technology, so that initial treatment can be provided and will help to combat the pandemic.

Social IoT (SIoT) has been gaining popularity with increasing use of social media. SIoT has been exploited in [8] to find out social relationships between mobile devices for the early identification of suspected COVID-19 cases and is helpful in controlling the infection rate. The authors applied minimum weight vertex cover (MWVC) and proposed an adaptive scheme by employing graph embedding technique for state representation and used reinforcement learning during the training phase. Digital contact tracing is a major way to prevent the spread of COVID-19; but in many parts of the world including Europe and US, its use is not preferred due to privacy issues. Garg et al. [9] employ trust-oriented decentralization of block chain for logging and retrieval of data. This model helps moving objects in sending or receiving notifications when they are close to a suspected or confirmed person or suspected object. So that infection spread networks model can be used to understand human connectivity.

Figure 2 shows how pandemic outbreak the globe worldwide. To stop the outbreak in future, Gupta et al. [10] proposes smart connected community scenarios. It gives an insight into how AI and IoT can be useful in COVID-19 patients tracking, enforcing social distancing, predicting the disease using symptoms, ensuring the efficient delivery of essential services and resources, and developing smart and intelligent infrastructures against COVID-19 to stop pandemic outbreaks in the future. To avoid the outbreak of COVID-19, many of the cities were having lockdowns. During the lockdown period for restricting public movement, IoT-based decentralized biometric face detection framework has been proposed by Kolhar et al. [11]. The framework uses three-layered edge computing architecture and deep learning for face detection.

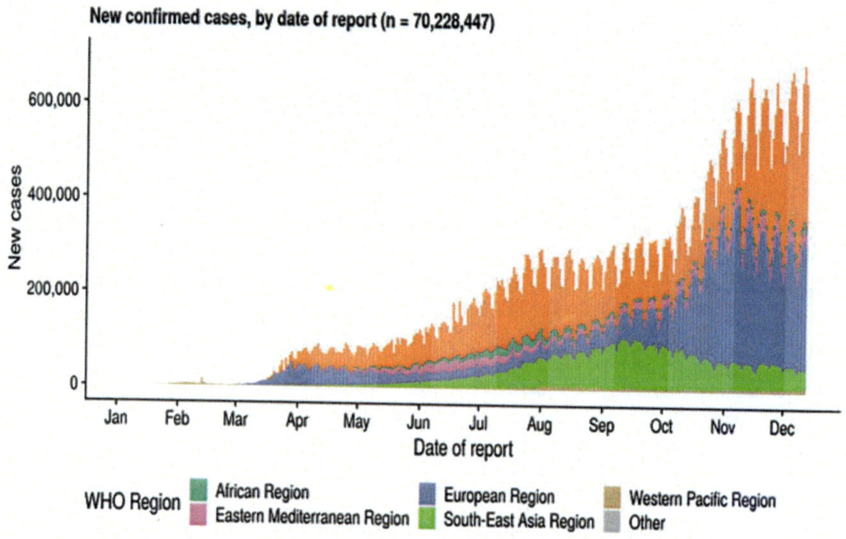

Fig. 2 WHO reported COVID-19 confirmed cases across the globe [18]

Drones or unmanned aerial vehicles (UAVs) are very helpful during this pandemic in remote areas having very limited Internet or wireless connectivity. It will also be very useful for handling the COVID-19 in very congested areas or containment zones where the spread of pandemic is high. Kumar et al. [12] propose architecture of the drone-based systems; to handle COVID-19 pandemic situations, wearable sensors are used to record the observations in body area networks (BANs). The system is useful for collecting and storing the significant amount of data in a short period, and accordingly, actions are taken when required. Real-time drone-based healthcare system has been implemented for COVID-19 operations such as thermal image collection, identification of patients, and sanitization in short time.

Most of the cities implement lockdown in phases. According to the growth or decay of COVID-19 cases, decision of closing malls, theaters and schools, sealing of national or international borders, and suspension or resume of traveling is taken. The accurate forecasting of the COVID-19 cases is a challenging task because of lack of knowledge about this disease and non-availability of real-time data samples. In [13], authors propose the multi-task Gaussian process (MTGP) regression model to predict novel coronavirus (COVID-19) outbreak. The model is useful for government authorities for planning and reducing the impact of this disease.

To combat COVID-19, several safety rules are suggested such as contactless temperature sensing should be used so that people, having high body temperature, are not allowed in public indoor spaces. Checks should be placed so that social distancing is maintained and people wear masks. Petrovic and Kocic [14] propose an IoT-based solution to impose these safety rules. For contactless temperature sensing, Arduino Uno has been used with infrared sensors or thermal cameras. For social distancing check and mask detection, computer vision techniques are used on camera-equipped

Raspberry Pi. Sanitization is another preventive measure for COVID-19. Pandya et al. [15] present a smart epidemic tunnel that uses ultrasonic sensors to detect people entering and then disinfect them by spraying sanitizer. For real-time detection of a person, an IoT-based sensor fusion assistive framework is proposed. This tunnel uses solar energy at day time, and at night, it uses a solar power bank for functioning. An LDR sensing unit has been used for this purpose.

To quarantine suspected covid individuals is another prevention approach that is helpful in reducing spread of the disease. Wearables can play an important role in imposing routines and restrictions on the quarantined people. In [16], an IoT-based wearable quarantine band (IoT-Q-Band) has been designed to figure out the absconding individuals. Along with the mobile app, Q-band is an economical solution for tracking and reporting absconding quarantines. To enforce home quarantine, signature home [17] provides a cost-efficient algorithm for monitoring quarantined people, so that they remain within a confined area. Signature home considers the identifiers of the network facilities such as Wi-Fi access points as the home signature.

Due to the highly contagious nature of Coronavirus, monitoring and taking care of the COVID patients and also their dependents is a challenging task. Akhund et al. [19] propose a cost-efficient IoT-based robotic agent for helping virus affected people and disabled persons by identifying their gestures and following their instructions. IoT-fog-cloud-based architecture and BPMN 2.0 extension are proposed in [20] that is helpful to monitor COVID-19 patients in hospitals or homes and autistic children in nursery.

To apply the social distancing rules, some countries including Germany and the UK have suggested the use of some documents to certify that a person has been infected and now are immune to COVID-19 disease. These documents are referred to as “immunity certificates” or “immunity passports” or “immunity licenses.” The individuals having the immunity certificates are exempted from various physical restrictions and can return to school, work, and other activities. To prevent forgery of information and spreading infection, a block chain technology has been proposed [21].

Initially without any specific cure or vaccine, social distancing and self-isolation play a significant role to reduce spread of COVID-19. Vedaei et al. [22] propose a framework based on IoT consisting of a lightweight IoT node for tracking health conditions (measuring temperature of body, oxygen level, etc.) and a mobile app to display these health parameters and provides notifications to maintain a distance of 2 m. Fog-based machine learning tools are used for data analysis and diagnosis. A fuzzy system has been proposed that considers the user health conditions and environment conditions to predict risk of infection. Table 1 provides the summary of the research contributions discussed in the paper.

Table 1 IoT research contributions to combat COVID-19

Research work	Focus	Algorithm/technique	How helpful
[7]	Survey of IoT wearable	Wearable for monitoring health parameters	Selecting the best device for monitoring temperature, respiration rate, and oxygen saturation
[8]	Social relationship between mobiles	Minimum weight vertex cover (MWVC) problem, reinforcement learning	Early identification of suspected person
[9]	Contact tracing	Trust-oriented decentralization of block chain	Alert when moving objects come in contact of infected person
[10]	Tracking, social distancing, and prediction	Smart-connected community using IoT and AI	Reduce pandemic outbreak
[11]	IoT-based biometric face detection framework	Three layered edge computing architecture and deep learning for face detection	For restricting public movement during lockdown
[12]	Drone-based systems	Wearable sensors are used to record the observations in body area networks	Thermal image collection, identification of patients, and sanitization in very short time
[13]	Prediction model	Multi-task Gaussian process (MTGP) regression model	Predicting the growth or decay in COVID-19 cases
[14]	Impose safety rule of coronavirus	Camera-equipped Raspberry Pi and Arduino Uno with thermal sensor	Imposing safety rules like contactless temperature measuring, social distancing, and mask
[15]	Smart epidemic tunnel	Uses ultrasonic and LDR sensors	Sanitization
[16]	IoT-Q-Band	IoT-Q-Band	For tracking and reporting absconding quarantines
[17]	IoT geofencing	Waterproof bluetooth low energy wristbands	For enforcing home quarantine
[19]	Robotic agent	Wireless gesture control robot	Helping virus affected and disabled person in daily activities
[20]	IoT aware business process modeling	IoT-fog-cloud-based architecture	To monitor autism and COVID-19
[21]	Security of immunity certificates	Block chain	To reduce forgery in immunity certificates

(continued)

Table 1 (continued)

Research work	Focus	Algorithm/technique	How helpful
[22]	Prediction of infection rate and notification of health parameters	Fuzzy system, fog-based machine learning	Predict the risk of infection by measuring health parameters and environment conditions

3 Conclusion

COVID-19 pandemic has affected the people worldwide, placing the world in economic slowdown. Medical scientists and researchers have put their best efforts for investigating vaccines and treatment for COVID-19. But till then, only ways to spread this pandemic are preventive measures such as quarantine, isolation, use of masks, sanitizer, and social distancing. The Internet of Things plays a vital role in healthcare systems including early diagnosis of illness, telemedicine, and real-time monitoring of the patients. During the COVID-19 pandemic also, researchers provided many IoT-based promising solutions for disease prevention, detection, and monitoring. In this paper, the research work done in the field of IoT to combat coronavirus disease has been reviewed. To prevent COVID-19, various IoT devices such as smart helmets, smart tunnels, drones, and wearables are used. For decision making about the lockdown phases to maintain social distancing, prediction of spread of pandemic is required. Different prediction approaches such as machine learning are used to forecast the spread. Many of the solutions will be helpful for any pandemic or can be used in disaster management also. Mobile crowd sensing applications and security approaches need to be explored further.

References

1. Perera C, Zaslavsky A, Christen P, Georgakopoulos D (2012) CA4IOT: context awareness for the Internet of Things. In: Proceedings—2012 IEEE international conference on green computing and communications, GreenCom 2012, Conference on Internet of Things, iThings 2012 and Conference on Cyber, Physical and Social Computing, CPSCom 2012, pp 775–782. https://doi.org/10.1109/GreenCom.2012.128
2. Baker SB, Xiang W, Atkinson I (2017) Internet of Things for smart healthcare: technologies, challenges, and opportunities. IEEE Access 5:26521–26544
3. Catarinucci L et al (2015) An IoT-aware architecture for smart healthcare systems. IEEE Internet Things J 2:515–526
4. Ashton K (2009) That 'internet of things' thing. RFID J 22:97–114
5. European Commission. Internet of Things in 2020 (2008) A roadmap for the future. RFID Work Gr Eur Technol Platf Smart Syst Integr 22, 1–29 (2008)
6. Chamola V, Hassija V, Gupta V, Guizani M (2020) A comprehensive review of the COVID-19 pandemic and the role of IoT, Drones, AI, Blockchain, and 5G in managing its Impact. IEEE Access 8:90225–90265
7. Islam MM et al (2020) Wearable technology to assist the patients infected with novel coronavirus (COVID-19). SN Comput Sci 1:1–9

8. Wang B, Sun Y, Duong TQ, Nguyen LD, Hanzo L (2020) Risk-aware identification of highly suspected COVID-19 cases in social IoT: a joint graph theory and reinforcement learning approach. IEEE Access 8:115655–115661
9. Garg L, Chukwu E, Nasser N, Chakraborty C, Garg G (2020) Anonymity preserving IoT-based COVID-19 and other infectious disease contact tracing model. IEEE Access 8:159402–159414
10. Gupta D, Bhatt S, Gupta M, Tosun AS (2020) Future smart connected communities to fight COVID-19 outbreak. arXiv 1–43 (2020)
11. Kolhar M, Al-Turjman F, Alameen A, Abualhaj MM (2020) A three layered decentralized IoT biometric architecture for city lockdown during covid-19 outbreak. IEEE Access 8:163608–163617
12. Kumar A et al (2020) A drone-based networked system and methods for combating coronavirus disease (COVID-19) pandemic. arXiv 14(4):337–339 (2020)
13. Ketu S, Mishra PK (2020) Enhanced Gaussian process regression-based forecasting model for COVID-19 outbreak and significance of IoT for its detection. Appl Intell. https://doi.org/10.1007/s10489-020-01889-9
14. Petrovic N, Kocic D (2020) IoT-based system for COVID-19 indoor safety monitoring. IcETRAN 2020 (2020)
15. Pandya S, Sur A, Kotecha K (2020) Smart epidemic tunnel: IoT-based sensor-fusion assistive technology for COVID-19 disinfection. Int J Pervas Comput Commun. https://doi.org/10.1108/IJPCC-07-2020-0091
16. Singh V et al (2020) IoT-Q-Band: a low-cost internet of things based wearable band to detect and track absconding COVID-19 quarantine subjects. EAI Endorsed Trans Internet Things 6:163997
17. Tan J et al (2020) IoT geofencing for COVID-19 home quarantine enforcement. IEEE Internet Things Mag 3:24–29
18. WHO (2020) COVID-19 explorer. COVID-19 Explorer 1. Available at: https://worldhealthorg.shinyapps.io/covid/
19. Akhund TMNU et al (2020) IoT based low-cost robotic agent design for disabled and Covid-19 virus affected people. In: Proceedings of world conference on smart trends systems, security and sustainability, WS4 2020, pp 23–26 (2020). https://doi.org/10.1109/WorldS450073.2020.9210389
20. Kallel A, Rekik M, Khemakhem M (2020) IoT-fog-cloud based architecture for smart systems: prototypes of autism and COVID-19 monitoring systems. Softw Pract Exp. https://doi.org/10.1002/spe.2924
21. Bansal A, Garg C, Padappayil RP (2020) Optimizing the implementation of COVID-19 "immunity certificates" using blockchain. J Med Syst 44:19–20
22. Vedaei SS et al (2020) COVID-SAFE: an IoT-based system for automated health monitoring and surveillance in post-pandemic life. IEEE Access 8:188538–188551

Chapter 17
Phylogenetic and Biological Analysis of Evolutionary Components from Various Genomes

Kshatrapal Singh, Manoj Kumar Gupta, and Ashish Kumar

1 Literature Review

As of now, mostly molecular phylogenetic analyzing builds on sequence comparison, the comparison of total genome/proteome sequence [1, 2]. The particular approaches can be subcategories in two types: alignment-based methods and alignment-free methods. As old phylogenetic as well as taxonomic approaches, alignment-based methods have been broadly adopted in vast analysis of species phylogeny [3, 4]. Yet, the latest advance in genome sequence techniques has helped the procurement of largely expanding amounts of sequence data gained from individual genomes as well as total genome [5]. Due to this fact, alignment-based approaches have grown into fewer useful as their defined phylogenetic information caliber. So, alignment-free methods based on total genome/proteome sequence might give highly powerful information for phylogenetic analyzing [6, 7]. Various alignment-free approaches were carried for figuring out the phylogeny of an organism: single value decomposition (SVD), feature frequency profiles (FFP) as well as the CV Tree method. The particular approaches, based on total genome analysis, have been largely as well as strongly applicable to phylogenetic analyzing as of their benefits as well as effectiveness.

The CV Tree approach has been strongly enforced for plants phylogenetic analyzing on the basis of complete chloroplast proteome as well as for prokaryote phylogenetics analyzing which is upon the basis of complete prokaryotic proteomes. CV Tree assumes a phylogenetic relationship between organisms based on the oligopeptides content (named as K-string) of protein sequence either from the

K. Singh (✉) · A. Kumar
Department of Computer Science and Engineering, ITS Engineering College, Greater Noida 201308, India

M. K. Gupta
School of Computer Science and Engineering, Shri Mata Vaishno Devi University, Katra, J&K 182320, India

A. Sheth et al. (eds.), *Intelligent Systems*, Algorithms for Intelligent Systems,
https://doi.org/10.1007/978-981-16-2248-9_17

oligonucleotides content of DNA sequence [8, 9]. As CV Tree applies calculation which involves a huge dataset containing either twenty (for protein sequence, where *K* is the count of *K*-string) or four (for DNA sequence) *K*-string for any species, this absorbs a huge quantity of processing duration as well as memory. This shows as not all *K*-strings are powerful mainly to develop phylogenetic tree. So, we do advance our main assumption: only a subset of *K*-string devotes mainly to tree generation as well as a particular subset consists of acceptable information for a phylogenetic analyzing faith on this string [10]. On the point of dimension reductions improve predictions efficiently, on using these main *K*-strings, we are able to lead the generation of extra phylogenetic trees. Hence, we are focused to identify as well as to collect particular special *K*-strings.

In this paper, we implemented dimensional contractions about a compilation of large-dimension sequence data and gained a batch of main *K*-strings from datasets of the complete metazoan genomes sequence of metazoan accessible. Now, by the use of key *K*-string, we rebuild the metazoans phylogenetic trees; further, we analyzed with the other trees drawn use of the composition vector tree method as well as with tree drawn used various other approaches. We too did again structural analysis of *K*-string to find the distribution patterns of amino acid composition. After that we have done the various analyses of the genetic characters as well as the appropriate pattern structures of the key *K*-string.

2 Methodology

From mt genomes for a group of protein sequences, CV Tree could be useful to get a composition vector matrix to encode the whole phylogenetic information as the group [11]. To gather the key *K*-string, we enforced a dimension contraction on particular composition vector matrices. We found the variance *V(X)* for all columns of the matrices to stands for the deviated amount of each dimensions; this devotes to the generation of a trees:

$$V(X) = \sum_{i=1}^{N} (x_i - \overline{x})^2 \tag{1}$$

where *N* represents count of row in the matrices that refers for the count of specie for the analysis. $\overline{x}$ represents mean amount of every columns of the matrices, named, for every *K*-strings. We arranged particular twenty deviation values as well as put them into a scatter plot. We then classified some critical points separating the point into two categories that devote either highly or lightly to tree generation. The L-shape corners have normally been treated as the best corners [12, 13]. The critical point, placed at the edge of the L-types curves, is established by a 90% devaluation of the maximal varied amount. The point with deviation value greater than the complex points gives our main point as well as the comparable K-strings generate the key K-string.

We created key K-strings trees on the basis of the cosine distance matrix that are gained by evaluating the cosine values of each two vectors in the composition vector matrix. Where $U(A, B)$ is the correlations among two species A and B, however, $V(A, B)$ stands for the distances among the two species. a and b are the vectors of the N-dimensional space of two species A and B.

$$U(A, B) = \frac{\sum_{i=1}^{N}(a_i * b_i)}{\sqrt{\left(\sum_{i=1}^{N} a_i^2 * \sum_{i=1}^{N} b_i^2\right)}} \tag{2}$$

$$V(A, B) = \frac{1 - U(A, B)}{2} \tag{3}$$

The neighbor-joining algorithm was applied to distance-based tree generation, as well as the visualization of the tree was done with the help of MEGA software, delivering the tree beyond each deliberation of branch length as the tree topologies could be openly seen [14, 15]. We enforced tree analogy to four trees: (i) a tree generated with our universal key K-string (the key K-string trees); (ii) an analogous tree generated from the ditto count of suddenly taken K-string, giving as a regulation; (iii) a tree generated from the whole sets of twenty K-string through CV Tree method (the CV Tree); and (iv) alignment-based tree generated with maximum likelihood approach. To examine the biological implication of our key K-string, we arranged them rear on the initial protein sequence. We analyzed 400 vast key K-strings in vertebrates which are the main critical for differentiating species belonging to the contrasting class. Later, selection of ten species randomly from vertebrate, with two species contained in one class, we have done a whole alignment of a particular protein sequence using the Clustal X software. We determined the conservative distribution patterns with sliding beyond sequence with a block size of five and steps size of one. The area of the 400 main K-strings was certainly shown on the protein sequence. So, we could analyze the points in which the main K-string compiled to a sustained region of protein sequence to declare a relationship between them.

Enforcing adaptive chi-square testing to certify the continuation of important differences among detected and conclude frequency of K-string. The detected frequency of a five-string ($K = 5$) is represented by $f(a_1a_2a_3a_4a_5)$; two analogous four-string are $f(a_1a_2a_3a_4)$ as well as $f(a_2a_3a_4a_5)$, as well as one three-string is $f(a_2a_3a_4)$. The estimated value of the five string is represented $f^0(a_1a_2a_3a_4a_5)$, as well as its chi-square test amount is represented by X^2:

$$f^0(a_1a_2a_3a_4a_5) = \frac{f(a_1a_2a_3a_4)f(a_2a_3a_4a_5)}{f(a_2a_3a_4)} \tag{4}$$

$$X^2 = \sum_{i=1}^{k} \frac{\left(f(a_1a_2a_3a_4a_5) - f^0(a_1a_2a_3a_4a_5)\right)^2}{f^0(a_1a_2a_3a_4a_5)} \quad (5)$$

where k refers to the number of K-strings verified.

3 Results and Discussion

For experiment, we get mt genomes metazoan protein sequence with the NCBI website. After the collection of key K-string, we have taken total 87 species for the generation as well as comparison of phylogenetic trees. We built a convergence intersection as well as a union set of the ten groups of vast key K-string, covering 400 and 1211 K-string, jointly. So, particular 400 K-string was fundamental for a tree generation among vertebrates as they presented in each group of vast key K-string, as well as the 1211 K-string from the union sets covered various vast key K-string from ten groups. With the use of particular 400 and 1211 K-string to generate phylogenetic tree, it is the observation that the final tree mainly categorized five class of vertebrates and failed to discuss the phylogenetic relationship of the species in every class in comparison to the tree generated by the CV Tree approach. So, particular two clusters of key K-string do not cover adequate information for tree generation, and they show the phylogenetic relationship between vertebrates at the class levels; hence, we consider them for our vast key K-string. As per expectation, the phylogenetic trees generated from particular main key K-string showed a topology analogous to that tree generated with the CV Tree approach (Fig. 1); on comparing with the CV Trees, we found that five species were separately placed. Figure 1a represents the phylogenetic tree generated from 23,223 key K-string, and Fig. 1b shows the phylogenetic tree generated from the all set of 20 K-string on use of the CV Tree approach.

Alignment-based phylogenetic analyzing is mainly done in old research as well as we generated a ML tree as the basis of alignments of all the mitochondrial protein's sequence (Fig. 2). Both main K-string trees as well as CV Trees show mostly similar to the ML trees (the topology values among K-string tree, CV Tree, and ML tree are 76.2% as well as 72.8%). Even statistically similarity, we have little difference among key K-string trees as well as ML trees, for example, CV Tree, ML tree potted branches of Coelenterat behind Placozo, on the contrary to key K-string trees. Moreover, the branch of vertebrates, the phylogeny points of Bird as well as Mammalian are exchanged among ML trees as well as the two alignment-free trees. Besides, more differences are also found among main K-string tree as well as ML tree, main in the member of arthropods, which phylogenie not shown at point. It is clear that the final output of the main K-string tree has similarity to alignment-base trees; also, a few differences are there.

Additionally, for analyzing traditionally alignment-base tree, we too compare main K-string tree to more alignment-free tree like as the SVDs Trees as well as three trees generated by Yue Zuguo: the dynamic languages models with correlation

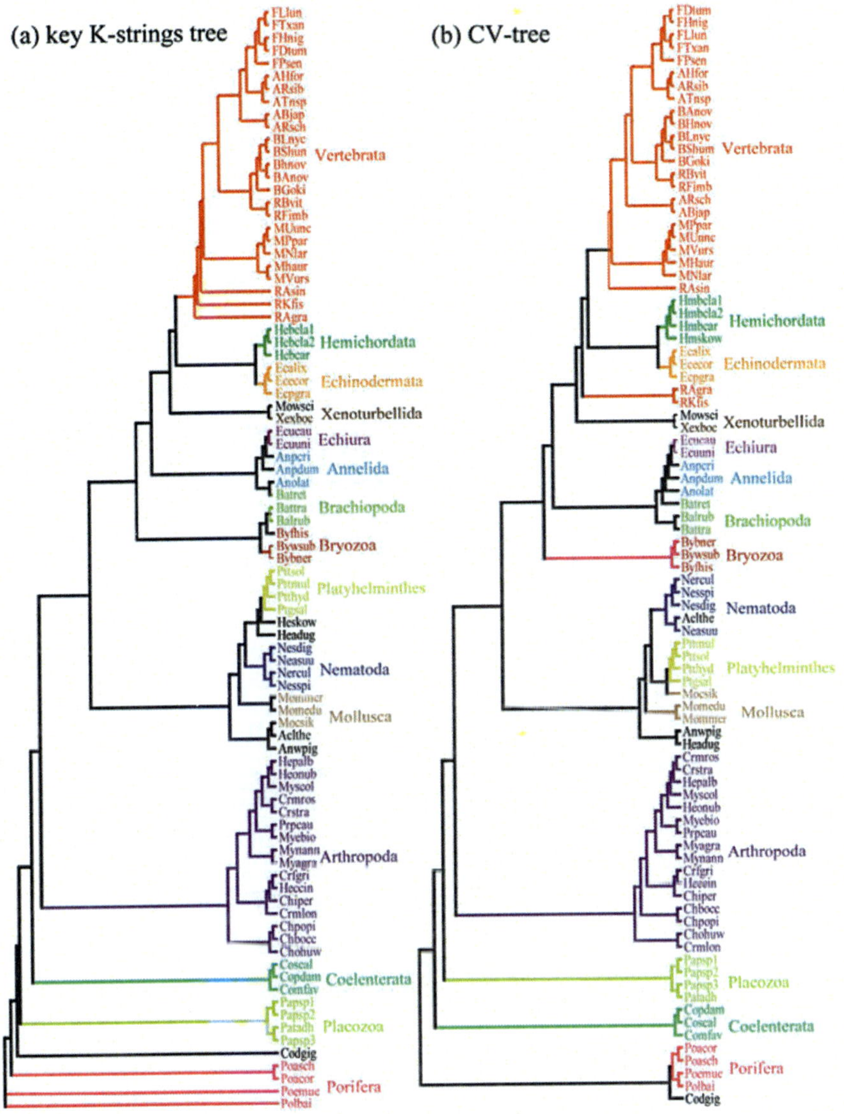

Fig. 1 Phylogenetic tree generated from our key *K*-strings and complete *K*-strings using the CV Tree approach

distances tree, the Fourier transform with KL divergence distances Tree as well as the log-correlation distances tree. Yu suggested the phylogenetic trees drawn by the CV Tree approach not cleared about performing separation of fish, bird, and reptile. Moreover, some divergence from the three trees of Yu was evidence in our key *K*-string trees (Fig. 3). In comparison, our tree with trees generated with KLD

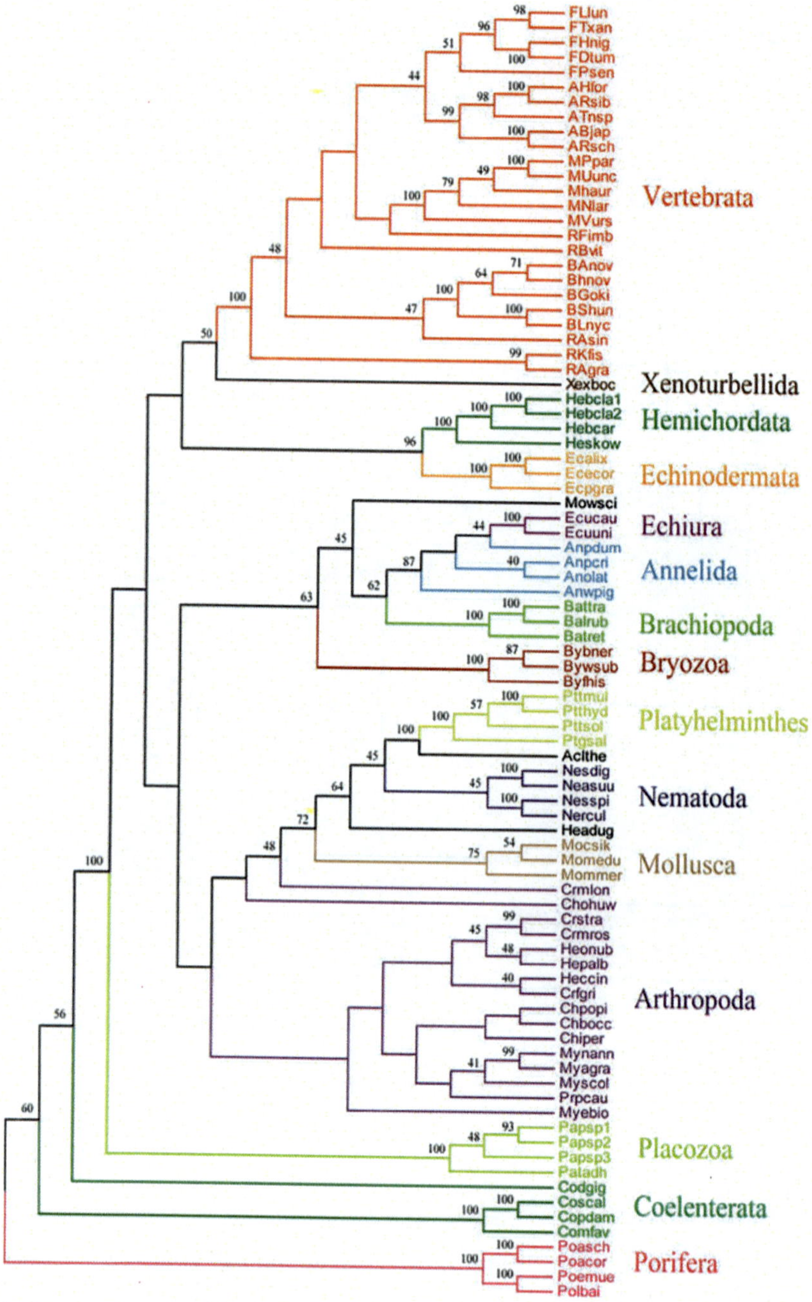

Fig. 2 ML phylogenetic tree

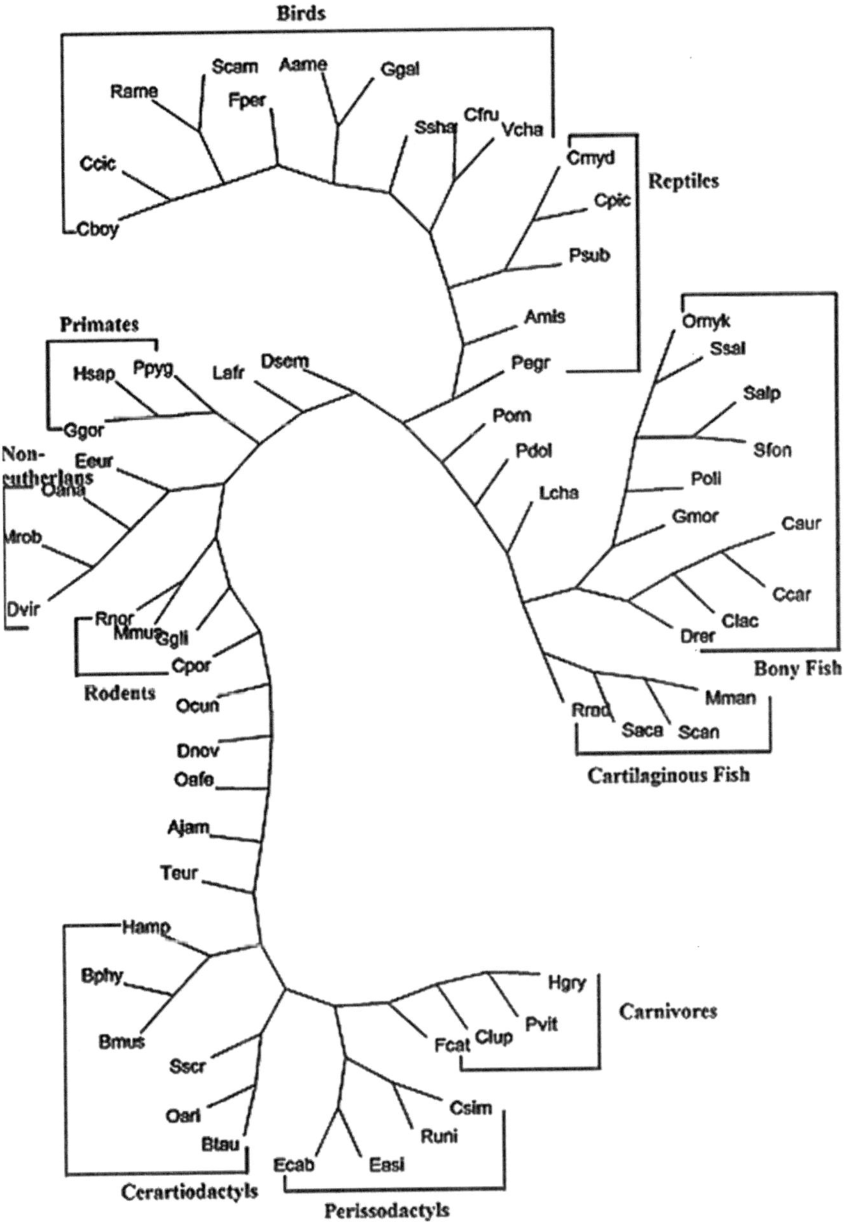

Fig. 3 Phylogenetic tree of vertebrate species generated with phylum-specific key *K*-string

distances, only two other arrangements were presented: Falco peregrinu and Danio rero. On use of DLM method, single species (Smithorni sharpeii) segregate unlike, and 3 species, Corvuss frugileegus, Falcoo peregrinuss and Smithorni sharpeii, were pointed different manner on use of LCD approach.

On other way, the KLD approach arranges the Cartilginous Fish subgroup of the Bonyy Fishes; the DLMs method assign Protpterus dolloie into the phylogroups of the Cartilginous Fishes; as well as LCDs approach place Sus scroofa into Carnivore as well as Protpterus dolloi into Cartilginous Fishes. Finally, we greatly applied the key K-string for evolutionary analysis of metazoan, and the trees generated from the 23,223 key K-string show high rational or higher topologies to the CV Tree as well as more other tree approaches.

4 Conclusions

The tree generated from the key K-string decreases computational time, as well as shown a high rational topology from CV Tree as well as various others alignment-base/alignment-free tree. Mark-ably, the key K-string inclines to gather at the reserved areas of homologous protein as well as have particular composition features which serve for the distortion of protein structures. Lastly, the key K-string has powerful diligence to the finding of phylogenies trajectory. The bauble and likely grandness of key K-string spark to conceive that they are necessary phylogenetic components. To our cognition, it is the main study to talk over the biological significance of this evolutionary element on integral protein sequence. They may play a major role in the procedure of species evolutions, as well as their recognition may, hence, edge to a new epoch of discovery in respect to the relationships among evolution as well as the functions of biological evolutions. The implication of this element might exceed the mt genomes, as well as the future work will disclose particular importance in plastids/chloroplasts, microbials as well as nuclear genomes.

References

1. Xie Q, Lin J, Qin Y, Zhou J, Bu W (2011) Structural diversity of eukaryotic 18S rRNA and its impact on alignment and phylogenetic reconstruction. Protein Cell 2:161–170
2. Gruning B et al (2018) Bioconda: sustainable and comprehensive software distribution for the life sciences. Nat Methods 15:475–476
3. Ondov BD et al (2016) Mash: fast genome and metagenome distance estimation using MinHash. Genome Biol 17:132
4. Li Q, Xu Z, Hao B (2010) Composition vector approach to whole-genome-based prokaryotic phylogeny: success and foundations. J Biotechnol 149:115–119
5. Smith SA et al (2011) Understanding angiosperm diversification using small and large phylogenetic trees. Am J Bot 98:404–414

6. Sankarasubramanian J, Vishnu US, Gunasekaran P, Rajendhran J (2016) A genome-wide SNP-based phylogenetic analysis distinguishes different biovars of *Brucella suis*. Infect Genet Evol 41:213–217
7. Lomsadze A, Gemayel K, Tang S, Borodovsky M (2018) Modeling leaderless transcription and atypical genes results in more accurate gene prediction in prokaryotes. Genome Res 28:1079–1089
8. Girault G, Blouin Y, Vergnaud G, Derzelle S (2014) High-throughput sequencing of *Bacillus anthracis* in France: investigating genome diversity and population structure using whole-genome SNP discovery. BMC Genom 15:288
9. Griffing SM et al (2015) Canonical single nucleotide polymorphisms (SNPs) for high-resolution subtyping of Shiga-toxin producing *Escherichia coli* (STEC) O157:H7. PLoS One 10:e0131967
10. Gardner SN, Slezak T, Hall BG (2015) kSNP3.0: SNP detection and phylogenetic analysis of genomes without genome alignment or reference genome. Bioinforma 31:2877–2878
11. Sahl JW et al (2015) Phylogenetically typing bacterial strains from partial SNP genotypes observed from direct sequencing of clinical specimen metagenomic data. Genome Med 7:52
12. Sahl JW et al (2016) NASP: an accurate, rapid method for the identification of SNPs in WGS datasets that supports flexible input and output formats. Microb Genom 2:e000074
13. Li PE et al (2017) Enabling the democratization of the genomics revolution with a fully integrated web-based bioinformatics platform. Nucleic Acids Res 45:67–80
14. Klenk H, Göker M (2010) En route to a genome-based classification of Archaea and Bacteria? Syst Appl Microbiol 33:175–182
15. Sims GE, Kim SH (2011) Whole-genome phylogeny of *Escherichia coli*/*Shigella* group by feature frequency profiles (FFPs). Proc Natl Acad Sci USA 108:8329–8334

Chapter 18
Indoor Positioning System (IPS) in Hospitals

Anant R. Koppar, Harshita Singh, Likhita Navali, and Prateek Mohan

1 Introduction

Positioning has been an important topic for many years since it is needed to locate people, guide them to places, and assist companies and organizations with their asset management. There has been humongous progress in this field. However, most of the current positioning techniques heavily depend on GPS for accurate location mapping. The global positioning system (GPS) is a satellite-based radio navigation system. It is a global navigation satellite system that provides geolocation and time information to a GPS receiver anywhere on or near the Earth as long as there is an unobstructed line of sight to four or more GPS satellites. The major point to note in the use of GPS is that it works best in an unobstructed line of sight. Obstacles such as mountains and buildings block the relatively weak GPS signals. So, we use the indoor positioning system as an alternative that overcomes the issues that GPS faces when used inside any closed environment, such as a multi-story building.

Indoor positioning holds value in a large number of fields, and it is convenient for positioning patients within buildings such as hospitals and nursing homes. There are two different types of mainstream indoor positioning systems—The first type is locating objects when the users and buildings are equipped with additional hardware. For example, RFID, NFC chip, or other hardware that is dependent on the receiving sensor arrangement present in the framework [1, 2]. These solutions fall under the

A. R. Koppar
Kushagramati Analytics Private Limited, RPC Layout, Bengaluru, Karnataka 560104, India

H. Singh
28/2, SH 35, Devasthanagalu, Gunjur Village, Bengaluru, Karnataka 560087, India

L. Navali (✉)
Kalakruti, Vidyanagar, Old Income Tax Office Road, Hubli 580021, India

P. Mohan
3C/10, Nandi Garden PH 1, JP Nagar 9th Phase, Bangalore, Karnataka 560108, India

A. Sheth et al. (eds.), *Intelligent Systems*, Algorithms for Intelligent Systems,
https://doi.org/10.1007/978-981-16-2248-9_18

fixed indoor positioning system. The second sort is locating clients who have the gear, for example, GPS when the structure does not have the framework introduced inside the given condition. This method is classified as pedestrian positioning.

In this paper, we try to look at the problem from another viewpoint, by using our android devices that are everybody's daily drivers, to utilize the sensors available in it for indoor positioning. This method falls under pedestrian positioning as it works without any stationary gear. In the hospital, a patient's well-being is always of the highest priority. Mishaps can happen in numerous ordinary circumstances that imperil the life of the patient. In such cases, the availability of a doctor affects the speed and quality of the first action taken. But what if the doctor is on another floor, or in another room? The staff must be able to easily locate the doctor and guide him to the patient. Indoor positioning can help locate the doctor in real time. This app can also be used by nurses and visitors to monitor their patient's movements.

2 Literature Survey

Paper [3] intends to furnish a correlation audit of various parameters that influence indoor restriction. There are three procedures known as angle of arrival (AoA), time of arrival (ToA), and fingerprinting methods to calculate indoor position. Four things that are used to compare the three metrics, those are—accuracy, cost, number of reference points/base station, and adaptiveness. Location detection technique and algorithm are as follows.

Time of arrival—Time of landing (ToA) frameworks depend on the definite estimation of the entry time of a flag transmitted from a convenient gadget to a few tolerating sensors. Since signals travel with a known speed (speed of light (c) or ~3 $\times 10^8$ m/s), the separation between the cell phone and each sensor receiving signals can be resolved from the time of the flag going between them.

Fingerprinting-based indoor localization—A one-of-a-kind imprint database is built by assembling imprints from every area in the intrigued territory. Table 1 shows comparisons of indoor positioning methods.

We see that the fingerprinting, which depends on the imprint, gives additional promising focal points contrasted with different procedures, and this procedure furnishes a phenomenal exactness contrasted with other frameworks. One can see

Table 1 Results of different algorithms

Method	Measurement type	Indoor accuracy	No. of beacons	Cost	Time synchronization
Direction (AoA)	Angle of arrival	Medium	Three	High	No
Time (ToA)	Time of arrival	High	Three	High	Yes
Fingerprinting	Received signal strength	High	Three	Medium	No

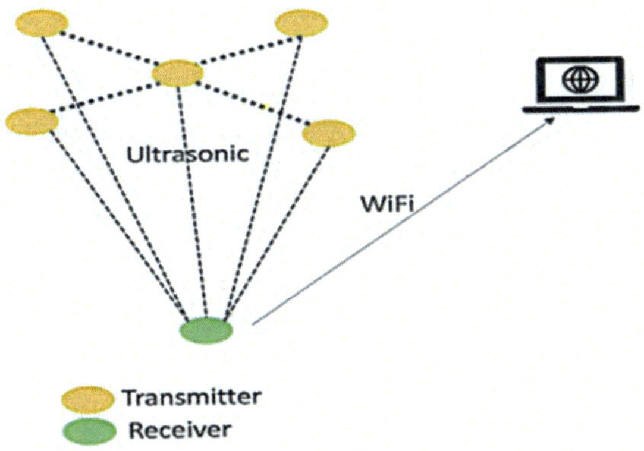

Fig. 1 Positioning system structure diagram

that the expense for this method is low since it does not need much foundation arrangement compared to the other two methods.

Paper [4] displays an ultrasonic indoor locating framework that locates objects exact to the centimeter. Transmitting hubs, accepting hubs, and display terminals establish the whole framework as shown in Fig. 1.

The system makes use of TOA to find the location of the receiver node. The transmitting nodes are fixed onto surfaces like the ceiling, and their spatial position is set. The receiver node captures the signals emitted by these transmitter nodes and calculates their 3D position. The values calculated by the receiver node are then shown onto the display nodes. While this solution claims to show promising results, the setup seems extremely cumbersome and costly.

Paper [5] surveys the advances in wireless approaches to indoor positioning. Different technological approaches are discussed and several trade-offs are brought to light. However, the survey also proves that, while there has been great advancement in the field of wireless indoor positioning systems, most of these technologies cannot keep up with the expected performance level of current-day systems.

Paper [6] deals with the survey of various indoor positioning systems and also puts them into different categories. The first being 'Type of Positioning' which is either fixed IPS or pedestrian positioning. Nuaimi and Kamel [6] also discuss received signal strength, angle of arrival, and time of arrival all of which use the proximity principle to calculate the distance. The paper concludes that the fixed indoor positioning system is saturated with errors. Pedestrian positioning on the other hand still lacks accuracy, and enhancements can be added to this system. A lot more techniques have been surveyed in [7], and similar concerns have been brought to light.

3 Method

This indoor positioning system technique takes advantage of the android devices and its sensors such as accelerometer and magnetometer to collect data streams, such as orientation, rotation, and acceleration. These chunks of data are then passed on to the server which calculates the location of the users. This section encapsulates the methodologies for data collection and calculation in depth.

3.1 Gathering Data

Data is procured using an application, which runs on any modern android device, supporting android 4.1 or greater version. The solution has the following dependencies:

Accelerometer and Magnetometer: In an android phone, the accelerometer (a type of motion sensor) and magnetometer (a type of position sensor) can utilize the Android SDK which enables us to get information of the speed and geomagnetic field quality, respectively, in X, Y, and Z heading.

Orientation: Hand-held devices such as mobile phones can use this information to automatically rotate the display to remain upright, presenting a wide-screen view of the content when the device is rotated so that its width is greater than its height. There are two specific JavaScript events that we use to handle the orientation information. DeviceOrientationEvent is used to acquire the accelerometer sensor data when there is a change in orientation of the device. DeviceMotionEvent is the second event, which is sent when a change in acceleration and magnetometer values are detected, portraying the changes in direction which allow us to track horizontal movements of the users.

3.2 Server-Side Data Pre-processing

The server hosts the algorithm to calculate the latitude and longitude using Haversine's formula (as shown in Fig. 2) [8] as the backbone. Data is collected from the accelerometer and magnetometer sensors using the android application. The values are collected in 3D-space as x, y, z coordinates.

The haversine formula determines the great-circle distance between two points on a sphere given their longitudes and latitudes. Next, using the direction between the two points, we can assume these points to be on a sphere (assuming the earth is spherical), and this allows us to use the property of a sphere to calculate the distance on its surface, hence getting the distance(d) using Eqs. (1) [8], (2) [8], and (3) [8] where ϕ is latitude, λ is longitude, and R is earth's radius (6371 km).

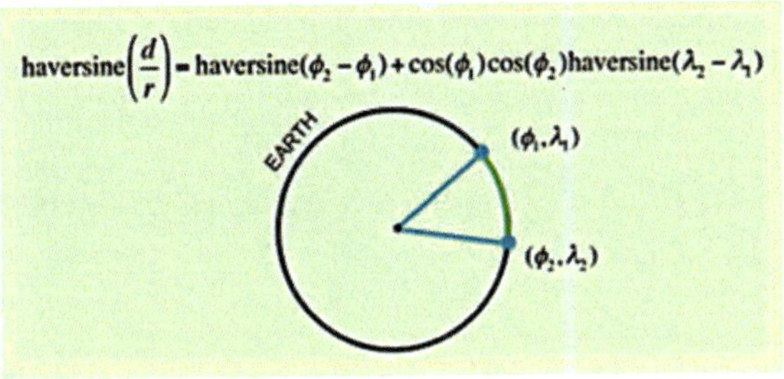

Fig. 2 Haversine formula

$$a = (\delta\phi/2) + \cos\cos\phi_1 \cos\cos\phi_2\left(\frac{\delta\lambda}{2}\right). \quad (1)$$

$$c = 22\left(\sqrt{a}, \sqrt{(1-a)}\right). \quad (2)$$

$$d = Rc. \quad (3)$$

The process of computing latitude and longitude includes computation of the number of steps, and the time at which the step is taken and computing angles using the Earth's magnetic field as shown in Eq. (4) [8] and (5) [8]. The steps are calculated using the root mean square of acceleration along *x*, *y*, and *z*.

$$\text{angle(radian)} = \frac{\tan\,(\text{Magnetic field} - y(\mu\text{T}))}{\text{Magnetic field} - x(\mu\text{T})}. \quad (4)$$

$$\text{angle(degree)} = \text{angle(radian)} * \left(\frac{180}{\pi}\right). \quad (5)$$

Indoor Positioning System Algorithm and Helper Functions: This code function reads the data from a server socket that collects all the values from the android device and dumps it into a CSV.

```
df = pd.read_csv("op.csv")
```

We correct the orientation, so that we can compensate for the way the phone is held in a user's hand.

```
for i in range(len(df)):
    orientation = (math.atan2(df1['ym'][i],\
    df1['xm'][i])*(180/math.pi))+270
    if orientation >360:
        orientation = orientation - 360
angle.append(orientation)
```

After this, the accelerometer, magnetometer, and the orientation values are sent to a recursive function that computes the new latitude and longitude using a haversine formula.

```
data=finder (lats, longs, steps, angles, len(angles), df, 0)

def finder (lats, longs, steps, direc, x, df, i):
    lat2 = degrees(lat2)
    lon2 = degrees(lon2)
    df1 = pd.DataFrame({'latitude':[lat2],\'longitude':
[lon2]})
    frame = [df, df2]
    new = pd.concat(frame)
    finder (lat2, lon2, steps, direc, x-1, new, i+1)
```

4 High Level System Architecture

This paper deals with two points of data collection and calculations, the first being the android device, in which its sensors are used to accumulate the data points for the users. This stream of data is sent to its final destination, the end-point Python server, which allows for a centralized calculation and display of the data points onto a map for a human-readable visualization.

4.1 Android Application and Data Acquisition

Accelerometer and magnetometer raw values are read from the sensors and displayed on the android app. The application user interface is shown in Fig. 3. The end-user installs the android application for data accumulation. The data gathered is shown in Fig. 4.

4.2 End-Point Server

We use Python 3 server as a backend, and it first receives the values from the android client and then calculates latitude and longitude. It is then displayed onto a map.

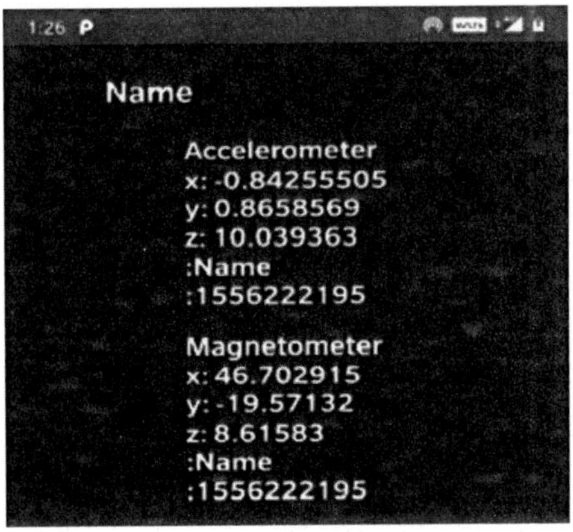

Fig. 3 Android application used to record accelerometer and magnetometer data

Acc	xa	ya	za	Mag	xm	ym	zm	Ori	xo	yo	zo	Name	Time
cc	0.084016	0.607145	9.949293	Mag	8.094788	33.93719	-6.07533	orien	346.618	356.508	359.5162	Name	1556252339
cc	0.589462	1.389509	10.11746	Mag	4.593681	34.12879	-6.56801	orien	351.8448	352.1931	356.6656	Name	1556252340
cc	0.120427	2.048744	9.560267	Mag	-1.28832	32.06915	-8.95682	orien	2.025359	347.9055	359.2783	Name	1556252340
cc	0.25122	1.588812	9.522418	Mag	-5.15377	31.65443	-7.82651	orien	8.643848	350.5307	358.4888	Name	1556252341

Fig. 4 Data point parsed after being processed in the server into a csv, readable format

5 Results

To summarize the workflow, each user's android app acts as a node that collects data points through the sensors present on the user's device and relays it to the backend server. This continuous data stream is fed into the algorithm, and the user's coordinates are updated on the monitoring system. The server receives this information in real time in the form of bytes as shown in Fig. 5.

For every 1000 rows of data collected, it translates to one step taken by the user. The accuracy of tracking is bolstered by the frequency at which the data points are logged in by the sensors. The user can then login and view their movements on our website which creates a visual map as shown in Fig. 6.

Traversing a set path to cross verify the data being recorded from the start till the end as seen in Fig. 6. These visual recordings have a high resemblance to the layout


```
b'\x00}Acc:-0.95993346:2.7362466:9.761487:Mag:-36.13514: -0.39484787: -19.707306:orien: 84.0643: 344.41278:5.6163435:Name:1556252227'
b'\x00~Acc:-1.1295334:2.9484859:8.999725:Mag:-36.13514: -0.39484787: -19.707306:orien: 83.627426: 341.99225:7.1536474:Name:1556252230'
b'\x00~Acc:-1.1295334:2.9484859:8.999725:Mag:-36.13514: -0.39484787: -19.707306:orien: 83.627426: 341.99225:7.1536474:Name:1556252230'
b'\x00|Acc:-1.1295334:2.9484859:8.999725:Mag:-36.474358: 1.0190353: -18.468983:orien: 82.26202: 341.99225:7.1536474:Name:1556252230'
```


Fig. 5 Data points for accelerometer and magnetometer are forwarded to the server every 1 s, pre-parsed

Fig. 6 Another dataset of coordinates displayed on the map in closeup-each pointer is a data point recorded per second

Table 2 Key characteristics of the proposed model

Measurement type	Multipath effect	Start Lat and Long	No. of beacons required	Cost	Need of time synchronization
Accelerometer versus magnetometer	No	Yes	Zero	Low	No

of the path traversed, thus ossifying the data collection and processing on both the client and server-side. Table 2 shows important attributes of our system.

6 Discussion and Summary

The outcome of this project is a novel indoor positioning system that can be used in a hospital setting by patients, doctors, and staff alike. This new solution makes use of accelerometer and magnetometer sensors that are present in most smartphones, along with the latitude and longitude of the starting position of the user to locate and track people on the map. This project serves as a more reliable, cheap alternative to GPS, or other proposed solutions that fall short at accurately tracking people in indoor settings at reasonable expenses.

There may be some possible limitations in this study. The first being the lack of data because the movements were recorded on a small set of devices. Because of this, we were not able to obtain data on a wide set of sensors from different sources. This can be fixed with manual tweaking before runs. Another limitation is due to the recursive nature of the calculation; if there are any slight errors from sensors, they might get multiplied in extreme conditions. This can cause issues with devices containing older generation sensors. To fix this, anytime the users stop, we can use it to break out of the recursive calculation and start afresh.

Due to the lack of ground truth of coordinates of the locations, accuracy cannot be measured in numerical percentage. Moreover, since accuracy, at present, is assessed through a visual map, it leaves room for future development which can be achieved with additional testing hardware. It must be noted that this system's usefulness is not confined to a hospital and can be used in other professional settings. For example—tracking shipments, locating professors in a university, and helping firefighters to locate victims stuck in a building fire.

Future work concerns deeper analysis of particular mechanisms, new proposals to try different methods, or simply curiosity. Currently using GPS, wherever accessible to improve accuracy, we can try other local beacons like Wi-Fi strengths to remove the dependency of GPS completely.

References

1. Zhou F (2017) A survey of mainstream indoor positioning systems. J Phys Conf Ser 910(1):012069. https://doi.org/10.1088/1742-6596/910/1/012069
2. Saab SS, Nakad ZS (2010) A standalone RFID indoor positioning system using passive tags. IEEE Trans Industr Electron 58(5):1961–1970. https://doi.org/10.1109/TIE.2010.2055774
3. Mrindoko NR, Minga LM (2016) A comparison review of indoor positioning techniques. Int J Comput (IJC) 21(1):42–49
4. Li J, Han G, Zhu C, Sun G (2016) An indoor ultrasonic positioning system based on TOA for the internet of things. Mobile Inf Syst. https://doi.org/10.1155/2016/4502867
5. Bonde GD, Barwal PU, Pal SR, Khan SI, Ablankar K (2015) Finding indoor position of person using wi-fi & smartphone: a survey. Int J Innov Res Sci Technol 1(8):202–207
6. Al Nuaimi K, Kamel H (2011) A survey of indoor positioning systems and algorithms. In: 2011 international conference on innovations in information technology, pp. 185–190. IEEE. https://doi.org/10.1109/INNOVATIONS.2011.5893813
7. Brena RF, García-Vázquez JP, Galván-Tejada CE, Muñoz-Rodriguez D, Vargas-Rosales C, Fangmeyer J (2017) Evolution of indoor positioning technologies: a survey. J Sens. https://doi.org/10.1155/2017/2630413
8. Info Geographical Information System Map. https://www.igismap.com/haversine-formula-calculate-geographic-distance-earth/

Chapter 19
A Review on Character Segmentation Approach for Devanagari Script

Manoj Sonkusare, Roopam Gupta, and Asmita Moghe

1 Introduction

The OCR system is made of four major steps, that is, preprocessing, segmentation, feature extraction, and classification. In this system, characters are recognized after perfect segmentation. So, the segmentation is an important phase, and recognition results of any OCR system are heavily dependent upon it. Segmentation is the step in which text images are subdivided into its constituent regions, these constituent regions are individual characters, and the classifier recognizes these characters only [1].

Due to improper scanning, adjacent characters touch each other in some documents and errors arise in the segmentation phase. Hence, inaccurate segmented characters are not properly recognized by the classifier. The text segmentation phase includes line, word, and character segmentation. Foremost, line segmentation is performed to find the number of lines and boundaries of each line. Segmentation of offline handwritten text in Devanagari script is more tedious because of different writing styles of individual writers. The presence of broken, overlapping, touching, half characters, vowels, and consonants with modifiers, and uneven header line makes this process more difficult.

Most of the segmentation and recognition algorithms described in the literature are implemented for multilingual scripts, and it is observed that the result of Indic script is inferior to Latin script [2, 3]. This review paper is related to various segmentation techniques for Devanagari script, and it presents an efficient approach at the character level.

M. Sonkusare (✉) · R. Gupta · A. Moghe
Department of IT, Rajiv Gandhi Proudyogiki Vishwavidyalaya, Bhopal 462033, India

R. Gupta
e-mail: roopamgupta@rgtu.net

A. Moghe
e-mail: aamoghe@rgtu.net

A. Sheth et al. (eds.), *Intelligent Systems*, Algorithms for Intelligent Systems,
https://doi.org/10.1007/978-981-16-2248-9_19

The paper is managed in a subsequent manner: In the next section, we have discussed the Devanagari script and about segmentation in Sect. 3. Section 4 presents various segmentation techniques and comparison between them is shown in Table 1. Finally, Sect. 5 contains conclusion and future scope of the research.

2 About Devanagari Script

A Devanagari word is made with core, top, and bottom strips [4]. The core and top strips are segregated by the header, while the lower modifier adjoining the core character is shown in Fig. 1.

2.1 Conjunct Characters

The difference between Hindi and other scripts is the existence of conjunct characters. It is the union of half and full consonants, where the left part is half and the right part is full consonant [2, 5], as shown in Fig. 2.

2.2 Overlapping Characters

Two or more characters in a word overlapped with each other in Fig. 3 [5].

Fig. 1 Three strips of Devanagari word

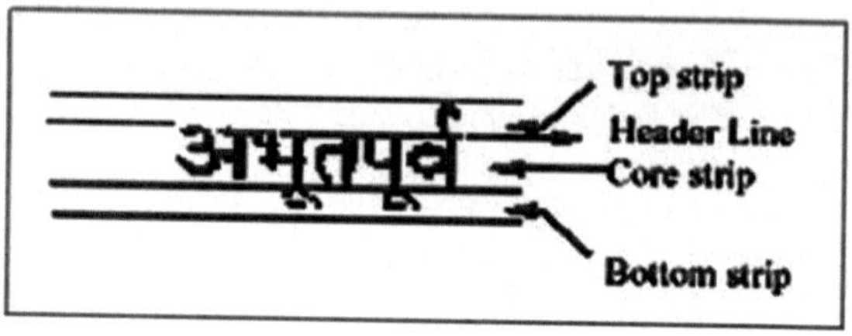

Fig. 2 Conjunct word

Fig. 3 Overlapped characters

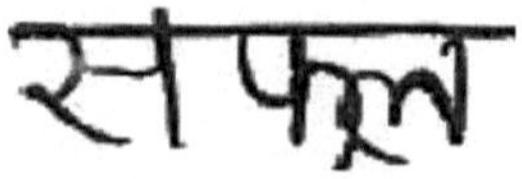

3 Segmentation

Segmentation is the process of segregating a digital image into multiple components. The aim of segmentation is the retrieval of a line or a word or even a character from the text-based images. Projection profile, connected components, structural properties, contour-based approach, edge detection, collusion dilation, water-reservoir, graph cut method, and machine learning-based method are the various techniques to segment words into characters.

The horizontal line (shirorekha), different writing style, shapes, cursive nature of handwriting in which characters connect to each other, characters sharing similar contours, variation of spaces between the characters, large character set which includes consonants, vowels, modifiers, compound characters, and device used for writing in handwritten Devanagari scripts make segmentation task more difficult [6]. Holistic approach reduces the accuracy results as compared to segmentation approach (Shaw, Parui, and Shridhar 2008) [7, 8]. Segmentation reduces the complexity of recognition. If a word is properly segmented, then the number of classes used in the recognition system will be equal to the number of characters.

4 Literature Survey

A trace and pixel plot-based system is developed in the recognition category to segment the characters using ANN [9–11]. A multi-oriented touched character segmentation method is presented in [12], where connected component analysis is carried out to segregate individual constituents, and a trained SVM is used to verify whether the segregated constituents are isolated or not. An important goal of the segmentation is to pull out each character from the text image. A printed text segmentation in Devanagari script is presented in [4], in which some modification is done in the algorithm proposed by Bansal [13] to obtain better results. Here, the bottom strip is considered with the core strip, and the projection profile method is used for separation. The performance of this method at character, top character, line, and word level is 99%, 97%, 100%, and 100%, respectively.

An approach based on a fuzzy multifactorial analysis is presented in [14], in which a word is segmented into characters by separating the header line from the word. A two-pass algorithm proposed in [13] was used for character separation, in which structural properties of the script are considered to remove the header line. All the characters that touched each other were left unseparated. Vertical and horizontal incline was used to predict the shape of the character, and the final choice of part of the image is based on the width and height of the text. A graph-based technique for character segmentation is presented in [15]. In this technique, the header line can be determined based on profile and run length. If the header line is separated from the word, individual characters can be categorized based on their distinct width, height, and top–bottom zone identification. A fuzzy multifactorial analysis for segmentation is also proposed in [16], where top to bottom of the word is considered as a standard size, and predictive algorithm is used for selection of cut points to the touching characters.

Algorithm for character segmentation is proposed for both Devanagari and Latin script by using multilingual documents consisting of both handwritten and printed texts in [2]. After preprocessing, key segmentation paths are obtained by using structural features, whereas overlapped and joined characters are segmented by using the graph distance and then correct and incorrect segmentations are distinguished by using SVM classifier. By using a proprietary database of Latin and Devanagari script, this method achieved the highest segmentation rate of 98.79% and 97.32%, respectively.

The process of segregation of handwritten text in Devanagari script is presented in [17]. The characters are segmented after removal of skewness. This technique allows skewness in documents within the range of $-5°$ to $5°$ only. Horizontal and vertical profile projections are applied in such a way that symbols in upper, middle, or lower zone are retained and segmented all through under one process. This technique achieves the accuracy of more than 90% and takes less than one second to execute the entire process which is faster than 70% with the related studies.

Segmentation of handwritten words of Hindi language is proposed in [18], where lines from text documents and words from lines are segregated by projection profile methods. The number of black pixels for each column of each line is counted, and the columns with null-black pixels are used as boundaries for word separation. The system is selected as the limiter with at least three continuous columns with null-black pixels for word segregation. The algorithm finds the header and base lines by estimating the average line height, and based on it, this system works for separation of words and shows word spacing and margin results. The segmented words are then stored as a result.

Edge-oriented filter for segmentation is used in a novel multilingual character recognition framework in [19]. The image document is downscaled and turned to grayscale image in the preprocessing phase. Grayscale image is then processed for edge enhancement followed by binarization through adaptive thresholding. At last, the character regions are segmented through edge density filter (EDF) in both vertical and horizontal directions. Experiments were conducted on the Telugu words dataset and Devanagari and Kannada words captured from the Chars74 dataset. Performance is measured through precision, recall, and accuracy and obtains outstanding results compared to the cutting-edge techniques.

A hybrid method is developed for segregation of handwritten Hindi words in [20] combining the vertical profile, horizontal profile, and clustering technique. The algorithm has been tested on a dataset of 300 handwritten words containing touching, broken, multiple touching, and isolated characters from different writers and gives segmentation accuracy of 96%.

A new shape-oriented segmentation (pihu) method for scene images of Devanagari words is proposed in [21]. This method focuses on the shape of the characters and does not separate the header line from the word to retain the structure of the character and overcome the limitation of under, over, and partial segmentation of the existing method. This method takes 2.70 s to segregate an image over the current method of 4.76 s and achieves a segmentation result of 92.11%.

A contour-based method for separation of handwritten Hindi words in bank cheques is proposed in [22]. This method points out the different zones of a word image and uses a fuzzy function to estimate the headline pixels. The estimated headline pixels along with the outer contour of the word are used to separate the meaningful character components along with upper and lower modifiers. The proposed method can be effectively used to segregate handwritten Hindi words in bank cheques.

A collusion dilation algorithm is developed for segregation of handwritten Devanagari words in [23]. The preprocessing step includes resizing, grayscale conversion, skew removal, etc. Header line is then removed based on collusion and dilation technique. The broken letters are identified and segmented in upper, middle, and lower zones. This approach has been applied on the database of 500 handwritten Devanagari words images and has achieved a segmentation rate of 96%.

5 Conclusion and Future Scope

This paper presents a comparison between various recent character segmentation techniques as shown in Table 1. The survey concludes that results of Indic script are inferior to Latin script, and segmentation accuracy of the hybrid approach (i.e., combination of projection profiles and any other segmentation method) is better than traditional approaches. In our survey, the two highest character segmentation accuracies achieved for Devanagari script is 97.32 and 96%. These hybrid approaches give better segmentation results, and recognition accuracy is highly dependent on the segmentation results. So, this paper finds an efficient approach from various segmentation techniques, and proposed hybrid segmentation approach helps the future researcher in segmentation as well as recognition of Devanagari characters.

Table 1 Comparison between various recent character segmentation techniques

S. No.	Author	Approach used	Type of input	Segmentation accuracy	Pros and Cons of the approach
1	Sahare et al. [2], (2018)	Hybrid approach (combining projection profile, graph distance theory, and SVM classifier)	documents containing Devanagari and Latin scripts images	Devanagari—97.32% Latin—98.79%	**Pros** 1. Two characters are segmented having tight joints with no connection 2. Missed, bad, and over-segmentation problems are resolved in post-processing and character validation steps **Cons**: Bad segmentation rates are too high as compared to over-segmentation rates
2	Sharma et al. [20], (2017)	Hybrid approach (combining horizontal and vertical profile and clustering technique)	Touching and broken characters in handwritten Hindi words	96%	**Pros**: This system gives promising results for isolated, conjuncts, and touching characters **Cons**: Words having uneven header lines, overlapping, or skewed characters are not segmented properly
3	Kaur et al. [23], (2017)	Collusion dilation	Handwritten broken letters of Devanagari scripts	96%	**Pros**: This technique is also applicable for skewed, broken text
4	Pramanik et al. [22], (2018)	Contour-based	Handwritten bank cheques in Hindi language	93.89%	**Pros:** Works superior to a system that employs a horizontal projection profile

(continued)

Table 1 (continued)

S. No.	Author	Approach used	Type of input	Segmentation accuracy	Pros and Cons of the approach
5	Jindal et al. [21], (2018)	Shape-oriented method	Devanagari words from natural scenic images	92.11%	**Pros**: Proposed method overcomes the limitation of under-segmentation, over-segmentation, and partial segmentation of the existing method **Cons:** Images that contain noisy pixels in background, unavailability of continuous vertical white pixel count, and skewed text are partially segmented
6	Bathla et al. [17], (2019)	Horizontal and vertical projections profile method	Handwritten Devanagari documents	90.35%	**Pros**: This technique achieves the accuracy in less than one second, it is faster than 70% with the related studies
7	Malpe et al. [18], (2018)	Horizontal and vertical projections profile method	Hindi and English Handwritten documents	Hindi: 85% English: 90%	**Cons:** Difficult to point out exact connecting points for separation of lower and upper modifier
8	Vishwanath et al. [19], (2020)	Contour- based (edge) approach	multilingual document images of Telugu, Devanagari, and Kannada words	Devanagari—77.70% Telugu—74.45% Kannada—76.37%	**Pros**: This technique gives better performance with overlapped words and noisy image

References

1. Singh R, Yadav S, Verma P (2010) Optical character recognition (OCR) for printed Devanagari script using artificial neural network. Int J Comput Sci Commun 1(1):91–95
2. Sahare P, Dhok SB (2018) Robust character segmentation and recognition schemes for multilingual Indian document images. IETE Techn Rev. https://doi.org/10.1080/02564602.2018.1450649
3. Mathew M, Singh AK, Jawahar CV (2016) Multilingual OCR for Indic scripts. In: Proceedings of 12th IAPR Workshop Document Analysis Systems, pp 186—191. IEEE Press, Santorini (2016)
4. Kumar V, Senger PK (2010) Segmentation of printed text in Devanagari Script and Gurmukhi Script. Int J Comput Appl 3(8):0975–8887
5. Thakral B, Kumar M (2014) Devanagari handwritten text segmentation for overlapping and conjunct characters—a proficient technique. In: Proceedings of 3rd International conference on reliability, Infocom Technologies and Optimization. https://doi.org/10.1109/ICRITO.2014.7014746
6. Kohli M, Kumar S (2020) Comparative analysis of segmentation and recognition techniques for offline handwritten words. Int Res J Adv Sci Hub (IRJASH) 02:41–48
7. Bag S, Krishna A (2015) Character segmentation of hindi unconstrained handwritten words. In: Proceedings of the 17th International Workshop on combinatorial image analysis, pp 247–260. Springer, Cham
8. Castro MJ, Jose M, Moya JG, Martinez FZ (2011) Improving offline handwritten text recognition with hybrid HMM/ANN models. IEEE Trans Pattern Anal Mach Intell 33:767–779
9. Sharma MK, Dhaka VP (2016) Pixel plot and trace-based segmentation method for bilingual handwritten scripts using feedforward neural network. Neural Comput Appl 27(7):1817–1829
10. Sharma MK, Dhaka VP (2015) An efficient segmentation technique for Devanagari off-line handwritten scripts using the feedforward neural network. Neural Comput Appl 26(8):1881–1893
11. Sharma MK, Dhaka VP (2016) Segmentation of English off-line handwritten cursive scripts using a feedforward neural network. Neural Comput Appl 27(5):1369–1379
12. Roy PP, Pal U, Lladós J (2012) Delalandre: multi-oriented touching text character segmentation in graphical documents using dynamic programming. Pattern Recogn 45(5):1972–1983
13. Bansal V, Sinha RMK (2002) Segmentation of touching and fused Devanagari characters. Pattern Recogn 35:875–893
14. Garain U, Chaudhuri BB (2001) Segmentation of touching characters in printed Devanagari and Bangla scripts using fuzzy multifactorial analysis. IEEE Trans Syst Man Cybern Part C Appl Rev, 805—809
15. Kompalli S, Nayak S, Setlur S, Govindaraju V (2005) Challenges in OCR of Devanagari documents. In: Proceedings of 8th conference on document analysis and recognition, pp 1–5. IEEE Press, Seoul
16. Garain U, Chaudhuri BB (2002) On OCR of degraded documents using fuzzy multifactorial analysis. In: Proceedings of AFSS international conference on fuzzy systems (AFSS-ICFS), pp 388–394. Springer, Calcutta
17. Bathla AK, Gupta SK, Jindal MK (2019) Character segmentation and skew correction for handwritten Devanagari scripts: a friends technique. Asian J Eng Appl Technol 8(1):50–54
18. Malpe K, Wankhade M (2018) Segmentation of Hindi handwritten words and personality prediction using word spacing and margin. Int J Innov Res Sci Eng Technol 7(2):1250–1260
19. Vishwanath NV, Manjunathachari K, Prasad KS (2020) Edge-oriented filter and feature invariant coding for multilingual character segmentation and recognition. J Crit Rev 7(14):610–620
20. Sharma P, Sachan MK (2017) A technique for character segmentation in the middle zone of handwritten Hindi words using hybrid approach. Int J Fut Revol Comput Sci Commun Eng 3(7):1–10

21. Jindal K, Kumar R (2018) A new method for segmentation of pre-detected Devanagari words from the scene images: Pihu method. Comput Electr Eng 70:754–763
22. Pramanik R, Bag S, Kumar R (2018) A fuzzy and contour-based segmentation methodology for handwritten Hindi words in legal documents. In: 4th Int'l conference on recent advances in information technology, pp 388–394. IEEE Press, Dhanbad
23. Kaur M, Bathla AK (2017) Segmentation of characters of Devanagari script documents. World Wide J Multidisc Res Dev 3(11):253–257

Chapter 20
Performance Characterization and Analysis of Bit Error Rate in Binary Phase Shift Keying for Future 5G MIMO Environment

Samarth Srivastava, Aman Gupta, Satya Singh, and Milind Thomas Themalil

1 Introduction

The future 5G communications system provides a higher level of performance than earlier generations of communication systems. The antenna technologies for future 5G have significant opportunities for performance enhancement over 4G. MIMO, used with 4G LTE, has been improved further. The earlier work on MIMO focused on space diversity; the MIMO system limits the signal impairment by multipath propagation [1, 2]. Later the system utilized the fading path propagations to favor, making the propagated signal pathways into additional channels to carry data.

By employing multiple antennas, MIMO technology can invariably enhance the efficiency of a given channel's capacity acting on Shannon boundary. Adding up the receiver and transmitter antenna numbers make it convenient to enhance the channel throughput/bandwidth. This makes MIMO technology an important wireless technique [3]. As spectral bandwidth is a valuable commodity, newest techniques are employed to maximize the available bandwidth effectively.

S. Srivastava · A. Gupta · S. Singh · M. T. Themalil (✉)
JK Lakshmipat University, Mahindra SEZ, Jaipur, Rajasthan 302026, India
e-mail: milind.themalil@jklu.edu.in

S. Srivastava
e-mail: samarthsrivastava@jklu.edu.in

A. Gupta
e-mail: amangupta@jklu.edu.in

S. Singh
e-mail: satyasingh@jklu.edu.in

A. Sheth et al. (eds.), *Intelligent Systems*, Algorithms for Intelligent Systems,
https://doi.org/10.1007/978-981-16-2248-9_20

Wireless Communication Systems have gained importance due to its high accuracy, greater extent of secure transfer and effectiveness. The major challenges faced by modern digital wireless communication is to satisfy the ever-growing demand of high-speed reliable communication in multimedia and Internet services with extremely limited frequency spectrum and limited power. Also, wireless channels suffer from fading. Wireless mobile communication has several technical challenges that must be overcome [4].

Transmitted signals undergo fading, shadowing, interference, path loss, etc. so the mechanics of transmitting modes are evaluated by Bit Error Rate (BER) against the signal-to-noise ratio (SNR) with respect to channel models like additive white Gaussian noise, Rayleigh fading and Rician fading channels. Investigations start by deriving the algebraic and geometric model for BER and SNR of AWGN, Rayleigh's and Rician's faded channels; then compare with AWGN channel and Rayleigh's faded channels as a noise channel and apply phase shift keyed modulation technique [5].

Section 1 contains the introduction of the MIMO environment and the related technological issues and challenges. Section 2 mentions the review of previous research work done over the last decade in the area of BER improvement for MIMO. Section 3 describes the work methodology while Sect. 4 shows the corresponding models and figures. Section 5 discusses the obtained results. Section 6 has concluding remarks and scope of future work. Section 7 contains Acknowledgement and Sect. 8 contains References.

2 Brief Review of Earlier Work

Wireless communication has gained a lot of trust in recent times due to its high accuracy, greater extent of secure transfer and effectiveness. The effectiveness of this communication system depends on the efficiency with which the input message reaches the output receiver end with very minimal distortion. The distortion in the message signal is added due to inefficiency of the encoder, attenuation in the channel or interferences in the channel, inefficiency of the decoder, etc., and efficiency of this channel lies in overcoming these errors [3]. The design issue for wireless systems is to optimize superior qualitative data at extremely high data rates. This issue is solved by applying MIMO technology [2]. Generally, in communication systems, the signal reaches the downlink from multipaths leading to intersymbol interferences enhancing the BER [6].

Modern devices and systems offer higher data rates, wider coverages and improved secured data reliance. To arrive at these optimized values in modern communication scenarios, the multiple input, multiple output systems are employed to efficiently increase the data transmission rates and coverages by taking into account multiple transmitters and receivers (antenna system) and geometric space dimension is used. As data transmission is done at higher rates, intersymbol interferences occur [4].

To efficiently utilize the communication system the frequencies reusing method is employed, but a severe setback is its leading to co and adjacent channel interferences due to several base stations accessing the same bands of frequencies having closer proximity to one other, however, design engineers need to reduce the channel interferences however, fail to completely do away with the issue of interference which in many cases become a dominating factor in computing the efficiency of the overall system [7].

3 Methodology

As a consequence of shifting the phases or keying of the signal's multipaths, there is creation of constructive and destructive interferences modeled by Rayleigh's faded channel. Rayleigh faded propagation channel models do not have direct paths between transmitter and receiver. The received signal is modeled as:

$$R(n) = \sum h(n, \tau)S(n - m) + w(n) \tag{1}$$

$w(n)$ is AWGN with unit variance and zero mean, $h(n)$ is impulse response for the channel. Firstly, MATLAB and Python programming was studied. In the proposed project work, characterization of Eb/No (Energy per bit/Noise) ratio [3] and comparison of the Bit Error Rates (BER) of BPSK was implemented [5]. Then BER was estimated for BER of BPSK for Rayleigh fading in comparison with AWGN channel. All these models were implemented using MATLAB and Python programs. Following this BER for Rayleigh Fading was implemented for fading channels using Python programming. Comparing performance parametric of channel capacity and BER analysis of the diversity for MIMO Environment for future 5G applications is formulated [8].

4 Different System Models

To mitigate multipath fading issues, diversity techniques are employed to enhance the secured reliable transmissions without further increase in the throughput/bandwidth or power outputs. Diversity schemes increase the quality of the baseband signal in which more than two communication channels are employed. This reduces the co channel interference and multipath fading of the signal since each channel has different levels of interference [4]. Various techniques used to classify the communication link depending on the number of transmit and receive signals are shown in Fig. 1. Single input and single output technique employs single transmitter and receiver but without diversity. Single input and multiple output technique employs single transmitter but multiple receivers hence providing only receiver diversity. Multiple input and single output technique employs multiple transmitters but single receiver and provides only transmitter diversity. Multiple input and multiple

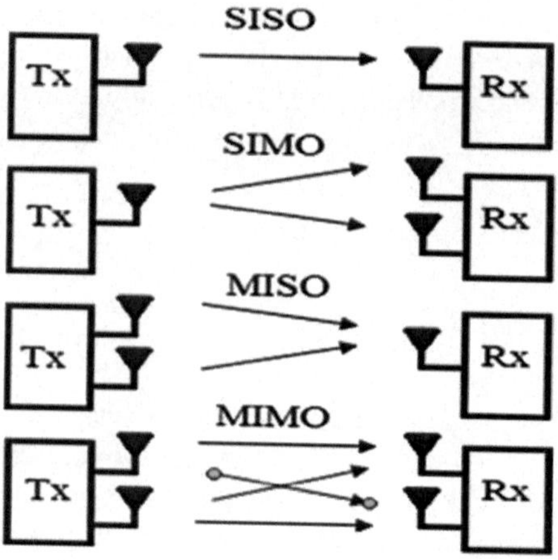

Fig. 1 Communication transmission techniques

output scheme employs several transmitters and multiple receivers and provide both transmitter and receiver diversities [9, 10].

A multiple input multiple output model of the communicating system is presented in Fig. 2. The inputs as a stream of bytes are fed into a binary phase shift keyed modulator (Fig. 3). The modulated signals are fed into an encoder in Orthogonal Space Time Block Code (OSTBC) format. Rayleigh fading channel is used here. Decoded signal (OSTBC) is fed into an equalizer. After that signal is fed into the BPSK demodulator and the approximated signal is recovered [10, 11].

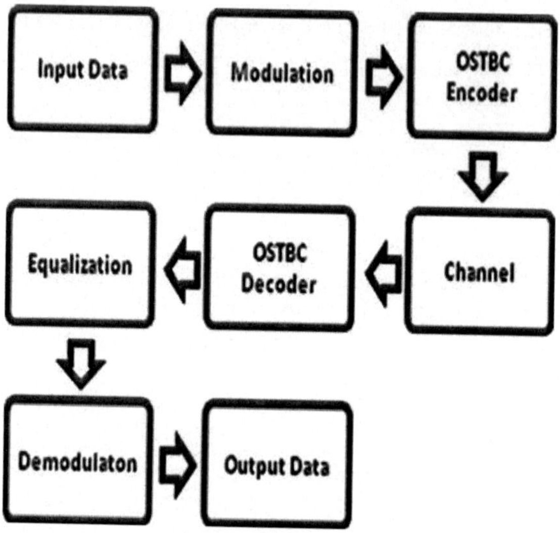

Fig. 2 MIMO system model

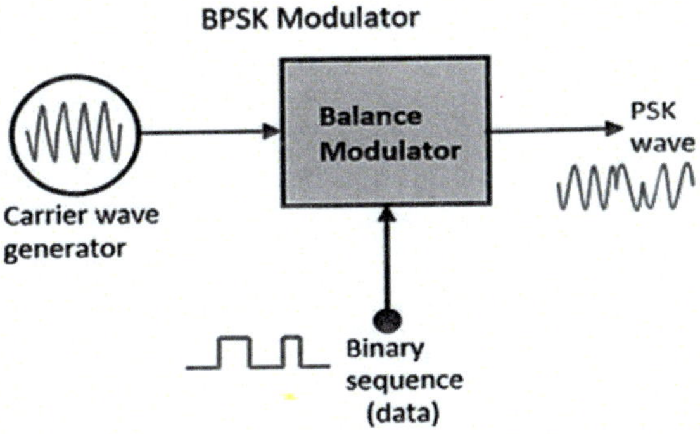

Fig. 3 BPSK system model

Table 1 Design framework for the MIMO system

Parameter	Value
No. of transmitter antennas	Two
No. of receiver antennas	Two
Modulator	BPSK
Channel type	Rayleigh fading
Noise type	AWG noise

BPSK signals are separated by 180° phase. Irrespective of the constellation vector points, at 0° and 180°, it can handle the highest noise level or distortion before the demodulator decision [12, 13]. Table 1 presents the parameters for the MIMO system [1, 14].

5 Results and Discussion

The BPSK outputs are shown in Fig. 4. The BPSK symbols are sampled with AWGN noise samples, randomly generated to characterize the channel. The BER for various Eb/No values are also shown in Fig. 5.

Figure 6 presents the BER analysis of single input and single output SISO systems. Results show the BER value variation from $10^{0.5}$ to 10^{-5} and Eb/No value variation from 0.5 to 35 for binary phase shifted modulation scheme using Rayleigh's fading. Figure 7 shows the BER characteristic of the SIMO system (along with the maximal ratio combining method—MRC). Results show the BER value variation from 10^{-1} to 10^{-4} and Eb/No value variation from 0.5 to 15.

Figure 8 shows the BER characteristics of multiple input single output MISO systems. Results show the BER value variation from $10^{-0.5}$ to 10^{-5} and Eb/No value variation from 0.5 to 18.5 with binary phase shifted modulation using OSTBC for Rayleigh's fading. Figure 9 presents the BER evolution of a 2 × 2 MIMO system. Results show the BER variation from $10^{-0.5}$ to 10^{-5} and Eb/No value variation from

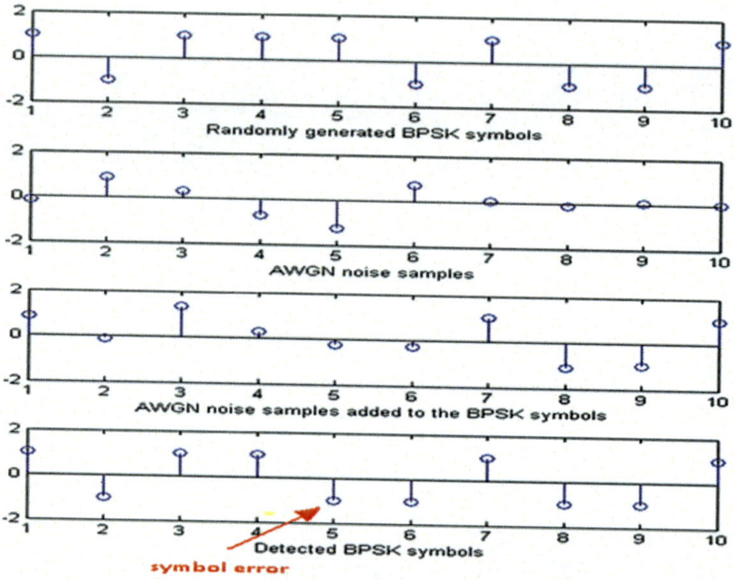

Fig. 4 BPSK symbols

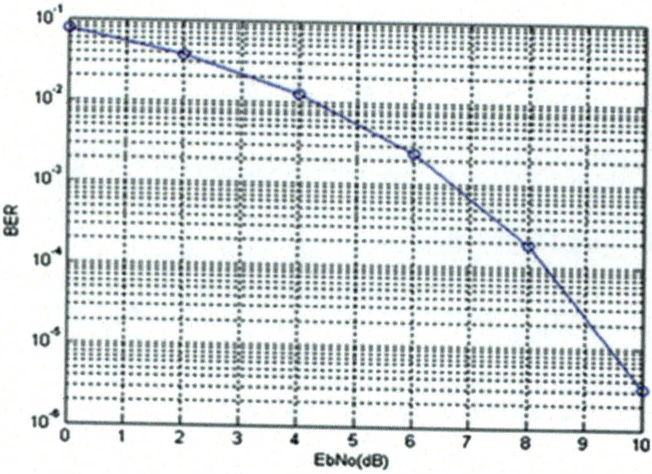

Fig. 5 BER versus Eb/N_0

0.5 to 10 with binary phase shifted modulation along with OSTBC for Rayleigh's fading, whereas Fig. 10 presents the BER comparative analysis for different schemes for transmission using the multiuser MIMO system.

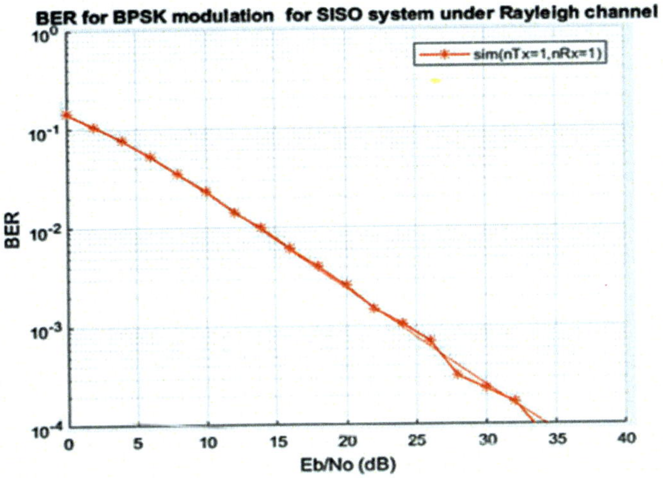

Fig. 6 BER of SISO system

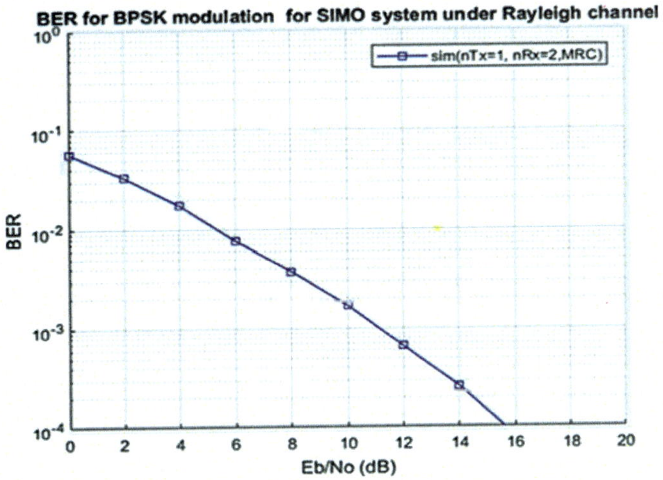

Fig. 7 BER of SIMO system

Analyses show the BER value variation from $10^{-0.5}$ to 10^{-5} and Eb/No value variation from 0.5 to 15 for binary phase shifted modulation along with MRC with Rayleigh's fading.

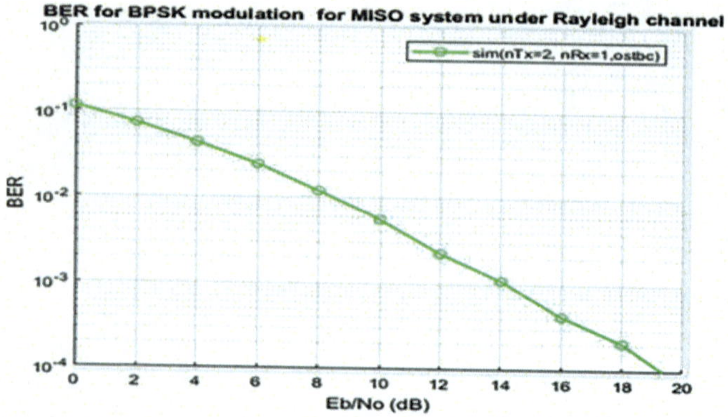

Fig. 8 BER of MISO system

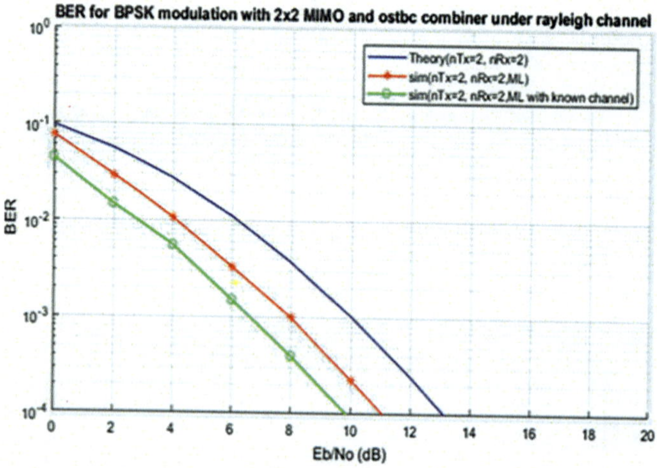

Fig. 9 BER of MIMO 2 × 2 system

6 Conclusions and Future Scope

In this work, a comparative parametric analysis of BER in different MIMO systems with binary phase shifted modulation for Rayleigh's faded channel is presented. The optimal performance of the system employing the 2 × 2 MIMO system with Eb/No 9.8 has a BER value of 10^{-4}. Further optimization to maximize channel capacity can be performed as a future work. Extending the scope of work to a larger MIMO array (say 4 × 4 or 8 × 8) may also be done (to validate the optimal results) as a future work. Rician fading model also can be performed for future MIMO indoor applications. Work is being continued to model the microstrip antenna performances in the new NR MIMO environment for future 5G infrastructure.

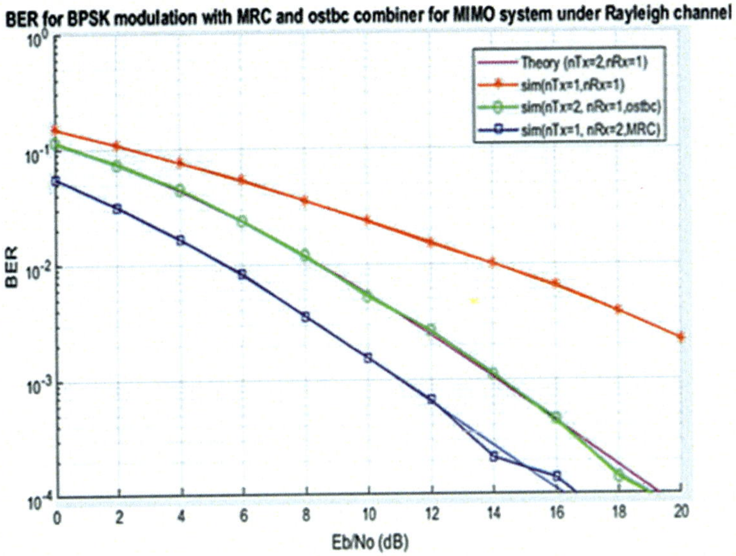

Fig. 10 Comparative OSTBC-MRC BER of different MIMO systems

Acknowledgements The authors thank the IET—Institute of Engineering and Technology of JK Lakshmipat University, Jaipur for providing their laboratory and library infrastructure.

References

1. Mo J, Schnitery P, González Prelcicz N, Heath RW Jr (2014) Channel estimation in millimeter wave MIMO systems with one-bit quantization
2. Nadeem Q-U-A, Kammoun A, Debbah M, Alouini M-S (2018) Design of 5G full dimension massive MIMO systems. IEEE Trans Commun 66(2)
3. Markosyan MV, Safin RT, Artyukhin VV, Satimova EG (2014) Determination of the Eb/N0 ratio and calculation of the probability of an error in the digital communication channel of the IP-video surveillance system
4. Rupa M, Sobha B (2019) BER performance improvement of MIMO. Intl J Recent Tech Eng
5. Vijaykumar K (2015) Comparison of MIMO OFDM system with BPSK and QPSK modulation. Int J Emerg Technol 188–192 (2015)
6. Dubey KK, Srivastava DK (2017) BER analysis of MIMO-OFDM system using BPSK modulation under different channel with STBC, MMSE and MRC. Int J Adv Res Comput Comm Eng 6(5)
7. Cheng Z, Chen B, Zhong Z (2012) A tradeoff between rich multipath and high receive power in MIMO capacity, 24 Nov 2012
8. Kharat S, Hanchate S (2015) Effect of BPSK and QPSK on MU-MIMO signal detection techniques. Effect of BPSK & QPSK on MU-MIMO signal detection techniques
9. de Lamare RC (2011) Massive MIMO systems: signal processing challenges and future trends. IEEE Trans Veh Technol 60(6):2482–2494
10. Ngo HQ (2015) Massive MIMO: fundamentals and system designs. Linköping 2015

11. Li T, Geyi W (2019) Design of MIMO beamforming antenna array for mobile handsets. Prog Electromagnet Res C 94:13–28
12. Gurdasani H, Ananth AG, Manjunath TC (2018) Performance of (2x2) MIMO communication systems for various PSK modulation schemes. IOSR J Electron Comm Eng. 13(4), Ver. I (July–Aug 2018), pp 34–40. e-ISSN: 2278–2834, p-ISSN: 2278–8735
13. Abur Khan M, Pal S, Ankita J (2015) BER performance of BPSK, QPSK & 16 QAM with and without using OFDM over AWGN, Rayleigh and Rician fading channel. Int J Adv Res Comput Comm Eng 4(7)
14. Singh MU, Kakkar S, Rani S (2016) BER performance analysis of OFDM-MIMO system using GNU radio. In: ICAET—2016

Chapter 21
Pose Estimation and 3D Model Overlay in Real Time for Applications in Augmented Reality

Pooja Nagpal and Piyush Prasad

1 Introduction

Previous approaches to pose estimation have involved depth cameras, body trackers, and multiple cameras in different views. Recent availability and accessibility to high-quality datasets like the COCO, MPII, and VGG have allowed fully convolutional neural networks to be trained with high accuracy. These have outperformed the older approaches. These high-quality datasets present us with numerous useful applications. One such application is in augmented reality.

Intrinsically, augmented reality is a motion tracking problem. Sensors like ultrasonic devices, inertial devices, and optical sensors have been considered. Vision-based techniques proved the other approaches to be suboptimal and non-scalable [1].

In this paper, an approach to 3D model overlay and orientation in accordance with key point information is presented. Rendering an image plane along with the 3D model is proposed. The model used in this work for pose inference is based on the paper titled Multi-Person Pose Estimation [2]. When seen at a higher level, the problem boils down to localization, key point detection, and model overlay. All of which, modern vision-based techniques can solve, provided high-quality datasets are available, which fortunately are not out of reach anymore.

P. Nagpal (✉) · P. Prasad
Computer Science Department, Amity University Gurugram, Panchgaon 122051, India
e-mail: pbnagpal@ggn.amity.edu

P. Prasad
e-mail: piyush.prasad@student.amity.edu

A. Sheth et al. (eds.), *Intelligent Systems*, Algorithms for Intelligent Systems,
https://doi.org/10.1007/978-981-16-2248-9_21

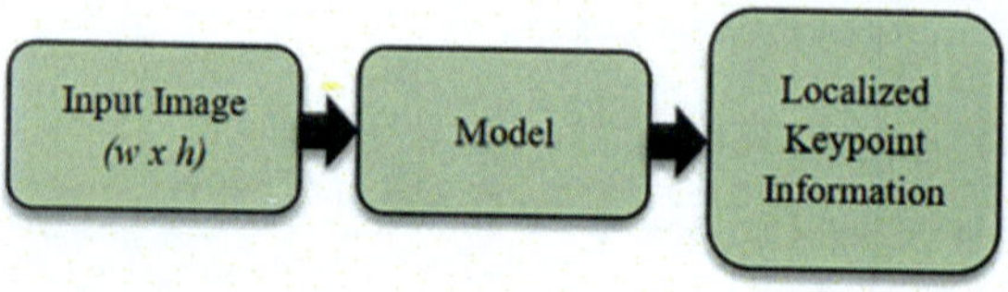

Fig. 1 High-level model pipeline, taking a $w \times h$ color image, and producing localized keypoint information for further processing

2 Related Work

Many researchers proposed various models in real-time applications. Cao et al. [1] done multi-person 2D pose estimation using part affinity fields. This approach uses a nonparametric representation. Further Lin et al. [3] proposed a model of augmented reality in 3D images which were human body pictures. Moa et al. [4] presented real-time simultaneous pose and shape estimation for articulated objects using a single depth camera. Main feature of this pose estimation was to embed an articulated deformation model using exponential-maps-based parameterization. This novel approach provided better accuracy and was less sensitive to complex motions. Next step of advancement in this area was work presented by Rosez et al. [5]. They dealt with the problem of 3D human pose estimation in the wild. Lack of training data was the biggest challenge for them, and they successfully resolved the concern. Toshey et al. [6] shifted to new technology and they implemented human pose estimation through deep neural networks. They benchmarked the process with better performance.

One recent approach proposed by Bin, Yanrui, et al. is Adversarial Semantic Data Augmentation (ASDA) [7] which exploits a generative network to dynamically predict tailored configurations with existing pose estimation networks acting as the discriminator.

3 Model Architecture Overview

The model is a deep neural network, which consists of three stages. The model takes a color image of size n worked upon, $w \times h$, and returns the localized pose key points from the image as shown in Fig. 1.

4 Methodology

Figure 2 illustrates the low-level pipeline of the system. The system takes color images as input, resizes them to 368 × 368, binary blobs are fed into the trained model, and keypoints containing 2D coordinates are extracted. These keypoints are

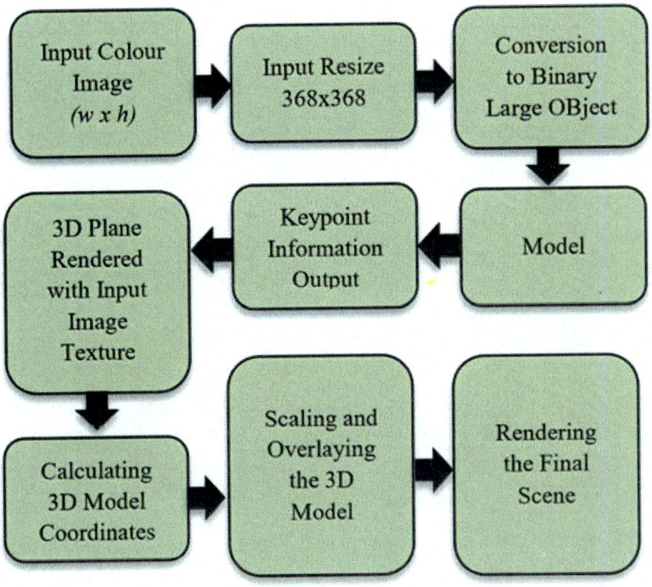

Fig. 2 Lower-level pipeline of the complete system. Each frame is loaded as a plane texture in the 3D scene

used to calculate the correct orientation for the 3D model [6, 8]. Finally, the image is rendered as a plane, and then the 3D model is rendered, intersecting the plane.

4.1 *Pose Inference*

A sample image resized to 368×368 is fed into the FCN and key points are detected as shown in Fig. 3. The ordered nature of key point information should be noted as this order will be important when we will deal with calculating the orientation and scaling factor of the 3D model [9].

4.2 *Intersecting Image Plane and 3D Model*

Overlaying the 3D model on the image is achieved by rendering the image as a plane in 3D space [10]. The plane is aligned with the camera orientation so as to cover the foreground of the scene completely. The input image is loaded as a texture in the plane as shown in Fig. 4.

Firstly, the input image is required to be rendered as a plane as shown in Fig. 5.

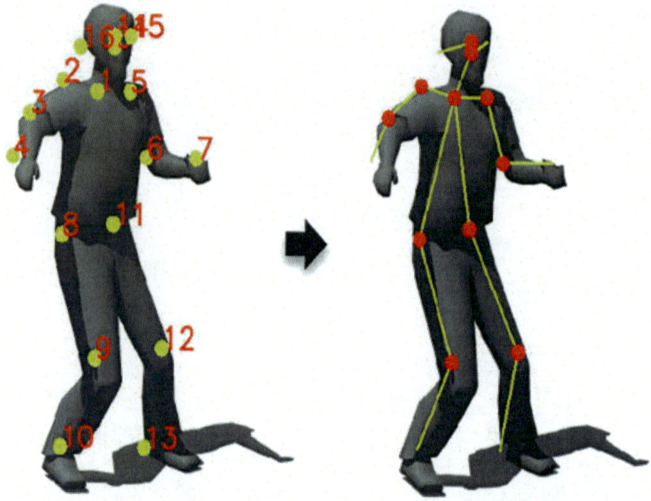

Fig. 3 Key point detection in a sample image (image is resized to 368 × 368). The model was trained on the COCO dataset, and it produces 18 key points which are labeled in order from 1, 2, 3, … 18

Fig. 4 3D model used in the work

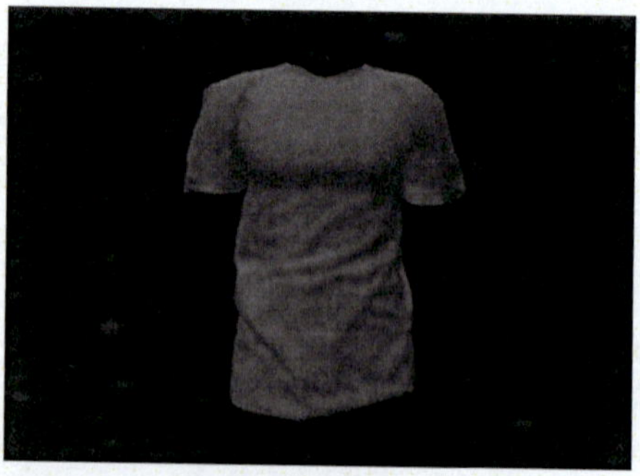

Fig. 5 Image rendered as a plane. The plane is loaded with the texture of the input image

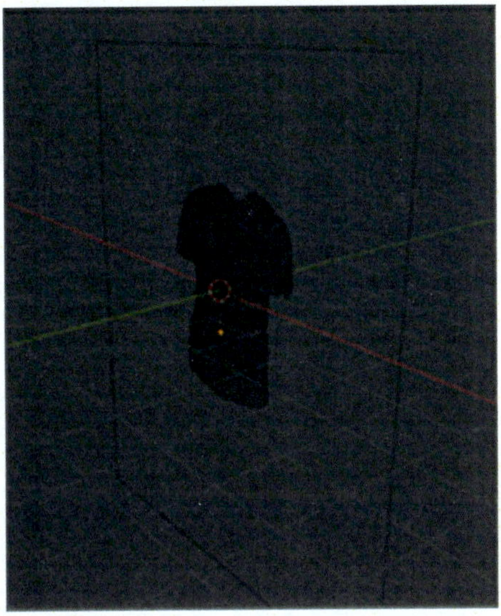

Fig. 6 Correct 3D model orientation and intersection with the image plane. Part of the problem is to calculate the correct coordinates and scaling factor for the 3D model, for a successful overlay

To overlay the 3D model in the right orientation, the key point coordinates are required. The 3D model must be scaled correctly, oriented and placed at the right coordinates. In Fig. 6, the image plane is intersecting with the 3D model and is rendered with the plane vertices coordinates (−4, − 4, − 8), (−4, 4, − 8), (4, 4, − 8), (4, − 4, − 8).

4.3 Overlay Calculation

To calculate the overlay coordinates, we will take two extracted key points, namely key point 1 and 5. Any two extreme key points (anchor points) can be chosen, provided the same corresponding anchor points are selected for the 3D model [11]. Let k_1, k_5 be the two keypoint vectors we are interested in and t_1, t_2 be the top-left most and top-right most vectors of the 3D model.

The first step is to superimpose t_1 on k_1 and translate the corresponding t_2, k_5 vectors. Once we have the two vectors superimposed, the only problem left is calculating the scaling factor for the model overlay.

This scaling factor S is given by Eq. (3).

$$d_1\left(\overrightarrow{k_5} - \overrightarrow{k_1}\right) = \left\|\overrightarrow{k_5} - \overrightarrow{k_1}\right\| = \sqrt{(k_{5i} - k_{1i})^2 + \left(k_{5j} - k_{1j}\right)^2} \quad (1)$$

$$d_2\left(\overrightarrow{t_2} - \overrightarrow{t_1}\right) = \left\|\overrightarrow{t_2} - \overrightarrow{t_1}\right\| = \sqrt{(t_{2i} - t_{1i})^2 + \left(t_{2j} - t_{1j}\right)^2} \quad (2)$$

$$S = \frac{\sqrt{(k_{5i} - k_{1i})^2 + \left(k_{5j} - k_{1j}\right)^2}}{\sqrt{(t_{2i} - t_{1i})^2 + \left(t_{2j} - t_{1j}\right)^2}} = \frac{d_1}{d_2} \quad (3)$$

$$S \in [0, \infty) \quad (4)$$

Now that we have **S**, the model scaling factor can be plugged in, and the model can be scaled. Note, that upon scaling the model, the vectors $\overrightarrow{k_5}$, $\overrightarrow{t_2}$ are automatically superimposed. This results in a perfect and successful model overlay.

5 Final Rendered Result

Figure 7 shows the final rendered result, with the 3D model overlaid with the correct orientation and scaling factor. Taking the overlay as a starting point, prospects for

Fig. 7 Correctly oriented overlay scaling factor and coordinates

ambient light inference from input image, ambient texture mapping for reflective surfaces are present. These can be integrated for seamlessly coloring, texturing virtual objects in input scenes.

6 Conclusion and Future Work

Previous works have approached the problem with single depth cameras which although provide precise input parameters are less accessible than a monocular camera. A significant challenge for pose estimation is the lack of training data, and some approaches to generate high-quality datasets have included using motion capture sensors on existing MoCap data for pose tracking and generating feature vectors. It can be concluded that using monocular, conventional cameras, are some of the feasible ways forward.

Techniques like SLAM, simultaneous localization and mapping, to localize and track camera displacement can be employed for ground-truth validation, camera tracking and seamless integration. Light mapping, shadow mapping, and inferring ambient light from an input image can prove instrumental in increasing the immersivity of the rendered result. Ambient texture mapping for reflective surfaces can also be implemented. This approach could be employed in AR applications, CGI, and or previsualization virtual mockups.

Future work can involve movement prediction and motion dampening to avoid jittery behavior. The performance can be improved with lower-level code although this work was mostly based in Python, OpenGL, and the C++ backend for OpenCV and rendering.

References

1. Cao Z et al. (2017) Realtime multi-person 2d pose estimation using part affinity fields. In: IEEE conference on computer vision and pattern recognition, pp. 1302–1310, July, 2017
2. Marchand E, Uchiyama H, Spindler F (2015) Pose estimation for augmented reality: a hands-on survey. IEEE Trans Vis Comput Graph 22(12):2633–2651
3. Lin H-Y, Chen T-W (2010) Augmented reality with human body interaction based on monocular 3D pose estimation. In: International conference on advanced concepts for intelligent vision systems. Springer, Berlin
4. Ye M, Yang R (2014) Real-time simultaneous pose and shape estimation for articulated objects using a single depth camera. In: Proceedings of the IEEE conference on computer vision and pattern recognition(CVPR), 2014, pp. 2345–2352
5. Rogez G, Schmid C (2016) Mocap-guided data augmentation for 3d pose estimation in the wild. In: Advances in neural information processing systems, NIPS'16: Proceedings of the 30th International Conference on Neural Information Processing Systems December 2016 Pages 3116–3124
6. Wohlhart P, Lepetit V (2015) Learning descriptors for object recognition and 3d pose estimation. In: Proceedings of the IEEE conference on computer vision and pattern recognition

7. Zhen J et al (2020) SMAP: single-shot multi-person absolute 3D pose estimation. In: European conference on computer vision. Springer, Cham
8. Newell A, Yang K, Deng J (2016) Stacked hourglass networks for human pose estimation. In: European conference on computer vision. Springer, Cham
9. Reitmayr G, Drummond TW (2006) Going out: robust model-based tracking for outdoor augmented reality. In: 2006 IEEE/ACM international symposium on mixed and augmented reality. IEEE
10. Schall G et al. (2009) Global pose estimation using multi-sensor fusion for outdoor augmented reality. In: 2009 8th IEEE international symposium on mixed and augmented reality. IEEE
11. Toshev A, Szegedy C (2014) Deeppose: human pose estimation via deep neural networks. In: Proceedings of the IEEE conference on computer vision and pattern recognition

Chapter 22
Comparative Analysis of Machine Learning Algorithms for Detection of Pulmonary Embolism—A Non-cardiac Cause of Cardiac Arrest

Naira Firdous, Sushil Bhardwaj, and Amjad Husain Bhat

1 Introduction

Human welfare is an indispensable task, holding the topmost priority on the medical bars. The major canon for the detection of diseases at proper time needs meticulous attention, which may not be available all the time due to paucity of medical facilities and scantiness of medical representatives at the desired destinations. The above-mentioned conditions take a catastrophic shape when cardiovascular diseases are encountered. So, it becomes important to automate the medical procedures in order to arrive at accuracy and precision on the output side. Previously, many attempts have been made in the form of expert systems but they did not come up with the desired expectations and hence were rejected. Tables were turned when machine learning was identified as a real benefit in the field of artificial intelligence. Cardiac arrest is the most common cause of hospitalization in developed countries for people aged 65 years or older and is the leading cause of death. Pulmonary emboli a blood clot in the pulmonary arterial system increases the blood pressure in the arterial system leading to pulmonary hypertension. Pulmonary hypertension can then lead to the right ventricle not being able to pump blood against this pressure, as the right ventricle was never designed to cope with that. Therefore, the formation of embolism leads to right heart failure resulting in sudden cardiac arrest. Pulmonary embolism can sit around for ages without being diagnosed as it mimics other symptoms. The only way to diagnose is to CT scan the patient. Doctors avoid this technique as it

N. Firdous (✉)
Department of Computer Science Engineering, RIMT University, Mandi Gobindgarh 147301, India
e-mail: nairafirdous@rimt.ac.in

S. Bhardwaj
Department of Computer Applications, RIMT University, Mandi Gobindgarh 147301, India

A. H. Bhat
Department of Computer Science, University of Kashmir, Srinagar 190006, India

A. Sheth et al. (eds.), *Intelligent Systems*, Algorithms for Intelligent Systems,
https://doi.org/10.1007/978-981-16-2248-9_22

is expensive and also a massive radiation dose. Therefore, patients can have these blocking arteries causing pulmonary hypertension for a significant period of time which can then lead to cardiac arrest.

Wong [1] has applied AI techniques to explore novel genotypes and phenotypes in existing diseases. Shrestha [2] has worked on detection of heart failure in coordination with preserved ejection fraction using machine learning. Chetankumar [3] has worked on "The heart rate variability (HRV) parameters" to detect cardiac arrest in smokers using machine learning. Alizadehsani [4] carried out a research with the help of a dataset called Z. Alizadeh Sani with 303 patients and 54 input variables. Hinto [5] produced expert systems and graphical models responsible for automation of the reasoning processes of experts. Matthews [6] has given a detailed review of the path physiology, by studying acute right ventricular failure in the setting of acute pulmonary embolism. Laher [7] has presented a paper on pulmonary embolism due to cardiac arrest with a mortality rate of 30%. Bizopoulos [8] has presented a review paper on deep learning on cardiology. Kim [9] has proposed a cardiovascular disease prediction model using the Six Korea National Health. Cano-Espinosa [10] observed CAD method for pulmonary embolism. Pawar [11] method showed results with improved efficiency of CA.

1.1 Abbreviations

Systolic blood pressure (*ap_hi*), diastolic blood pressure (*ap_lo*), pulse pressure (*Pp*), stroke volume (*sv*), pulmonary embolism (*pe*), support vector machine (*SVM*), naive Bayes (*NB*), *K*-nearest neighbor (*K*NN), hidden layers (HL), neural network (NN), cardiac arrest (CA).

1.2 Dataset and Data Processing

In this paper, we worked on modified UCI dataset, as UCI dataset for cardiology was not ample to carry out our work. We derived extra input features from the pre-existing features by following all the medical protocols, details of which are mentioned in phase-I of methodology. This dataset is a newly created dataset which is formed with the help of UCI cardiology dataset. This is the first study where this database has been used to establish a link between cardiac arrest and pulmonary embolism using machine learning. Previous work has been done on cardiovascular diseases without its association with pulmonary embolism with the accuracy of 99%.

We first of all effectuated data cleaning for carrying out our research; in this step, we processed our data for analysis by removing the inconsequential data, and the data which was improperly formatted was made unerring.

2 Methodology

Sections 2.1 and 2.2 are the sub-topics under Sect. 2. Since we have divided our methodology into two parts, so sub-topic 2.1 shows the establishment of a medical algorithm, and Sect. 2.2 shows the applicability of machine learning on the above-formed algorithm.

2.1 Phase-I (Establishment of Connectivity Between Pulmonary Embolism and Cardiac Arrest)

The major condition of pulmonary embolism is the drop of stroke volume (volume of blood pumped out of the left ventricle of the heart during each systolic cardiac contraction).

The stroke volume is directly proportional to pulse pressure difference (between systolic blood pressure and diastolic blood pressure:

Equation (1) represents pulse pressure which is equal to the systolic pressure–diastolic pressure.

$$Pp = (ap\_hi) - (ap\_lo) \tag{1}$$

Equation (2) represents the relationship between pulse pressure and stroke volume.

$$Pp = sv/C \tag{2}$$

Pulse pressure is considered to be critically low if it is less than 25% of systolic blood pressure (*ap_hi*). Using these above-mentioned biological equations, we formulated a medical algorithm. With the addition of new variables, the attributes in the dataset were increased to 18 with 70 K records.

We propose our algorithm as shown in Fig. 1, for describing a link between pulmonary embolism and cardiac arrest.

- In the first step, we calculated the pulse pressure (*Pp*) with the help of systolic and diastolic blood pressures. [$Pp = (ap\_hi) - (ap\_lo)$].
- *Pp* is directly proportional to the stroke volume (*sv*). So, decrease in stroke volume will cause decrease in the amount of blood ejected out of the left ventricle during each systolic cardiac contraction, thereby leading to the formation of emboli which may result in cardiac arrest.
- The value of stroke volume may be either 0 (low) or 1(high) depending upon the two conditions of *Pp*:

  (1) If $Pp < 25\%$ of *ap_hi*, $sv = 0$ (low), it indicates the cause of cardiac arrest which is pulmonary embolism.

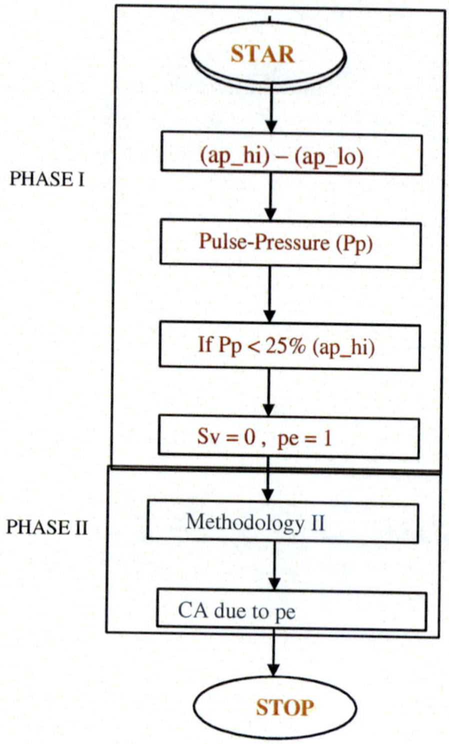

Fig. 1 Flowchart shows the connectivity between cardiac arrest and pulmonary embolism

(2) If $Pp > 25\%$ of *ap_hi*, $sv = 1$ (high), it indicates pe which is not a cause of CA.

2.2 *Phase-II (Implementation of Machine Learning)*

We now explain the various algorithms and techniques that we have used in our proposed algorithms:

(a) Adaboost: The basic principle behind the functioning of the Adaboost is to engender weak learners and to coalesce their prognosis to form one strong rule.

In case of boosting algorithms, the default weak learners are decision trees.

- First of all, we assigned equal weights to data points, and we drew out a decision stump (a single-level decision tree that tries to classify the data point) for a single input feature.
- The results that we got from the first decision stump were examined, and highest priority was assigned accordingly.
- In the next step, another decision stump was drawn that tried to classify the data points with high priority.

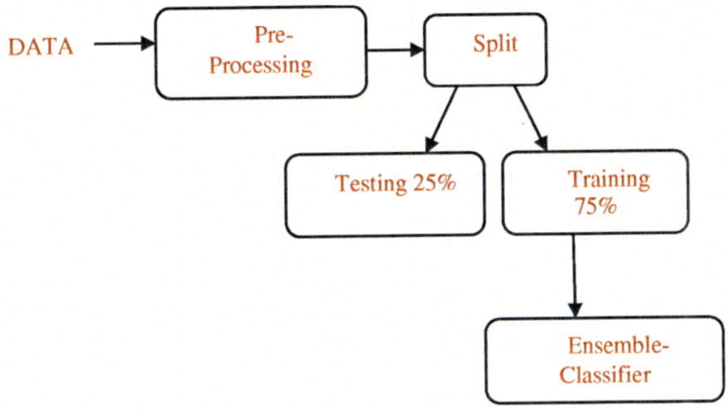

Fig. 2 Shows ensemble classifier

- Steps were repeated until all the observations were seen to have fallen at their arrant positions.

We carried out our work using 100 estimators, and the learning rate was kept as 1.

(b) Ensemble Method: While working with boosting algorithms, we encountered a problem of overfitting. So, in order to get rid of this problem, we replaced the decision trees in the Adaboost with strong machine learning models.

An ensemble model shown in Fig. 2 is described by making use of multiple machine learning algorithms, namely SVM, *K*NN, and naive Bayes. Ensemble learning is used to aggrandize the performance of machine learning models by combining several learners. When compared to a single model, this type of learning builds model with improved efficiency and accuracy with proper precision in hand.

(c) Neural Networks: Neural networks mimic our human brain.

(1) In first step, we fed our novel dataset to the input layer of our neural network.
(2) Hidden layer of the neural network, which is the next layer after the input layer, performs two functions:

- Performs sum of the product of weights with input data, which is represented in Eq. (3)

$$\sum WiXi \tag{3}$$

where $i = 1$ to n.
With the addition of bias (bias allows activation functions to be shifted to the left or right, to better fit the data).

- Application of activation function.
Equation (4) shows the multiplication of weights with respective inputs.

$$\mathrm{Act}\left(\sum WiXi\right) \tag{4}$$

where $i = 1$ to n.

Finally, the result obtained from the hidden layer is multiplied by the weight and passed to the output layer where activation function called as sigmoid is used as it transforms values between 0 and 1.

This phase is forward propagation.

At the end of the forward propagation, we obtain our predicted output represented as $\overline{Y}$. In the next step, we compared the predicted value $\overline{Y}$ with the actual value or the target value y. This is done by means of a loss function that will optimize the parameter values present in our model.

Equation (5) represents the expression for calculating the loss function.

$$\text{Loss} = (y-)2 \tag{5}$$

Squaring is performed to get the positive results.

where y is the actual value, and $\overline{Y}$ is the predicted value. If our $\overline{Y} = Y$, then our model is working correctly, but if $\overline{Y} \neq Y$, then we have to back propagate.

The main aim of back propagation is to reduce the loss function. Loss function should be reduced to a minimal value. During the training phase, weights should be updated in such a way that loss function should get reduced to a greater extent. Loss function is reduced by making use of an optimizer.

3 Results

Finally, we enumerate the results on the basis of our research are as follows:

- On the implementation of the Adaboost algorithm on our novel dataset, we encountered the problem of overfitting. Since the default weak learners in Adaboost are decision trees which are considered prone to overfitting when working with deep datasets. The reason is that the model in the training phase not only considers the original data but also comprehends noise as the useful data. Implementation was performed using 100 estimators, and learning rate was kept as 1.
- In order to get rid of this overfitting problem, we proposed an ensemble method in which we combined many machine learning models by using a voting classifier. We trained our KNN for $k = 5$ values, and our SVM model was trained by using a linear classifier followed by Gaussian naïve Bayes.

The accuracy of ensemble classifier on replacement of decision trees with KNN, NB, and SVM is shown in Table 1.

Table 1 Shows the accuracy of ensemble model

Classifier	Accuracy
Ensemble classifier	99.99%

Table 2 Shows the accuracy values of neural network by invoking different hidden layers and activation functions on the novel dataset

No. of hidden layers	Activation function	Optimizer	Accuracy (%)
2	relu	Adam	99.33
2	tanh	Adam	99.32
2	softmax	Adam	99.13
5	relu	Adam	99.23
5	tanh	Adam	99.22
5	softmax	Adam	99.32
10	relu	Adam	99.33
10	tanh	Adam	99.31
10	softmax	Adam	99.21

- While working with neural networks, we checked the accuracy by invoking different hidden layers and activation functions, keeping the optimizer the same. In the first step, we divided our modified dataset into 80% training and remaining 20% into testing phase. Next step was to instantiate the model which was done by making use of a sequential model. After instantiating our sequential model, we added layers to our model by making use of dense function. We first of all specified the number of neurons in the dense function which was put = 256. Since this was the first layer of our model, it was necessary for us to assign it something very important called as input dimensions, which will obviously be the number of features that we have in our dataset. In order to incorporate the number of features in our dataset, we used len() method. We also trained our network with the help of a Kernel initializer. An initializer defines a way to set the initial random weights of the Keras layers. Kernel initializer generated tensors with a normal distribution (Table 2).

We checked the results for 10 hidden layers, and we observed that the accuracy remained the same (Fig. 3).

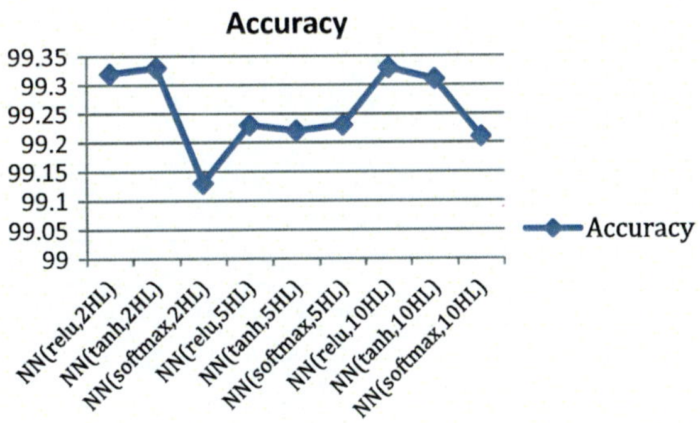

Fig. 3 Graph represents the accuracy values of a neural network on a novel dataset

Table 3 Previous results on pulmonary embolism

Author	Technique	Accuracy (%)
Rucco et al. [12]	Introduced an approach for the analysis of partial and incomplete dataset based on *Q* analysis	94
Agharezaei et al. [13]	Used ANN for prediction of pulmonary embolism	93.23
Chen et al. [14]	Deep learning CNN model	99
Weifang et al. [15]	Deep learning CNN model	92.6
Remy et al. [16]	Deep learning CNN model	99

Table 4 Results of previous research on cardiac arrest

Author	Technique	Accuracy (%)
Kannan et al. [17]	Logistic regression, random forest, stochastic gradient boosting, SVM	86, 80, 84, 79
Rahma et al. [18]	Hard voting ensemble method	90
Divya et al. [19]	Random forest, decision tree, *K*NN	96.80, 93.33, 91.52
Liaqat et al. [20]	Linear SVM	90
Ashier et al. [21]	RSA-based random forest	90

3.1 Results of Previous Research

Previous work has been done on cardiac arrest but without its association with pulmonary embolism. Results of cardiac arrest and pulmonary embolism have been displaced in Tables 3 and 4. Authors have attained an accuracy of 99%. Work has been carried on CSV dataset as well as on image files for detection of cardiovascular diseases and pulmonary embolism.

3.2 Overall Results of Proposed System

In this work, two medical fields have been combined in order to extract out the cause-and-effect relationship between them using machine learning (Table 5; Fig. 4).

Table 5 Overall results of proposed system

Classifier	Accuracy
Ensemble classifier	99.99
Neural network (relu)	99.33
Neural network (tanh)	99.33
Neural network (softmax)	99.32

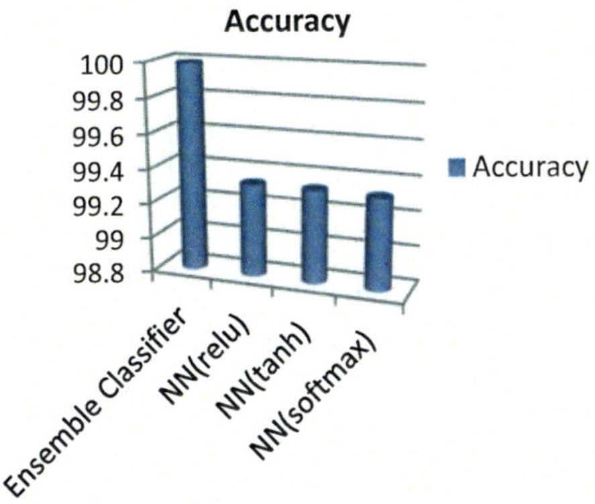

Fig. 4 Graph shows the comparison of accuracies of ensemble and neural network

4 Conclusion and Future Scope

This paper contributes the correlative application and analysis of distinct machine learning algorithms in the Python software. The aim of this work was the application of machine learning algorithms with divergent operational metrics and to ameliorate the efficiency by addressing the overfitting problem due to the Adaboost algorithm. We concluded that the ensemble classifier along with neural networks performed well and showed promising results as compared to single machine learning algorithms. Implementation of machine learning for identifying pulmonary embolism as a cause of cardiac arrest can greatly prove helpful in saving a number of lives. People particularly belonging to the lower section of the society need not to go for further investigations viz CT scans and other expensive medical examinations. So that the poor people who are not in a position to undertake such expensive medical treatments or investigations are saved from incurring heavy financial expenditure. Automating the medical procedure will also help medical representatives from maintaining bulk of data in the form of records of patients on this account. In the future, we will implement deep learning models on this novel dataset.

References

1. Chayakrit K, Zhang H (2017) Artificial Intelligence in Precision Cardiovascular Medicine. J Am Coll Cardiol 69(21):2657–2664
2. Shrestha S, Sengupta PP (2018) Machine learning for nuclear cardiology: the way forward
3. Shashikant R, Chetan Kumar P (2019) Predictive model of cardiac arrest in smokers using machine learning technique based on heart rate variability parameter. J Appl Comput Inf (2019)
4. Alizadehsani R, Habib J, Javad M, Hosseini MJ, Mashayekhi H, Boghrati R (2013) A data mining approach for diagnosis of coronary artery disease
5. Hinton G (2018) Deep learning: a technology with the potential to transform health care. JAMA 320(11):1101–1102
6. Matthews JC., McLaughlin V (2018) Acute right ventricular failure in the setting of acute pulmonary embolism or chronic pulmonary hypertension. Bentham Science Publication

7. Ebrahim LA (2018) Cardiac arrest due to pulmonary embolism. Indian Heart J 70(5):731–735
8. Bizopoulos P, Koutsouris D (2019) Deep learning in cardiology. IEEE Rev Biomed Eng 12:168–193
9. Kim J, Kang U, Lee Y (2017) Statistics and deep belief network based cardiovascular risk prediction. Healthc Inf Res 23(3):169–175
10. Cano-Espinosa C, Cazorla M, Gonzalez G (2020) Computed aided detection of pulmonary embolism using multi-slice multi-axial segmentation. MDPI
11. Singh S, Pandey S, Pawar U, Janghel RR (2018) Classification of ECG arrhythmia using recurrent neural networks. Proc Comput Sci 132:1290–1297
12. Rucco M, Sousa-Rodrigues D, Merelli E, Johnson JH (2015) A Neural hyper network approach for pulmonary embolism diagnosis. BMC Res Notes 8(1):617
13. Agharezaei L, Agharezaei Z, Nemati A, Bahaadinbeigy K, Keynia F, Baneshi MR (2016) The prediction of the risk level of pulmonary embolism and deep venous thrombosis through artificial neural network. Acta Inf Med 24(5):354–359
14. Chen MC, Ball RL, Yang L, Moradzadeh N, Chapman BE, Larson DB, Langlotz CP, Amrhein TJ, Lungren MP (2017) Deep learning to classify radiology free-text reports. Radiology 286(3):845–852
15. Liu W, Liu M (2020) Evaluation of acute pulmonary embolism and clot burden on CTPA with deep learning. In: Imaging informatics and artificial intelligence. Springer
16. Jardin R, Martin F (2020) Machine learning and deep neural network application in thorax. J Thorac Imaging 35(Suppl 1):S40–S48
17. Kannan R, Vasanthi V (2018) Machine learning algorithms with ROC curve for predicting and diagnosing heart disease. In: Springer briefs in applied science and technology
18. Atallah R, Mousa A (2019) Heart disease detection using machine learning majority voting ensemble method. IEEE
19. Krishnani D, Kumari DA (2019) Prediction of coronary heart disease using supervised machine learning algorithm. IEEE
20. Ali L, Khan SU (2019) Early detection of heart failure by reducing the time complexity of machine learning based predictive model. In: 1st international conference on electronics and computer engineering
21. Ashier Zhou S, Yongjian, L (2019) An intelligent learning system based on random search algorithm and optimized random forest model for improving heart disease detection. IEEE Xplore

Chapter 23
Conditional Generative Adversarial Network with One-Dimensional Self-attention for Speech Synthesis

Yash Javeri, Nirav Jain, and Sudhir Bagul

1 Introduction

Using computers to generate natural-sounding audio has been a research topic for quite a while now, but no satisfactory solution has been found yet. This research has applications in a wide variety of fields like entertainment, hospitals, etc. However, generating human-like audio is not an easy task because audio is a complex data format. It involves several intricacies like frequency and amplitude, which govern different aspects of an audio clip like pitch, loudness, etc. Also, the high temporal resolutions of an audio ranging from 16 to 192 kHz make the task even more difficult. Furthermore, the network must understand and establish meaningful long-term and short-term dependencies in an audio file.

Most approaches usually model a low-resolution representation like mel-spectrogram to overcome the challenges of generating raw audios. Even in the text to audio, the process is generally divided into two parts, converting text to an intermediate form and converting it to raw audio. Various methods have been tried to produce raw audio from intermediate forms like mel-spectrograms or aligned linguistic features. The method proposed in the paper combines ideas from MelGan [1] and Self-Attention GAN [2]. However, many different methods have been tried previously. The most generic way of making raw audio from such formats is by using signal processing methods like Griffin-Lim [3]. However, the problem with such methods is that they introduce a lot of noise, disturbances, and unclear sound. Another approach would be to use neural networks for the same. Models like WaveNet [4] produce state-of-the-art results but at the cost of slow inference speed. Similarly, parallel WaveNet [5] produces good results at the cost of high memory. Such solutions cannot be used in real-time applications and hence need improvement. One

Y. Javeri · N. Jain (✉) · S. Bagul
Computer Department, Dwarkadas J. Sanghvi College of Engineering, Vile Parle, Mumbai, India

S. Bagul
e-mail: sudhir.bagul@djsce.ac.in

A. Sheth et al. (eds.), *Intelligent Systems*, Algorithms for Intelligent Systems,
https://doi.org/10.1007/978-981-16-2248-9_23

model that introduces such improvements is MelGan, which is lightweight and easy to train. It uses GAN's to achieve its high and reliable performance. The use of GAN's to generate audio has never been explored before effectively, but the performance of GAN's in producing high-quality images motivates the research in this particular area.

The proposed paper focuses on improving the performance of MelGAN by introducing self-attention in MelGAN architecture. Self-attention has been used before in GANs to generate images. The addition of the self-attention layer has shown that GANs can produce images with improved quality. We try to achieve similar results for one-dimensional audio data by adding a self-attention layer in the generator of MelGAN. Self-attention GAN was used to establish long-range dependencies within an image. Such dependencies are present in a much more considerable amount in an audio file because human language is structured in such a way that each part of a sentence is related to some other part of the same sentence. Hence establishing such dependencies is an essential task in the audio generation and can be easily achieved using a one-dimensional self-attention layer. We train the proposed network on LJ Speech Dataset for 1000 epochs and then compare the results obtained with MelGAN.

2 Literature Review

In "MelGAN: Generative Adversarial Networks for Conditional Waveform Synthesis," the paper proposes to generate raw audio waveforms from mel-spectrograms by training GANs. Mel-spectrogram is given as an input to the generator, a fully convolutional feed-forward network, which produces the output. The generator consists of upsampling layers followed by residual blocks. These upsampling layers with transposed convolutional layers help in upsampling the input by 256×. The researchers adopt a multiscale architecture with three discriminators, each operating on different scales of audio. In order to test the significance of the various architectural decisions, the model is trained and unit-tested by removing the various key parts of the proposed architecture one-by-one. The observations show that dilated convolutional stacks, weight normalization, and the multiscale discriminator architecture (instead of a single discriminator) play a vital role in the performance of MelGAN. MelGAN can potentially be a plug-and-play replacement in various higher-level audio generation tasks like the [6] Universal Music Translation Network or the [7] vector-quantized VAEs.

In self-attention generative adversarial networks, the authors propose self-attention GANs to generate high-quality images. Normal GANs cannot establish long-range dependencies as they focus on only spatially local points of an image. The paper tries to overcome this limitation by mapping long-range dependencies in an image by using attention maps. In self-attention, the value at any position considers the features at all the positions. Hence, it provides a better representation of the overall image. The paper proposes to use self-attention in both the generator as well as the discriminator. The paper also uses spectral normalization in both the discriminator and the generator and helps improve the training dynamics by stabilizing the generator. It assists in preventing unusual gradients in the generator. The paper shows that the trained model performed best when the attention mechanism was applied to larger feature maps compared to smaller ones. It becomes easier to map dependencies as it has more evidence to choose from the image. The model shows promising results compared to previous state-of-the-art models with an Inception score of 52.52 and a Frechet Inception distance of 18.65.

The authors of the paper "WaveNet: A generative model for raw audio" propose a probabilistic and autoregressive model whose output depends not only on the current input but also on all the previous inputs. The model takes a mel-spectrogram as input and provides the corresponding audio for each time step as the output. As proposed by the authors, the most crucial aspect of WaveNet is the dilated causal convolutions, which help to correctly predict the audio at any given time steps without considering mel-spectrograms from any future time steps. The problem with causal convolutions is that they have a small receptive field. To overcome this obstacle, the authors propose using stacked dilated convolutions to increase the receptive field by orders of magnitude with just a few layers. The paper also proposes the use of softmax distributions with nonlinear quantization for better results. To further improve the results and training speed, the paper proposes gated activation units and residual connections. The model was used for three tasks: multi-speaker speech generation, text to speech, and music audio modeling. In terms of the naturality of audio, the model outperforms all its previous methods of text-to-speech generation and shows promising results in the other two tasks as well.

In WaveGlow [8], the researchers propose a flow-based network to generate audios from mel-spectrograms by combining ideas from Glow [9] and WaveNet. Unlike WaveNet, the paper posits and shows that the autoregressive model is unnecessary for synthesizing speech. WaveGlow is a generative model that samples a simple input distribution. The authors have used zero mean spherical Gaussian distribution with dimensions equal to the desired output's dimensions. In order to make the training stable and more straightforward, the model uses only a single network with one cost function that is "maximizing the negative likelihood of the training data." WaveGlow uses twelve coupling layers and twelve invertible convolution layers, with each having eight dilated convolution layers. Only with slight differences, WaveGlow has the maximum mean opinion score among other speech-synthesis architectures like WaveNet, Griffin-Lim. In terms of speed of inference speed, even the unoptimized implementation of WaveGlow is slightly faster than WaveNet with a synthesis rate of 500 kHz on an NVIDIA V100 GPU. The model has an advantage over others with respect to the speed of inference and simplicity of training.

3 Proposed Methodology

Figure 1 illustrates the proposed architecture of the generator of SA-MelGAN, designed by combining ideas from MelGAN and SAGAN, following the general

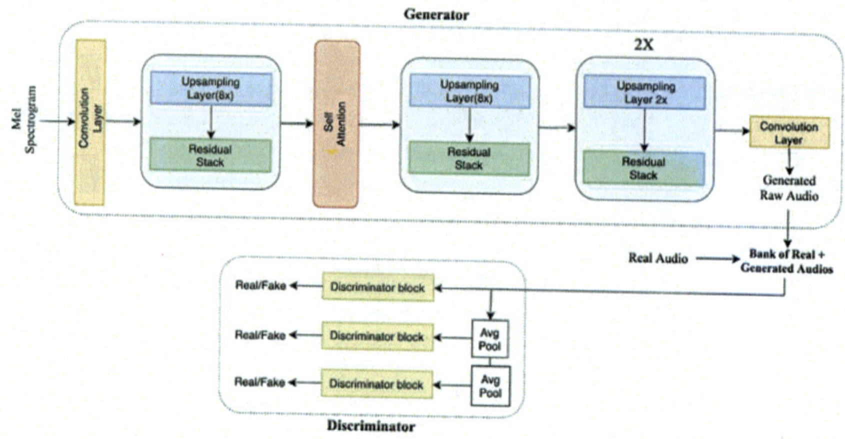

Fig. 1 Proposed SA-MelGAN architecture

adversarial game between the generator and the discriminator. As shown, we introduce a self-attention block in the model architecture in order to improve its performance. The mel-spectrogram is given input to the first one-dimensional convolution layer with a kernel of size 7. This output then undergoes a total of 256× upsampling in the consecutive blocks. The upsampling is done in four stages 8×, 8×, 2×, 2× by adjusting the stride values. Each upsampling layer consists of a one-dimensional transposed convolution layer with stride values depending upon the stage of upsampling and the kernel sizes equal to twice that of the stride. LeakyReLU has been used as the activation function in all the upsampling blocks. Each upsampling output is given to a residual stack block containing dilated convolutional layers. All the levels of the generator and the discriminator use 1D reflection padding to ensure that the outputs are of the required size.

3.1 *Residual Stack*

Each block consists of three dilated convolutional layers with dilation values equal to 3, 6, and 9, respectively, for the three layers. The dilated convolution later is followed by a regular convolutional layer with a kernel size of 1. This output is coupled with the initial input passing through a bunch of convolution layers. Dilated convolutions help increase the receptive field significantly without any extra computation power or loss of resolution.

3.2 Self-attention

The self-attention layer, as shown in Fig. 1, is added after the first residual block. We have used 1D self-attention, which takes an input having 256 channels and 80 × 1 dimensions obtained from the first residual stack. This input is first passed through three 1 × 1 convolutions, as shown in Fig. 2, to reduce the dimensionality of the input received from the previous layer.

Since the data is one-dimensional, we use 1D convolutions with kernel size 1 to achieve the same effect that a 2D 1 × 1 convolution has. More precisely,

$$\text{Query} = f(x) = W_f x \tag{1}$$

$$\text{Key} = g(x) = W_g x \tag{2}$$

$$\text{Value} = h(x) = W_h x \tag{3}$$

where $x \epsilon R^{CxN}$, $W_f \epsilon R^{\underline{C}xC}$, $W_g \epsilon R^{\underline{C}xC}$, $W_h \epsilon R^{\underline{C}xC}$.

$\underline{C} = C/8N$ is the dimension of the previous layer.

That is $N = L \times 1$ where L is the length of the 1-D array.

From Eqs. (1), (2), and (3), we have three outputs used to calculate the self-attention viz. query, key, and value. First, we transpose the query and matrix-multiply it with the key. Then, we pass this multiplied matrix through a softmax function to obtain the attention map.

$$s_{ij} = f(x_i)^T g(x_j) \tag{4}$$

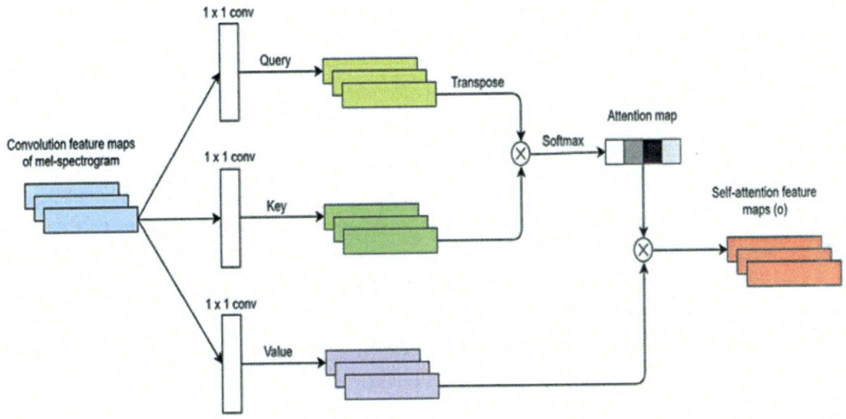

Fig. 2 One-dimensional self-attention

$$\beta_{j,i} = \frac{\exp(s_{ij})}{\sum_{i=1}^{N} \exp(s_{ij})} \tag{5}$$

The attention map obtained from Eqs. (4) and (5) is then matrix-multiplied with the value matrix to obtain the self-attention feature maps. Furthermore, this output is multiplied with a scalar (learned during training) and added back to the input as shown in Eqs. (6) and (7). The network then uses these feature maps to reconstruct the audio.

$$o_j = \sum_{i=1}^{N} \beta_{j,i} h(x_i), \quad o \epsilon R^{CxN} \tag{6}$$

$$y_i = \gamma o_i + x_i \tag{7}$$

All the convolutions used to calculate self-attention in our implementation are 1D because of the nature of the input from the previous layers. Also, the self-attention layer is located at the beginning to efficiently map the long-range relationships in the input mel-spectrogram in a memory-efficient way. For example, if mel-spectrogram of dimension (80, 150) is given as the input, the input dimension that the self-attention layer in the proposed model gets is (16, 256, 1200) [the format used is (batch_size, channels, dimensions)], which then converts to three matrices of size (16, 32, 1200) when we apply 1×1 convolutions. These three matrices are the query, key, and value matrices. After transposing the query and matrix multiplying it with the key, the size of the input (150) increases by eight times (1200), resulting in a matrix of the dimension (16, 1200, 1200), which corresponds to utilization of 171.66 MB of the GPU. On the other hand, if self-attention is embedded after the second residual stack, the input size increases by 64 times after matrix multiplication of query with the key, resulting in a matrix of dimension (16, 9600, 9600). This corresponds to 10.98 GB of GPU memory. Hence, the self-attention layer's proposed location is after the first residual stack, where matrix multiplications for calculating attention can be carried out more efficiently.

3.3 Normalization Technique

Which normalization technique to use is a vital decision. Image generation GAN architectures like [10] use instance normalization. However, according to MelGAN, instance normalization removes essential information related to the pitch and generates metallic audios. Spectral normalization also generates poor results, which affects the generator's feature mapping objective. Weight normalization works best in this case as it does not mitigate the discriminator's capacity in any way as it simply reparametrizes the weight matrices. Therefore, we use the weight normalization technique for the discriminator as well as for the generator.

3.4 Dataset and Preprocessing

We have used the LJ Speech Dataset [11], consisting of 13,100 audio clips, each varying in length from 1 to 10 s. All the clips are read by a single speaker from seven non-fiction books. We use a mel-spectrogram as an input to our model. It is a representation in which the audio signal in the time domain is mapped onto the mel-scale. First, the audio is converted into a frequency domain, and then, these frequencies are mapped to the mel-scale using a nonlinear transformation. The crucial information from audio, such as its loudness and amplitude, varies over time at different frequencies and can be efficiently obtained from a mel-spectrogram. To convert the audio into a mel-spectrogram, we have used a window length of 1024 and a hop length of 256, according to which the audio is sampled for computing the fast Fourier transform. The frequency spectrum is then divided into 80 mel channels that are then mapped to the mel-scale to obtain the mel-spectrogram. The proposed methodology uses audios with a sampling rate of 22,050 Hz. In order to have all the audios of the same length while training, we pad the wav file with zeros. Also, the frequencies of the generated mel-spectrogram are restricted between 0 and 8000 Hz. We divide the dataset into a 4:1 ratio for training and validation, respectively. Hence, out of 13,100 audio-mel pairs, 10,480 instances are used for training, while the remaining 2620 are used for validation.

3.5 Training

We use the hinge loss version of the GAN objective [12] for training as shown in Eqs. (8) and (9).

$$\min_{D_k} E_x[\min(0, 1 - D_k(x))] + E_{s,z}[\min(0, 1 + D_k(G(s, z)))], \quad \forall k = 1, 2, 3 \tag{8}$$

$$\min_{G} E_{s,z}\left[\sum_{k=1,2,3} -D_k(G(s, z))\right] \tag{9}$$

where x denotes the raw waveform, s denotes the conditioning information, and z represents the Gaussian noise vector.

To minimize the L1 distance between the discriminator feature maps of the real and synthetic audio, we also use feature matching objectives for the generator's training along with the discriminator's signal. L1 loss is not added in the audio space. This is because it introduces noise in the audio and affects its quality, as mentioned in work on MelGAN.

$$L_{\mathrm{FM}}(G, D_k) = E_{x,s\sim p_{\mathrm{data}}}\left[\sum_{i=1}^{T} \frac{1}{N_i} \left\| D_k^{(i)}(x) - D_k^{(i)}(G(s)) \right\|_1\right] \tag{10}$$

where $Dk(i)$ denotes the ith layer's feature map output of the kth discriminator block and N_i represents the number of units in each layer. We use the feature mapping from Eq. (10) at all the intermediate layers of each discriminator block.

In conclusion, to train the generator, we use the objective function as described in Eq. (11).

$\lambda = 10$ as in [13].

$$\min_G \left(E_{s,z} \left[\sum_{k=1,2,3} -D_k(G(s,z)) \right] + \lambda \sum_{k=1}^{3} L_{\mathrm{FM}}(G, D_k) \right) \quad (11)$$

Both the architectures are trained with a batch size of 16 on Google Colab, which provides an Nvidia K80 GPU with 12 GB of memory, 0.82 GHz clock speed, and 4.1 TFlops of performance. We use 'adam' as the optimizer with a learning rate of 0.001, $\beta 1$ as 0.5, and $\beta 2$ as 0.9.

4 Results

We carry out qualitative and quantitative analysis between our proposed architecture and MelGAN. From Figs. 3 and 5, it can be seen that as the discriminator loss decreases, i.e., as it becomes more accurate in distinguishing real and generated audio, the generator loss increases (Figs. 4 and 6), forcing the generator to produce even better results. This indicates that the architectures are correctly implemented, and the GANs are getting trained on the right track. Observing the change in loss values, as the models train progressively, it is seen that MelGANs loss values oscillate relatively more as compared to that of SA-MelGAN. From these observations, it can be concluded that SA-MelGAN trains are more stable than MelGAN.

In the case of MelGAN, at around 545, 100th step, the loss of the discriminator while training again starts increasing. This gradual increase in discriminator loss

Fig. 3 Discriminator loss for MelGAN

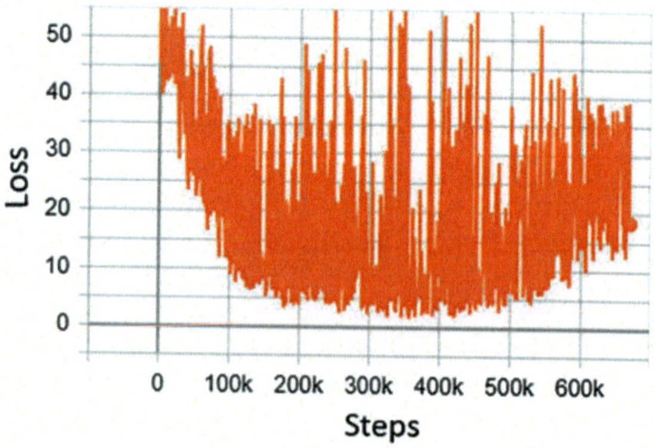

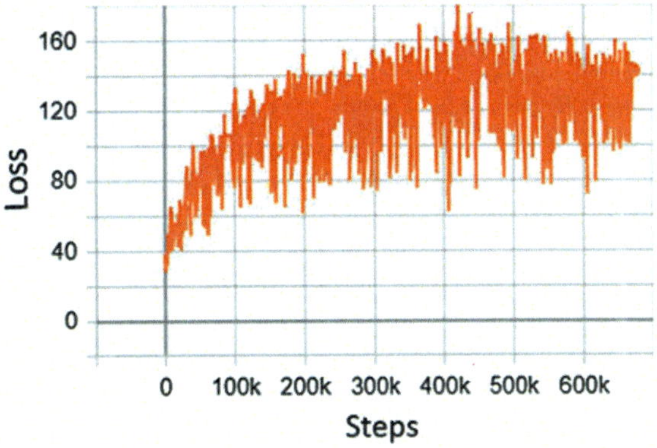

Fig. 4 Generator loss for MelGAN

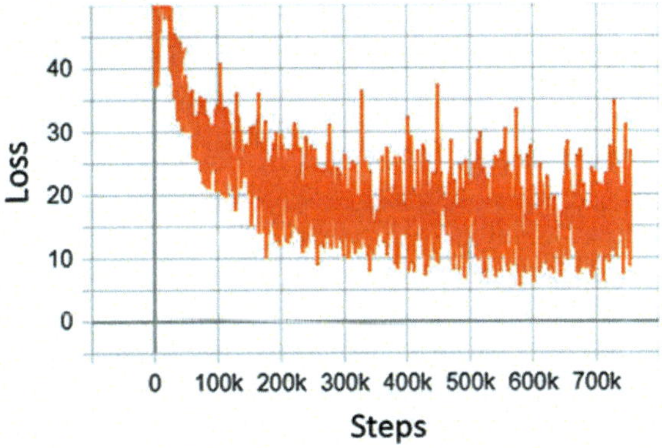

Fig. 5 Discriminator loss for SA-MelGAN

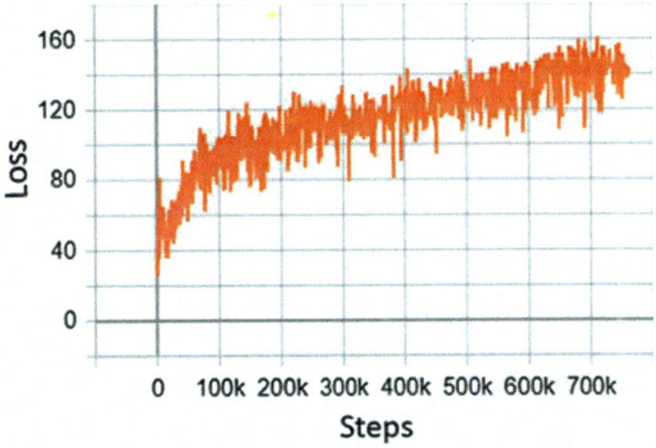

Fig. 6 Generator loss for SA-MelGAN

leads to some fluctuations in the generator as well. This increases the time taken by the GAN to reach the state of equilibrium. On the contrary, such changes are not seen in SA-MelGAN (Fig. 5).

Table 1 Mean opinion score (human evaluation) and mean opinion score-listening quality objective (MOS-LQO)

Category	Mean opinion score	MOS-LQO
Original	4.502	4.732
SA-MelGAN	3.646	3.298
MelGAN	3.346	3.207

For the qualitative analysis, we randomly select five audio samples from each category, viz. the reconstructed audios by MelGAN, reconstructed audios by SA-MelGAN, and original raw audios. These 15 audio clips are mixed and rearranged. We then ask 120 raters to rate these audio clips on a scale of 1–5, keeping the label (original or generated) of the audio hidden. In order to calculate the mean opinion score, we average the ratings within each category. The MOS for every category is mentioned in Table 1. Table 1 shows that the MOS for SA-MelGAN is greater than MelGAN by almost 8.94%. While few of the audios generated by SA-MelGAN sound less naturalistic than those generated by MelGAN, most of them generated by SA-MelGAN seem very close to the original ones.

We also calculate a mean opinion score-listening quality objective (MOS-LQO) score that ranges from 1 (worst) to 5 (best). It is an objective and full-reference metric for perceived audio qualities. We calculate this metric using the ViSQOL, i.e., Virtual Speech Quality Objective Listener [14], an open-source tool by Google. The main idea behind ViSQOL is that it calculates the spectro-temporal measure of similarity between the real and generated audios. Table 1 demonstrates the results received by averaging the MOS-LQO values of 20 random audio samples each generated by both architectures. Although SA-MelGANs MOS (human evaluation) was 0.3 greater than that of MelGAN, we did not observe a significant increase in the computer-generated MOS-LQO.

From Figs. 7 and 8, it can be posited that the amount of overall noise present in the audio generated by SA-MelGAN is not significantly different from that of audio generated by MelGAN architecture. However, the maximum of all the noise values throughout the audio in the case of SA-MelGAN is less than that in MelGAN

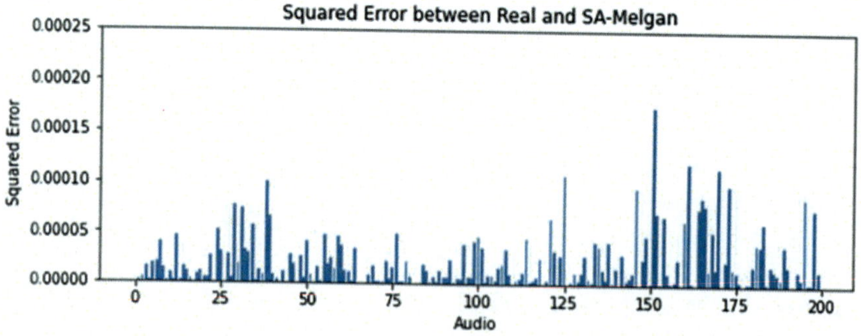

Fig. 7 Squared error plot for SA-MelGAN

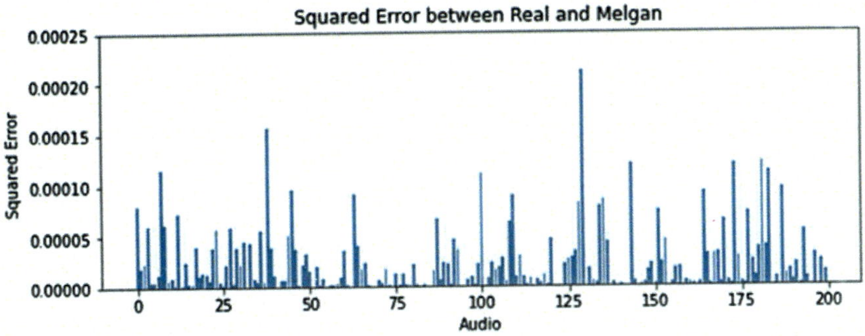

Fig. 8 Squared error plot for MelGAN

generated audios. Hence, the noise is not much noticeable to human ears in the case of SA-MelGAN. This attributes to the reason behind our better human-evaluated MOS compared to the MOS of MelGAN and also the reason behind the insignificant difference in MOS-LQO between the two architectures.

5 Conclusion

In this paper, we have introduced a modification to MelGAN. We have added a self-attention layer after the first residual stack, and we see an improvement in the mean opinion score, thus obtained compared to MelGAN. Although there is a notable increase in the mean opinion score, we observe only a slight increase in MOS-LQO. Also, the higher peaks in squared error in the case of MelGAN are the reasons for its low human-evaluated MOS but almost equal MOS-LQO. Hence, we conclude that the addition of a self-attention layer for audio generation in MelGAN improves its overall quality with minor errors. However, MelGAN produces audio with higher and noticeable fluctuations without this layer when trained till the 1000th epoch. Self-attention also facilitates a smooth and stable convergence with the addition of only a few parameters. Although there is a negligible difference in the inference time between the proposed model and MelGAN, it should be noted that the proposed model uses memory intensive calculations. Hence, further research is required in this field to reduce the memory requirements and improve the quality of the audio produced.

References

1. Kumar K, Kumar R, Boissiere TD, Gestin L, Teoh WZ, Sotelo J, Brébisson AD, Bengio Y, Courville AC (2019) MelGAN: generative adversarial networks for conditional waveform synthesis. ArXiv abs/1910.06711

2. Zhang H, Goodfellow IJ, Metaxas D, Odena A (2019) Self-attention generative adversarial networks. ArXiv abs/1805.08318
3. Griffin D, Lim J (1984) Signal estimation from modified short-time Fourier transform. IEEE Trans Acoust Speech Signal Process 32:236–243
4. Oord A, Dieleman S, Zen H, Simonyan K, Vinyals O, Graves A, Kalchbrenner N, Senior A, Kavukcuoglu K (2016) WaveNet: a generative model for raw audio. ArXiv abs/1609.03499
5. Oord A, Li Y, Babuschkin I, Simonyan K, Vinyals O, Kavukcuoglu K, van den Driessche G, Lockhart E, Cobo LC, Stimberg F, Casagrande N, Grewe D, Noury S, Dieleman S, Elsen E, Kalchbrenner N, Zen H, Graves A, King H, Walters T, Belov D, Hassabis D (2018) Parallel WaveNet: fast high-fidelity speech synthesis. ArXiv abs/1711.10433
6. Mor N, Wolf L, Polyak A, Taigman Y (2018) Autoencoder-based music translation
7. Oord A, Vinyals O, Kavukcuoglu K (2017) Neural discrete representation learning. NIPS
8. Prenger R, Valle R, Catanzaro B (2019) WaveGlow: a flow-based generative network for speech synthesis. In: ICASSP 2019—2019 IEEE international conference on acoustics, speech and signal processing (ICASSP), pp 3617–3621
9. Kingma DP, Dhariwal, P (2018) Glow: generative flow with invertible 1×1 convolutions. ArXiv abs/1807.03039
10. Isola P, Zhu J-Y, Zhou T, Efros AA (2017) Image-to-image translation with conditional adversarial networks. In: 2017 IEEE conference on computer vision and pattern recognition (CVPR), pp 5967–5976
11. Ito K, Johnson L (2017) The LJ speech dataset. https://keithito.com/LJ-Speech-Dataset/
12. Larsen ABL, Sønderby SK, Larochelle H, Winther O (2016) Autoencoding beyond pixels using a learned similarity metric. ArXiv abs/1512.09300
13. Wang T, Liu M-Y, Zhu J-Y, Tao A, Kautz J, Catanzaro B (2018) High-resolution image synthesis and semantic manipulation with conditional GANs. In: 2018 IEEE/CVF conference on computer vision and pattern recognition, pp 8798–8807
14. Chinen M, Lim FSC, Skoglund J, Gureev N, O'Gorman F, Hines A (2020) ViSQOL v3: an open-source production ready objective speech and audio metric. In: 2020 twelfth international conference on quality of multimedia experience (QoMEX), pp 1–6

Chapter 24
Comparative Study of Various Dehazing Algorithms

Nithish Praveen, Samridhi, and Manisha Chahande

1 Introduction

For various image processing projects and research activities, the image in discussion should be available in its pure form, i.e., free from any noise. Noise in discussion can be in any form, atmospheric light, smoke, fog, improper lighting conditions, etc. These noises called 'haze' can cause the processing techniques to provide us with incorrect readings or prevent us from getting the desired results. One of the most prominent image processing applications, 'object detection,' gets deeply affected with the addition of haze. The paper covers and compares various dehazing algorithms based on MSE [1], PSNR [1], and SSIM [1].

Outdoor images always gets disturbed by the addition of various noises, mainly atmospheric light scattering, fog, and smoke concepts. Therefore, to perform processing techniques, we need a haze free image. Dehazing algorithms can be classified on the amount of data required. Multiple image dehazing algorithms take 1 or more reference images and perform the algorithms on it. This algorithm may produce less errors but the fact that the reference image must be in its pure form is an obstacle as it is difficult to realize it practically. Multiple image dehazing algorithms cannot be an ideal algorithm when dealing with videos as sufficient data may always not be available. Taking the above problems into consideration, single image dehazing algorithms are proposed. Single image dehazing algorithms work by the comparisons of characteristics of a pure image with the target image. It may be obtained either by estimating the transmission map, atmospheric light estimation, or using neural networks.

N. Praveen (✉) · Samridhi · M. Chahande
Amity University Uttar Pradesh, Noida, Uttar Pradesh 201313, India

M. Chahande
e-mail: manisha_157@rediffmail.com

A. Sheth et al. (eds.), *Intelligent Systems*, Algorithms for Intelligent Systems,
https://doi.org/10.1007/978-981-16-2248-9_24

The comparison of these algorithms will be done by the parameters which are used to evaluate the quality of an image after an image processing algorithm has been performed.

The most used parameters to measure the quality of a processed image are:

Mean Square Error (MSE) [1]: Mean square error gives the value of the cumulative error between the original image and the processed image. It is the square of the average of the differences between the predicted value and the actual value. For an image to be of greater quality, its MSE should be as small as possible. The MSE of an image is calculated by the expression,

$$\mathrm{MSE} = \frac{1}{MN}\sum_{n=1}^{M}\sum_{m=1}^{N}[\hat{g}(n.m) - g(n,m)]^2 \tag{1}$$

where $\hat{g}$ (n, m) and $g(n, m)$ are two images and M and N are their sizes, respectively.

Peak Signal to Noise Ratio (PSNR) [1]: The PSNR measures the ratio of the signal power to its corresponding noise power in decibels (dB). In image processing, it is a performance measure of the reconstructed image. It determines the quality of an image, and for an image to be of better quality, PSNR should be high.

The expression to compute PSNR of an image from its MSE is given as,

$$\mathrm{PSNR} = 20\left(\frac{\mathrm{MAX_I}}{\sqrt{\mathrm{MSE}}}\right) \tag{2}$$

where $\mathrm{MAX_I}$ is the maximum pixel value possible, $\mathrm{MAX_{-I}}$ is given by,

$$\mathrm{MAX_I} = 2^B - 1 \tag{3}$$

where B is the number of bits per sample, usually images are encoded with 8 bits per sample, hence $\mathrm{MAX_I}$ becomes 255.

Structural Similarity Index Measure (SSIM) [1]: Structural similarity index measure predicts the perceived quality of images, and it measures the similarity between two images which has a strong role in image processing. The SSIM is given by,

$$\mathrm{SSIM}_{(x,y)} = \frac{(2\mu_x\mu_y + c_1)(2\sigma_{xy} + c_2)}{(\mu_x^2 + \mu_y^2 + c_1)(\sigma_x^2 + \sigma_y^2 + c_2)} \tag{4}$$

where

μ_x *is the average of* x_i
μ_y *is the average of* y_i
σ_x^2 *is the variance of* x_i
σ_y^2 *is the variance of* x_i
σ_{xy} *is the covariance of* x_i*and* y_i.

c_1 and c_2 are variables used to stabilize weak denominator division.

In this research, we compare three existing single image dehazing algorithms based on its values of MSE [1], PSNR [1], and SSIM [1] for post image processing applications.

Since the research is for dehazing images for various image processing projects, an analysis for computational time will also be done.

Some of the algorithms are dependent upon the system configuration, and the system configuration used during the project execution is:

Intel Core i3 6006U @2 GHz Processor.
8 GB Memory.
Intel HD Graphics 520 integrated graphics processing unit.
Windows 10 Home, 64 Bit as the operating system.

2 Image and Atmospheric Light [2, 4]

An image is a matrix of order $m \times n$, where m and n are the dimensions of the image, and pixels are the matrix values which takes the value between 0 and 256. Combination of different RGB levels gives the values of the pixels. Mathematically, an image is represented as,

$$I(x) = J(x)\mathrm{e}^{-\beta d(x)} + A\left(1 - \mathrm{e}^{-\beta d(x)}\right) \tag{5}$$

where I is the image obtained, x represents the coordinates of the image matrix. J is the haze free image. A is the global atmospheric light, β is the scattering coefficient of the image, and d represents the scene depth.

The term $\mathrm{e}^{-\beta d(x)}$ represents the transmission map t which is given by,

$$t(x) = \mathrm{e}^{-\beta d(x)} \tag{6}$$

Equation (5) can be rewritten as

$$I(x) = J(x)t(x) + A(1 - t(x)) \tag{7}$$

When the image under consideration is subjected to no haze then $I(x) = J(x)$, which implies β becomes 0. But in normal conditions, β becomes non-negligible and causes the degradation of the image. When the scene depth increases, the A (Atmospheric light) increases but the term $J(x)t(x)$ (Direct Attenuation) decreases. Dehazing of the image aims to obtain J from I; to attain this, we need to estimate the values of t and A, after this J can be obtained by,

$$J(x) = \frac{I(x) - A}{t(x)} + A \tag{8}$$

3 Dehazing Algorithms [4]

As discussed, for most image processing projects or applications, the images are to be converted into their true form, one such process includes the elimination of noise components from the image, image dehazing algorithms work to reduce/eliminate the noise components commonly referred to as 'haze.' The algorithms work in different methods, mostly manipulating the characteristics of hazy images to the ones observed in haze free images. The algorithms discussed as.

- Dark Channel Prior [2–4]
- Multi-scale Fusion [5]
- Color Attenuation Prior [6–8].

The reason for choosing these algorithms is they are simple to implement and can be used in any image processing project or application directly.

3.1 Dark Channel Prior (DCP) Dehazing Algorithm [2–4]

Dark channel prior (DCP) algorithm works with the characteristics of the dark channel of a haze free image, and it was noted in the literature survey that in the dark channel, the pixels take values which are close to zero, the pixels in which the pixel value is zero is due to presence of shadows and other dark colored objects, other objects take values which are close to zero. Atmospheric light tends to take values greater than zero, and DCP algorithms work to keep those pixels into a near zero value range. The DCP algorithm works in the following stages.

- Dark Channel Construction [2, 3]
- Atmospheric Light Estimation [2, 3]
- Transmission Map Construction [2, 3]
- Transmission Map Refinement [2, 3].

3.1.1 Dark Channel Construction [2, 3]

In order to start with the image dehazing with DCP, we need to find the dark channel of the input hazy image. The dark channel of a hazy image is given by:

$$J^{\text{dark}} = \left(J^{c}(y)\right) \quad (9)$$

where J^{dark} is the dark channel and J^{c} is the intensity of the color channel $c \in \{r, g, b\}$.

3.1.2 Estimating Atmospheric Light [2, 3]

When the dark channel is constructed, the pixels with above zero values consist of atmospheric lighting and bright objects in the scene. Estimation of bright pixels from the dark channel is given by,

$$A = I\left(\arg\max_{x}\left(I^{\text{dark}}(x)\right)\right) \tag{10}$$

where I^{dark} is the dark channel of the image. However, this results in a situation where the bright objects in the scene also get counted in the atmospheric light. To prevent this, we use a local entropy $E(x)$ which is given by,

$$E(x) = \sum_{i=0}^{N}(p_x(i) \times (p_x(i)) \tag{11}$$

where $p_x(i)$ represents the probability of a pixel value i in the image and N is the maximum pixel value. During the literature survey, it was noted that local entropy $E(x)$ is low for regions of smooth variations, i.e., the condition of haze-opaque elements is present in the image. Therefore, the pixels with low entropy are calculated for the estimation of atmospheric light.

3.1.3 Transmission Map Construction [2, 3]

After the estimation of the dark channel and the atmospheric light, according to Eq. (8), we need to construct the transmission map from the constructed dark channel to obtain the pure image. The construction of the transmission map from the dark channel of the image is given by,

$$\hat{t}(x) = 1 - \left(\frac{I^c}{A^c}\right) \tag{12}$$

But since the complete elimination may cause image quality to be of displeasing value to the eyes, we need to keep a small amount of haze by including a factor $\omega(0 < \omega \leq 1)$ in Eq. (12).

$$\hat{t}(x) = 1 - \omega\left(\left(\frac{I^c}{A^c}\right)\right) \tag{13}$$

3.1.4 Transmission Map Refinement [2, 3]

If the transmission map estimation is performed incorrectly, this can yield errors in the final image. Using Eq. (9), the final image may become blurry. To avoid this, several methods are devised to sharpen the transmission map. Common methods to achieve this are to pass through some filters. In the project, we used Laplacian filter [9] to refine our transmission map.

The dark channel dehazing algorithm is suitable for outdoor images as its dehazing algorithm is dependent on the estimation of atmospheric lighting. In our research, we find out the DCP algorithm is suitable when the image contains no white or bright element as the algorithm can assume those components to be a part of atmospheric light and can filter it out. Although the MSE [1], PSNR [1], and SSIM [1] values are above average, this algorithm is very much dependent on hardware. Several papers we studied stated the runtime was about 15–20 min, 0.5 s per second, to complete the project scaling down of images was required. Choice of filter can be a solution to reduce the computational time (Fig. 1).

3.2 *Multi-scale Fusion [5]*

In a fusion-based algorithm, we derive multiple inputs from a single image and fuse them, keeping only the important characteristics from it.

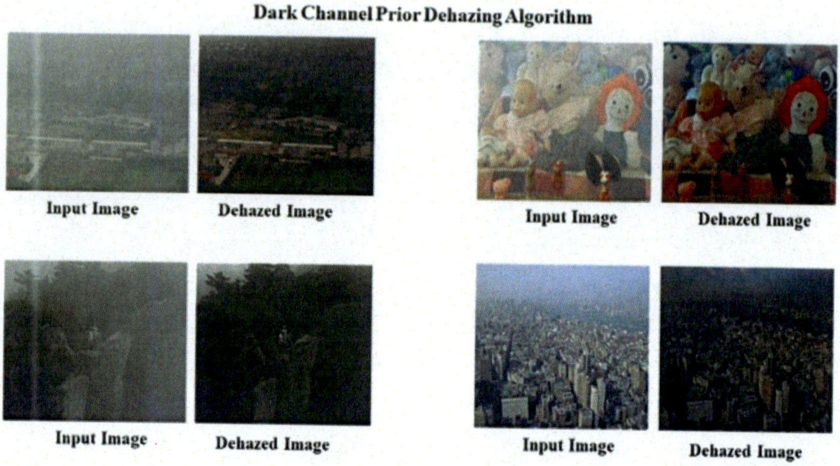

Fig. 1 Dark channel prior dehazing algorithm, the following figures shows the input hazy images and its corresponding output dehazed images

3.2.1 Inputs [5]

Input I_1 is derived from white balancing the original hazy input image, by doing this, we aim to eliminate chromatic casts which are the results of atmospheric color.

Input I_2 is derived by subtracting the average luminance of the image from the given image. Mathematically, it is given by,

$$I_2 = \gamma\left(I(x) - \tilde{I}\right) \tag{14}$$

where $\tilde{I}$ represents the average luminance of input hazy image $I(x)$.

Where γ represents a factor, which is used to increase the luminance of the recovered hazy image regions. By default, the value of γ is taken as 2.5.

3.2.2 Weight Maps [5]

In the literature survey, we find if using these two inputs only the images suffer from poor visibility in dense haze regions. The main reasons for it are the fact that optical density of hazy regions varies unevenly, and its effect on each pixel value is different. Another limitation is that the convolutional contrast enhancement techniques perform the same operation in a single image. To prevent this limitation, three weight maps are introduced. The objective of these three is to balance the input contribution and make sure regions with high contrast and high saliency in an input get higher values.

Luminance Weight Map [5]

This weight map measures the visibility of each pixel and assigns high values to the pixels with high visibility and low to the rest. One way to achieve this by calculating the loss of colorfulness as hazy regions tend to produce low saturation, the weight map using loss of colorfulness can be mathematically expressed by,

$$W_L^k = \sqrt{\frac{1}{3}\left[\left(R^k - L^k\right)^2 + \left(G^k - L^k\right)^2 + \left(B^k - L^k\right)^2\right]} \tag{15}$$

where k represents the indices of the derived inputs and the luminance weight map is the measure of deviations from the R, G, B channels.

Chromatic Weight Map [5]

This weight map controls the saturation gain in the resultant image, and it is since human eyes prefer high level of saturation, mathematically this weight map is calculated by,

$$W_C^k = \exp\left(-\frac{S^k(x) - S_{\max}^k}{2\sigma^2}\right)^2 \tag{16}$$

where S is the saturation level, k represents the indices of the derived inputs, and σ is the value of the standard deviation. $S_{\max}^k$ a constant which is dependent on the type of colour space employed.

Saliency Weight Map [5]

It is the degree of conspicuousness compared with respect to the neighborhood regions. Saliency tends to estimate the contrast of image regions by comparing the factors of neighboring regions based on orientation, and intensity, color.

The saliency weight at pixel position represented as (x, y) is given by,

$$W_S^k = \left\| I_k^{\omega hc}(x) - I_k^u \right\| \tag{17}$$

where I_k^u is the arithmetic mean pixel value of the input image I_k where k is the indices of the input image. $I_k^{\omega hc}(x)$ is the blurred version of the input by employing a small binomial kernel of value 5×5 with a high frequency cut-off which is given by,

$$\omega hc = \frac{\pi}{2.75} \tag{18}$$

When the values of $I_k^{\omega hc}$ and I_k^u are computed, then the saliency is computed as a per pixel fashion.

3.2.3 Multi-scale Fusion [5]

After the inputs and the weights have been calculated, fusion takes place. The final image output is represented by $F(x)$. This $F(x)$ is calculated by,

$$F(x) = \sum_k W^{-k}(x) I_k(x) \tag{19}$$

In the literature survey, we understand by using only the weight maps and input images, we get some halo effect which degrades the image quality, to prevent this from happening, we decompose the input image with a Laplacian pyramid [5, 10] and the weight map with a Gaussian pyramid [5, 10]. After applying the pyramid models, Eq. (19) can be rewritten as,

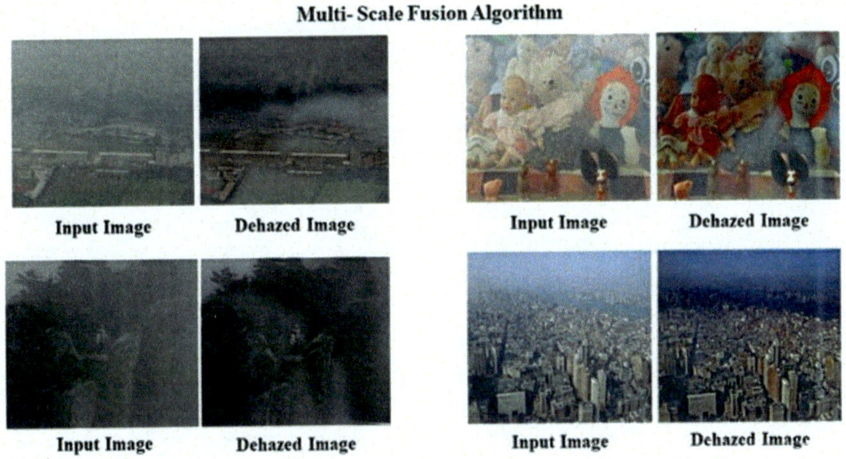

Fig. 2 Multi-scale fusion algorithm, the following figures shows the input hazy images and its corresponding output dehazed images

$$F_l(x) = \sum_k G_l\{W^{-k}(x)\}L_l\{I_k(x)\} \tag{20}$$

where l represents the level of the number of levels in the pyramid. The final output haze image is represented by $J(x)$ is the summation of the contributing of the resulting input pyramids. $J(x)$ is given by (Fig. 2)

$$J(x) = \sum_l F_l(x) \uparrow^d \tag{21}$$

3.3 *Color Attenuation Prior (CAP) [6–8]*

Color attenuation prior (CAP) algorithm is quite like dark channel prior algorithm; but in DCP algorithm, the hazy regions are obtained from the dark channel, and the concept in CAP algorithm is that image saturation and image brightness vary greatly in the regions of haze and haze free areas of an image. In CAP algorithm,

- We find the atmospheric scattering model (ASM) [6].
- Then, we use haze lines to estimate the atmospheric light A [6].
- Estimate the exponential model of dynamic atmosphere scattering coefficient β [2–4].
- Use these parameters in our ASM to get the haze free image [6].

In CAP-based algorithms, the image equations are the same as discussed in dark channel prior algorithm [2–4]. The ASM model is represented by Eq. (5). From the definition given above, we can infer that,

$$d(x) \propto c(x) \propto v(x) - s(x) \tag{22}$$

where $d(x)$ is the scene depth, $c(x)$ is the concentration of haze, $v(x)$ is the intensity of brightness, and $s(x)$ is intensity of saturation.

In CAP [6]-based algorithm, the scene depth is calculated by the equation

$$d(x) = \theta_o + \theta_1 v(x) + \theta_2 s(x) + \varepsilon(x) \tag{23}$$

where θ_o, θ_1, and θ_2 are some unknown parameters and $\varepsilon(x)$ is some random noise which follows the Gaussian distribution model $N(0, \sigma^2)$ where σ^2 is the variance.

3.4 Haze Lines [6]

For the estimation of atmospheric light, CAP algorithm employs the use of haze lines, the haze lines are the pixels in a hazy region in the *RGB* color space, and these points are modeled as some lines, and their point of intersection is the value of atmospheric light A. This approach was found more accurate rather than the estimation of top $p\%$ [2–4, 6] values for atmospheric light estimation.

3.5 Dynamic Atmospheric Scattering Coefficient Β [6–8]

The ASM model described in Eq. (5) consists of a term β which is known as dynamic atmospheric scattering coefficient. In most of the cases, β is considered as a constant, but in cases when this β is smaller than the actual situation, we might experience incomplete haze removal. To prevent this, the value of β must be calculated, and β is calculated by the expression.

$$\beta = a.e^{b.d(x)} \tag{24}$$

where a and b are two unknown constants. We find that β depends on screen depth.

After the values of β and A are estimated, the scene radiance and transmission map are estimated, we put these values in the ASM model which is described in Eq. (5), and from it, we can get the haze free image which is described in Eq. (8) (Fig. 3).

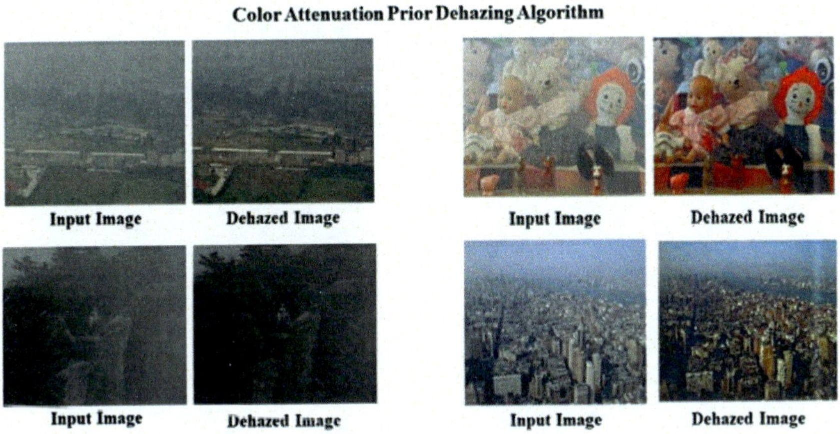

Fig. 3 Color attenuation prior dehazing algorithm, the following figures shows the input hazy images and its corresponding output dehazed images

4 Quantitative Analysis

After studying some of the image dehazing algorithms, we implemented them on MATLAB 2019a and found out the values of MSE [1], PSNR [1], and SSIM [1] of the processed images. We ran the above-mentioned algorithms [9–11] on a set of 12 hazy images and found out those values. The values of MSE [1], PSNR [1], and SSIM [1] are calculated by Eqs. (1), (2), and (4), respectively.

The dataset we used is the MATLAB implementation of single image haze removal using dark channel. He et al. [12], the images used in this paper are the standard images used in various papers of single image dehazing (Table 1).

Dark channel prior [4] offers good values of MSE [1], PSNR [1], and SSIM [1]; but due to its high dependence on hardware and high computational time, this dehazing algorithm is not suitable for real-time data or further projects with it. We find the CAP algorithm very efficient, with its suitable values and less computational times, but it has limited enhancement as it uses hard thresholding. We find that the methods used in these algorithms does not have a single approach but multiple approaches to it, take for example the DCP-based algorithm in the transmission map which works

Table 1 Average values of MSE, PSNR, and SSIM obtained after running the image dehazing algorithms on a set of 12 hazy images

Algorithm	MSE	PSNR (dB)	SSIM
DCP	0.033858333	64.04938	0.686554005
Multi-scale fusion	0.051325	61.60114167	0.439371239
CAP	0.035266667	63.04431667	0.748788575

differently based on the choice of filters, same goes for multi-scale fusion algorithm when we have a choice of methods for image input decomposition and weight map decomposition. The computational time issue of the dark channel prior image dehazing algorithm can be resolved by having a better choice of hardware, probably using dedicated graphic processing units (GPU) and solid state drives (SSD).

5 Conclusion

We performed quantitative analysis of three single image dehazing algorithms and emphasized the need of single image dehazing over multi-image dehazing. Apart from quantitative analysis, we performed some visual analysis of the above-mentioned algorithms and found out about the different approaches for a particular image dehazing algorithm and learnt how parameters vary from the choices of the filters and decomposition pyramids.

All the codes and materials are used in accordance with fair use, and proper citations have been given.

References

1. University of Edinburgh homepage. http://homepages.inf.ed.ac.uk/rbf/CVonline/LOCAL/COPIES/VELDZEN/node18.html
2. Lee S, Yun S, Nam J et al (2016) A review on dark channel prior based image dehazing algorithms. J Image Video Proc 2016:4. https://doi.org/10.1186/s13640-016-0104-y
3. He K, Sun J, Tang X (2011) Single image haze removal using dark channel prior. IEEE Trans Pattern Anal Mach Intell 33(12):2341–2353. https://doi.org/10.1109/TPAMI.2010.168
4. Parihar AS, Gupta YK, Singodia Y, Singh V, Singh K (2020) A comparative study of image Dehazing algorithms. In: 2020 5th international conference on communication and electronics systems (ICCES). Coimbatore, India, pp 766–771. https://doi.org/10.1109/ICCES48766.2020.9138037
5. Ancuti CO, Ancuti C (2013) Single image Dehazing by multi-scale fusion. IEEE Trans Image Process 22(8):3271–3282. https://doi.org/10.1109/TIP.2013.2262284
6. Aiswarya Menon N, Anusree KS, Jerome A, Sreekumar K (2020) An enhanced digital image processing based Dehazing techniques for haze removal. In: 2020 fourth international conference on computing methodologies and communication (ICCMC). Erode, India, pp 789–793. https://doi.org/10.1109/ICCMC48092.2020.ICCMC-000147
7. Wang Q, Zhao L, Tang G, Zhao H, Zhang X (2019) Single-image Dehazing using color attenuation prior based on haze-lines. In: 2019 IEEE international conference on big data (big data). Los Angeles, CA, USA, 2019, pp 5080–5087. https://doi.org/10.1109/BigData47090.2019.9005603
8. Zhu Q, Mai J, Shao L (2015) A fast single image haze removal algorithm using color attenuation prior. IEEE Trans Image Process 24(11):3522–3533. https://doi.org/10.1109/TIP.2015.2446191
9. Sreekuttan AM (2020) Image dehazing.zip. MATLAB Central File Exchange. https://www.mathworks.com/matlabcentral/fileexchange/47147-image-dehazing-zip. Retrieved 4 Oct 2020
10. Github homepage. https://github.com/JiamingMai/Color-Attenuation-Prior-Dehazing

11. Stoica A (2020) Single image haze removal using dark channel prior. MATLAB Central File Exchange. https://www.mathworks.com/matlabcentral/fileexchange/46147-single-image-haze-removal-using-dark-channel-prior. Retrieved 4 Oct 2020
12. He PK, Sun J, Tang X (2011) MATLAB implementation of single image haze removal using dark channel. IEEE Trans Pattern Anal Mach Intell 30(12):2341–2353

Chapter 25
Real-Time Model for Preventive, Emergency and Restorative Controls of Grid-Based Power System Network

Youssef Mobarak, Nithiyananthan Kannan, and Mohammed Almazroi

1 Introduction

AC systems have been placed on DC systems since the massification of electricity usage due to factors such as their ease of transforming voltage levels and the simplicity and economy of AC machines against DC machines. In order to maintain safety and reliability in the operation of the system [1, 2], the behavior of the frequency and voltage control systems is important. The efficient operation of interconnected power systems includes the combination of total generation with total load demand and system losses associated with it. The operating point of a power system varies over time, and so, the system can experience fluctuations in the nominal frequency of the system and scheduled power exchanges to other areas that may cause undesirable effects [3, 4]. Currently, the regulation of generator excitation is the regulation of generator output power and also affects the stability of the entire electric power system. Exciter is the electrical power source for the generator's field winding and is realized as a separate DC or AC generator. AVR [5] is regulated by eexciter. VAR compensation requires reactive power control for enhancing the efficiency of the electric power system. Reactive power control addresses power quality issues such as the maintenance of the flat voltage profile at all stages of power transmission, power factor enhancement, transmission reliability, and system stability [6]. To alter the natural electrical properties of electric power systems, series and shunt

Y. Mobarak · N. Kannan (✉) · M. Almazroi
Electrical Engineering Department, Faculty of Engineering, Rabigh, King Abdulaziz University, Jeddah 560 037, Saudi Arabia
e-mail: nmajaknap@kau.edu.sa

Y. Mobarak
e-mail: ysoliman@kau.edu.sa

M. Almazroi
e-mail: mhommedalmazroi@stu.kau.edu.sa

A. Sheth et al. (eds.), *Intelligent Systems*, Algorithms for Intelligent Systems,
https://doi.org/10.1007/978-981-16-2248-9_25

VAR compensation techniques are used. Series compensation modifies the transmission or delivery system reactance parameter, while shunt compensation modifies the corresponding load impedance. In both cases, it is possible to control the reactive power of the line efficiently, thereby improving the efficiency of the overall electrical power system.

2 System Modeling

This section discusses the static state estimation procedure for power systems, handling errors, and bad data in static state estimation for power system network and normal/alert state in an intelligent power system control.

2.1 *Static State Estimation Procedure for Power Systems*

In an informal fashion, it has been tried to understand the procedure with a basic power system with two buses as shown in Fig. 1. Consider the simple system shown below: The following measurements are acquired. P_1, Q_1, P_2, Q_2, V_1, V_2: i.e., the real and reactive flows on both ends of the line and the voltage magnitudes at the two ends. If $\delta_1 - \delta_2$ is known, then the system is fully estimated. It has been assumed that the parameters of the line are known fairly accurately. If all measurements are accurate, then it should be possible to obtain $\delta_1 - \delta_2$ quite easily by the following formula as shown in Eqs. (1) and (2) for real and reactive power of the one bus and Eqs. (3) and (4) for real and reactive power for bus 2:

$$P_1 = \frac{V_1 V_1}{Z} \cos\cos\theta - \frac{V_1 V_2}{Z} \cos\cos(\theta + \delta_1 - \delta_2) \tag{1}$$

$$Q_1 = \frac{V_1 V_1}{Z} \sin\sin\theta - \frac{V_1 V_2}{Z} \sin\sin(\theta + \delta_1 - \delta_2) \tag{2}$$

$$P_2 = \frac{V_2 V_2}{Z} \cos\cos\theta - \frac{V_1 V_2}{Z} \cos\cos(\theta + \delta_2 - \delta_1) \tag{3}$$

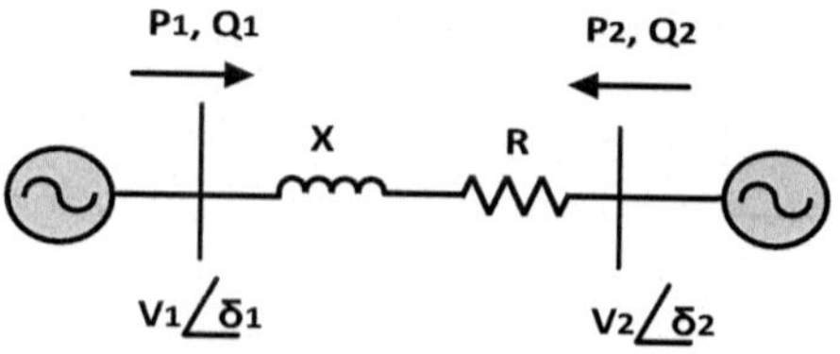

Fig. 1 Two bus system

$$Q_2 = \frac{V_2 V_2}{Z} \sin \sin \theta - \frac{V_1 V_2}{Z} \sin \sin(\theta + \delta_2 - \delta_1) \tag{4}$$

where $Z = |R + jX|$ and $\theta = \arctan(X/R)$. However, if there is a measurement error in P_1, V_1, or V_2, then $\delta_1 - \delta_2$ will be inaccurate. So, it is best to cross-check the results with other formulae.

2.2 *Handling Errors and Bad Data in Static State Estimation for Power System Network*

One possibility is to take the average value of $\delta_1 - \delta_2$ obtained from formulae for P_1 and P_2. This way, one does not place too much faith on one measurement. Alternatively, one could try to obtain a value of $\delta_1 - \delta_2$ such that plugging that value will minimize the following error J as shown in Eq. (5):

$$\begin{aligned} J = {} & k_1\left(P_1 - \frac{V_1 V_1}{Z} \cos \cos \theta + \frac{V_1 V_2}{Z} \cos \cos(\theta + \delta_1 - \delta_2)\right)^2 \\ & + k_2\left(Q_1 - \frac{V_1 V_1}{Z} \sin \sin \theta + \frac{V_1 V_2}{Z} \sin \sin(\theta + \delta_1 - \delta_2)\right)^2 \\ & + k_3\left(P_2 - \frac{V_2 V_1}{Z} \cos \cos \theta + \frac{V_1 V_2}{Z} \cos \cos(\theta + \delta_2 - \delta_1)\right)^2 \\ & + k_4\left(Q_2 - \frac{V_2 V_2}{Z} \sin \sin \theta + \frac{V_1 V_2}{Z} \sin \sin(\theta + \delta_2 - \delta_1)\right)^2 \end{aligned} \tag{5}$$

where k_1, k_2, k_3, and k_4 are positive weights.

Note that J is always greater than or equal to zero. Minimization of J, results in the "least square" deviation from the measured values. The main problem, however, is to choose the weights k_1, k_2, k_3, and k_4 appropriately. For similar quantities (power, reactive power), one can choose weights to be equal. Minimizing J requires using an optimization procedure as outlined in the previous module. Both the approaches [(1) averaging and (2) minimizing J] are likely to work well if errors are small in magnitude. If there exist gross errors in the data (which occur, say, due to failure of communication), then the above approaches may fail.

2.3 *Normal and Alert State in an Intelligent Power System Control*

To distinguish among a standard state and a warning state, a device operator brings out the following cases using the network formation, capacity, and generation standards attained as of a static state estimation procedure. Static security study includes inspection for apparatus boundary damages, if one of the elements of the network/load/generation configuration existing at that point of time were to be tripped due to some contingency. Note that this element is not actually tripped by an operator, but only simulated using a computer program (essentially a load-flow study which computes the steady-state power flows in transmission lines, generator real and reactive power output, and voltages at various nodes for such a tripping). Dynamic security analysis: this involves checking the stability of the system, if one of the elements of the network/load/generation configuration existing at that point of time was to be tripped due to some contingency. The exact nature of the contingency can impact the transient behavior. For example, the contingency could be due to a single phase to ground fault which results in protective action (circuit breakers disconnecting the faulted element) within, say, 0.1 s. Note again, that this element is not actually tripped by an operator, but only simulated using a computer transient analysis program (which essentially does a numerical integration of the differential equations which describe the system).

A computer program which checks for angular stability requires a significantly large amount of computation time. Figure 2 indicates the entire procedure on power system security procedure. First step is to collect data from the actuation of stored data, and it will run the state estimation for that data. After that contingency analysis will be carried initially with static data and later with dynamic data received through remote terminals. Based on the outages, the list of critical contingencies will be listed based on the severity of the outages. Therefore, it is not implemented in most load dispatch centers at present. It is important to carefully choose the element

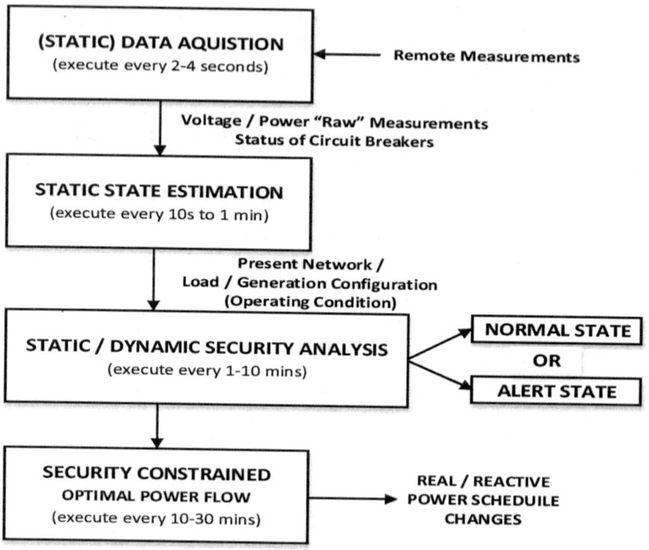

Fig. 2 Schematic of security assessment procedure

whose outage is to be simulated since the number of elements in a power system are too numerous for all of them to be considered one by one. Usually, a set of critical elements are chosen by some rough screening based on an operator's experience, and the security analyses are carried out for the outage of these elements. If the security analysis shows that the system is secure, it is classified as a normal state. If the state is normal, then a system operator may wish to do some minor changes in real and reactive scheduling (from an economic perspective), if such flexibility exists. However, any such change should not bring the system out of the secure state. If the system is not secure (alert), then the operator has to try to steer it into the secure state by real or reactive power re-scheduling (preventive control). However, note that this re-scheduling is done to improve security and may result in higher cost if cheaper generators are asked to "back down" their generated power, while costlier ones are ramped up. Therefore, even if preventive control is to be done, it should be done in a way which will minimize any cost increase while simultaneously ensuring security. This is done using a security constrained optimal power flow program. A schematic of the procedure is shown in Fig. 2. It is clear that operating condition 2 is not a secure operating condition (i.e., the system is in "alert state") Therefore, in order to bring the system back to normal state, a system operator has to reschedule the generation. System security cannot be assessed by only considering post-contingency steady-state power flows. A system could be unstable for a disturbance even if a post-disturbance steady state exists and power flows and voltages for that steady state are within equipment limits. If a system is unstable, it will not settle down to that steady state. The assessment of dynamic security (stability) is a more complex task as it requires numerical integration of the system dynamic equations (e.g., swing equations of all generators). This is computationally intensive, and many probable contingencies have to be considered. Direct numerical integration of differential equations can be avoided if one uses criteria like "equal area criterion" to adjudge system angular stability. However, equal area criterion cannot be extended in a straightforward manner for multi-machine systems with detailed models of all system components. Therefore, quick assessment of dynamic stability is still a challenge to system engineers.

3 Angular Instability and Emergency Control of a Power System

A real-time situation has been considered, wherein an emergency is caused by loss of synchronism between generators. Consider a four generator system as shown in Fig. 3. It has been assumed that the loads at the two buses shown in Fig. 3 are not voltage dependent and that losses in the system are negligible. Suppose that a fault occurs (e.g., a short circuit between phase to ground) on one line which is carrying 500 MW, this is detected by relays at both ends of the line, and they send a trip signal to the circuit breakers which disconnect this line. This is usually done in a

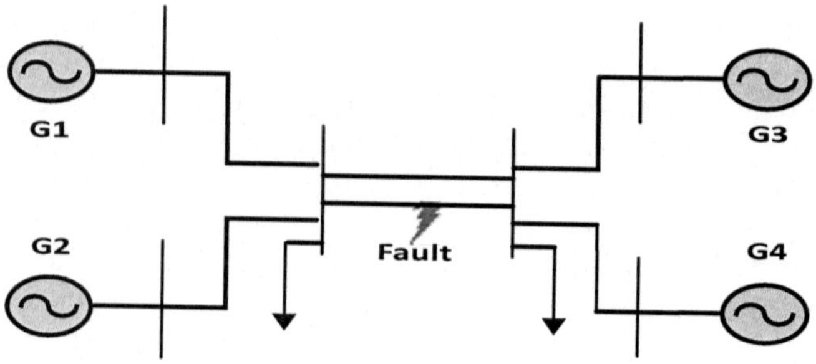

Fig. 3 Four generator system

very short time (about 100 ms). The loss of this line causes power to get diverted to the parallel line. While the steady-state scenario following loss of line does not cause violation of equipment limits, the question arises: The worrisome aspect of this disturbance is that the transient behavior of the relative generator speeds after the disturbance may not be stable. During the fault, generator speeds deviate due to a sudden change in the electrical power due to the fault. After the line is tripped, the generator speeds do not directly go into steady state because of the deviation caused due to a fault. However, if the relative angular differences between the machines are not too large, then the electrical torques try to pull all the generator speeds together in synchronism, i.e., the system is angular stable. The typical waveforms of generator speeds, power flow (*P*), and the phase angular difference between the bus voltages at both ends of the line are shown below. The power oscillates and eventually settles at a constant value after some time (damper windings in generators can contribute to damping these oscillations). Also, all generator speeds reach the same value. The typical waveforms have been seen when a system is angular stable after a large disturbance like a fault. If relative speed and angular deviations are excessive, then the restorative torque may become negative. Under this situation, the generators are unable to pull back into synchronism (angular instability).

While the generators are all connected to one another, the speeds do not settle down to a common value, as shown below. This causes large changes in power and voltages in the interconnecting lines which cannot be tolerated for long. It becomes necessary to split (island) the system such that generators with large differences in speeds are not connected to each other. Usually, distance relays mistake these large power and voltage swings for a fault and cause interconnecting lines to trip. Thus, a "natural" uncontrolled splitting occurs. It would be desirable if the evolving transients were swiftly and correctly diagnosed as being unstable, and system splitting is done "gracefully," i.e., in a controlled fashion by tripping a preselected set of lines. If wide-area synchronized measurements are available, then they can aid the diagnosis of the evolving transients. For example, which it has been considered, the tripping of the line shown splits the system into two islands. One of the islands has less generation. In order to prevent a large and sudden drop of frequency, some load is shed by sensing the frequency (under-frequency load shedding). Stabilizing the voltage and frequency in an island is not easy due to the reduced cumulative generator inertia (due to lesser number of generators) and possibly lower real and reactive power reserve margins. In an over-generated island, generation has to be reduced quickly to stabilize frequency.

4 Emergency Control of a Studied Power System

Consider the two machine systems shown in Fig. 4 and there is a fault on one of the connecting lines. This fault is clean by tripping the lines using circuit breakers which are activated by defending relays.

In order to understand the various possibilities, the system considered is shown below which is modeled as follows: A generator in this example is modeled as a voltage source (constant magnitude) behind a reactance. The angle dynamics are described by the swing Eqs. (6) and (7):

$$\frac{\mathrm{d}(\omega_i - \omega_0)}{\mathrm{d}t} = \frac{\mathrm{d}\omega_i}{\mathrm{d}t} = \frac{\omega_B}{2H_i}(P_{mi} - P_{ei}) \tag{6}$$

$$\frac{\mathrm{d}\delta_i}{\mathrm{d}t} = (\omega_i - \omega_0) \tag{7}$$

where $i = 1$ and 2 for generator 1 and 2, respectively. $(\omega_i - \omega_0)$ is the speed deviation from nominal, $(P_{mi} - P_{ei})$ is the difference in mechanical input and electrical output power. The electrical power is obtained from the circuit solution for the figure shown above. ω_B is the base (nominal frequency), and H_i is the inertia constant. The value of the rotor angle and speed is obtained by numerical integration (e.g., Runge Kutta method) of the equations. Caution: In actual practice, a generator model is much more complicated due to the dynamics of stator and field fluxes and the excitation system. The loads are assumed to be resistance type (no frequency dependence and unity power factor). The values of various parameters are shown in the figure, and the swing equations are $P_{L1} = 0.63$ pu, $P_{L2} = 1.27$ pu, $x_{e1} = x_{e2} = 0$, $x_g = 0.25$ pu,

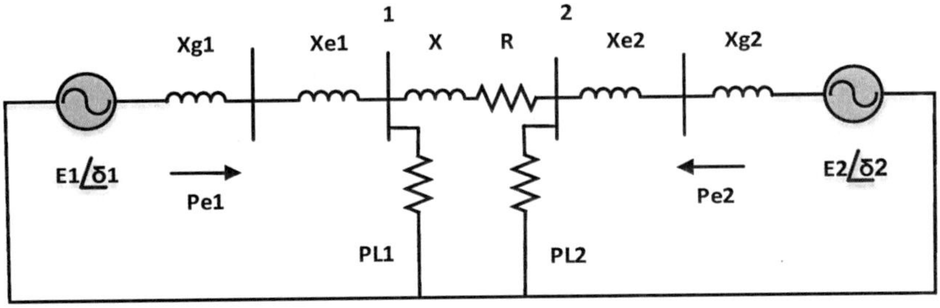

Fig. 4 Two machine systems

$H_1 = H_2 = 6$ MJ/MVA, and $w_o = w_B = 2p50$ rad/s. Initial operating conditions: $V_1 = V_2 = 1.0$ and $P_{m1} = P_{m2} = 0.95$ pu. Note that initially power flows from bus 2 to bus 1 via two parallel lines (shown in the figure above as one equivalent line with impedance $R + jX$). Prefault: Two identical lines in parallel: $R = 0$, $X = 0.5$. Three Phase Fault: On one of the parallel lines at bus 1 and lasts for T_{clear} seconds. Post Fault: The faulted line is tripped. Therefore after the fault, $R = 0$ and $X = 1.0$.

5 Results and Discussion

Real-time Case 1: T_{clear} = 0.1 s: In this case, the disturbance does not lead to a loss of synchronism. However, power, frequency, and voltage undergo "swings" which are caused by relative motions of the generator rotors. These swings are acceptable and will usually die down due to the effect of damper and field windings in a generator, these swings do not die down but grow with time. This is due to the effect of large gain feedback controllers—like voltage regulators in a generator excitation system as shown in Fig. 5. Note: Due to slightly lower voltages at the buses 1 and 2 after tripping of the line, the total power drawn by the loads decreases.

Since the mechanical power is not changed, the frequency of both generators keeps increasing. The frequencies will settle down to a value greater than 50 Hz if the loads are frequency dependent or if governors adjust the mechanical input to the generators (not considered here).

Real-time Case 2: T_{clear} = 0.6 s: In this case, the disturbance causes the two generators to lose synchronism. The frequencies of the two generators "separate out," and electrical power and voltage undergo violent pulsations as shown in Fig. 6 (normally, this will not be allowed to continue).

Real-time Case 3: T_{clear} = 0.6 s, T_{isl}= 0.3 s: In this case, the disturbance leads to a loss of synchronism, but the two generators are separated by disconnecting the remaining line connecting them 0.3 s after the faulted line is cleared as shown in Fig. 7. Frequency in the islands, one of which has excess load and the other which has surplus generation, changes very rapidly due to the large imbalance, calling for quick measures.

Real-time Case 4: T_{clear} = 0.6 s, T_{isl}= 0.3 s, Frequency Controls enabled: The effect of two (idealized) emergency frequency control schemes is shown here: Shedding of some load when frequency <48.5 Hz, in the under generated island, and fast acting overspending control to reduce mechanical power at frequency >52 Hz in the over-generated island, prevents large frequency deviations as shown in Fig. 8.

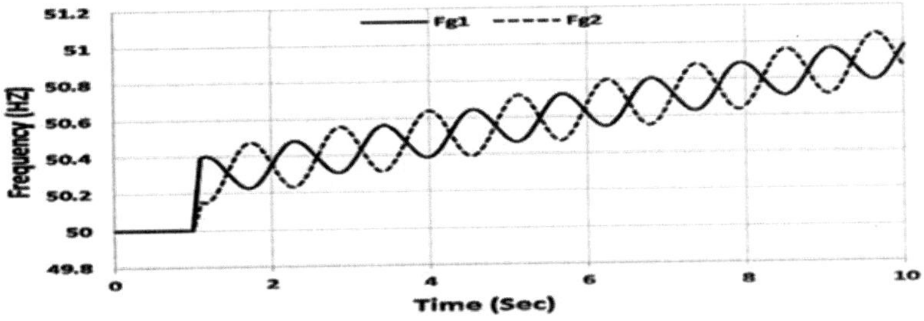

(a) Generator frequency (Hz)

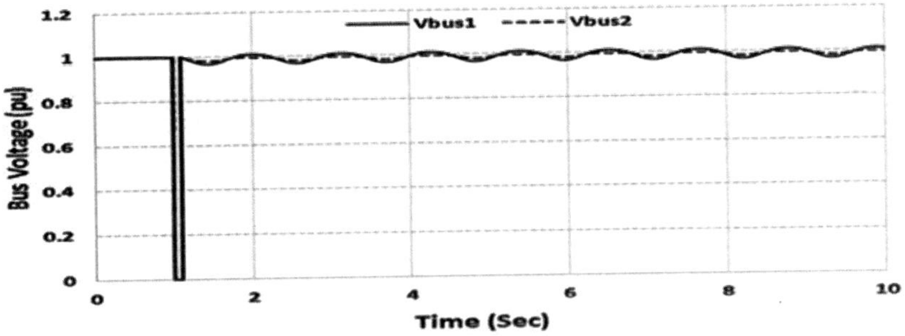

(b) Bus voltage (Pu)

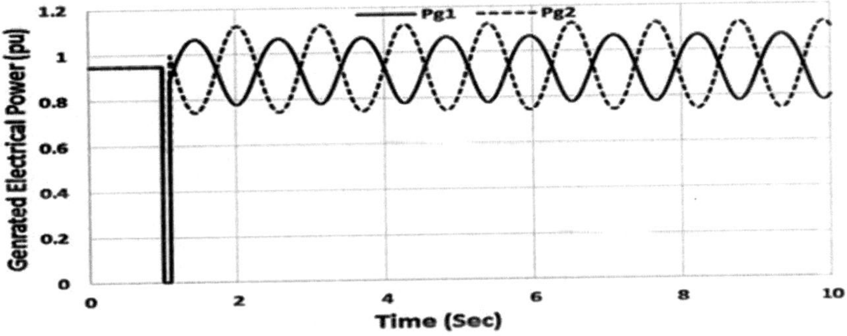

(c) Generator output power (Pu)

Fig. 5 Three phase fault on one of the parallel lines at bus 1 and $T_{\text{clear}} = 0.1$ s

6 Conclusion

Preventive control, restorative control and emergency control, and the role of the load dispatch center and what it comprises of power system state estimation is a method by which data can be formed into a collection of accurate data for control and recording purposes, from network measuring points to a central computer, and

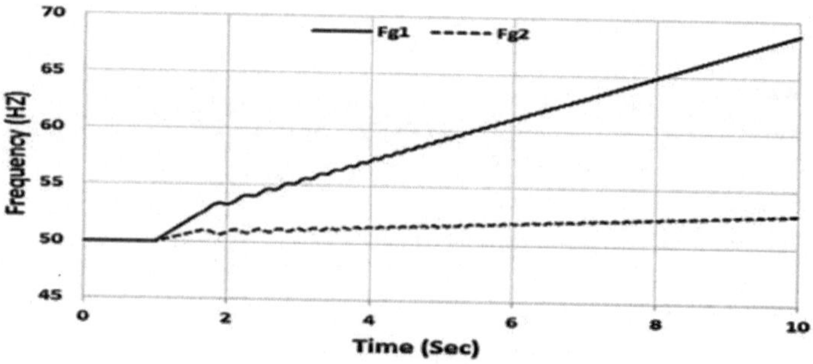

(a) Generator frequency (Hz).

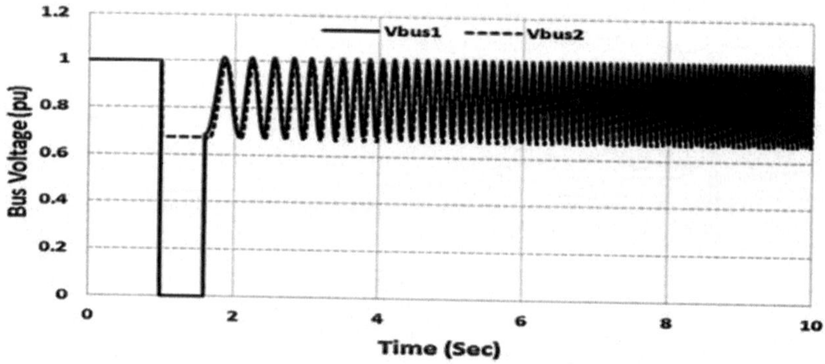

(b) Bus voltage (Pu).

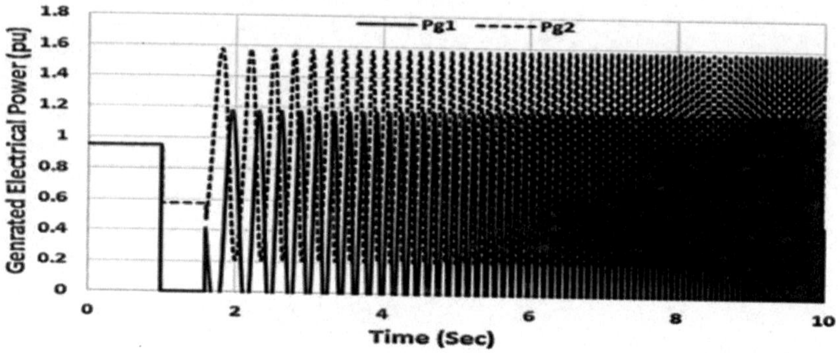

(c) Generator output power (Pu).

Fig. 6 Three phase fault on one of the parallel lines at bus 1 and $T_{clear} = 0.6$ s

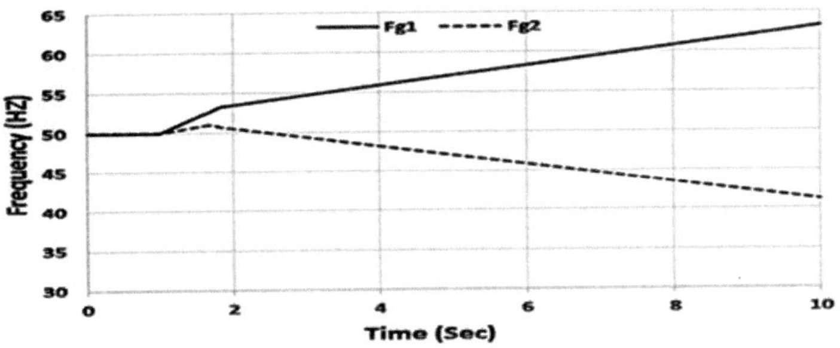

(a) Generator frequency (Hz).

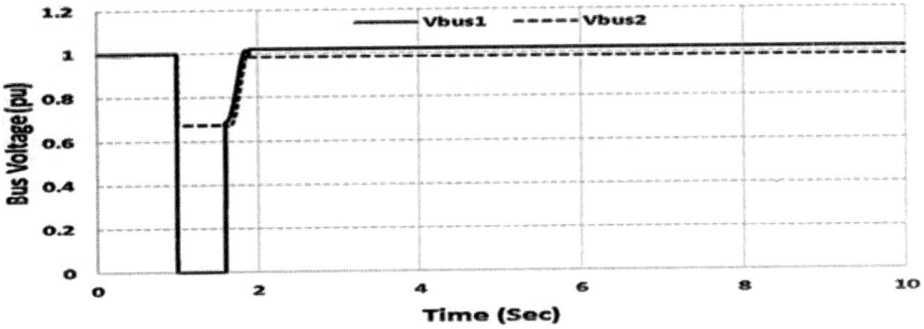

(b) Bus voltage (Pu).

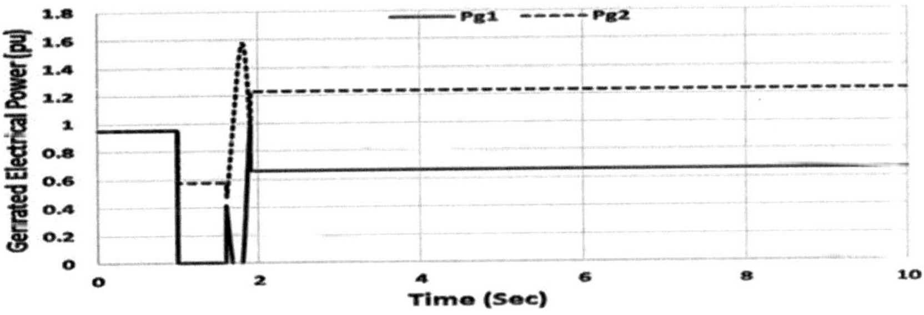

(c) Generator output power (Pu).

Fig. 7 Three phase fault on one of the parallel lines at bus 1 and $T_{\text{clear}} = 0.6$, $T_{\text{isl}} = 0.3$ s

dynamic state estimation allows time-synchronized data acquisition at a faster rate. Classification of systems into normal or alert state, is carried out with an example to illustrate preventive control. An example to illustrate the system angular instability and islanding, if generators within a grid lose synchronism, then they have to be disconnected from each other, and the separate subsystems (islands) may not survive if adequate frequency control measures are not in place.

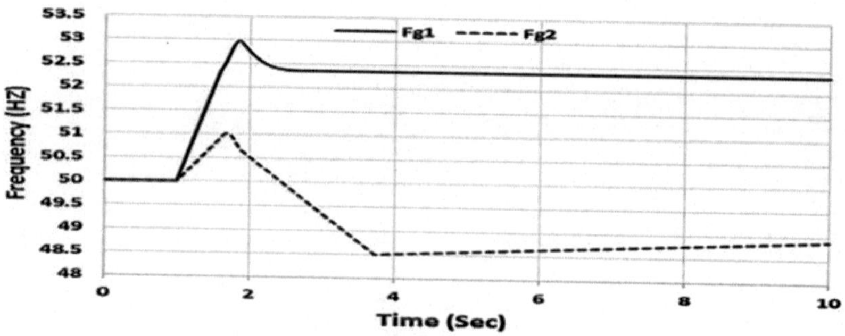

(a) Generator frequency (Hz).

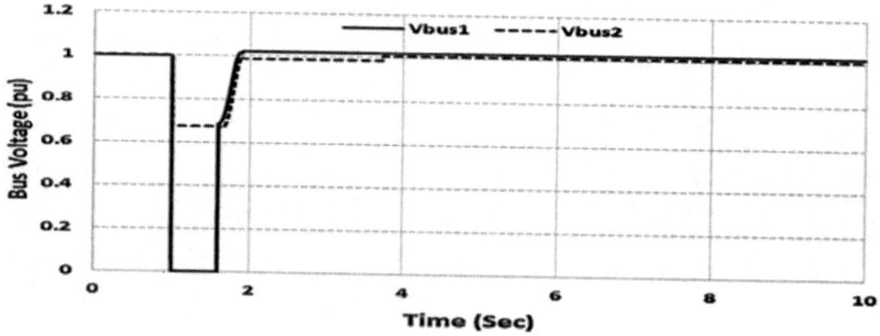

(b) Bus voltage (Pu).

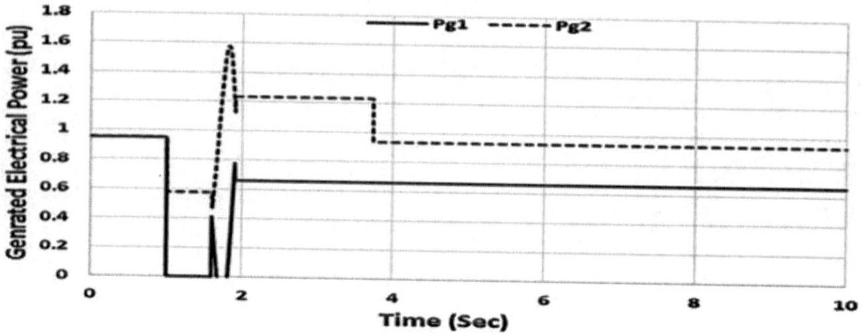

(c) Generator output power (Pu).

Fig. 8 Three phase fault on one of the parallel lines at bus 1 and $T_{\text{clear}} = 0.6$ s, $T_{\text{isl}} = 0.3$ s and frequency control enable

References

1. Sun S, Cong W, Sheng Y, Chen M, Wei Z (2020) Distributed power service restoration method of distribution network with soft open point. In: 2020 IEEE/IAS industrial and commercial power system Asia (I&CPS Asia). Weihai, China, pp 1616–1621. https://doi.org/10.1109/ICPSAsia48933.2020.9208641

2. Ai H, Wang H (2020) Evaluation method of static voltage stability in multi-infeed HVDC system during power system restoration. In: 2020 5th Asia conference on power and electrical engineering (ACPEE). Chengdu, China, pp 528–532. https://doi.org/10.1109/ACPEE48638.2020.9136383
3. Seid B, Aazami R (2019) Intelligent operation and automation management in distribution networks—case study of Ilam-Iran. In: 2019 international power system conference (PSC). Tehran, Iran, pp 5–14. https://doi.org/10.1109/PSC49016.2019.9081568
4. Alassi A, Ellabban O (2019) Design of an intelligent energy management system for standalone PV/battery DC microgrids. In: 2019 2nd international conference on smart grid and renewable energy (SGRE). Doha, Qatar, pp 1–7. https://doi.org/10.1109/SGRE46976.2019.9020679
5. Shabib G, Mobarak YA, El-Ahmar MH (2009) Combined SVC and SSSC controllers for power system transient stability improvement. In: 13th international Middle East power systems conference. Assiut University, Egypt, vol 21–23, pp 573–577
6. Mobarak YA (2012) Fault duration for voltage instability and voltage collapse initiation as influenced by generator voltage magnitudes GVM. J Eng Sci 40(3):846–866 (Assiut University, Egypt)

Chapter 26
Outage and Sum Rate Analysis of Half-Duplex Relay Assisted Full-Duplex Transmission Based NOMA System

P. Bachan, Aasheesh Shukla, and Atul Bansal

1 Introduction

Due to growing demand of number of smart devices and their related utilization in Internet of Things (IoT) and spectrum sharing via cognitive decision-based applications, the next generation wireless networks are becoming pervasive in our daily life which results in ever-increasing demand for large data traffic support and massive connectivity among wireless networks [1–3]. This demand is very challenging, and the conventional multiple access technique such as OMA which has the limitation of excessive channel usage is not suitable to fulfill the massive connectivity requirement of 5G. As with NOMA, multiple devices can share the same channel resource so it is considered as the most prominent multiple access technique with superior spectral efficiency, throughput, and user fairness [4].

In the recent literature, authors have explored the cooperative NOMA (C-NOMA) systems where the reliability of the system can be enhanced by exploiting the information of weaker users' message feature by encouraging strong users to relay the decoded message to weaker users' [5, 6]. In the literature [7, 8], authors presented the analysis of HD user relaying for based NOMA with Rayleigh channels in the form of feasible sum rate and the outage probability. However, the HD relaying-based NOMA has the limitation of excess bandwidth requirement and lower spectral efficiency (SE). To overcome these limitations, references [9, 10] have analyzed full-duplex relaying in C-NOMA systems in which relay transmits and receives the message concurrently. In [11, 12], authors have studied and analyzed system performance

P. Bachan (✉) · A. Shukla · A. Bansal
Department of Electronics and Communication, G. L. A. University, Mathura 281406, India
e-mail: p.bachan@gla.ac.in

A. Shukla
e-mail: aasheesh.shukla@gla.ac.in

A. Bansal
e-mail: atul.bansal@gla.ac.in

A. Sheth et al. (eds.), *Intelligent Systems*, Algorithms for Intelligent Systems,
https://doi.org/10.1007/978-981-16-2248-9_26

of half-duplex (HD) C-NOMA systems where reception of signal at near user and forwarding of signal from near user to far user takes place in two different stages. To further enhance the system performance, full-duplex (FD) C-NOMA systems were also investigated in [13].

However, FD relaying suffers from self-interference problems which reduce spectral efficiency. To overcome the loss of SE in FD-NOMA, distributed HD relays can be used in place of FD relay which is known as sequential relaying [14].

In this work, we study and analyze the outage performance and sum rate of HD relay-assisted FD transmission-based NOMA system [HAF-NOMA]. In this system, two HD relay-based secondary users receive the message from BS sequentially in consecutive time slots and relay the message to the primary user by employing decode and forward strategy.

The organization of this paper is as follows: The system model, SNR and SINR models for HAF-NOMA, is discussed in Sect. 2. In Sect. 3 and Sect. 4, we derive the closed-form analytical expressions outage probability and sum rate for secondary and primary users. In the subsequent Sect. 5, mathematical analysis of probability of outage and sum rate for conventional FD-NOMA system is provided for comparative analysis. Section 6 provides the simulation-based numerical results of SR-NOMA system. Finally, in Sect. 7, we conclude the work.

2 System Model

The system model for sequential HD relaying-based NOMA for cognitive radio network is shown in Fig. 1. In cognitive radio (CR) scenario, the users are arranged as per their priority of service, e.g., higher priority is given to serve the users availing voice services to attain the best quality of service, whereas lower priority is given to users accessing a website [3, 15]. This system consists of a single base station (BS) and three users PU, S_1, S_2 attaining different services in the CR network. User PU

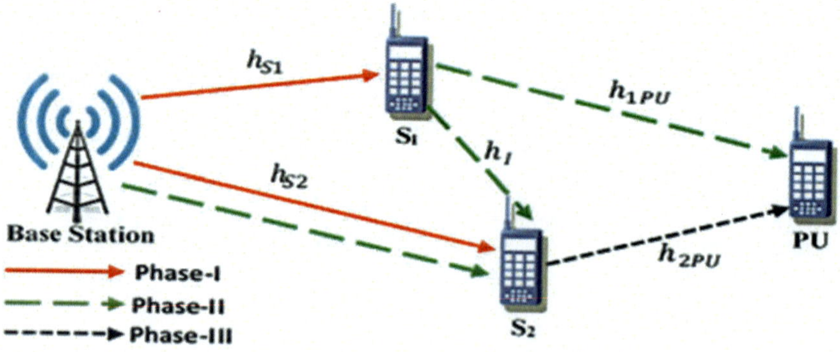

Fig. 1 System model for HAF-NOMA

is considered as the primary user, and S_1 and S_2 are half-duplex (HD) relay-assisted secondary users. Due to heavy shadowing, P is not directly connected to the BS, and it receives the messages via S_1 and S_2.

As shown in Fig. 1, the complete system model is divided into three phases. In phase-I, the base station broadcasts the superimposed signal to S_1 and S_2. S_1 relays the decoded signal to primary user PU, and simultaneously, BS transmits new superimposed signal to S_2 in phase-II. In the final stage, S_2 relays the decoded signal to primary user PU. Here, all the channel coefficients are modeled as quasi-static Rayleigh distributed and considered as reciprocal. The channel parameters between various nodes are as follows: h_{S1} and h_{S2} represent the channel between BS to S_1 and BS to S_2, respectively, with channel variance λ, h_I represents the channel between S_1 and S_2, and h_{1PU} and h_{2PU} represent the channel between secondary users and primary user. In the following subsection, we provide and analyze the signal model for the HAF-NOMA system.

2.1 Signal Model

In phase-1, the base station broadcasts the superimposed composite NOMA signal to secondary users S_1 and S_2 which is given by,

$$m_C(t_1) = \sqrt{aP_B} m_S(t_1) + \sqrt{bP_B} m_{PU}(t_1), \tag{1}$$

where $m_S(t_1)$ and $m_{PU}(t_1)$, contain the messages for S_1 and PU, respectively, and a and b are chosen as the power distribution coefficients of secondary user and primary user, respectively. The coefficients are distributed such that $a + b = 1$ and $a < b$, to serve the primary user with higher priority as compared to the secondary user S as mentioned in [10]. The signals received by S_1 and S_2 are,

$$r_{S_1}(t_1) = h_{S_1} m_C(t_1) + n_1 \tag{2}$$

$$r_{S_2}(t_1) = h_{S_2} m_C(t_1) + n_2 \tag{3}$$

where n_1 and n_2 represent the noise at S_1 and S_2, respectively. Both n_1 and n_2 are Gaussian noise zero mean and variances σ^2.

In the same phase, S_1 and S_2 decode their own message $m_S(t_1)$ after decoding and canceling $m_{PU}(t_1)$ using successive interference cancelation (SIC) technique.

In phase-II, the decoded signal $m_{PU}(t_1)$ is relayed by S_1 to PU, and simultaneously, BS broadcasts a new composite NOMA signal to S_2.

The message received at PU is given by,

$$r_{PU}(t_2) = \sqrt{P_S} h_{1PU} m_{PU}(t_1) + n_{PU} \tag{4}$$

where P_S and n_{PU} represent the transmit power of secondary user S and noise at PU, respectively. n_{PU} is Gaussian noise with zero mean and variances σ^2.

The composite NOMA signal received at S_2 is given as,

$$m_C(t_2) = \sqrt{aP_B}m_S(t_2) + \sqrt{bP_B}m_{\text{PU}}(t_2), \tag{5}$$

The signal received by S_2 from BS is given by,

$$r_{S2}(t_2) = h_{S2}m_C(t_2) + \sqrt{P_S}h_I m_{\text{PU}}(t_1) + n_2 \tag{6}$$

The second part in (6) is the inter-user interference signal from S_1 to S_2. By estimating the value of $\sqrt{P_S}h_I$ using previously received $m_{\text{PU}}(t_1)$, this interference can be mitigated at S_2.

In the final phase-III, the decoded signal $m_{\text{PU}}(t_2)$ is relayed by S_2 to PU. The message received at PU is given by,

$$r_{\text{PU}}(t_3) = \sqrt{P_S}h_{2\text{PU}}m_{\text{PU}}(t_2) + n_{\text{PU}} \tag{7}$$

2.2 *SINR Model*

At Si interference, cancelation technique is applied to decode the message $m_{PU}(t_i)$, where $i = \{1, 2\}$. Thus, SINR for detecting this message $m_{\text{PU}}(t_i)$ at Si is given by,

$$g_{S_i,\text{PU}} = \frac{bP_B|h_{S_i}|^2}{aP_B|h_{S_i}|^2 + \sigma^2} \tag{8}$$

The secondary users S_1 and S_2 subtract the successfully decoded primary user signal $m_{\text{PU}}(t_i)$ from their received signals $r_{Si}(t_i)$ to decode their own message. Thus, SINR for detecting the message $m_S(t_i)$ at Si is given by,

$$g_{Si} = \frac{aP_B|h_{Si}|^2}{\sigma^2} \tag{9}$$

Finally, the relayed message $m_{PU}(t_i)$, from Si to PU will be detected at PU. Thus, the SINR for successful detection of $m_{\text{PU}}(t_i)$ at PU is,

$$g_{\text{PU}} = \frac{P_S|h_{i\text{PU}}|^2}{\sigma^2} \tag{10}$$

3 Probability of Outage

Probability of outage is one of the prominent metrics to estimate the message transmission and detection quality in accordance with the fixed message transmission rate of the user. This section presents the analytical expressions of probability of outage for HD-assisted FD relaying (HAF)-based NOMA systems.

To begin, let us first define the outage events that may take place at different nodes:

O_1, O_2: When S_1 fails to detect $m_{PU}(t_1)$, S_1 fails to detect $m_S(t_1)$.
O_3: When S_2 fails to detect $m_{PU}(t_1)$ and $m_{PU}(t_2)$.
O_4: When S_2 fails to detect $m_S(t_2)$.
O_5, O_6: When PU fails to detect $m_{PU}(t_1)$, PU fails to detect $m_{PU}(t_2)$.

Let R_S, R_{PU} denote the desired data rates for secondary user (Si) and primary user (PU), respectively. PO_{S1}, PO_{S2}, PO_{PU} denote the outage parameters for S_1, S_2, and PU, respectively. So, on the basis of outage events, the closed for expressions of probability of outage are given as,

$$\begin{aligned} PO_{S_1} &= 1 - \Pr[O_1] \cap \Pr[O_2] \\ &= 1 - \exp\left(-\frac{(2^{R_{PU}} - 1)\sigma^2}{(1 - a2^{R_{PU}})P_B\lambda}\right)\exp\left(-\frac{(2^{3R_S} - 1)\sigma^2}{aP_B\lambda}\right) \end{aligned} \tag{11}$$

$$\begin{aligned} PO_{S_2} &= 1 - \Pr[O_1] \cap \Pr[O_2] \\ &= 1 - \exp\left(-\frac{(2^{R_{PU}} - 1)\sigma^2}{(1 - a2^{R_{PU}})P_B\lambda}\right)\exp\left(-\frac{(2^{3R_S} - 1)\sigma^2}{aP_B\lambda}\right) \end{aligned} \tag{12}$$

$$\begin{aligned} PO_{PU} &= 1 - Pr[O_1] \cap Pr[O_5] \cap Pr[O_3] \cap Pr[O_6] \\ &= 1 - exp\left(-\frac{(2^{3R_{PU}} - 1)2\sigma^2}{(1 - a2^{3R_{PU}})P_B\lambda}\right)exp\left(-\frac{(2^{3R_{PU}} - 1)2\sigma^2}{P_S\lambda}\right) \end{aligned} \tag{13}$$

4 Ergodic Rate of HAF-NOMA

Ergodic rate is the sum rate of all the users in a particular network considering the impact of time variant nature of channel on the instantaneous rate of each user. Here, the ergodic rate of HD-assisted FD-NOMA system is the average sum of rates achieved by low-priority (S_1, S_2) and high-priority users (PU) which is given by,

$$\underline{R}^e = \underline{R}^{S_1} + \underline{R}^{S_2} + \underline{R}^{PU} \tag{14}$$

where $\underline{R}^{S_1}$, $\underline{R}^{S_2}$, $\underline{R}^{PU}$ are the sum rate of S1, S2, and PU, respectively.

As the complete message transmission of the HAF-NOMA system from BS to PU is performed in three time slots, therefore 1/3 is the rate constant for sum rate of all three users.

On the basis of successful detection of $m_{\mathrm{PU}}(t_1)$ and $m_{\mathrm{PU}}(t_2)$ at S1 in first and third phase, respectively, the sum rate of secondary users S1, S2 is given by,

$$\underline{R}^{S_1} = \underline{R}^{S_2} = -\frac{1}{3\ln 2}\exp(\Phi).Ei(-\Phi) \tag{15}$$

where $\Phi = \frac{\sigma^2}{aP_B\lambda}$ and $Ei(-\Phi)$ are the exponential integral operator.

During the second phase, BS broadcasts a new composite NOMA signal to S2 and S2. Also decodes $m_{\mathrm{PU}}(t_2)$ so the ergodic rate is the same as indicated in Eq. (15).

Combining the messages received from S1 and S2, the ergodic rate of PU is given by,

$$\underline{R}^{\mathrm{PU}} = -\frac{2\log_2\exp(1)}{3\ln 2}\{\exp(2\Phi)\cdot Ei(-2\Phi) - \exp(2a\Phi)\cdot Ei(-2a\Phi)\} \tag{16}$$

On the basis of successful detection of $m_{\mathrm{PU}}(t_1)$ and $m_{\mathrm{PU}}(t_1)$ at PU in phase-2 and phase-3, respectively, the ergodic sum rate of HAF-NOMA from Eq. (15) and Eq. (16) is given by,

$$\begin{aligned}\underline{R}^{S-HD} &= \underline{R}^{S_1} + \underline{R}^{S_2} + \underline{R}^{\mathrm{PU}} \\ &= -\frac{2\log_2\exp(1)}{3\ln 2}\{\exp(2\Phi)\cdot Ei(-2\Phi) - \exp(2a\Phi)\cdot Ei(-2a\Phi)\} \\ &23In2\exp\Phi\cdot Ei - \Phi\end{aligned} \tag{17}$$

5 Full-Duplex NOMA

This section presents the analytical expressions of outage probability and ergodic rate for conventional full-duplex NOMA as given in [9].

In FD-NOMA scheme, the SINR for detecting the message of primary PU at secondary user S is given as,

$$g_{S,\mathrm{PU}} = \frac{bP_B|h_S|^2}{aP_B|h_S|^2 + kP_S|h_I|^2 + \sigma^2} \tag{18}$$

where h_I is the channel for self-interference at user relay, and k is self-interference cancelation factor.

The SINR to detect the message of S at S user is given by,

$$g_S = \frac{a P_B |h_S|^2}{k P_S |h_I|^2 + \sigma^2} \tag{19}$$

If h_{PU} represents the channel between S and PU, then the SINR to detect the message of PU at PU is given by

$$g_{\mathrm{PU}} = \frac{P_S |h_{\mathrm{PU}}|^2}{\sigma^2} \tag{20}$$

The probability of outage for user-assisted relaying-based full-duplex NOMA [9] at S and PU is given as

$$\mathrm{PO}_S = 1 - \Pr\left[\log_2\left(1 + g_{S,\mathrm{PU}}\right) \geq R_{\mathrm{PU}}\right] \cdot \Pr\left[\frac{1}{T_S}\log_2(1 + g_S) \geq R_S\right] \tag{21}$$

$$\mathrm{PO}_{\mathrm{PU}} = \Pr\left[\frac{1}{T_S}\log_2\left(1 + \min\left(g_{S,\mathrm{PU}}, g_S\right)\right) \geq R_{\mathrm{PU}}\right] \tag{22}$$

where the requisite time slot Ts = 2, for complete transmission in FD-NOMA.

On the basis of Eqs. (18), (19), and (20), the ergodic rate for FD-NOMA system is given by

$$\underline{R}^{FD} = \frac{2}{T}\log_2(1 + g_S) + \frac{2}{T}\log_2\left(1 + \min\left(g_{\mathrm{PU}}, g_{S,\mathrm{PU}}\right)\right) \tag{23}$$

6 Simulation Results

In this section, we validate the performance of HD relaying-assisted FD-NOMA (HAF-NOMA) model through numerical simulations. To perform the numerical analysis, we divide the simulation procedure into two steps: Firstly, we carry out simulations of outage probability with respect to varying SNR values for secondary users and primary users. Secondly, we present the numerical analysis of ergodic rate of S_1, S_2, and PU in HAF-NOMA. In both the analysis, we compare the performance of HAF-NOMA system with full-duplex NOMA system for the same scenario. The simulation parameters for comparative numerical analysis are provided in Table 1.

6.1 Outage Probability Results

In Fig. 2, we demonstrate the comparative probability of outage for secondary users (S1 and S2) in the HAF-NOMA and FD-NOMA schemes with low and high target data rates, $R = 0.1$ and 0.9 bits/s/Hz, respectively.

Table 1 System parameters for numerical simulations of HAF-NOMA and FD-NOMA

Simulation parameters	Numerical values
Channel realizations	10^5
SNR	0 to 40 dB
Noise variance (σ^2)	1/SNR
Channel variance (λ)	1
Power distribution coefficients (a, b)	0.08, 0.92
Base station power (PB)	1 W
Relay user power (PS)	0.5 W

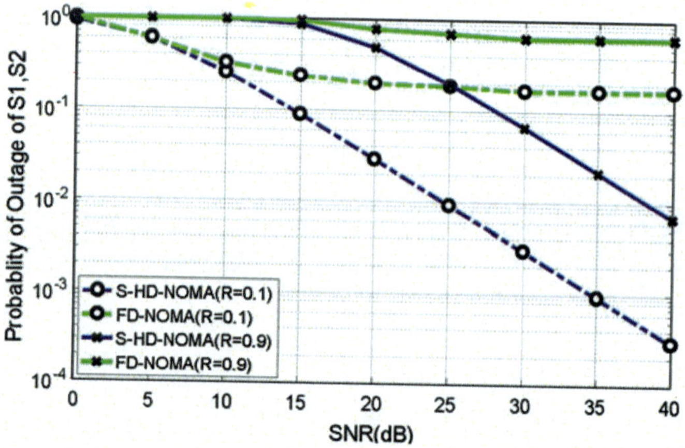

Fig. 2 Probability of outage of S1, S2 in HAF-NOMA and FD-NOMA

Based on the mathematical analysis discussed in Sect. 3, the probability of outage is same for S1 and S2 as given in (11) and (12), respectively. From the figure, it may be seen that self-interference by the user relay causes poor outage probability in FD-NOMA when compared with HAF-NOMA for both lower and higher data rate cases.

In Fig. 3, we have shown the probability of outage for primary users (PU) in the SR-NOMA and FD-NOMA schemes with low and high predetermined target data rates, $R = 0.1$ and 0.9 bits/sec/Hz, respectively. Here, we observe that the theoretical outage performance described in (13) with the numerical values generated through Monte Carlo-based simulations. From the figure, it may be seen that for a higher range of SNR values, i.e., SNR > 17 dB HAF-NOMA outage performance is better than FD-NOMA when R = 0.1 b/s/Hz. Similarly, for higher data rates ($R = 0.9$ b/s/Hz) also for HAF-NOMA provides lower probability of outage than FD-NOMA when SNR is greater than 20 dB.

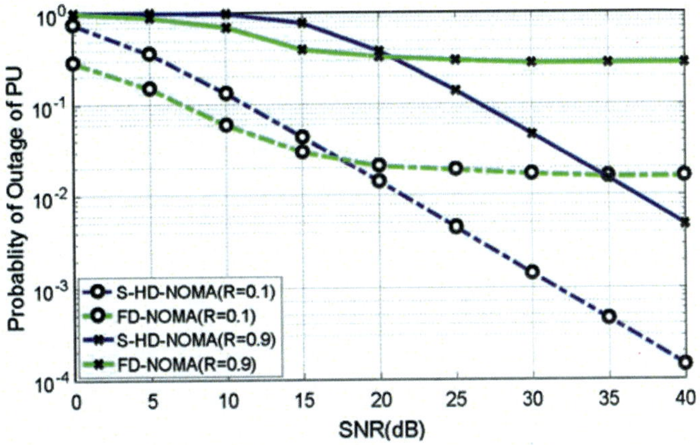

Fig. 3 Probability of outage of PU in HAF-NOMA and FD-NOMA

6.2 Ergodic Rate Results

In Fig. 4, we demonstrate the comparative ergodic rate for secondary users (S1 and S2) in the HAF-NOMA and FD-NOMA schemes considering variable power distribution coefficient $a = 0.08, 0.05$ and $b= 0.92, 0.95$ for secondary and primary user, respectively. Based on the mathematical analysis discussed in Sect. 4, the ergodic rate is the same for S1 and S2 as given in Eq. (15). We can observe that the numerical analysis values generated through Monte Carlo-based simulations perfectly validate the exact mathematical analysis. From the figure, it may be seen that the ergodic rate of HAF-NOMA is significantly high as compared to FD-NOMA when SNR is greater than 15 dB for both power distribution coefficient 0.05 and 0.08, respectively.

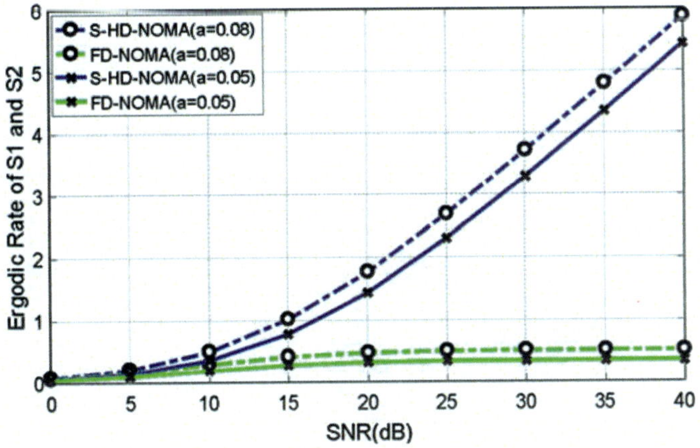

Fig. 4 Ergodic rate of S1, S2 in HAF-NOMA and FD-NOMA

Figure 5 represents the numerical analysis of primary user's (PU) ergodic rate for HAF-NOMA and FD-NOMA transmission schemes. Though in the SNR range 10 dB to 20 dB, FD-NOMA achieved slightly better rate but in majority of lower (SNR < 10 dB) and higher SNR regions (SNR > 20 dB), HAF-NOMA achieves better rate for PU while considering both power allocation confidents. One important observation that is evident from Fig. 4 is PU's ergodic rate which is improved by lowering the value of power distribution coefficient in both the schemes.

In Fig. 6, we have shown the ergodic sum rate of HAF-NOMA as given in Eq. (17) and compared the performance with the FD-NOMA scheme as given in Eq. (23). Because of the residual self-interference from the relay user present in FD-NOMA, HAF-NOMA achieves a significantly higher sum rate than its counterpart for the

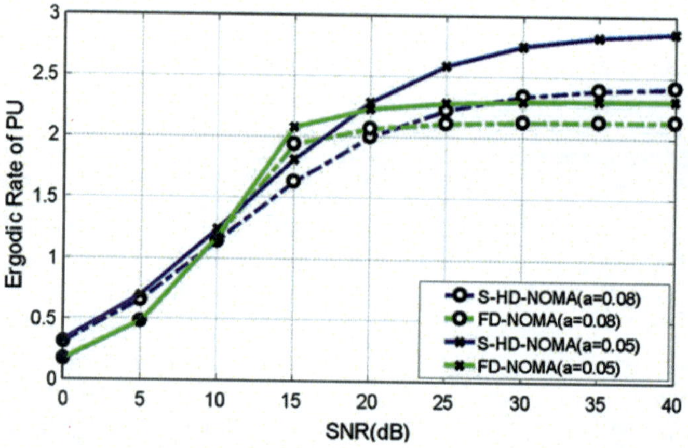

Fig. 5 Ergodic rate of PU in HAF-NOMA and FD-NOMA

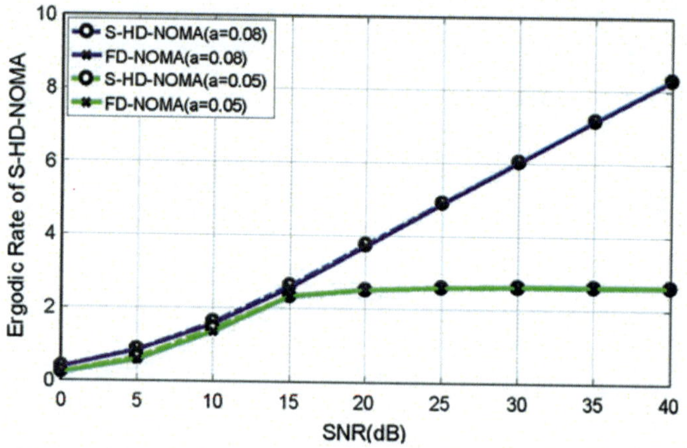

Fig. 6 Sum rate of HAF-NOMA and FD-NOMA transmission

SNR values greater than 15 dB. It may also be observed here that change of power distribution coefficient does not have a significant impact on the sum rate comparable schemes.

7 Conclusions

In this work, we have investigated sequential HD transmission-assisted cooperative NOMA systems for cognitive radio networks where one primary user is placed away from the base station and two secondary users placed closer to BS relay the message from BS to PU in successive time slots. Firstly, we have analyzed the closed expressions of SINR and probability of outage HAF-NOMA systems. Secondly, we have also discussed the SINR and outage expressions of FD-NOMA for comparative analysis. We have also analyzed the ergodic rates of users individually and in combined sum rate form for both the transmission schemes. The simulation-based numerical results validate the analytical results presented for HAF-NOMA and FD-NOMA systems. The probability of outage of primary user and secondary users is verified in sequential HD-NOMA, and numerical analysis clearly shows that sequential HD-NOMA outperforms FD-NOMA. The ergodic sum rate numerical analysis also proves that the performance of HAF-NOMA scheme is significantly higher over FD-NOMA due to perfect cancelation of interference between secondary users of the network. Results also verify that the change in power distribution coefficient does not have much of an impact on the sum rate of both the transmission schemes.

References

1. Andrews JG et al (2014) What Will 5G Be? IEEE J Sel Areas Commun 32(6):1065–1082. https://doi.org/10.1109/JSAC.2014.2328098
2. Shirvanimoghaddam M, Dohler M, Johnson SJ (2017) Massive non-orthogonal multiple access for cellular IoT: potentials and limitations. IEEE Commun Mag 55(9):55–61. https://doi.org/10.1109/MCOM.2017.1600618
3. Ghosh SK, Bachan P (2017) Performance evaluation of spectrum sensing techniques in cognitive radio network. IOSR J Electron Commun Eng (IOSR-JECE), 2278–2834. https://doi.org/10.9790/2834-1204051721
4. Dai L et al (2015) Non-orthogonal multiple access for 5G: solutions, challenges, opportunities, and future research trends. IEEE Commun Mag 53(9):74–81. https://doi.org/10.1109/MCOM.2015.7263349
5. Chang Z et al (2018) Energy-efficient and secure resource allocation for multiple-antenna NOMA with wireless power transfer. IEEE Trans Green Commun Netw 2(4):1059–1071. https://doi.org/10.1109/TGCN.2018.2851603
6. Zheng B et al (2018) Secure NOMA based two-way relay networks using artificial noise and full duplex. IEEE J Sel Areas Commun 36(7):1426–1440. https://doi.org/10.1109/JSAC.2018.2824624
7. Ding Z, Peng M, Poor HV (2015) Cooperative non-orthogonal multiple access in 5G Systems. IEEE Commun Lett 19(8):1462–1465. https://doi.org/10.1109/LCOMM.2015.2441064

8. Liu Y, Pan G, Zhang H, Song M (2016) Hybrid decode-forward & amplify-forward relaying with non-orthogonal multiple access. IEEE Access 4:4912–4921. https://doi.org/10.1109/ACCESS.2016.2604341
9. Alsaba Y, Leow CY, Abdul Rahim SK (2018) Full-duplex cooperative non-orthogonal multiple access with beamforming and energy harvesting. IEEE Access 6:19726–19738. https://doi.org/10.1109/ACCESS.2018.2823723
10. Zhang L, Liu J, Xiao M, Wu G, Liang Y, Li S (2017) Performance analysis and optimization in downlink NOMA systems with cooperative full-duplex relaying. IEEE J Sel Areas Commun 35(10):2398–2412. https://doi.org/10.1109/JSAC.2017.2724678
11. Su B, Ni Q, Yu W (2019) Robust transmit beamforming for SWIPT-enabled cooperative NOMA with channel uncertainties. IEEE Trans Commun 67(6):4381–4392. https://doi.org/10.1109/TCOMM.2019.2900318
12. Tian Fy, Chen Xm (2019) Multiple-antenna techniques in non-orthogonal multiple access: a review. Front Inform Technol Electron Eng 20:1665–1697. https://doi.org/10.1631/FITEE.1900405
13. Li G, Chen H, Cai J (2019) Joint user association and power allocation for hybrid half-duplex/full-duplex relaying in cellular networks. IEEE Syst J 13(2):1145–1156. https://doi.org/10.1109/JSYST.2018.2850861
14. Hong S, Caire G (2015) Virtual full-duplex relaying with half-duplex relays. IEEE Trans Inf Theory 61(9):4700–4720. https://doi.org/10.1109/TIT.2015.2453942
15. Thakur P et al (2019) Frameworks of non-orthogonal multiple access techniques in cognitive radio communication systems. China Commun 16(6):129–149. https://doi.org/10.23919/JCC.2019.06.011

Chapter 27
Machine Learning-Based Approach for Nutrient Deficiency Identification in Plant Leaf

Parnal P. Pawade and A. S. Alvi

1 Introduction

The precision farming focuses on various topics among that nutrient content identification in plants is determining task. This task is very complex if performed without any destruction. Most of the existing research does not identify the most of the nutrient deficiencies in the plant leaves. For providing an automatic and trustworthy economic solution for nutrient insufficiency detection, it is proposed in [1]. For real-time texture detection, RGB color feature extraction, edge detection, etc., the dataset for healthy leaves and deficient leaves are generated using image processing methodology. In demand to apply precautionary measures to increase the yield, this generated dataset is given to supervised machine learning models as training dataset for additional identification and detection of precise nutrient insufficiency in unhealthy plants and to separate fit plants.

The examination of tonalities and geometric physiognomies of the leaves of coffee plantation is carried out by the proposed algorithm in [2] for identification of nutritional insufficiencies of coffee plantations. By visual perception, the algorithm tried to minimize the subjectivity in the analysis. The prescription plan of nutrients and fertilizers smeared by producers is affected by the errors in this analysis. A procedure for contrast improvement from the luminance is initially applied in the algorithm, and then, a scale invariant feature transform (SIFT) algorithm is used which delivers the vital points for the creation of the corresponding descriptors. In comparison with obtaining Fourier and Hu descriptors, the enhanced image is exposed to thresholding. For the detection of particular nutritional deficiency, a particular neural network is trained independently with the three types of descriptors. For comparing the results, Kappa index was used with those occupied by pictorial inspection. The results are

P. P. Pawade (✉)
1CSE Department, PRPCEM, Amravati, Maharashtra 444602, India

A. S. Alvi
IT Department, PRMIT&R, Badnera, Maharashtra 444701, India

A. Sheth et al. (eds.), *Intelligent Systems*, Algorithms for Intelligent Systems,
https://doi.org/10.1007/978-981-16-2248-9_27

found pleasing, with a Kappa coefficient of 0.92 for boron deficiency and 0.96 for nitrogen and potassium deficiencies.

In this paper, a proposed machine learning-based approach is presented which utilizes random forest, voting classifier, linear support vector classifier, and k nearest neighbor classifier for identification of nutrient deficiencies in plant leaves. In the proposed approach, features are extracted from gray-level co-occurrence matrix and gray-level co-occurrence properties of leaf images. The extracted features comprise contrast, energy, homogeneity, correlation, and dissimilarity. The extracted features are utilized for identifying nitrogen, potassium, copper, and magnesium nutrient deficiencies in plant leaf. The subsequent sections present existing literature in nutrient deficiency identification, proposed machine learning-based algorithm for nutrient deficiency identification in plant leaves, and evaluation of proposed machine learning-based approach.

2 Related Work

To increase both the quantities and the qualities of crops, plant nutrient insufficiency classification is important for the agricultural industry. To solve the various kinds of complex problems, deep learning and computer vision methodologies, specifically convolutional neural networks, play an important role in biological and agricultural divisions. Macronutrients and micronutrients [3] are two categories of plant nutrients.

For providing an automatic and dependable economic solution for nutrient insufficiency detection, it is proposed in [4]. For real-time texture detection, RGB color feature extraction, edge detection, etc., the dataset for healthy leaves and deficient leaves are generated using image processing methodology. In order to apply protective measures to the amount of the production, this generated dataset is applied to supervised machine learning models as training dataset for additional identification and of rigorous nutrient insufficiency in unhealthy plants and to separate well plants.

The analysis of geometric features and tonalities of the leaves of coffee plantation is carried out by the proposed algorithm in [5] for discovery of nutritional insufficiencies of coffee plantations. By visual perception, the algorithm tried to minimize the subjectivity in the analysis. The prescription plan of nutrients and fertilizers smeared by producers is affected by the errors in this analysis. A procedure for contrast improvement from the luminance is initially applied in the algorithm, and then, a scale invariant feature transform (SIFT) algorithm is used which offers the imperative points for the creation of the corresponding descriptors. In parallel to obtain Fourier and Hu descriptors, the enhanced image is exposed to thresholding. For the detection of particular nutritional deficiency, a precise neural network is trained distinctly with the three types of descriptors. For comparing the results, Kappa index was used with those engaged by graphical examination. The results found satisfactory, with a Kappa coefficient of 0.92 for boron deficiency and 0.96 for nitrogen and potassium deficiencies.

For any fruit for its complete growth and development, all the nutrients are necessary. But deficiency disorder may be found due to certain reasons such as waterlogging, parched soils, and further natural catastrophe factors. So, for detecting such insufficiency and to provide relief in decreasing manual interference of inspection and detection, there must be an automated system. Hence, an application-based computer vision tool, a different approach, is used in [6] that deals with a relative investigation of double nutrient, i.e., calcium and boron, insufficiency recognition system for apple fruit. A graphical user interface (GUI) has been demonstrated in order to offer a pretty communication among program and a user, which has been implemented with the assistance of MATLAB. The approach used by [6] mainly offers the user to identify the insufficiency existent in a fruit, which assists the user to determine the numerous usages of image processing method overall and also choose the discriminating measures over the identified insufficiency. A very acceptable performance of this tool is observed from the results. Hence, it can be considered as a bendable and vigorous method by spontaneous usage.

For recognizing nutrient insufficiencies in plants grounded on its leaf, a novel image analysis method is proposed in [7]. An input leaf image is distributed into small blocks by the proposed method. To a set of convolutional neural networks (CNNs), every block of leaf pixel is fed. For identification of a nutrient shortage, each CNN is specifically trained and it is used to agree if a block is showing any symptom of the conforming nutrient deficiency. For producing a single answer for the block, the replies from all CNNs are combined using a winner-take-all strategy. To generate a last response for the complete leaf, the responses from all blocks are combined into single using a multilayer perceptron. A group of black gram plants developed underneath nutrient-controlled environments is used for validating the performance of the proposed method. A cluster of plants with all nutrients and five types of insufficiencies, i.e., Mg, Ca, K, Fe, and N, were studied in [7]. For experimentation, a dataset comprising 3000 leaf images was composed and utilized. The supremacy of the proposed technique over trained humans from experimental results can be observed in nutrient deficiency identification.

Mango tree leaves are immensely exaggerated by numerous nutrient insufficiencies like potassium, copper, nitrogen, and iron. The natural color of mango leaves can be changed due to deficiency of these nutrients. The numerous nutrient insufficiencies of mango leaves are detected by the work presented in [8]. The diverse features of mango leaves are extracted for creating the dataset using digital image processing. The texture and RGB color features are extracted from the leaves. The unsupervised machine learning model is fed with this dataset for clustering and further nutrient deficiency detection. An early detection of nutrient deficiency helps the farmers to take the measures for reducing the risk of unhealthy growth of plants. The work can be extended to recognize the nutrient deficiency in various agricultural plants and crops not only in mango.

The usage of numerous deep convolutional neural networks (CNNs) is investigated in [9] with transfer learning to determine nutrient insufficiencies from a leaf image. A dataset comprising 4088 images of black gram leaves is used for experimental evaluations developed in seven diverse treatments, i.e., comprehensive nutrient treatment and six nutrient shortage treatment, including phosphorus (P), nitrogen (N), calcium (Ca), magnesium (Mg), iron (Fe), and potassium (K) deficiencies. A deep CNN model recognized as ResNet50 was the top among all investigated models with an accuracy of 65.44% and an F-measure of 66.15% as observed from experimental results. The ResNet50 model performs better than the block-based technique and the human performance described in literature. The nutrient insufficiencies on plants that are lacking more than one nutrient in the same time is not identified in [9].

One of the vital indicators of plant health is leaf chlorophyll content (LCC). Too little or too much nutrients, i.e., nitrogen in specific, received by the plant can be identified by using LCC. Yield reduction can be prevented with remedial measures based on primary identification of nitrogen shortage. For determining the reflectance of leaves, i.e., close-range detecting, or canopy, i.e., remote sensing, optical approaches with a spectrophotometer or a hyperspectral sensor can be used to estimating LCC. LCC estimation using color cameras has recently gained interest due to the cost of optical sensor-based devices. But their usefulness in practice is limited because present methodologies are mostly aimed for very nearby range sensing, i.e., they permit the measurement of LCC for a particular plant at a time. Furthermore, the anisotropic reflective properties of canopies and leaves cannot be explored with these devices. A simulation grounded on a plant canopy reflectance model is presented in [10] to examine the possibility of RGB imaging for remote sensing-based assessment of canopy LCC. For generating RGB images with numerous reference illuminants and from ten dissimilar observing angles, an RGB camera model was utilized. For predicting LCC from RGB values deprived of white balance, linear and neural network-based regression approach is used. For remote sensing, LCC estimation results indicated a noteworthy prospective to use RGB sensors, predominantly with a multi-angle methodology. The camera wants to be positioned at an observer zenith angle near 90° when the solar zenith angle is 40° to achieve the best estimation. The assessment accuracy can be improved by allowing two viewing angles instead of one. Solitary simulation data is used in [10], and further real data can be used for evaluating the model of chlorophyll content assessment with. Possibility of recovering spectral reflectance from RGB pixel values can be further investigated.

Most of these approaches only identify the specific nutrient deficiencies in specific plant leaves. The proposed machine learning-based approach in this paper can identify nitrogen, potassium, copper, and magnesium nutrient deficiencies in a variety of plant leaves.

3 Proposed Machine Learning Approach for Nutrient Deficiency Identification

The proposed machine learning-based approach for nutrient deficiency identification in plant leaves uses random forest classifier, voting classifier, linear support vector classifier, and k neighbors classifier to identify nutrient deficiency in plant leaf. The input given to the proposed machine learning model comprises mean, variance, standard deviation, contrast, energy, homogeneity, correlation, dissimilarity, and angular second moment (ASM) features extracted from plant leaf. These features are extracted from the plant leaf images using gray-level co-occurrence matrix method. The proposed machine learning model predicts the nutrient deficiency in the plant leaf. For training the proposed model synthetic dataset comprising mean, variance, standard deviation, contrast, energy, homogeneity, correlation, dissimilarity, angular second-moment features, and their nutrient deficiency class. The nutrient deficiency classes used are healthy, nitrogen, potassium, copper, and magnesium. If there is no nutrient deficiency in plant leaf, then it is classified into healthy class. Table 1 shows the dataset used to train and test the proposed machine learning model. More than 800 plant leaves features are stored in the dataset.

In the proposed approach initially, plant leaf image is taken as an input. The input image is converted from RGB to LAB where the Euclidean distance between two colors is equal to their perceptual distances. The LAB image is divided into l, a, and b components. The obtained l, a, and b components are used to segment the image. By default, the image is segmented into 30 segments. The Gaussian blur of image is computed in 5 × 5 dimensions, and the image is reshaped. The label of each segment in the image is identified using k nearest neighbor classifier. These labeled segments are utilized to identify which segments comprise nutrient deficiency. If any segment comprises nutrient deficiency, then type of deficiency will be identified.

For obtaining the gray-level image, the identified image comprising nutrient deficiency segments is converted from LAB to RGB, and then RGB image is converted to gray-level image. From the gray-level image, gray-level co-occurrence matrix and properties are computed. The gray-level matrix is used to compute features vectors which comprises mean, variance, standard deviation, contrast, energy, homogeneity, correlation, dissimilarity, and angular second moment features. This feature vector is used to predict the type of nutrient deficiency in the plant leaf using SVC, random forest, and voting classifiers. Finally, the accuracy of the proposed model is computed. The proposed algorithm used in the proposed machine learning model is as follows:

Proposed Machine Learning-based Algorithm

Usage: Random Forest Classifier, Voting Classifier, Linear SVC, K Neighbors Classifier, Plant Leaves Dataset.

Input: Plant Leaf (Features Vector).

Output: Gray-Level Co-occurrence Matrix, Gray-Level Co-occurrence Properties, Identified Nutrient Deficiency, Accuracy.

1. Define number of segments. Default segments = 30
2. Take plant leaf image
3. Convert the leaf image from RGB to LAB
4. Split image into l, a, and b component
5. Segment the images into number of segments specified

Table 1 Dataset used for training and testing the proposed machine learning model

Mean1	Mean2	Stddev1	Stddev2	Variance	Contrast	Energy	Homogeneity	Correlation	Dissimilarity	ASM	Class
154.835	100.921	23.0723	43.472	3094.55	76.529	0.0497	0.329497772	0.985894627	4.313488871	0.00247	Nitrogen
107.501	115.148	15.5017	31.2156	2336.94	91.9984	0.02988	0.356123467	0.985742991	4.122450782	0.00089	Nitrogen
158.32	113.616	19.9381	43.2896	3441.05	348.953	0.03952	0.309305461	0.96865512	7.467932605	0.00156	Nitrogen
123.765	127.926	26.5703	30.7493	2434.97	180.896	0.02242	0.198705727	0.9981432041	7.753455343	0.0005	Nitrogen
130.244	116.2	14.0709	24.1101	3236.83	149.624	0.03722	0.355810791	0.989967826	4.597266174	0.00139	Nitrogen
136.96	104.696	29.2842	45.8272	2253.42	57.0903	0.06403	0.40169009	0.979950997	3.374338235	0.0041	Nitrogen
149.468	106.001	29.3411	46.3389	3655.05	72.6069	0.05913	0.413465399	0.992948732	3.84098675	0.0035	Nitrogen
145.527	104.76	25.4108	45.4684	3341.67	415.124	0.03834	0.191284013	0.942579323	10.25432605	0.00147	Nitrogen
159.285	108.654	18.7764	31.3684	2166.48	372.475	0.02744	0.152769949	0.893113389	11.00031328	0.00075	Nitrogen
139.455	105.856	29.2936	45.4208	2454.26	803.3	0.04935	0.11782022	0.82627009	18.0722469	0.00244	Potassium
144.46	98.5402	26.5937	37.9203	2232.36	1352.47	0.03096	0.103800242	0.663426454	22.36966787	0.00096	Potassium
79.4002	125.991	17.0283	29.6659	3036.45	103.862	0.23643	0.514788759	0.988953219	4.766080417	0.0559	Potassium
108.926	104.205	28.0173	30.309	2051.06	360.95	0.03056	0.152785843	0.939641011	12.22251143	0.00093	Potassium
111.791	112.447	13.0917	16.5484	823.841	446.111	0.01946	0.124105429	0.807670541	11.52531844	0.00038	Potassium
119.47	102.943	24.8808	30.3927	1893.72	1140.03	0.02488	0.106324262	0.690168424	19.92278481	0.00062	Potassium
98.0628	107.431	33.567	43.0395	2647	116.396	0.07825	0.329311664	0.981595412	5.101719996	0.00612	Potassium
122.764	90.7449	49.254	62.2616	4101.62	459.484	0.22338	0.41181741	0.911790107	7.114197715	0.0499	Potassium

6. Compute Gaussian blur of image with 5 × 5 dimension
7. Reshape the image
8. Convert the obtained result in vector
9. Using k nearest neighbors classifier, classify image components into specific segments
10. Return the labels of each segment
11. Extract the required image components from the labeled segments needed for nutrient deficiency identification
12. Display the segmented and extracted image components
13. Compute mean, variance, and standard deviation
14. Convert the extracted image component from LAB to RGB
15. Convert RGB results to gray-level image
16. Compute gray-level co-occurrence matrix and gray-level co-occurrence properties
17. Extract features comprising contrast, energy, homogeneity, correlation, dissimilarity, and angular second moment
18. Obtain the features vector comprising mean, variance, standard deviation, contrast, energy, homogeneity, correlation, dissimilarity, and angular second moment features
19. Use leaves classification dataset and obtain train set and test set
20. Train the model using SVC, random forest, and voting classifiers
21. Predict the class of the features vector
22. Compute accuracy.

4 Evaluation Results

The proposed machine learning-based approach for nutrient deficiency identification is implemented using Anaconda 3.8 Python module on platform i3 with 4 GB RAM. The opencv, numpy, skimage, pandas, and sklearn libraries of Python are used for the implementation. Figure 1 shows the two cases evaluated on the proposed machine learning model comprising a screenshot of input image, RGB image, and extracted nutrient deficiency segments.

Figure 2 shows the screenshot of results obtained on the anaconda prompt during performance evaluation of the proposed machine learning approach. In Table 2, the features value extracted from the images shown in Fig. 1 and their identified nutrient deficiencies are shown.

5 The Conclusion & Future Scope

In this paper, a proposed machine learning-based approach is presented which utilizes random forest, voting classifier, linear support vector classifier, and k nearest neighbor

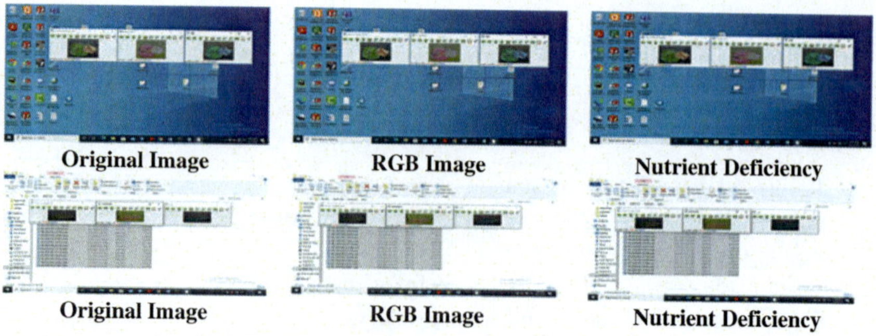

Fig. 1 Two cases evaluated on the proposed machine learning model comprising screenshot of input image, RGB image, and extracted nutrient deficiency segments

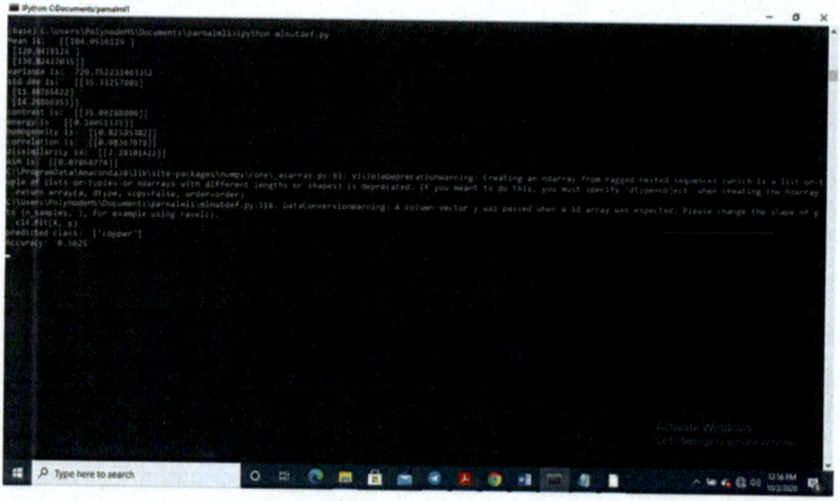

Fig. 2 Screenshot of results obtained on the anaconda prompt during performance evaluation

Table 2 Extracted features and identified nutrient deficiency and accuracy for two cases shown in Fig. 1

Contrast	Energy	Homogeneity	Correlation	Dissimilarity	ASM	Nutrient deficiency	Accuracy
35.0928	0.28051	0.82585	0.98368	2.28101	0.07869	Copper	0.5625
5.67581	0.19457	0.80241	0.99046	0.91663	0.03786	Magnesium	0.59375

classifier for identification of nutrient deficiencies in plant leaves. In the proposed approach, features are extracted from gray-level co-occurrence matrix and gray-level co-occurrence properties of leaf images. The extracted features comprise mean, standard deviation, variance, contrast, energy, homogeneity, correlation, dissimilarity, and angular second moment. The extracted features are utilized for identifying nitrogen, potassium, copper, and magnesium nutrient deficiencies in plant leaf. The evaluation results and accuracy of the proposed approach are presented. From the evaluation results, it can be concluded that the average accuracy of the proposed approach is 60%. In future, modification in the proposed machine learning-based approach can be explored for improving the overall accuracy of nutrient deficiency identification. In future, the proposed approach can be evaluated for identifying the nutrient deficiencies in other parts of plants.

References

1. Shah A, Gupta P, Ajgar YM (2018) Macro-nutrient deficiency identification in plants using image processing and machine learning. In: Proceedings of 2018 3rd IEEE international conference for convergence in technology (I2CT), November 2018
2. Sosa J, Ramírez J, Vives L, Kemper G (2019) An algorithm for detection of nutritional deficiencies from digital images of coffee leaves based on descriptors and neural networks. In: Proceedings of 2019 XXII IEEE symposium on image, signal processing and artificial vision (STSIVA), June 2019
3. Chen Q et al (2019) Autophagy and nutrients management in plants. Cells 8(11):1–17
4. Latte MV, Shindal S (2016) Multiple nutrient deficiency detection in paddy leaf images using color and pattern analysis. In: Proceedings of 2016 IEEE international conference on communication and signal processing
5. Tomas JF, Zitova SB (2017) 2D and 3D image analysis by moments. Wiley
6. Makkar T et al (2019) A computer vision based comparative analysis of dual nutrients (boron, calcium) deficiency detection system for apple fruit. In: Proceedings of 2018 4th IEEE international conference on computing communication and automation (ICCCA), July 2019
7. Watchareeruetai U, Noinongyao P, Wattanapaiboonsuk C, Khantiviriya P, Duang S (2019) Identification of plant nutrient deficiencies using convolutional neural networks. In: Proceedings of 2018 IEEE international electrical engineering congress (iEECON), May 2019
8. Merchant M, Paradkar V, Khanna M, Gokhale S (2018) Mango Leaf Deficiency Detection Using Digital Image Processing and Machine Learning. In: Proceedings of 2018 3rd IEEE international conference for convergence in technology (I2CT), April 2018
9. Han KAM, Watchareeruetai U (2019) Classification of nutrient deficiency in black gram using deep convolutional neural networks. In: Proceedings of 2019 16th IEEE international joint conference on computer science and software engineering (JCSSE), Oct 2019
10. Chang Y, Le Moan S, Bailey D (2000) RGB imaging based estimation of leaf chlorophyll content. In: Proceedings of 2019 IEEE international conference on image and vision computing New Zealand (IVCNZ), Jan 2020

Chapter 28
Software Quality Enhancement Using Hybrid Model of DevOps

Pooja Batra and Aman Jatain

1 Introduction

In today's era, progressive adoption of software development reaches the world of DevOps. Traditional software development process was suffering from many issues like unreliable products, risk occurrence, unsatisfied customers, delayed projects, and many more. DevOps is a mitigation strategy for all issues facing existing development methods, and this is the reason that software organizations are ready to adopt DevOps practices [1]. In spite of many benefits, software firms that are adopting DevOps processes are facing many challenges while implementing it such as non-standardization of process, not defined performance metrics, existing systems are not evaluated [2]. To overcome these types of problems, a hybrid model is proposed that not only defines a standard for DevOps processes but also validates the process by evaluating performance of the overall system. Hybrid models consist of two approaches like test-driven development and DevOps which are discussed in detail in the following section.

1.1 Test-Driven Development

Test-driven development consists of two main concepts, unit tests and refactoring. While writing unit tests programmers are allowed to write it in a detailed manner as compared to functional tests. A small piece of code can be tested at application interface level that is even easier to execute since they do not require the entire production environment for execution. The process of test-driven development (TDD) is

P. Batra (✉)
Career Point University, Kota, Rajasthan 325003, India

A. Jatain
Amity University, Gurugram, Haryana 122051, India

A. Sheth et al. (eds.), *Intelligent Systems*, Algorithms for Intelligent Systems,
https://doi.org/10.1007/978-981-16-2248-9_28

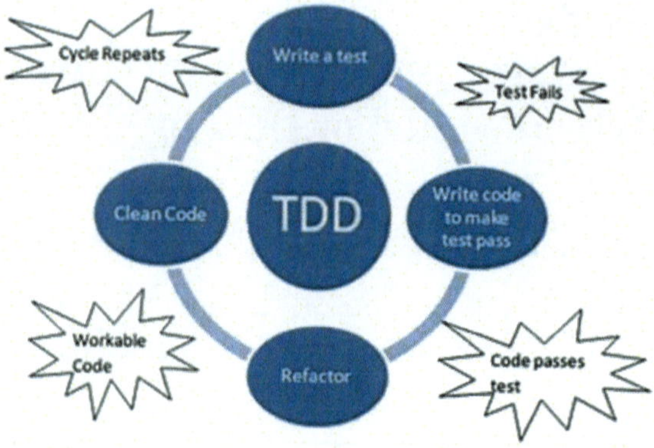

Fig. 1 Process of test-driven approach

overviewed in Fig. 1. Initially, a test is added with the intention of failing the test. The next step test is run but no complete suite is executed for the sake of run time. While executing the code developer ensures the failing of the test as refactoring of code is done to pass the new test. Now if test passes, then development continues until whole test cases are executed.

1.2 DevOps Practices

DevOps is a combination of two terms, development and operations. In almost all software firms, there is a different department for individual tasks. For example, developers' team is made of developers, while QA team is made of software testing personnel and IT executives are part of operations team. While different teams are working on a single project, then conflicts are obvious between teams as they have to deliver work to each other [3]. Better communication and collaboration between development team and operation team is one of the drivers of DevOps processes that yield happy and satisfied customers by fast and reliable delivery of projects [4]. DevOps practices involve build formation, continuous integration, continuous delivery, continuous deployment, and monitoring as shown in Fig. 2.

Paper is organized in the following manner. First section introduced the concepts of test-driven development and DevOps along with the purpose of present research. Main idea of research is also discussed in this section. Second section discusses the related work in the field of TDD and DevOps. Third section describes the detailed methodology of work. Fourth section details the implementation process which includes description of dataset and experimental set-up. Next section outlines results achieved by implementation, and comparative analysis is done to show the usefulness of our approach. Last section concludes the research work and delineates the future aspects of work.

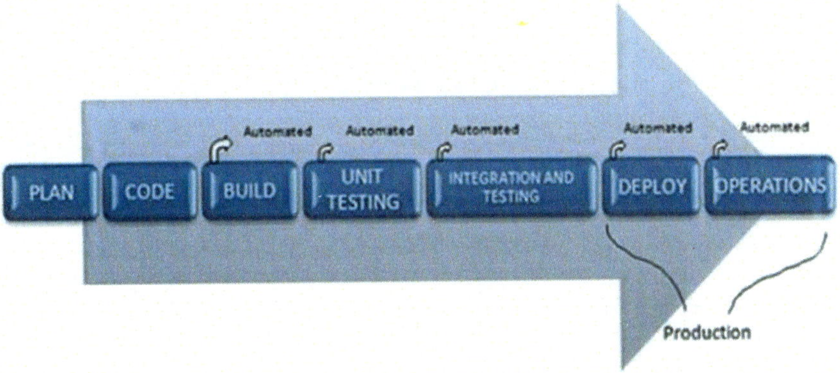

Fig. 2 DevOps process flow

2 Related Work

In related work subsections, two main pillars of hybrid architectures are discussed, i.e., test-driven development (TDD) and DevOps.

Beck et al. [5] propose TDD approach and at that time emerged as a very successful methodology in terms of efficiency [6]. This approach supports write test first technique and also known as test first approach. In this approach, recurrence of steps is followed in a sequential manner. Initially, a test is added before writing a functional code or to just see the failure of code. Further, test cases are executed and succeeded test cases will go next by implementing through code and failed test cases are responsible for updation of code. Test cases are executed again until it passes the process and continues, so on [7, 8]. TDD approach is generally supported by frameworks such as JUnit [9, 10], NUnit [11], PyUnit, and XMLUnit [12]. Some researchers and practitioners described the process of TDD in detail in their research work [13, 14].

DevOps process came from an agile methodology which supports iterative processing. The term DevOps is a permutation of two terms, development (Dev) and operations (Ops) [15]. This approach evolved as a solution to various challenges faced by traditional software development [16, 17]. This approach not only provides faster delivery of products but also produces happy and satisfied customers [18]. According to DevOps concept, it fills the communicational and technological gap between developer and operations. It automates the processes like continuous development, continuous testing, and continuous integration [3]. In traditional and agile methodologies, development and operation teams were working in isolation and yielded communicational gaps. Not even any single process was automated. In DevOps, every process from build creation to monitor tasks every process is automated without any human intervention [19]. In the next section, detailed methodology is explained to scale the process to DevOps.

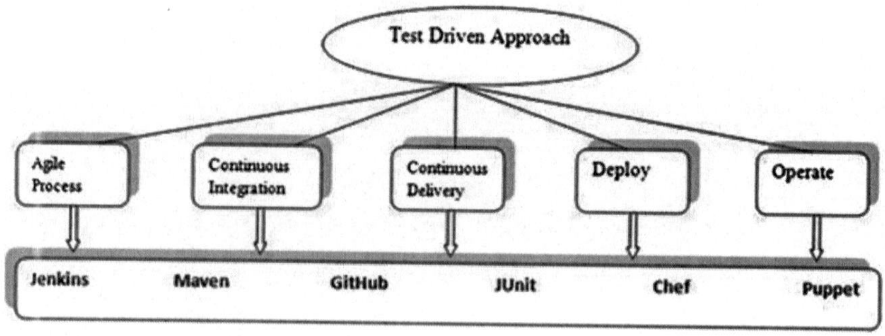

Fig. 3 Hybrid model of proposed approach

3 Proposed Methodology

In the proposed hybrid model, test-driven development is integrated into the DevOps process. Figure 3 shows the architecture of the proposed model.

The hybrid approach model is divided into three layers where the first layer test-driven development approach is applied to every practice of DevOps. In the second layer, DevOps practices are aligned to produce efficient products. Every process in this layer comprises a test-driven approach that incorporates writing of test cases before implementation of each process. In the third layer, various supporting tools are present which will be used to implement integrated models. Although the model consists of three layers, implemented two phases as test-driven development is executed in first phase where test cases are written before execution of those, until it passes all tests, whereas in second phase, DevOps practices are implemented through various mentioned tools and results are achieved.

4 Implementation of Process

Integrated model is implemented through Jenkins tool with customized dataset as discussed in the next section.

4.1 Dataset and Experimental Setup

An application is designed which issues a token for meal coupons in mess in a school. Size of application is 1680 LOC, and number of classes identified are 21, whereas recognized components are 10. All classes have different methods and various attributes. Application is developed in Java environment. Jenkins tool is used to implement DevOps practices, and test cases are written in JUnit framework.

JUnit is used as a plug-in to Jenkins tool. Cobertura tool is used to show coverage percentage of test cases.

4.2 Metrics Evaluation

Two products metrics are chosen to evaluate hybrid models. These metrics are complexity and test coverage. Test coverage can be defined as the percentage of features required by a test strategy that have been exercised by a given test [20]. Statement coverage technique is used while validations of TDD approach. In statement coverage technique, all the executable statements in the source code are executed at least once. Purpose of this technique is calculation of the number of statements in source code which have been executed, and all possible path, lines and statement should be covered statement coverage [20]. Complexity metrics talks about maintainability and understandability of software. It depends upon the line of code metric. The general scale table of complexity shows that it lies in between 1 and 10 on scale. If scale table shows value 1, i.e., complexity is low, and if value is 10, then complexity is at higher level. Mid-value represents the medium complexity of a project.

5 Results and Discussion

When both of the metrics are evaluated on an integrated model, the following table shows the results. Table 1 depicts the area covered under test in different packages. More coverage leads to better quality of software. Although 100% coverage does not guarantee a defect-free product, still test coverage is the best method to analyze quality of software with test-driven development. Better test coverage shows the quality of the product by covering its risk factor.

While the complexity of different packages is nominal, a single package shows complexity of mid value. Less complexity yields the better maintainability of software, and the following table concludes that by using an integrated model approach, software complexity is reduced, and hence, software is more maintainable.

Table 1 Analysis report of test coverage of application

S. No.	Packages	Number of classes	Line coverage (%)	Branch coverage (%)	Complexity
1	edu.ips.admin	16	95	100	5
2	edu.ips.messstaff	2	100	100	1
3	edu.ips.domain	1	98	100	1
4	edu.ips.utility	2	100	100	1

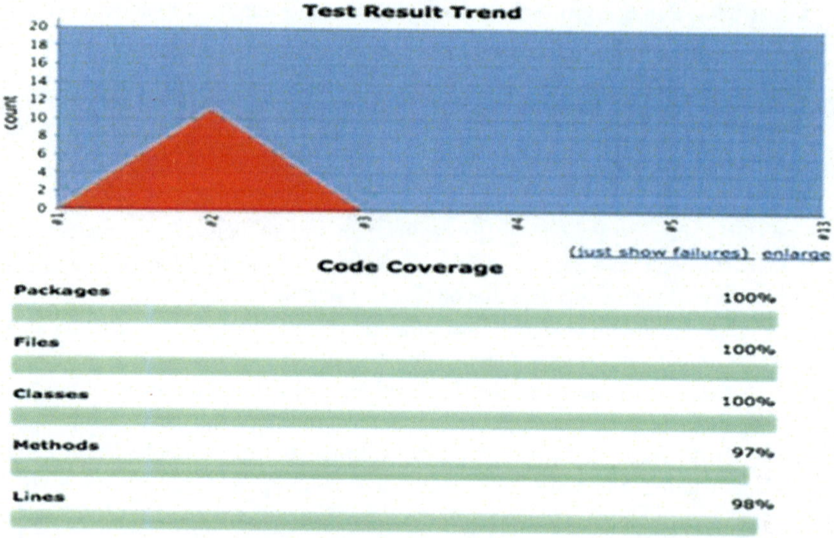

Fig. 4 Jenkins code coverage report with Cobertura plugin

Line coverage and branch coverage area show the good strength as branch coverage presents 100% results and complexity of application is also not high.

Figure 4 shows the glimpse of coverage metric report in Jenkins when Cobertura is used as a plug-in. Results that are displayed in Table 1 are derived through the following report by implementing the above-mentioned dataset. Tool provides individual methods coverage reports also that information will be used to calculate quality of software. Jenkins also provide the functionality of checking coverage reports at every milestone that even enables to check quality concerns at each iterative phase.

5.1 Comparative Analysis of Proposed and Existing Approach

Figure 5 shows comparative analysis of code coverage and code complexity metrics of both approaches and clearly depicts that proposed approach is better than the existing approach on the defined metrics in terms of software quality and maintainability.

6 Conclusion and Future Scope

In this research work, hybrid model for DevOps was presented and evaluated through metrics. As DevOps system is facing challenges of standardization of process along

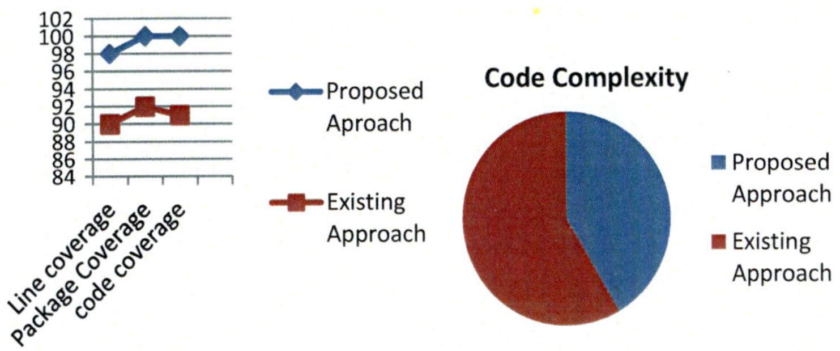

Fig. 5 Comparative analysis of code coverage and complexity of proposed and existing approach

with non-availability of performance defining system. This research fulfills the mentioned research gap by defining standard procedure for DevOps through introduction of integrated models. Results show that software quality is enhanced through test coverage metrics evaluation, and software is more maintainable than existing one. In this way, performance of DevOps systems can be measured easily by measuring software quality and complexity. Although test coverage metrics does not assure a defect less software and can be considered a limitation of this research work, this metric is able to validate a performance measuring system.

In the future, other variants of test-driven approach such as behavior-driven functionality can also be integrated to produce better results and more metrics can be defined and evaluated to create a huge performance measuring tool for DevOps.

References

1. Gottesheim W (2015) Challenges, benefits and best practices of performance focused DevOps. In: LT 2015—Proceedings of the 4th international workshop on large-scale testing, Conjunction with ICPE 2015, vol 3. https://doi.org/10.1145/2693182.2693187
2. Riungu-Kalliosaari L, Mäkinen S, Lwakatare LE, Tiihonen J, Männistö T (2016) DevOps adoption benefits and challenges in practice: a case study. In: Lecture Notes in Computer Science (including subseries Lecture Notes in Artificial Intelligence and Lecture Notes in Bioinformatics). https://doi.org/10.1007/978-3-319-49094-6_44
3. Huttermann M (2012) Integrate development and operations, the agile way. In: DevOps for Developers, pp 196–223. Apress
4. Salmikangas E (2019) An approach to Software deployment information and communications technology, software engineering
5. Beck K (2003) Test-driven development: by example
6. Astel D (2003) Test driven development: a practical guide. Prentice Hall Professional Technical Reference
7. Janzen D, Saiedian H (2005) Test-driven development: concepts, taxonomy, and future direction. Computer (Long. Beach. Calif). https://doi.org/10.1109/MC.2005.314
8. Kollanus S (2010) Test-driven development—still a promising approach? In: Proceedings—7th international conference on the quality of information and communications technology, QUATIC 2010. https://doi.org/10.1109/QUATIC.2010.73

9. Gamma & Beck 2013 JUnit - Google Scholar, https://scholar.google.com/scholar?hl=en&assdt=0%2C5&q=Gamma+%26+Beck+2013+JUnit&btnG=. Last accessed 24 Oct 2020
10. Tahchiev P, Leme F, Massol V, Gregory G (2010) JUnit in action
11. Osherove R (2009) The art of unit testing: with examples in. Net
12. Hamill P (2004) Unit test frameworks: tools for high-quality software development
13. Freeman S, Aaddison-Wesley NP (2009) Growing object-oriented software, guided by tests
14. Mäkinen S, Münch J (2014) Effects of test-driven development: a comparative analysis of empirical studies. In: Lecture notes in business information processing. https://doi.org/10.1007/978-3-319-03602-1_10
15. Ebert C, Gallardo G, Hernantes J, Serrano N, Software technology DevOps.
16. Ravichandran A, Taylor K, Waterhouse P (2016) Practical DevOps. In: DevOps for digital leaders, pp. 125–137. Apress. https://doi.org/10.1007/978-1-4842-1842-6_8
17. Ravichandran A, Taylor K, Waterhouse P (2016) Practical DevOps. In: DevOps for digital leaders, pp. 27–47. Apress. https://doi.org/10.1007/978-1-4842-1842-6_3
18. Hosono S (2012) A DevOps framework to shorten delivery time for cloud applications. Int J Comput Sci Eng 7:329–344. https://doi.org/10.1504/IJCSE.2012.049753
19. Wettinger J, Breitenbücher U, Leymann F (2014) DevOpSlang—bridging the gap between development and operations. In: Lecture Notes in Computer Science (including subseries Lecture Notes in Artificial Intelligence and Lecture Notes in Bioinformatics). 8745 LNCS, pp 108–122. https://doi.org/10.1007/978-3-662-44879-3_8
20. Memon AM, Soffa M, Lou, Pollack ME (2001) Coverage criteria for GUI testing. https://doi.org/10.1145/503209.503244

Chapter 29
Prediction Model for Cervical Cancer in Female Patients Using Machine Learning

Pooja Nagpal and Palak Arora

1 Introduction

Cervical cancer is cancer arising from cervix; basically, it happens due to abnormality of cells in a woman's cervix. Cervical cancer grows slowly in our body but can spread to other parts of our body, often to the lungs, liver, vagina. As per research women under the age 35–50 are more prone to cervical cancer [1]. There is no sign and symptoms at an early stage but with time symptoms start coming up. We can use machine learning techniques to predict cancer.

Machine learning is an application of artificial intelligence (AI) that provides systems the ability to automatically learn and improve from experience without being explicitly programmed. Machine learning algorithm basically used for classification and prediction problems. In our research, we used various risk factors like number of pregnancies, age, smoking, HPV, contraceptives, STDs that lead to cervical cancer, on that basis we are predicting biopsy. It is a diagnostic procedure which finds precancerous cell or cervical cancer. In our paper, we are using the most popular classification algorithm random forest classifier along with python libraries.

1.1 Symptoms of Cervical Cancer

- Abnormal menstrual—irregular bleeding between periods or even after menopause.
- Pain—pelvic pain or pain during intercourse and other body pain.

P. Nagpal (✉) · P. Arora
Computer Science Department, Amity University, Gurgaon, Haryana 122051, India
e-mail: pbnagpal@ggn.amity.edu

P. Arora
e-mail: palak.arora6@student.amity.edu

A. Sheth et al. (eds.), *Intelligent Systems*, Algorithms for Intelligent Systems,
https://doi.org/10.1007/978-981-16-2248-9_29

- Groin—abnormal vaginal discharge or bleeding.
- Fatigue—loss of weight and appetite.
- Difficulty in urinating—pain while urinating or blood in urine.
- Others—swelling in legs, swollen abdomen, nausea, vomiting and constipation.

2 Literature Review

In recent years, the machine learning model achieved a lot in the medical diagnosis field for its prediction system. As it makes things easier on both sides for patients and for doctors, also many research works are still going on for better prediction models. Some of the previous research works are mentioned below.

Kahng et al. [2] developed a web-based prediction tool using the SVM algorithm, and it is basically for the patient who tested positive for HPV and on the edge of developing cervical cancer. The main aim for the application is to visualize patients with the higher risk of cervical cancer.

Vidya and Nasira [3] proposed a model using data mining algorithms such as CART, random forest classifier and K-means learning tree for prediction of cervical cancer. They found the combination of K-means and RFT achieved high accuracy.

Zhou et al. [4] proposed a model for predicting cervical squamous cell carcinoma using a regression algorithm. In their study, they found a linear equation between hTERC, HR-HPV viral load, MCM5 and observe high accuracy 98.5%.

Menon et al. [5] designed a model for cervical cancer prediction. In their work, they compared machine learning algorithms K-nearest neighbor (KNN), decision tree (DT), and random forest. Among them, KNN shows best results in terms of accuracy, precision, F1-score. Also, for KNN false negative value was zero. Hence, KNN algorithm is used for prediction.

Kumar et al. [6] compared various machine learning algorithms on the basis of accuracy, performance matrix and confusion matrix for prediction of biopsy samples of cancer patient. After the execution, the Bayes Net classifier yields better results among others.

Singh et al. [7] designed a model for diagnosing the current stage of cervical cancer, and they used six powerful algorithms and the outcome was decision tree which gives appropriate stage prediction with higher precision and F measure also less false negative value.

Abdullah et al. [8] predicted cervical cancer based on gene expression, and they used two popular classification algorithms of machine learning random forest classifier and support vector machine (SVM). In their work, random forest achieved 94.21% accuracy higher than SVM.

Seera [9] built a prediction model for cervical cancer using machine learning and neural network using behavioral dataset to obtain the pattern according to the dataset and used these patterns to predict risk factor of cervical cancer.

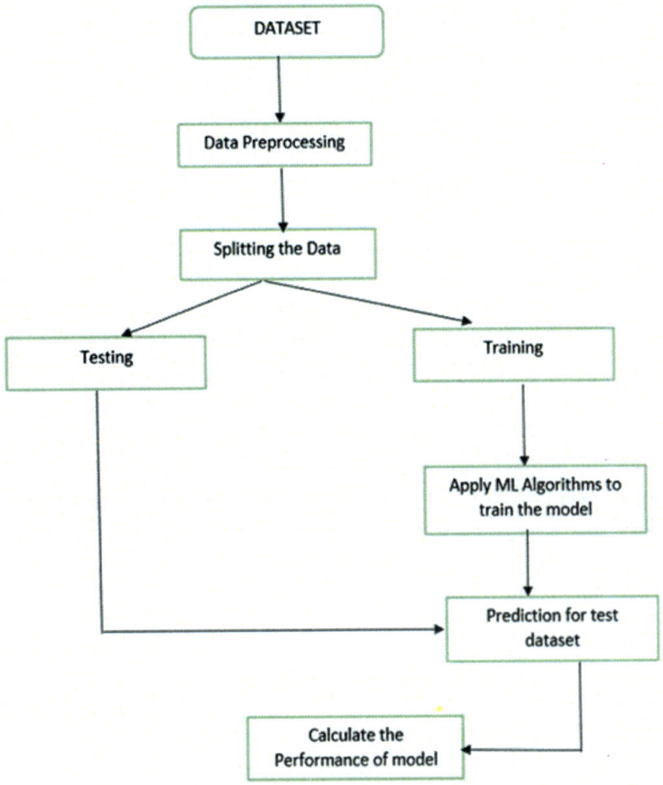

Fig. 1 Implementation of prediction model and its performance

3 Process Flow

Figure 1 shows the brief information for our implementation of our prediction model and its performance; first we will clean the dataset, then preprocessed the dataset by filling all the missing and null values and extracting the main features, and in second step, we will split the dataset into two training and testing phases or input and output. After splitting the dataset, we will train the model using various machine learning algorithms, then will predict the output for the testing set, and then calculate the accuracy of the prediction model and its performance.

4 Methodology

Our main aim is to predict cervical cancer based on a risk factor which is done using machine learning algorithms and then finding out accuracy and performance matrix.

The algorithm used, will evaluate the dataset, and make accurate and necessary predictions. In this paper, we used the random forest classifier algorithm.

4.1 Algorithm Used

Random forest classifier: Random forest is a supervised machine learning algorithm based upon the concept of ensemble learning, used for both classification and regression problems. It is a collection of decision trees used to classify a dataset. It is an updated form of decision tree as it iteratively asks a series of questions and based on that answer it will pass another series of questions to classify data [9] .

Random forest is a collection of multiple decision trees which merge together to give much accurate prediction and less false negative cases; also, more the number of trees will be the accuracy of the prediction. Logic behind the RF is that multiple uncorrelated models perform much better as a group then they do alone. In RF, each tree gives classification as vote and RF considers the classification with the majority of votes.

Below steps explain the working of random forest algorithm [10]:

Step1: Select random data points (say K) from the training set.
Step2: According to the selected data points build the decision trees.
Step3: Select the number for decision trees (say N) that you want to build.
Step4: Repeat the Step number 1 and 2.
Step5: For new data points, find the predictions of each decision tree, and assign the new data points to the category that wins the majority votes.

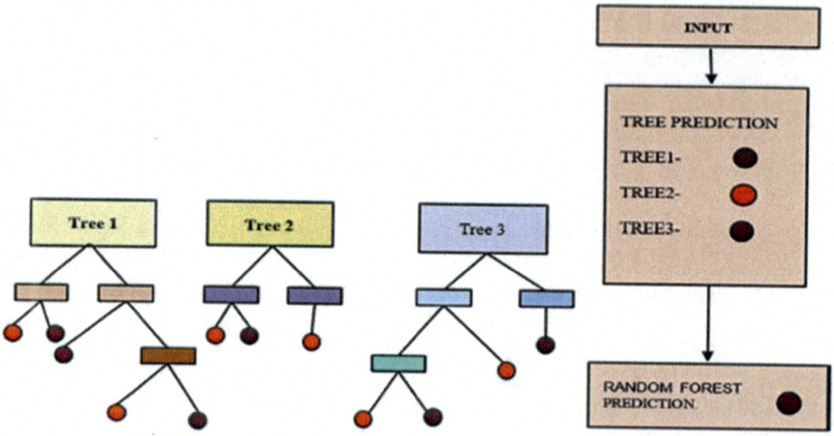

Fig. 2 Visualization of a random forest classifier as mentioned in definition it is a collection of multiple decision trees. It makes predictions from each decision tree and considers the best solution from them by means of voting

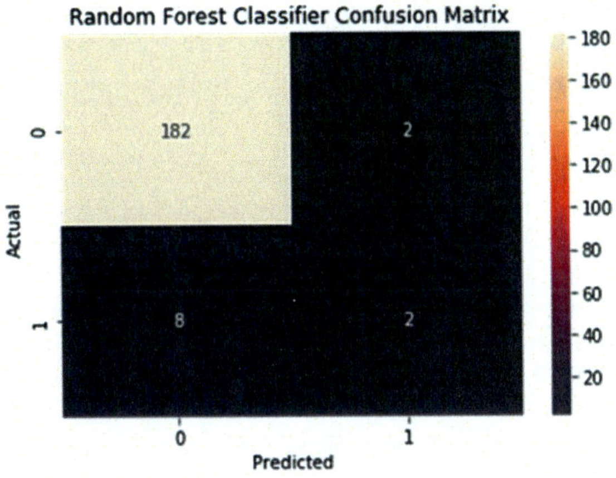

Fig. 3 Confusion matrix, the matrix is formed using the seaborn library of python. It displays the True Positive (TP), True Negative (TN), False Positive (FP), False Negative (FN) value in a precise manner

4.2 Confusion Matrix

Confusion matrix is also known as error matrix. It basically measured the performance of classification problems (Fig. 3).

Terms of Confusion Matrix:

- True Positive (TP)—Refers to when the model predicts positive, and it is actually positive.
- False Positive (FP)—Refers to when the model predicts positive but it is actually negative.
- True Negative (TN)—Refers to when the model predicts negative but it is actually negative.
- False Negative (FN)—Refers to when model predicts negative but it is actually positive.

4.3 Classification Report

Classification reports evaluate the precision, recall, and F1 score values; they basically measure a model's predictive power. It is important to have a classification report of a model as it calculates the performance of classification problems.

Classification report terms:

- Precision—Precision calculates the accuracy of positive prediction means when a model predicted true/positive, then how often was it right?
- Recall—Recall is also known as sensitivity or true positive rate. It is the ability to find all positive samples.
- F1 score—F score or F measure calculates the test's accuracy. F score is the harmonic mean of precision and recall.

5 Implementation

This section discusses the dataset and tools used for our research.

5.1 Dataset

We have taken the dataset from kaggle. Link for the same is given here. https://www.kaggle.com/loveall/cervical-cancer-risk-classification. Dataset was collected at 'Hospital Universitario de Caracas' in Caracas, Venezuela. Dataset is classified into various risk factors for cervical cancer leading to biopsy test. Dataset has a record of 968 patients with 36 different risk factors with some missing values due to personal reasons; hence, we modify the dataset into some useful data with no null values and 21 risk factors.

5.2 Tools Used

For implementation of the model, we used Anaconda, an open-source distribution for Python and R language use for scientific computing. We further worked using the Jupyter notebook as it supports the python programming language.

6 Results

The reliability and accuracy of cervical cancer prediction using RF algorithm are 94.84% higher than the past studies as shown in Fig. 9. Random forest gives us higher precision value (0.93), recall value (0.95), and F1 score (0.94) refer to Fig. 4.

```
report_rfc=classification_report(y_test,model_predict)
print(report_rfc)
```

	precision	recall	f1-score	support
0	0.96	0.99	0.97	184
1	0.50	0.20	0.29	10
accuracy			0.95	194
macro avg	0.73	0.59	0.63	194
weighted avg	0.93	0.95	0.94	194

Fig. 4 Classification report of the proposed model having precision, recall and F1 score based on support—a number of actual occurrences in a class

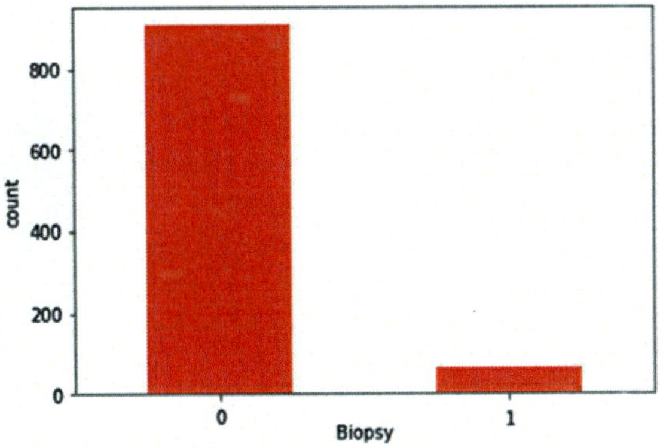

Fig. 5 Number of patients who must go for a diagnostic procedure biopsy through a bar graph. Here we can clearly see most of the patients are healthy

```
df.head(10)
```

	Age	Numberofsexualpartners	Firstsexualintercourse	Numofpregnancies	Smokes	Hormonal_Contraceptives	IUD	STDs	STDs_condylomatosis	STDs_syphilis
0	18	4	15	1	0	0	0	0	0	0
1	15	1	14	1	0	0	0	0	0	0
2	34	1	17	1	0	0	0	0	0	0
3	52	5	16	4	1	1	0	0	0	0
4	46	3	21	4	0	1	0	0	0	0
5	42	3	23	2	0	0	0	0	0	0
6	51	3	17	6	1	0	1	0	0	0
7	26	1	26	3	0	1	1	0	0	0
8	45	1	20	5	0	0	0	0	0	0
9	44	3	15	2	1	0	0	0	0	0

10 rows × 21 columns

Fig. 6 Attributes of the dataset after data preprocessing used to predict diagnosis test biopsy

PREDICTING BY VALUES

```
model.predict([[17,2,15,1,0,0,0,0,0,0,0,0,0,0,0,0,0,0,0,0]])
```

```
array([0], dtype=int64)
```

```
model.predict([[51,3,17,6,1,0,1,0,0,0,0,0,0,0,0,0,0,0,0,0]])
```

```
array([1], dtype=int64)
```

Fig. 7 Prediction of diagnostic procedure biopsy using predict function

From the confusion matrix (Fig. 3), RF shows less FN cases means when a patient should undergo biopsy, the machine considered as fine on the basis of the dataset provided.

On compiling the above values, we can say that RF suited best for the proposed model (Fig. 8).

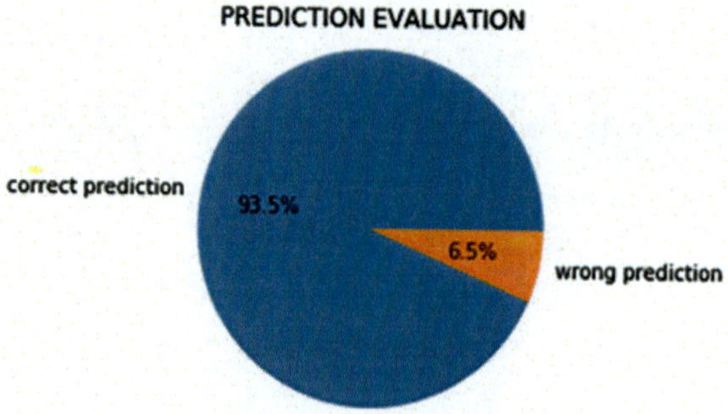

Fig. 8 Evaluation for the model the percentage is on the basis of the test set which we generated using splitting and the accuracy we got. It clearly shows model efficiency as the correct prediction rate is quite high. Hence, it gives us a clear vision that random forest classifier is best of the model we implemented

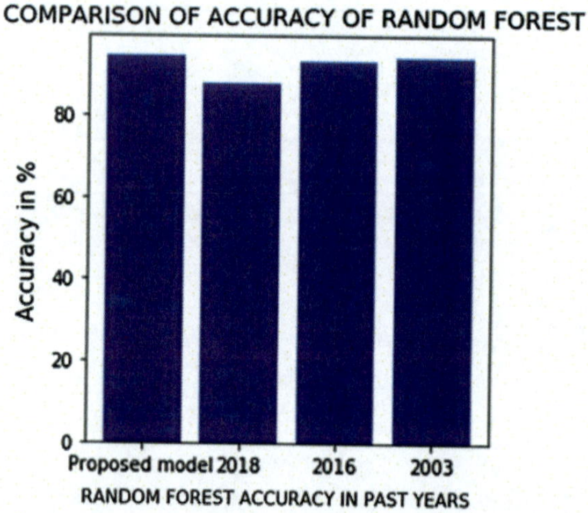

Fig. 9 Comaparison of Accuracy of Random Forest Algorithm

6.1 Comparison of Accuracy of Random Forest

Accuracy comparison of the proposed model and past studies is shown in figure 9. We can see random forest gives better accuracy, and for the proposed model, the accuracy is slightly more compared to past studiesSee Fig. 9.

6.2 Conclusion and Future Work

The main motive of this research is to predict the cervical cancer in women which take inputs of female behavioral and their medical terms and consider the risk factors for cervical cancer and then the model predicts the diagnostic procedure.

We worked with a single machine learning algorithm, but in the future, we can extend our learning and can work with a hybrid model of various algorithms that can be prepared to generate better precise values. Furthermore, risk factors can be identified that can be supported through multidimensional data.

References

1. Saha A, Chaudhury AN, Bhowmik P, Chatterjee R (2010) Awareness of cervical cancer among female students of premier colleges in Kolkata, India. Asian Pac J Cancer Prev
2. Kahng J, Kim E-H, Kim H-G, Lee W (2015) Development of a cervical cancer progress prediction tool for human papillomavirus-positive Koreans: a support vector machine-based approach. Res Rep J Int Med Res 43:518–525. https://doi.org/10.1177/0300060515577846
3. Vidya R, Nasira GM (2016) Prediction of cervical cancer using hybrid induction technique: a solution for human hereditary disease patterns. Indian J Sci Technol 9. https://doi.org/10.17485/ijst/2016/v9i30/82085. www.indjst.org
4. Zhou Y, Fan W, Deng J, Xi HL (2017) Establishment and analysis of the prediction model for cervical squamous cell carcinoma
5. Parikh D, Menon V (2019) Machine learning applied to cervical cancer data. Int J Math Sci Comput 1:53–64. https://doi.org/10.5815/ijmsc.2019.01.05
6. Kumar Suman S, Hooda N (2019) Predicting risk of Cervical Cancer: a case study of machine learning. J Stat Manage Syst 22:689–696.https://doi.org/10.1080/09720510.2019.1611227
7. Singh J, Sharma S (2019) Prediction of cervical cancer using machine learning techniques
8. Abdullah AA, Abu Sabri NK, Khairunizam W, Zunaidi I, Razlan ZM, Shahriman AB (2019) Development of predictive models for cervical cancer based on gene expression profiling data. iopscience.iop.org. 557. https://doi.org/10.1088/1757-899X/557/1/012003.
9. Seera M, Lim CP (2014) A hybrid intelligent system for medical data classification. Expert Systems with Applications. Elsevier

Chapter 30
A Comprehensive Study of Feature Selection Techniques for Evaluation of Student Performance

Randhir Singh and Saurabh Pal

1 Introduction

Data mining, statistical methods, and machine learning are applied on EDM results to get information from instructive or education conditions. As of now, it is popular and increasing more consideration in view of increment in the instructive data of e-learning frameworks, and in any event, advancing conventional training. The developing methods for finding the unmistakable sorts of data present in academic conditions try to separate important results to progress and value learning measures from immense measures of crude data [1]. There is proper monitoring to foresee future students' conduct to improve educational program plan and plan meditations for scholastic help and direction on the educational plan. For this purpose, data mining [2] becomes an integral factor. DM strategies dissect datasets and extricate facts to change it into justifiable structures for some time in the future. Machine learning (ML), collaborative filtering (CF), etc., are computer-assisted methodologies for an. The examination led by Mueen et al. [3] is one of the underlying investigations which explored the use of analyzing the outcomes of students enrolled in any academic course. They may also calculate the rate of failure of students. The most critical commitment by this investigation was that it was a pioneer and cut the way for a few such examinations. The financial achievement of any nation exceptionally relies upon making advanced education more reasonable and that thinks about one of the primary worries for any legislature. One of the components that add to the instructive costs is the examining time spent by students to graduate. For instance, the financial loan of the American students has been expanded because of the disappointment of many students in getting graduated on time [4, 5]. Advanced education is given to the students in Iraq by the legislature. However, coming up short of graduating on time costs the administration other miscellaneous costs. To evade these costs, the administration needs to guarantee that the student graduates on schedule.

R. Singh (✉) · S. Pal
VBS Purvanchal University, Jaunpur, UP 222001, India

A. Sheth et al. (eds.), *Intelligent Systems*, Algorithms for Intelligent Systems,
https://doi.org/10.1007/978-981-16-2248-9_30

1.1 Analyzed Component

Different components were chosen in this research work and comparative analysis; their performance has been done using an analysis tool. Educational dataset was pre-processed therefore MLP, SVM, K-NN classifiers, decision tree, training, and testing were performed in ten different folds resulting in an accurate model. The obtained results from the proposed model were measured in different terms like accuracy and ROC curve area.

The first step is gathering the dataset from the data sources. For our situation, the data has been gathered by utilizing a review given to the students and the student's evaluation book. The later step is the pre-processing the observations to get a standardized dataset and afterward marking the data lines. In the third step, the preparation and testing of dataset are done through the machine learning calculation. The machine learning algorithm constructs a model utilizing the preparation data and tests the model utilizing the test data. At last, the machine learning algorithm creates a model or a prepared classifier that can take another data as an information column and predicts its name.

2 Literature Review

Fernandes [1] used analysis algorithm to evaluate the scores of the graduate degree and current performance in post-graduate courses which will help in uncovering information that queries and reports would not be able to offer. It presumed that Naive Bayes for sure performed in a way that is better than some other machine learning calculation. She likewise inferred that "Instead of demographic characteristics of students, using first attendance and homework grades produce better prediction rate at earlier stages." The study demonstrated important aspect in affirming the uniqueness of the proposed application [6]. The work also showed that although there are techniques available to find rate of people leaving a course but less work has been done in the field of performance analysis of enrolled students. A unique way of measuring the performance of an understudy is by analyzing the daily activities that he or she does on Moodle platform. Based on this idea, predicting frameworks have been made by random forest and support vector machine. Random forest gave much reliable and good result [7, 8]. These strategies are further followed by coding and overall give significant results [3, 9]. It is conceivable to expect last students' performance before because of conduct data enhanced with other more significant data [10]. Moreover, the data acquired from records seem to be significant, surprisingly better than applying course subordinate recipes to foresee performance.

3 Research Experiment

This part deals with the parameters that are instrumental in foreseeing the academic performance of an understudy. The features along with the methodologies of prediction are the two pivotal things on which this entire process depends. Figure 1 shows the block diagram of performance foreseeing methods which are computer based. Firstly, the student data is organized and used as an input. This is followed by selecting features, and for this purpose, various machine learning models are employed. Further training and testing of the machine learning models are done, and then, we get the results. Data mining is the most important thing that is used in this process. The detailed description of the significance of data mining has been explained in the subsequent section.

3.1 Feature Selection

The prediction model performance is incredibly relying upon the alternative of selection of most significant features from the rundown of features which is utilized in the student dataset. This can be accomplished by methods for using different feature selection techniques on datasets. Truth be told, level of exactness is generally not picked for characterization, as estimations of precision are profoundly as per the base paces of various classes. Moreover, numerous elements influence the accomplishment of data mining calculations on a given errand. The value of the data is characterized as the boundary of investigating the data is unessential or repetitive, and at that point, information revelation during preparation is more troublesome. As a rule, attribute subset selection is the way toward identifying and dispenses with both unimportant and excess features as hard as could be expected under the

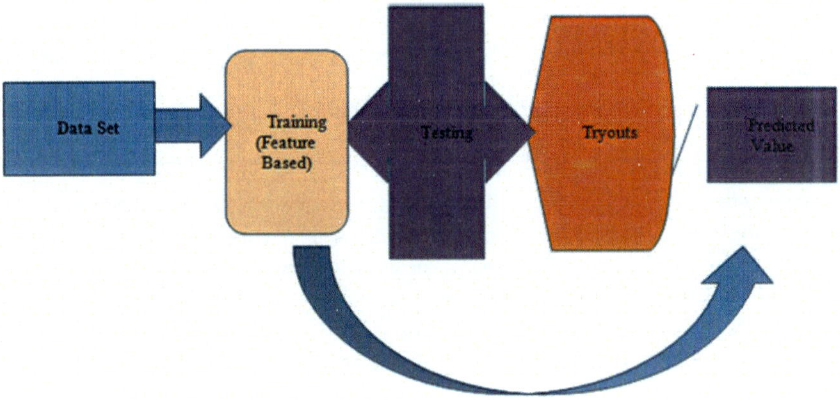

Fig. 1 Process flow diagram

circumstances. Feature selection techniques are used to recognize which feature has the best impact on our yield variable (scholarly status). There are a broad scope of trait selection techniques that can be gathered in different practices. One well-known arrangement is one in which the methodologies contrast in the manner in which they gage credits and are named: channels, which select and examine features independently of the learning cycle and coverings, which use the classifier performance to appraise the attractive quality of a subset. Feature selection has been a functioning and productive field of examination zone in design acknowledgment, machine learning, measurements, and data mining networks [11, 12]. Feature selection principle is to choose a subset of erasing information factors targeting features, which are superfluous or of no prescient data. Feature selection has demonstrated in both hypothesis and practice to be powerful in improving learning productivity, expanding prescient precision, and diminishing intricacy of educated outcomes [6, 13]. Feature selection in directed learning has a primary objective of finding a feature subset that produces higher classification precision or accuracy (Table 1).

4 Results and Discussion

In this paper, an algorithm for predicting student performance as well as earlier 2 years record for same course and behavioral analysis has been proposed. The performance of student has been calculated through various approaches. This is then saved in the database for further evaluation. Some fixed no. of instances for behavioral analysis of student has been used. Average of this prediction leads to calculation of impact factor. The impact factor calculated for the method is used for calculating impact from average values of obtained accuracy for comparative study about methods. The current examination centers on different feature selection strategies that are habitually used in data pre-processing for data mining. The overall systems on feature selection about filter strategy are followed with the impact of feature selection methods on a produced database on higher secondary students. Viability of the calculations or algorithms is introduced about various estimates like PCA values and F1-measure esteems. At first, all feature selection techniques are used on the first feature set, and the features have been positioned by their benefits in rising trends. Since no understanding was found among the feature positioning techniques, we performed the student performance assessment based on PCA value, and F1-measure esteems on many subsets of feature vectors. Indeed, the assessment based on PCA value and F1-measure was completed iteratively on the different subsets beginning from two with one as an augmentation from the position or rank-based list.

Initially, all feature variables were applied on the original feature set, and the features were valued according to their merits in ascending or descending order. Since comparison was found among the feature statistics, we performed student performance evaluation in terms of values on multiple subsets of feature vectors (Figs. 2, 3, and 4). In fact, the evaluation on the basis of measure was carried out

Table 1 Feature selection criteria

Feature	Attribute	Domain
S1	Sex of students	2 = Female, 3 = Male
S2	Students category	2 = General, 3 = OBC, 4 = SC, 5 = ST, 6 = Minority
S3	Discussion at home	2 = Always, 3 = Almost always, 4 = Sometimes, 5 = Never
S4	Own computer/laptop	2 = Yes, 3 = No
S5	Laptop shared with family	2 = Yes, 3 = No
S6	Study desk at home	2 = Yes, 3 = No
S7	Own mobile phone	2 = Yes, 3 = No
S8	Own gaming system	2 = Yes, 3 = No
S9	Heating/cooling systems at home	2 = Yes, 3 = No
S10	Absent from school	2 = Once a week or more, 3 = Once every two weeks, 4 = Once a month, 5 = Never or almost never
S11	How often use computer/laptop at home	2 = Every day or almost every day, 3 = Once or twice a week, 4 = Once or twice a month, 5 = Never or almost never
S12	How often use computer at school	2 = Every day or almost every day, 3 = Once or twice a week, 4 = Once or twice fifteen days, 5 – Once or twice in a month, 6 = Never or almost never
S13	Access textbooks	2 = Yes, 3 = No
S14	Completed assignments	2 = Yes, 3 = No
S15	Collaborate with classmates	2 = Yes, 3 = No
S16	Communicate with teacher	2 = Yes, 3 – No
S17	Students grade in senior secondary education	2 = 90–100%, 3 = 80–89%, 4 = 70–79%, 5 = 60–69%, 6 = 50–59%, 7 = 40–49%, 8 ≤ 40%
S18	Fathers qualification	2 = elementary, 3 = secondary, 4 = graduate/post-graduate, 5 = doctorate
S19	Mother's qualification	2 = elementary, 3 = secondary, 4 = graduate/post-graduate, 5 = doctorate
S20	Father's occupation	2 = Service, 3 = business, 4 = not applicable
S21	Mother's occupation	2 = Housewife, 3 = Service, 4 = business, 5 = not applicable
S22	Grade obtained in B.C. A	2 ≥ 60%, 3 ≥ 45 & < 60%, 4 ≥ 36 & < 45%, 5 ≤ 36%

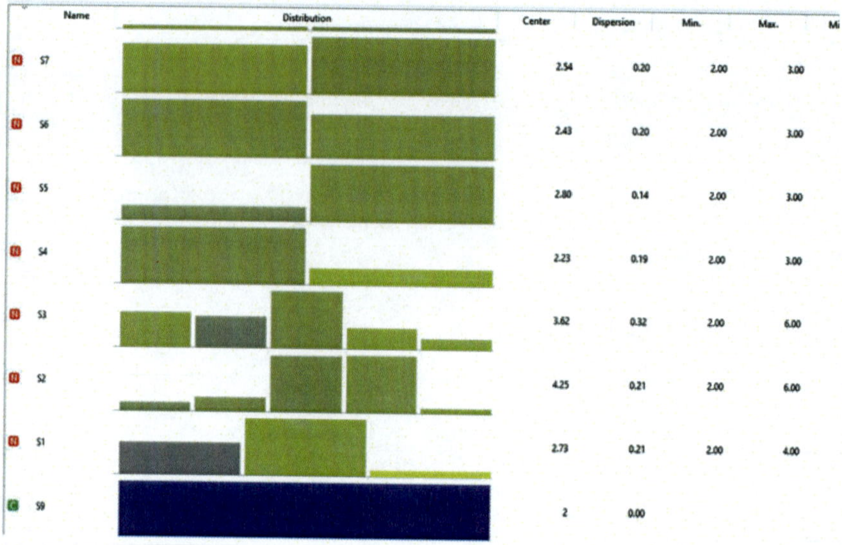

Fig. 2 Feature statistics_1

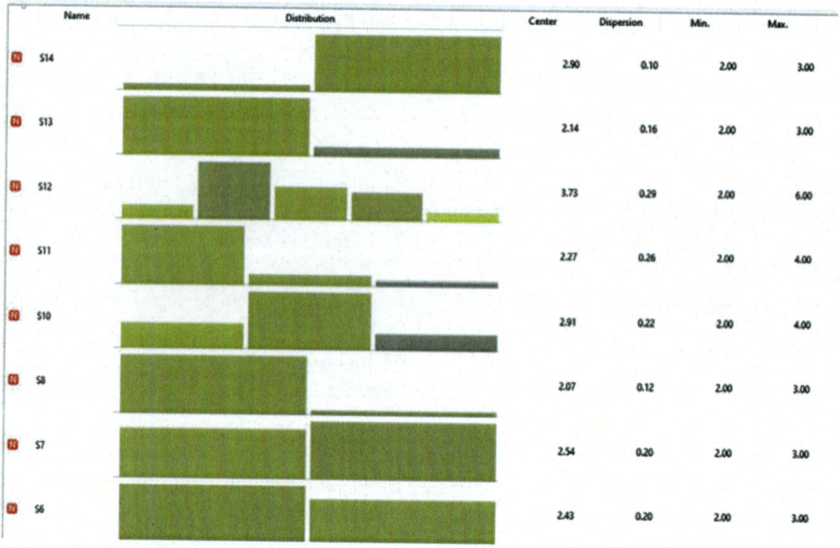

Fig. 3 Feature statistics_2

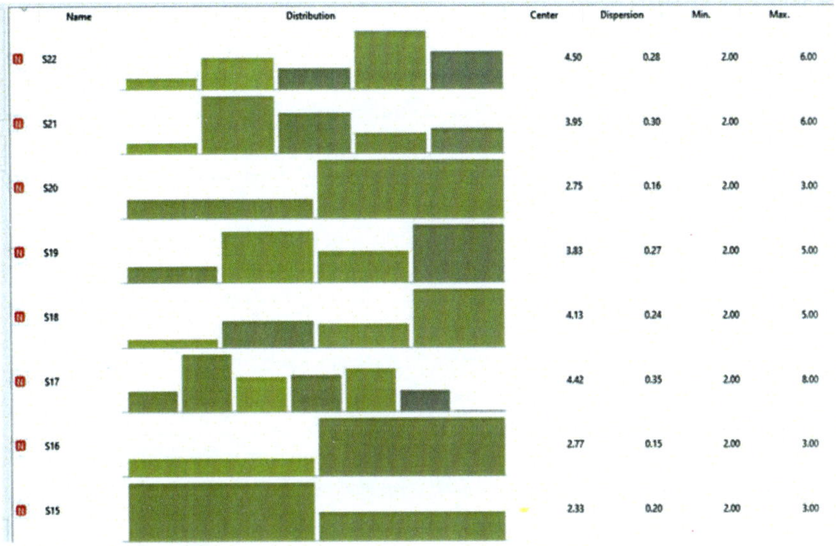

Fig. 4 Feature statistics_3

iteratively on the multiple subsets starting from two with one as an increment from the feature values.

Principal component analysis is an unaided feature reduction technique for anticipating high-dimensional data into another lower-dimensional portrayal of the data that depicts however much of the variance in the data as could be expected with least remaking error. In Fig. 5, principal component analysis is a quantitatively rigorous

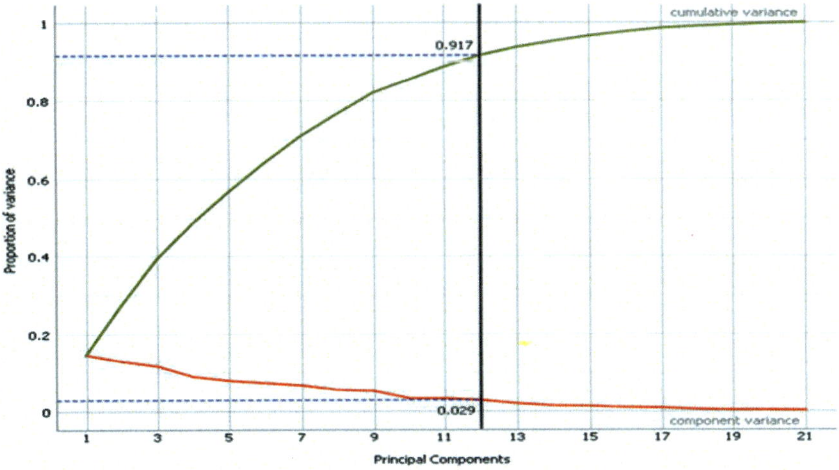

Fig. 5 PCA observations

method for achieving X-axis (component variances) and Y-axis (cumulative variances). This method creates a new set variable, called principal component. Each major component is a linear combination of the original variables which is depicted in Fig. 7. All the essential segments are symmetrical to one another, so there is no repetitive data. Figure 6 presents the consequences of rank utilizing both crisp-valued and real-valued informational collections. At first, we have assessed the algorithm on a dataset, which is accessible in the UIM.

The experiments combining training and testing data, classifiers match columns by names, not by their positions (e.g., first column in one table to the first column in another). Since it does not do so in general, it also does not do it for targets; even if you have just a single target in each data instance. Figure 8 presents the comparative study with various models and corresponding train time, test time, MSE, RMSE, MAE, and R2 values. R2 values illustrate the accuracy level of prediction method and dataset errors level.

A decision tree (or tree diagram) is a decision support instrument that utilizes a tree-like chart or model of decisions and their potential results, including chance occasion results, asset expenses, and utility. Decision tree induction method has

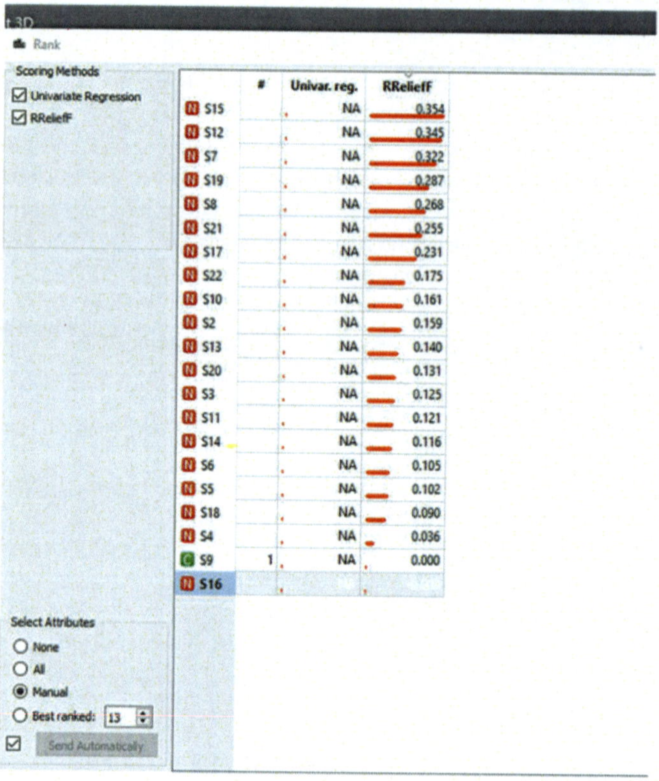

Fig. 6 Rank statistics

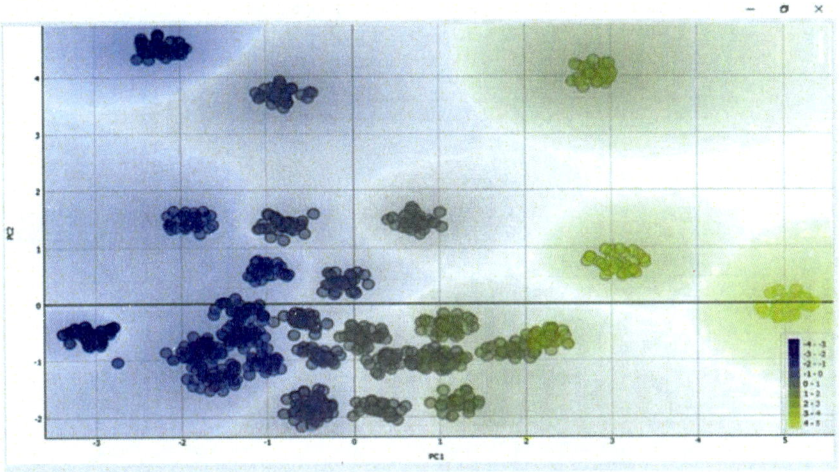

Fig. 7 Scatter plot analysis of instances

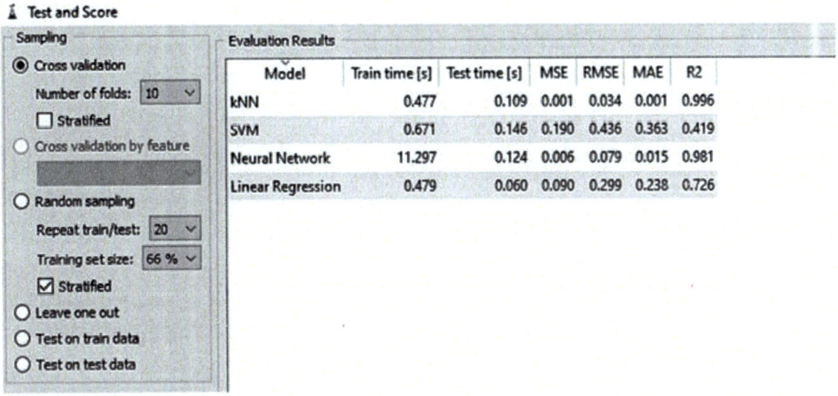

Model	Train time [s]	Test time [s]	MSE	RMSE	MAE	R2
kNN	0.477	0.109	0.001	0.034	0.001	0.996
SVM	0.671	0.146	0.190	0.436	0.363	0.419
Neural Network	11.297	0.124	0.006	0.079	0.015	0.981
Linear Regression	0.479	0.060	0.090	0.299	0.238	0.726

Fig. 8 Comparative study of various methods

been effectively utilized in master frameworks in catching the information. Decision tree enlistment method is useful for various attribute datasets. The proposed tree characterization algorithm is depicted in Fig. 9 with five phases, and it has been divided into low-level forecasts.

5 Discussion

Education system data mining is very relevant to the analysis of the students' performance in academics by considering various performance factors and for prediction

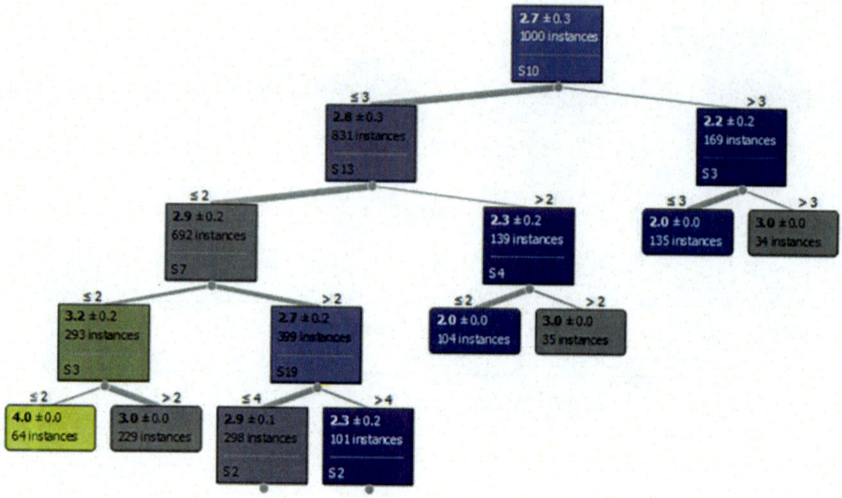

Fig. 9 Tree classification

purposes. It will play a major role in their care at the right time for creative development and direction of study students, after predicting and analyzing the instances and hopefully would lead to the decline in dropouts. Several strategies have been evaluated and implemented enabling the educational institutions to transform bulk information into information having good predictability, stability, and profits. Feature choice technique is applied for arranging the performance of the understudy in the educational institutions. This evaluation has shown the instances features and eliminated the redundant features present in the relevant features as much as possible. This has improved the accuracy of the outcomes acquired from various methods. Our outcomes are empowering and have to be approved by bigger examples of courses from various offices and projects. A fascinating errand is to apply a model for a particular assignment, such as predicting the understudy's performance, for another connected undertaking, such as the prediction of understudy's dropout or for regression tasks (e.g., for predicting student's background). This can help to analyze the results in much effective way that actually contributes to improvement of methods with various parameters related to the student performance.

References

1. Hussain S, Dahan NA, By-Law FM, Ribata N (2018) Educational data mining and analysis of students' academic performance using WEKA. Indonesian J Electr Eng Comput Sci 9
2. Hasan R, Palaniappan S, Mahmood S, Abbas A, Sarker KU, Sattar MU (2020) Predicting student performance in higher educational institutions using video learning analytics and data mining techniques. Appl Sci 10:3894. https://doi.org/10.3390/app10113894

3. Mueen B, Zafar A, Manzoor U (2016) Modeling and predicting students' academic performance using data mining techniques. Int J Modern Educ Comput Sci 8:36
4. Fernandes E, Holanda M, Victorino M, Borges V, Carvalho R, Erven GV (2019) Educational data mining: predictive analysis of academic performance of public-school students in the capital of brazil, vol 94, pp 335–343., IEEE
5. Baker R, Yacef K (2009) The state of educational data mining in 2009: a review and future visions. JEDM 1(1):3–17
6. Iqbal Z, Qadir J, Mian AN, Kamiran F (2017) Machine learning based student grade prediction: a case study
7. Motohashi H, Teraoka T, Aoki S, Ohwada H (2018) Regression models and ranking method for p53 inhibitor candidates using machine learning. In: International conference on bioinformatics and biomedicine (BIBM)
8. Sivakumar S, Venkataraman S, Selvara R (2016) Predictive modeling of student dropout indicators in educational data mining using improved decision trees. Indian J Sci Technol 9
9. Altujjar Y, Altamimi W, Al Turaiki I, Al-Razgan M (2016) Predicting critical courses affecting students' performance: a case study. Procedia Comput Sci 82:65–71
10. Jalota C, Agrawal R (2019) Analysis of data mining using classification. In: IEEE international conference on machine learning, big data, cloud and parallel computing (COMITCON)
11. Serra A; Perchinunno P; Bilancia M (2018) Predicting student dropouts in higher education using supervised classification algorithms. Lect. Notes Comput. Sci., 10962 LNCS, pp 18–33
12. Zaffar M, Hashmani MA, Savita KS (2017) Performance analysis of feature selection algorithm for educational data mining. In: IEEE conference on big data and analytics (ICBDA)
13. Longadge R, Dongre SS, Malik L (2013) Class imbalance problem in data mining: Review. Int J Comput Sci Netw 2(1):83–87

Chapter 31
Computational Modeling and Governing of Standalone Hybrid Electric Power Generation System

Raviprabhakaran Vijay and B. Deepika

1 Introduction

The hasty rise in energy demand and intensifying apprehensions in the direction of the ecological impact owing to the high degree of dependability on fossil fuels, renewable power production, and sparkling power techniques play a crucial part in the future tolerable power system. The dispersed non-conventional power systems and small-production systems such as solar photovoltaic array (PVA) and wind farms are offered to reduce power demand from the utility grid. Since non-conventional energies are from the organic environment, it is dependent on all seasons this constructs them insecure once a renewable resource is applied to shape a constant power system. With an amalgamation of numerous renewable resources for instance PVA and wind supplementing separately, someplace sunshine is accessible over the daytime, and wind at dark in wintertime and summer seasons.

The combination of variable renewable resources is allowed for soft, steady, and authentic output to power grids aimed at the protection, consistency, and constancy of dispatch power, which is inexpensive than capitalizing in sole non-conventional technology [1]. Recently, eco-friendly solutions are gaining importance for energy generation to overcome environmental problems, but uncertainty in nature are the main constraints of non-conventional sources to produce uninterrupted power. To overcome these disadvantages, different renewable resources such as PVA, wind farm, and battery remain to be included. The biggest challenge of this incorporation is to control and handle the power flow [2]. The best design and performance of a connectionless PVA/fuel cell/diesel alternator energy system have been explained as follows.

The chief idea is to propose an energy scheme through a lofty non-conventional portion; truncated greenhouse gas discharges, besides less cost of energy. The main

R. Vijay (✉) · B. Deepika
Department of Electrical Power Engineering, CVR College of Engineering, Hyderabad 501510, India

A. Sheth et al. (eds.), *Intelligent Systems*, Algorithms for Intelligent Systems,
https://doi.org/10.1007/978-981-16-2248-9_31

aim is to control a grid-connected fossil fuel energy system to a non-conventional and good power system [3]. Standalone photovoltaic (PV) systems are mostly used to produce electrical energy in agricultural areas. Migrants in PV radiation influence electrical energy produced by solar standalone PV systems. The PV systems strictly need energy-saving units like batteries to balance the solar energy deficiency. The batteries have the pleasure to store electric strength with an excessive-electricity ratio; however, their potential is constrained. Supercapacitors have a high energy density with a low-electricity density. A combination of each power storage gadget is required to be related in a suitable configuration. So that it caters to the power call for and strength exceptional concerns of the sun standalone PV systems [4].

A Hybrid Power Storage Device (HPSD) is a promising solution for mitigating these strength fluctuations. The electricity that the HPSD additives must deliver/absorb, the Electricity Control Method (ECM) impacts the dimensions/capacity of the Energy storage system. Due to this consideration, sizing and ECM of a battery/supercapacitor (SC) HPSD is mutually optimized employing the use of a deep reinforcement mastering-based total technique. The proposed technique splits the electricity between the HPSD additives such that even as the operational constraints are happy, electricity garage length and losses are minimized [5, 6]. A techno-economic evaluation based totally on incorporating exhibiting, replication, and optimization techniques is accustomed to layout an off-grid hybrid solar/fuel cell control device. The major goal is to enhance the plan and broaden dispatch, manage techniques of the impartial hybrid renewable electricity device to satisfy the anticipated electrically powered load of a residential area positioned in a barren region.

The results of warmness and dirt buildup at the solar panels in the plan and overall concert of the fusion power machine in a wasteland location are examined. The intention of the projected off-grid hybrid renewable electricity system is to increase the penetration of renewable power in the power combination. Lesser the greenhouse fuel discharges after fossil gasoline incineration and decrease the rate of power from the transmission structures [7].

The feasibility and ideal sizing layout of a stand-on wind/hydrogen hybrid power machine for a residence are considered with no connection to the electricity grid lines. The designed device ensures uninterrupted, reliable, continuous electricity to the residence at any time. The wind strength, and the considered hybrid PV-wind and the fuel cell system is a good alternative for providing the power [8]. Off-grid packages (i.e., users not connected to a country's primary electric grid) are assuming an increasing number of essential functions inside the future energy structures [9, 10].

In contrast, almost the complete transportation region (excluding trains) may be considered as made of off-grid [11] structures (e.g., vehicles, trucks, planes, and ships). More typically, numerous elements contribute to a renewed interest in small-scale technology. Among them are the price and public opposition to new transmission lines and large power vegetation [12, 13]. The need of reducing the vulnerability of the supply chain in centralized structures, and the elevated overall performance of small electricity technologies [14, 15]. Hybrid renewable electric powered energy era devices emerge as crucial to most electrical networks and standalone systems such as the water impelling and cable systems. Renewable resources generally are essential garage devices owing to variation in the electrical outputs in the day.

As a result of a surge in the demand for the use of batteries, the indicting procedure of battery gadgets desirable to be suitably controlled over an adaptive managed power handling machine. Fuel Cell (FC) strength plants are electrochemical devices that change the chemical strength of a reaction unswervingly into electricity. It generates power through an electrolytic reaction, not combustion. In a fuel cell, hydrogen and oxygen are used to produce electricity, heat, and water. Fuel cells are cast off nowadays in a variety of applications, to provide power to homes and businesses, and to supply power to critical facilities like hospitals, supermarkets, and computer centers. These are used to move a variety of vehicles including cars, buses, goods vehicles, excavators, trains, and more.

This paper deals with the different kinds of energy resources such as solar PVA and wind farms. As well as storage batteries, and diesel generators, or fuel cells with different types of control techniques to contribute and manage the power in off-grid. The remaining of this paper is outlined as follows: Sect. 2 defines the prototype of the hybrid system. Section 3 is about energy monitoring methods. The modeling and simulation outcomes are revealed in Sect. 4. Conclusions are presented in Sect. 5.

2 Hybrid Energy System

The wind and solar systems are needed to supply the power to the load. But if they fail to meet the load demand, then the backup devices, i.e., storage battery and diesel generator/fuel cell are needed to supply the power to the load.

2.1 Hybrid System Configuration

The hybrid system consists of the following characteristics.

(1) The various energy sources are interlinked in parallel.
(2) Removal of diversion load by using an individual diversion energy control focused on avoiding battery excess charging.
(3) A high-swiftness line is unnecessary for battery modern-day/voltage report statistics communication.
(4) Extension of capacity is easy through a parallel connection of extra power sources to manage with forthcoming load growth.

Concentration has also been specified to Phase-Locked Loop (PLL) control methods via workshop tests and examines the performance of the current/voltage effect of input real-reactive energy parameters into an energy control network test and established suitable energy to manage the result. The proposed self-power composite electricity production system is, as shown in Fig. 1.

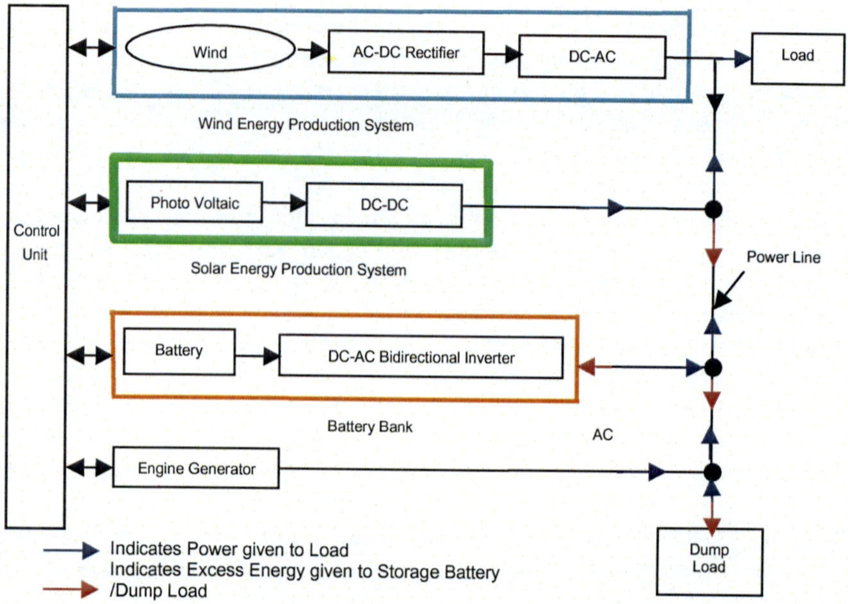

Fig. 1 Hybrid energy system with backup devices

2.2 *Hybrid System Operation*

Chief maneuver drifts of the planned amalgam arrangement are explained below.

Once leftover battery potential is satisfactory: Engine Generator (EG) action halts and whole inverters operated in parallel. Energy excess and shortage, rendering to the equilibrium among the output and load are best used via battery indicating or squaring.

Once the leftover battery potential is unsatisfactory: EG and entire inverters work in parallel. As soon as the energy produced by wind and PV energy production systems is inadequate to reach load demand, EG balance is intended for the shortage. Simultaneously, EG charges the battery through the two-way inverter. This inverter improves indicating energy for the battery with the intention of EG run at the better load factor matches with high proficiency and the subsequent expertise from the monitoring element.

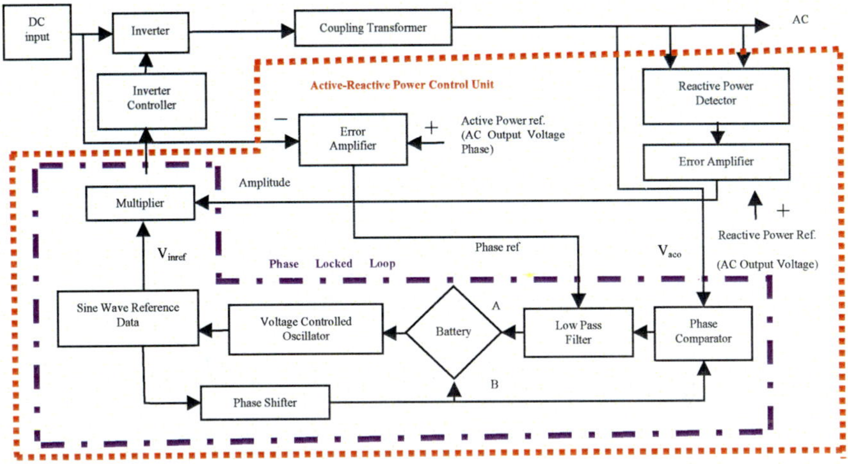

Fig. 2 Block diagram of an elementary power control mechanism of the inverter

3 Hybrid Energy System Monitoring Methods

In the projected composite system, it fixated on monitoring real-reactive energy on load distribution in parallel inverter operations and phase management. Through further study energy diversion control method is considered without a diversion load.

3.1 *Real-Reactive Energy Balance*

The auto-master–slave control practice is pragmatic in entire inverters. When EG is in maneuver, the switch of the respective inverter is locked. The following switches are in AC-synchronized operation with all inverters that operate as slaves and with EG as the master. When EG operation stops, the switch of the battery bank two-way inverter is closed. This inverter works as a master and is under the Constant Voltage Constant Frequency condition. The switch of each remaining inverter that acts as a slave is closed which is shown in Fig. 2. Then, the AC-synchronized operation is undergone. The proposed composite system concentrates on the method of the PLL in the real-reactive energy control.

3.2 *Parallel Inverter Operation*

In this assembly a sample self-power composite electricity production system and conducted experiments. X_1, X_2 and X_3 are interlinked reactors arranged in WT

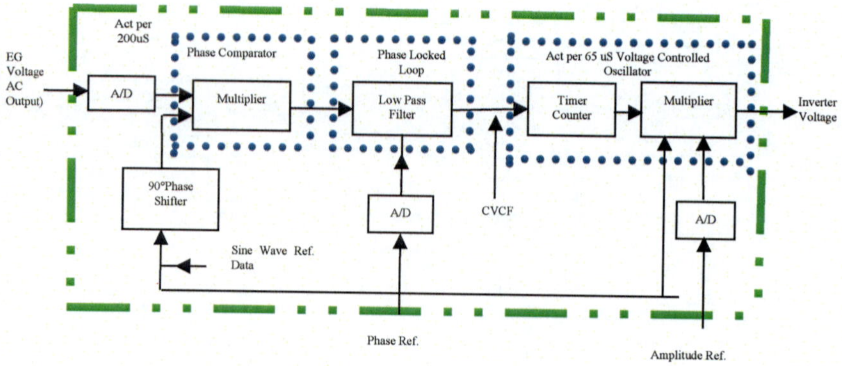

Fig. 3 Phase-locked loop control

inverter, PV inverter, and two-directional inverters, respectively. This research finds the optimal real-reactive energy parameters for each inverter to improve the output under the conditions that each inverter capacity is 3 kVA (with a power factor of 0.8) and the output voltage is single-phase 100 V, 60 Hz.

3.3 *Phase-Locked Loop/Feedback Control Loop*

The PLL, which turns as a phase synchronization regulator consists of a phase comparator, low-pass filter (LPF), phase shifter, multiplier, and Voltage-Precise Oscillator (VPO). The phase comparator performs to expand the AC output voltage by the cosine wave pass on hence found from the reference sine wave which passes over the phase shifter. The expanded wave is changed to DC voltage for VPO frequency control through the LPF.

PLL consists of two control elements, the phase comparator output, and phase reference signal which is introduced into the LPF and shown in Fig. 3. The phase comparator output is introduced as harmonization data (i.e., the variance equated to reference frequency). The phase reference signal is introduced as the quantity of phase shift in inverter output voltage at the same time by balancing the synchronization in opposition to voltage in the marketable energy system. Therefore, the real power changes with the alteration in the phase reference sign.

3.4 *Parallel Inverter Operation*

In this construction a sample self-power composite electricity production system and conducted laboratory experiments. X_1, X_2, and X_3 are interlinked reactors arranged in wind turbine inverter, photovoltaic inverter, and two-directional inverters correspondingly. The investigation finds the optimum real-reactive energy limitations for the respective inverter to improve the output underneath the circumstances that separate inverter size is 3 kVA (at 0.8 power factor) energy storage system, battery, the fuel cell.

Assume V_{se} is demarcated as the input side voltage, V_{re} as the load side voltage, and the angle of phase difference is δ, each of P_{se}, Q_{se}, P_{re}, and Q_{re} is embodied as trails.

$$P_{se} = P_{at} = \frac{V_{se} V_{re}}{X_e} \sin\delta \tag{1}$$

$$Q_{se} = \frac{V_{se}^2 - V_{se} V_r \cos\delta}{X_e} \tag{2}$$

$$Q_{re} = \frac{V_{se} V_r \cos\delta - v_r^2}{X_e} \tag{3}$$

where

P_{se}—active power

Q_{se}—reactive power

P_{at}—sending end active power

Q_{re}—reactive power at the load side

X_e—reactance of the interlinked reactor.

Here, e specifies the number of power sources that function in parallel. It is attractive to normalize the voltage amplitude variation and the phase variation angle to be inside the period from 5 to 15 V, and from 5° to 10°, correspondingly, depends on the behavior of the real-reactive energy regulator unit as shown.

When δ is 7.5° and V_r is 100 V, V_{sm} turns into 109 V by using (1) and (2). The reactance of the interlinked reactor develops 1.57 mH.

Each inverter current and voltage waveform. 50% of real energy is sent toward the load by separate wind turbines and photovoltaic inverters. Underneath this state reactive energy of the battery bank, two-directional inverters turn out to be zero. The size of the load and the power factor is 3 kVA and 0.8 respectively in this scenario.

The evaluation of the respective battery module is 12 V, 24 Ah 24 battery units are allied in series in two analogous rows. The balanced indicting voltage is regulated as 331.2 V and the indicting current as 10 A.

3.5 Diversion Power Control

The real-reactive energy control method is developed for the successful governor of diversion power. Once either power production from WT or PV becomes better than load, the engine generator stops, and the 2-directional inverter as a grasp has functioned underneath the Constant Voltage Constant Frequency situation. At that time, the remaining share was attained subsequently by removing production power load (diversion power), which is routine as an indicating power for the battery-operated bank. During battery indication, a progressive practice to avert battery swindle is required. Diversion load (e.g., resistive load or radiator), is used to utilize diversion power, is usually placed in analogous through the battery or AC production output.

The diversion power changes continuously, it is challenging to adjust the diversion load by merely input diversion load, as these delays supply regulate the battery indication. To alleviate together the battery bank's indicting current and voltage, the improved matchless advanced dump power control method is factorized by diversion energy guidelines without any diversion load and their difference is shown in Figs. 4 and 5. This method allows for immediate response to diversion energy as well as a decrease in unwanted diversion power governor and contributes to further successful practice of renewable energy resources.

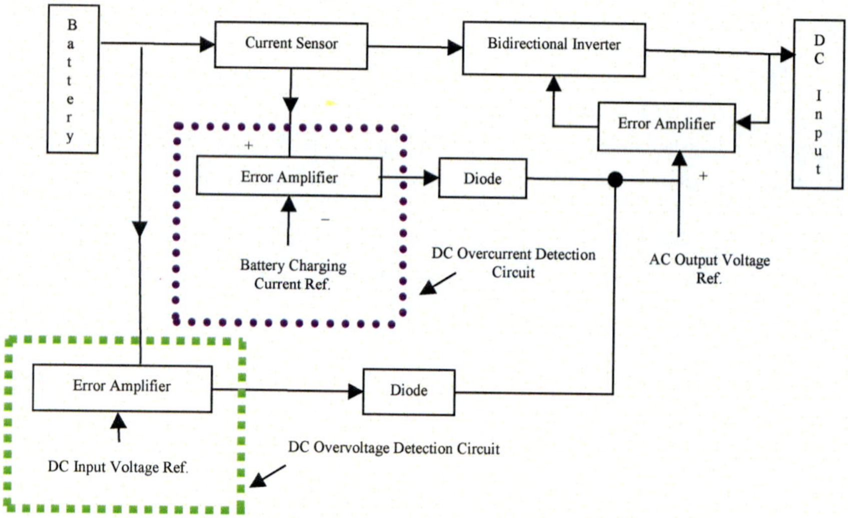

Fig. 4 Bidirectional inverter dump power control

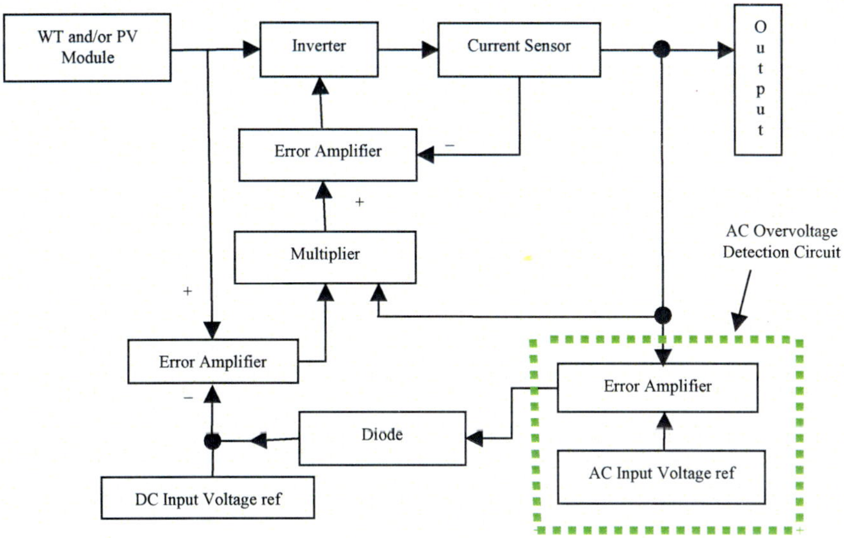

Fig. 5 WT and PV inverters dump power control

4 Simulation Results and Discussions

The proposed hybrid system consists of four energy sources such as PVA, wind farm, storage battery, and diesel generator. The solar PVA and wind farm are directly connected to the load through a transmission line, whereas the battery and diesel generator is connected to the transmission line through a circuit breaker (CB). For every system, the 5 kVA rating of the transformer is used to step up the voltages. All the above four systems are operated in parallel by synchronizing their phase angles and magnitudes.

The synchronization is done by using the Synchronous Reference Frame (SRF) controller. The SRF controller has been executed in all four systems to maintain synchronization. In this system total, three loads are introduced. Each load consumes 5 kw power, i.e., the total power consumed by the load is 15 kW. Hence, it is needed to supply the 15 kW power to the load by synchronizing all four sources. The results of different energy systems are simulated in a MATLAB environment as shown below. But in this simulation, there is one diesel generator (non-renewable energy source) is connected.

The PVA generates 3 kW power, but the power given to the load is 2.650 kW only. The 350 W power is lost due to the conversion and transformation losses in the system. The above graph shows the solar output power, initially, transients are present in the system. Then the system came into a stable state and supply 2.650 kW to the load as shown in Fig. 6.

The actual power generated by the wind farm is 4 kW, but due to losses, the power inoculated to the load is only 3.30 kW. The remaining 700 W power is lost due to the

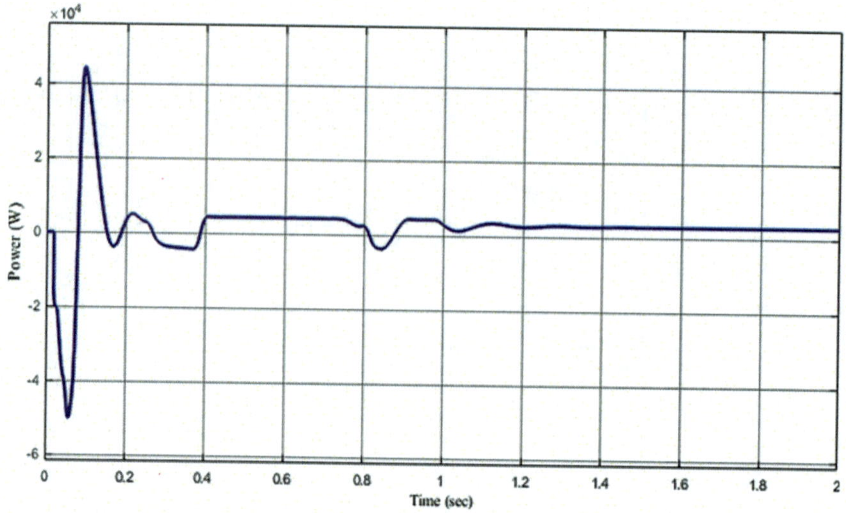

Fig. 6 Solar PVA output power

rotating parts of the wind turbine. The Permanent Magnet Synchronous Generator (PMSG) is cast off in the wind energy system. Initially, there are transients in the system, then the system came into a stable position and supplies 3.30 kW to the load. The simulation result is shown in Fig. 7.

The lead-acid battery is used with the rating of $E = 450$ V with the capacity of 24 Ah (Ah). There are two states in the battery, i.e., charging and discharging state.

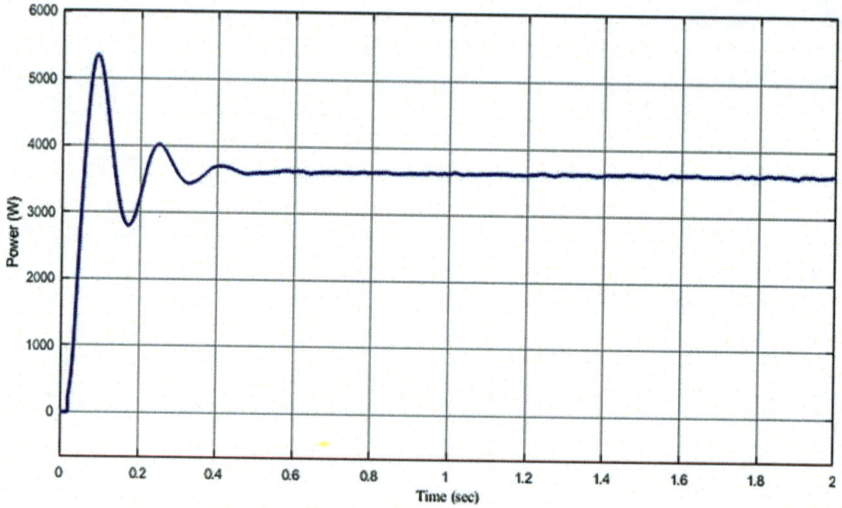

Fig. 7 Wind farm output power

Figure 8 shows there is a transient (harmonics) in the starting and the negative part of the graph shows the charging of a battery and the positive part shows the discharging of a battery. If the State of Charge (SoC) is 80% it indicates the discharging of a battery and 10% indicates the charging of a battery.

The diesel generator generates only 2 kW to the load which is used as a backup device. Initially, there are transients in the system then it becomes the stable state and supplies power to the load shown in Fig. 9. is a non-renewable energy resource. The cost of fuel is high, so, instead of using diesel, one more renewable energy resource, i.e., the fuel cell is introduced.

In this system, as shown in Fig. 10, three loads are added each load consists of 5 kW. Depends upon the load demand the renewable resources are activated. If the first load demand is 5 kW then PVA and wind farm supply power to the load. If the second load is added into the system, then the total load demand is 10 kW Then the power supplied by the wind farm and solar didn't adequate to meet the load demand.

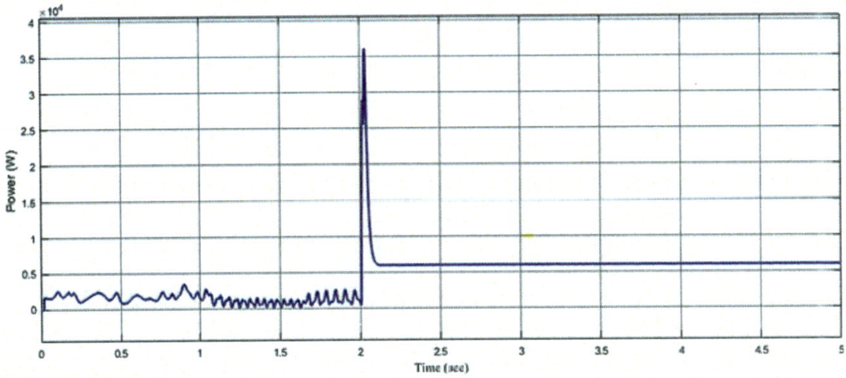

Fig. 8 Storage battery output power

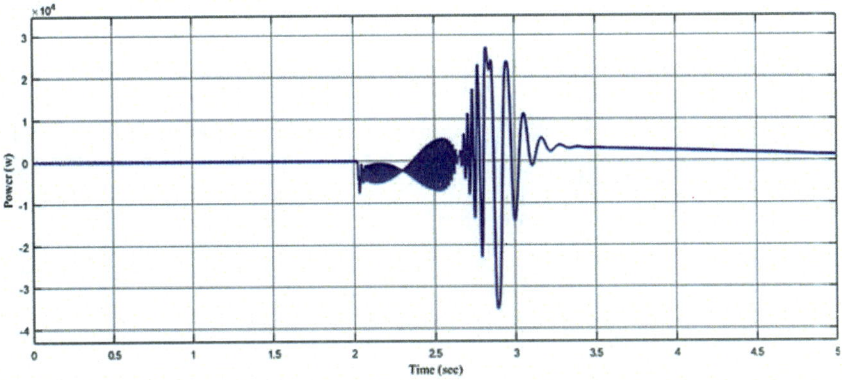

Fig. 9 Diesel generator output power

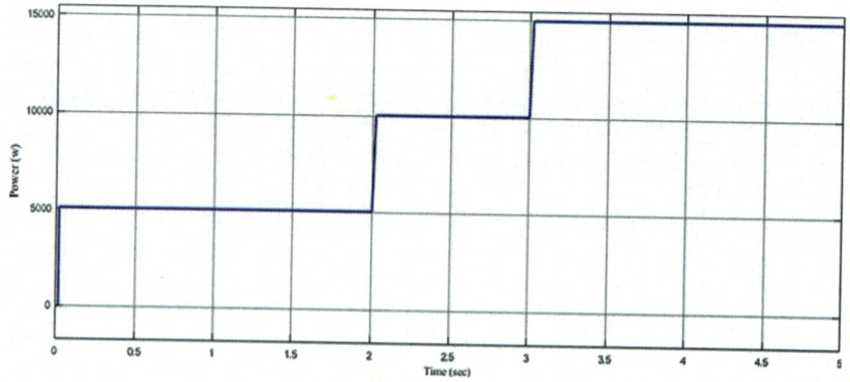

Fig. 10 Load output power

Hence the storage battery supplies the power to the load demand. If the third load is added, then the power supplied by these three devices is not enough. Hence in this situation, all four energy sources are activated and supply power to encounter the load demand.

Figure 11 shows the simulation diagram of four energy sources, but all these are renewable energy sources only. Here diesel generators are removed, and fuel cells are introduced to reduce the cost so that the system efficiency is improved.

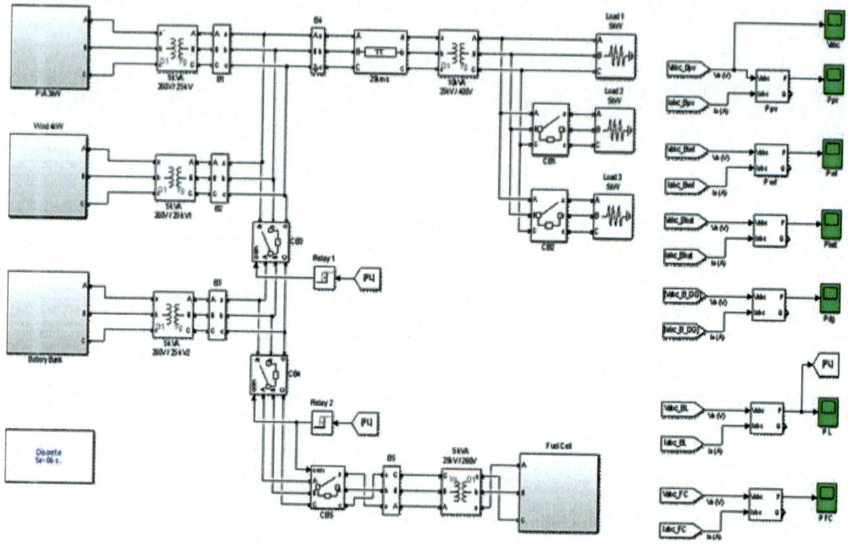

Fig. 11 Simulation diagram of hybrid power system with PVA, windfarm, battery and fuel cell

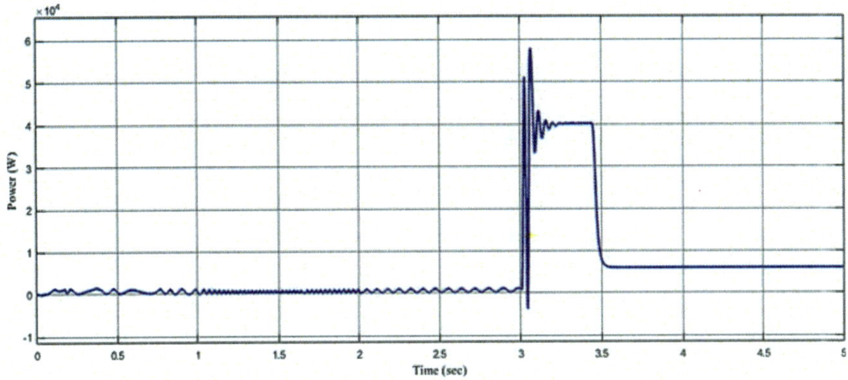

Fig. 12 Fuel cell output power

In Fig. 12, a fuel cell generates 6 kW. This power is supplied to the load when PVA, windfarm, and the storage battery are not able to fulfill the load demand. The effective use of fuel cells is also presented in the paper for optimal operation.

Renewable energy sources like Solar PVA and wind farms are used as main power suppliers, whereas non-renewable energy sources like a storage battery, diesel generator, and fuel cell are used as backup devices. The output of the energy from the solar PVA is 3 kW and from the wind turbine is 4 kW. The backup device lead-acid battery will generate 4 kW power, the diesel generator will generate 2 kW and the fuel cell will generate 6 kW output power. If the main sources are failing to supply the load demand, then the backup devices will supply the total load demand.

The total power being supplied by solar PVA is 42% and the wind farm is supplying 58% of the power to the load. If there is any dump load in the system, then the main energy sources will fail to supply the load demand. In such a case the backup devices will supply power to the load.

The storage battery will supply 33.34% of the power, the diesel generator will supply 16.67% of the power and the fuel cell will supply 50% of the power to the load. All the backup devices are not activated at a time, depending upon the load requirement, it will be operated optimally in the standalone system.

5 Conclusion

This paper is equipped with a self-power composite electricity production system, composed of PLL and diversion energy control. The diversion energy controller permits the arrangement of a return path in the scheme, there is no requirement intended for a devoted ultra-speed communication line to deliver battery loading voltage and current information. If the energy line is utilized as a medium for the transfer of data, the line voltage magnitude is also used as the transmission of information. So, there is no usage of a fiber optic communication line or the energy line transporter system through the harmonic signals are pragmatic to the energy line. Moreover, neither the diversion load nor diversion load control device is required. Further, the proposed diversion energy control, management of output is completed without the battery fully charged, and the current use of excess energy is made possible. This furnishes the extension of battery life and comprehension of a less-price system. The system, via the AC system interlinked, has allowed the stretchy expansion of the system in the future. Energy sources together with EG have interconnected flexibly any location with the same energy line, and the energy quality constant is sustained through governing the AC output voltage magnitude and phase angle. It is estimated that the considered composite arrangement with natural sources of energy makes use of a variety of power control techniques, is validated. The system will also help other countries in context protection through an appeal on remote islands with no dependence on marketable energy systems.

References

1. Uzunoglu M, Alam MS (2006) Dynamic modeling, design, and simulation of a combined PEM fuel cell and ultracapacitor system for stand-alone residential applications. IEEE Trans Energy Conv 21(3):767–775 (2006). https://doi.org/10.1109/TEC.2006.875468
2. Ahmed NA, Miyatake M, Al-Othman AK (2008) Power fluctuations suppression of stand-alone hybrid generation combining solar photovoltaic/wind turbine and fuel cell systems. Energy Convers Manage 49(10):2711–2719. https://doi.org/10.1016/J.ENCONMAN.2008.04.005
3. Kim SK, Jeon JH, Cho CH (2008) Dynamic modeling and control of a grid-connected hybrid generation system with versatile power transfer. IEEE Trans Ind Electron 55(4):1677–1688. https://doi.org/10.1109/TIE.2007.907662
4. Joos G, Li W, Belanger J (2010) Real-time simulation of a wind turbine generator coupled with a battery super capacitor energy storage system. IEEE Trans Ind Electron 57(4):1137–1145. https://doi.org/10.1109/TIE.2009.2037103
5. Karakoulidis K, Mavridis K, Bandekas DV, Adoniadis P (2011) Techno-economic analysis of a stand-alone hybrid photovoltaic-diesel–battery-fuel cell power system. Renew Energy 36(8):2238–2244. https://doi.org/10.1016/J.RENENE.2010.12.003
6. Bratcu AI, Munteau I, Bacha S, Picault D, Raison B (2011) Cascaded dc–dc converter photovoltaic systems: power optimization issues. IEEE Trans Ind Electron 58(2):403–411. https://doi.org/10.1109/TIE.2010.2043041
7. Dursun E, Kilic O (2012) Comparative evaluation of different power management strategies of a stand-alone PV/Wind/PEMFC hybrid power system. Int J Electr Power Energy Syst 34(1):81–89. https://doi.org/10.1016/J.IJEPES.2011.08.025
8. Ravi Prabhakaran V, Subramanian RC (2018) Enhanced ant colony optimization to solve the optimal power flow with ecological emission. Int J Syst Assur Eng Manag 9(1):58–65. https://doi.org/10.1007/S13198-016-0471-X
9. Belfedhal SA, Berkouk EM, Messlem Y (2019) Analysis of grid connected hybrid renewable energy system. J Renew Sustain Energy 11(1):014702. https://doi.org/10.1063/1.5054869
10. Vijay R (2018) Quorum sensing driven bacterial swarm optimization to solve practical dynamic power ecological emission economic dispatch. Int J Comput Methods 15(3):1850089–1850124. https://doi.org/10.1142/S0219876218500895

11. Benlahbib B, Bouarroudj N, Mekhilef S, Abdeldjalil D, Abdelkrim T, Bouchafaa F (2020) Experimental investigation of power management and control of a PV/wind/fuel cell/battery hybrid energy system microgrid. Int J Hydrogen Energy 45(53):29110–29122. https://doi.org/10.1016/J.IJHYDENE.2020.07.251
12. AmarBensaber A, Benghanem M, Guerouad A, AmarBensaber M (2019) Power flow control and management of a hybrid power system. Przegląd Elektrotechniczny 9(1):189–190. https://doi.org/10.15199/48.2019.01.46
13. Shaqour A, Farzaneh H, Yoshida Y, Hinokuma T (2020) Power control and simulation of a building integrated stand-alone hybrid PV-wind-battery system. In: Kasuga City, Japan. Energy reports vol 6, pp 1528–1544. https://doi.org/10.1016/J.EGYR.2020.06.003
14. Ravi Prabhakaran V (2018) Optimal and stable operation of microgrid using enriched biogeography based optimization algorithm. J Electr Eng 17(4):1–11
15. Ghenai C, Salameh T, Merabet A (2020) Technico-economic analysis of off grid solar PV/Fuel cell energy system for residential community in desert region. Int J Hydrogen Energy 45(20):11460–11470. https://doi.org/10.1016/J.IJHYDENE.2018.05.110

Chapter 32
Sensor Fusion of Camera and Lidar Using Kalman Filter

Reshma Kunjumon and G. S. Sangeetha Gopan

1 Introduction

Autonomous vehicles have become an important field of interest due to their capability of autonomous navigation. The autonomous (or ego) vehicles can operate and react to the environment without the help of the driver as they can make decisions based on the information perceived through the sensors. Sensors are the vital components of the self-driving vehicles and the robust sensing and fusion of information passed down from these sensors and their interpretation are key for the control of the vehicle.

The Intelligent Transport Systems that make use of various technological aspects have largely benefited from the recent rapid technological advancements. These systems are mostly based on sensor technologies like RADAR, Lidar or Camera. Certain systems are designed such that they alert the driver through audio, visual or haptical signal about the danger. Also, some systems are designed such that they can take over the control of vehicles during emergencies, which is intended to avoid accidents that result from mistakes and carelessness in complex traffic [1]. Furthermore, these sorts of systems enable a better driving environment.

Though the driving experience and vehicle safety measure has increased drastically over the years, the ever-changing traffic culture has sparked an extreme interest in the research and development of systems with a high level of autonomy. Rather than just providing driving assistance, an autonomous car can drive itself to reach the desired location. Such a high level of autonomy requires the vehicle to perceive the environment as done by humans, i.e., the vehicle will have to sense the environment and identify distance to the obstacle, location, movement of pedestrians, etc. Also, it will have to carry out a decision-making process like humans. This requires a new array of sensing devices, information processing algorithms, computing units and an entirely new electrical and electronic architecture.

R. Kunjumon (✉) · G. S. Sangeetha Gopan
College of Engineering Trivandrum, Sreekaryam PO, Trivandrum 695016, India

A. Sheth et al. (eds.), *Intelligent Systems*, Algorithms for Intelligent Systems,
https://doi.org/10.1007/978-981-16-2248-9_32

The main feature that sets the autonomous vehicle different from the conventional vehicle is the devices that are used for perceiving and understanding the surrounding environment of the vehicle. Such systems consist of devices like RADAR, Lidar, GPS, Ultrasonic, Camera, Inertial Measurement Unit (IMU), etc. These devices help in integrating information processing algorithms like signal processing, machine learning, encryption/decryption and decision making. The combination of all devices and associated algorithms helps in determining the autonomy level [2] of the autonomous vehicles.

The sensors play a vital role in autonomous driving as they help in better perception of the environment around the vehicle. The selection of sensors is essential for the design of an efficient autonomous driving system. This research is focusing on the use of Lidar and Vision sensors. The use of these sensors will reduce the complexity that is involved in computation. The higher number of sensors will require a performance intensive processor. Another problem with using more than two sensors is writing the library function to read the data from the sensors. It is possible to write a library for a Lidar or the Vision sensor, but if the system has a greater number of sensors, writing libraries could take a large amount of time. This is the reason for choosing two sensors.

2 Sensor Fusion

Each of these sensors has its own advantages and disadvantages. Among all perception functions, visual perception is one of the most important components. When we consider a camera, it has better performance when the object detection is considered. The advantage of a camera is that it can provide specific information related to the target. It interprets visual data and performs critical tasks such as vehicle and pedestrian detection. In an autonomous driving system, cameras are essential because they mimic human eyes and most traffic rules are designed by assuming the ability of visual perception. For example, many traffic signs share similar physical shapes and they are differentiated by their colored patterns that can only be captured by a visual perception system. The better resolutions of the camera at both lateral and elevation are also an added advantage. However, the camera's functioning is largely affected by the weather. The night conditions or bad weather can affect its performance. This is totally against the idea of a safety sensor. In such cases, a RADAR or Lidar can be implemented. A comparison between RADAR and Lidar shows that a Lidar-based sensor can provide good resolution about the position [5]. The short wavelength of Lidar lets us detect small objects and a Lidar can build an exact 3D monochromatic image of an object. The Lidar also has its working principle that facilitates the calculation of distance and velocity to an object. The camera can measure the distance only when it is used in the stereo vision configuration. There is a lot of limitation in the amount of data about a physical quantity, procured by one sensor [3]. This is because one sensor has many of the following disadvantages:

- Field of View (FOV) coverage is limited.
- Temporal coverage is limited because the rate of data acquisition by sensor is limited.
- The malfunction of one sensor affects the reliability of the entire system.
- The precision of the entire system is determined by the precision of the sensing element.
- When some features are missing (e.g., occluded objects), measured data becomes uncertain.

Thus, a probabilistic fusion of radar and video data improves the estimation accuracy by compensating the weaknesses of one sensor with the strengths of another. Combining information from a multi-sensor system introduces new challenges [4]. One of the important challenges is a spatio-temporal task: the spatial part is the alignment of frame sensors while the second is handling the update rates of sensors. In the alignment process, a relation is found between the different coordinates in each sensor frame to ensure the transformation from one frame into another. Another major challenge is the timing of operation. This applies to both homogeneous and heterogeneous sensors. The operation frequencies of the sensors are different. For above-mentioned reasons, sensor fusion algorithms have to tackle temporal, noisy input and output a probabilistically sound estimate of kinematic state.

3 System Design

The proposed system as represented in Fig. 1 consists of the following features:

- Targets Data Acquisition: Taking input from the sensors, i.e., camera and Lidar in the respective formats and storing them.
- Camera Image Processing: Detecting targets in images captured by camera, classification and boundary of targets and the centroids are obtained.

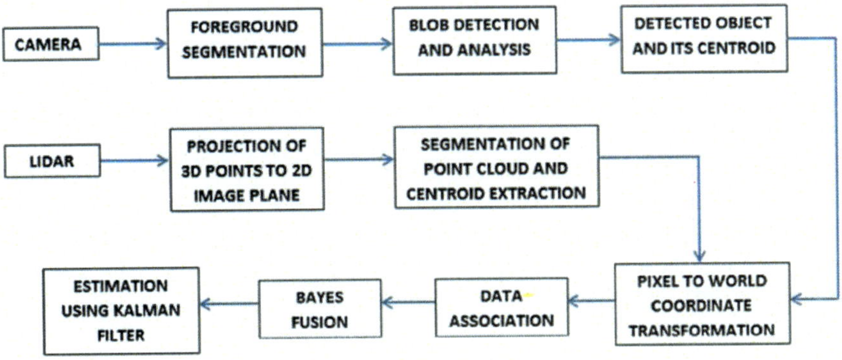

Fig. 1 Block diagram of the proposed system

- Projection: The cloud of points obtained from Lidar is projected from the 3-dimensional spaces onto the 2-dimensional images procured by camera.
- Lidar Frame Processing: Segmentation of the Lidar point cloud and extraction of centroid from segmented cloud.
- Data Association: This step is being done to ensure that the same obstacle is being detected by the 2 sensors.
- Bayes Fusion: To create a less noisy resultant measurement.
- Estimation Filter: Obstacle Path Estimation using Kalman filter.

4 Data Acquisition and Pre-processing

In order to prove my hypothesis practically, an experimental setup is needed with practical real-time measurement of the parameter values. But such a setup is outside the scope of this project tenure due to pandemic restrictions. Hence, I had to rely on an online data set called KITTI VISION BENCHMARK SUITE.

4.1 KITTI Vision Benchmark Suite

It is a joint venture of Karlsruhe Institute of Technology, Germany and Toyota Technological Institute, Chicago for Sensor Fusion and Robotic Analysis purposes. It has been made available for academic purposes. For the KITTI Database, a standard station wagon like Volkswagen Passat was fitted with 4 cameras: 2 high resolution color cameras and 2 Gy scale video cameras, an HDL-64E Velodyne Laser Scanner and a GPS localization system along with an Inertial Measurement Unit. The location chosen was a highway in Karlsruhe that contains up to 15 cars and 30 pedestrians. Both raw images and benchmarked images were made available of which I have chosen the latter subject to evaluation metrics and real-world benchmarks. The calibration files and equations have also been made available by the authors.

The cameras provide images at the rate of 10 frames per second. There were 360 images provided per camera. Each grayscale frame is an 8-bit PNG image consisting of 1392 × 512 pixels. In effect, each image was around 5.6 MB in size. While each frame of the RGB image is a 16-bit PNG image, amounting to double the size. The HDL-64E Velodyne Lidar is a 360° counterclockwise rotating Lidar at 10 Hz. The sensors were time synchronized by the authors. It procures the 3-dimensional coordinates of about 1.3 million points per second and their reflectance values in the single path mode. These points are available in the form of text files and BIN files. The values stored as binary float matrices in BIN files which allows them to have greater accuracy. Hence, they were chosen for my project. Also, due to the storage limitations, the number of points was rectified to about 21,947 points per frame.

4.2 Camera Image Processing

Once the sensor data has been procured, relevant obstacles have to be detected from the data. In the case of a camera, we know that the sensor data consists of two-dimensional images. Advanced image processing techniques in object detection can be used to identify and recognize the obstacles. There are many object detection algorithms like HOG detector, R-CNN, SSD, YOLO, etc. which can be implemented. In this case, considering hardware limitations, I have opted for a simple object detection based on foreground segmentation and blob analysis technique. This technique is based on Gaussian Mixture Models and Blob Analysis. It consists of four steps:

- Foreground segmentation
- Blob Detection
- Blob Analysis
- Blob Tracking.

The vision toolbox of MATLAB is made use of for this purpose. Using the command vision. ForegroundDetector(), the image frames are compared to a mixture of built-in Gaussian mixture models. The parts of the frame that match the model are considered as background and the rest of the frame is considered as foreground. Depending on this conclusion, the image is binarized with pixel intensity 0 or black given for background pixels while intensity 255 or white given for foreground pixels. This step is followed by a morphological opening operation to remove the noise and fill the gaps in detected objects. This is done using the imopen() command making use of a structural element (strel()) of size 1 percent of video's width. Following this, the minimum area of the foreground required to be detected as a blob is specified as 8 percent of the video's frame area. Using the vision. BlobAnalysis() command, the object is detected and using insertShape() a green bounding box is drawn around the detected obstacle and its centroid is displayed (Fig. 2).

It is worth noting here that when multiple objects are detected, we have to go for object recognition algorithms and prioritization steps, in order to assign priorities to the detected obstacles in further tracking. In order to reduce complications, I have stuck to the main obstacle captured by the sensors, i.e., the vehicle in front of the ego vehicle and skipped the prioritization step. In future scope of the project, these steps can be included to maximize the output (Fig. 3).

4.3 Multi-Sensor Data Fusion Network

After the process of data acquisition, before doing fusion, we need to make sure that the same object is being detected by the sensors. For this, we need to do the process of data association. This is discussed in the next section. For data association, we need to make sure that the data from different sources belong to the same frame of reference. This process is called sensor alignment. We can extend the proposed approach to n

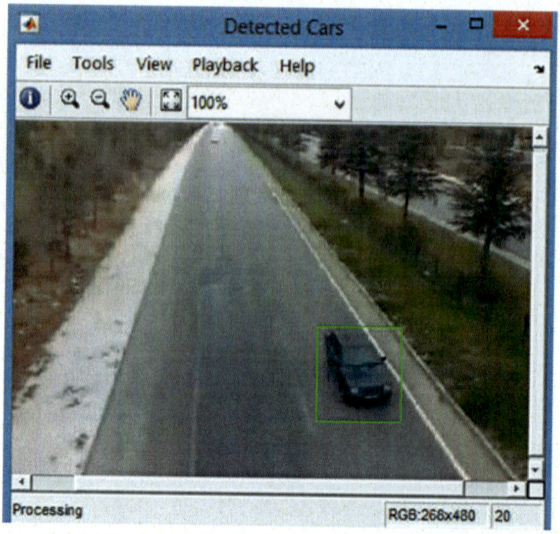

Fig. 2 Output of the object detection code

sensors and append the alignment process. These sensors are homogeneous and/or heterogeneous and their task is to measure the position of the detected obstacles. It consists of two main processes:

- Sensor alignment process (off-line). The inputs of this process are sensor data while the outputs are the calibration parameters (rotation matrix and translation vector). This process is an extrinsic calibration between different sensors (source and targets). It allows estimating the relative position of point p in a common frame.
- Object detection process (online). In this step, there are n processing chains. Each of them provides a list of the detected objects.

With the conjunction of the calibration parameters obtained from the sensor alignment process, we are able to fuse the data together to better detect the objects in the surroundings. In this work, focus is made on Lidar and camera sensors. In the following sections, each task will be detailed.

4.4 Sensor Alignment: Calibration Parameters

To efficiently perform sensor fusion, the sensors should be calibrated. The calibration process is an alignment procedure of the given sensor frames. That is to say, to find the relation between the coordinates of sensor frames to ensure the transformation from a frame into another. To carry out this process, we need two calibration parameters, i.e., the rotational matrix and the translation matrix. In this case, these parameters were already provided in the KITTI data set. Else, they have to be experimentally

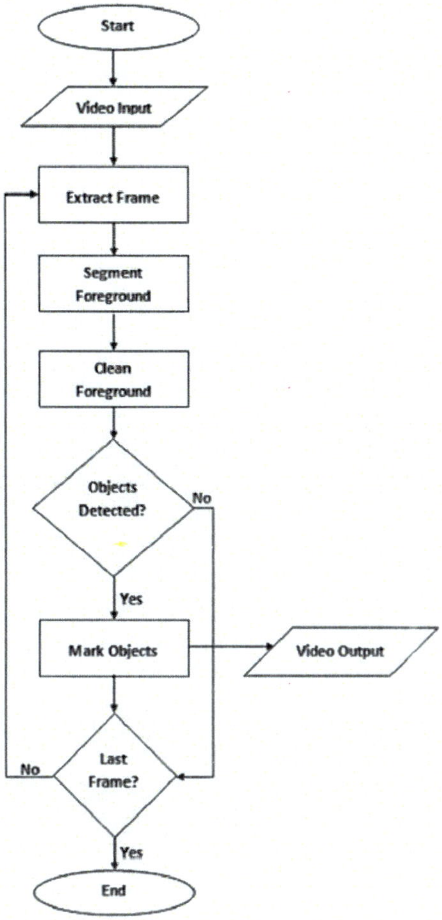

Fig. 3 Algorithm for object detection using blob analysis

calculated. The data set contains 3 calibration files: camera to camera, Velodyne to camera and IMU to Velodyne. Of these, the third file is of no use to this project as the IMU sensor is not included. The other two files contain the various parameters needed to calibrate the camera and project the Lidar points from the 3-dimensional space to the image plane. Of these parameters, the ones that we need in order to calculate the transformation or projection matrix are R_rect and P_rect. Also, each of these parameters has been listed for the four cameras used in the experiment. We have considered only one-color camera for this project. Hence, only the first camera's parameters will be used.

The second file we need is called Tr_velo_to_cam. txt. It is dependent on the camera's external parameters. It is a 4×4 double matrix. This is used for velodyne to camera transformation. It contains the rotational matrix values in the first 9 cells, translation matrix in the fourth column and 2 small delta values in the fourth row. We define a 4×4 double matrix called R_cam_to_rect. This contains the R_rect values from the first file copied to a 4 × 4 matrix and the remaining values appropriately filled. This is done to correct the order of matrices for proper multiplication. The screenshot of the required parameter values in MATLAB workspace is given in Fig. 5.

The next step is to calculate the Projection Matrix given by Eq. 1. The variable name given is P_velo_to_img. It is a 3 × 4 matrix obtained by multiplying the cameras internal parameters (P_rect), rectified rotational matrix (R_cam_to_rect) and external parameters (Tr_velo_to_cam). The point cloud in the Lidar plane can be converted to the 3D camera plane by using Eq. 1.

$$\text{P\_velo\_to\_img} = \text{P\_rect}\{\text{cam} + 1\} * \text{R\_cam\_to\_rect} * \text{Tr\_velo\_to\_cam}. \quad (1)$$

Figure 6 illustrates the values of the projection matrix as calculated using the above equation. Once the projection matrix is calculated, we can easily convert the point cloud from Lidar plane to camera plane by multiplying the points with the projection matrix. As mentioned already, the Lidar data provided by the KITTI data set contains the x, y and z coordinates of the points along with a fourth column containing their reflectance values. The number of points has been rectified by KITTI authors from 1.3 million to a useful set of 21,947 points. These are stored per frame in BIN files. The BIN files are read using a file descriptor into a 21,947 × 4 double matrix called velo. In order to further reduce the computation time, we have scaled the original set by choosing a set of relevant 6535 points. Also, reflectance values have been neglected because it is outside the scope of this project. The projection equation is given in Eq. 2. Here, P_t contains the transformed points to the camera plane. It is a 6535 × 3 matrix.

$$\text{P\_t} = \text{P\_velo\_to\_img} * \text{velo}. \quad (2)$$

Now the transformed points are in the camera plane but they are still in 3 dimensions. They need to be converted to two-dimensional points to project them to the image plane. Dividing the x and y coordinates with the z coordinates will project the 3D point to a 2D plane. This is contained in the P_p matrix which is of size 6535 × 2 and contains only x and y coordinates. This is plotted on the image plane as shown in Fig. 4.

4.5 Lidar Frame Processing

Next step is to segregate the cloud of points such that points belonging to relevant obstacles can be isolated and segmented. This process is called segmentation. To extract the projected points of the source sensor on the calibration board, the automatic extraction approach by differentiation of the measurements and background data in static environments is often used. However, since this work is not based on experimental setup, this approach cannot be used. Hence, this portion is out of scope

Fig. 4 Lidar points projected onto camera plane

Fig. 5 Calibration parameters

calib.R_rect{1, 1}

	1	2	3	4
1	0.9999	0.0098	-0.0074	
2	-0.0099	0.9999	-0.0043	
3	0.0074	0.0044	1.0000	
4				

calib.P_rect{1, 1}

	1	2	3	4
1	721.5377	0	609.5593	0
2	0	721.5377	172.8540	0
3	0	0	1	0
4				

Tr_velo_to_cam

4x4 double

	1	2	3	4
1	0.0075	-1.0000	-6.1660e-04	-0.0041
2	0.0148	7.2807e-04	-0.9999	-0.0763
3	0.9999	0.0075	0.0148	-0.2718
4	0	0	0	1
5				

Fig. 6 Projection matrix

VARIABLE SELECTION

P_velo_to_img

3x4 double

	1	2	3	4
1	609.6954	-721.4216	-1.2513	-123.0418
2	180.3842	7.6448	-719.6515	-101.0167
3	0.9999	1.2437e-04	0.0105	-0.2694
4				

Fig. 7 Coordinates of obstacle's centroid

Clipboard Font

J11 *fx*

	A	B	C	D
1	CamX	CamY	LidX	LidY
2	586.25	221	622	237
3	587.75	222.1	622.5	237.5
4	588.25	223.25	623	238
5	590.75	225.75	623.5	238.5
6	590.75	227.75	624	239
7	592.75	228.25	624.5	239.5
8	593.75	229.25	625	240
9	595.25	231.75	625.5	240.5
10	596.25	232.75	626	241
11	597.25	233.75	626.5	241.5

of this project. Rather, what has been done is the manual selection of the centroids from the given frames. The coordinates of the centroid of the obstacle obtained from image frames and Lidar points were collected to an Excel sheet which has been displayed in Fig. 7.

4.6 Pixel to World Coordinate Transformation

The coordinates of the centroid are available in pixel values. It needs to be converted in terms of real-world distances because the points of Lidar correspond to real-world distances. For this, we need the actual measurement of road width, size of the obstacle, etc. to find the scale between pixel value and real-world distance. Since that information was not available in the online data set, an approximated scaled-down transformation was done considering the real-world length of the image frame and the obstacle size. This data is recorded and shown in Fig. 8.

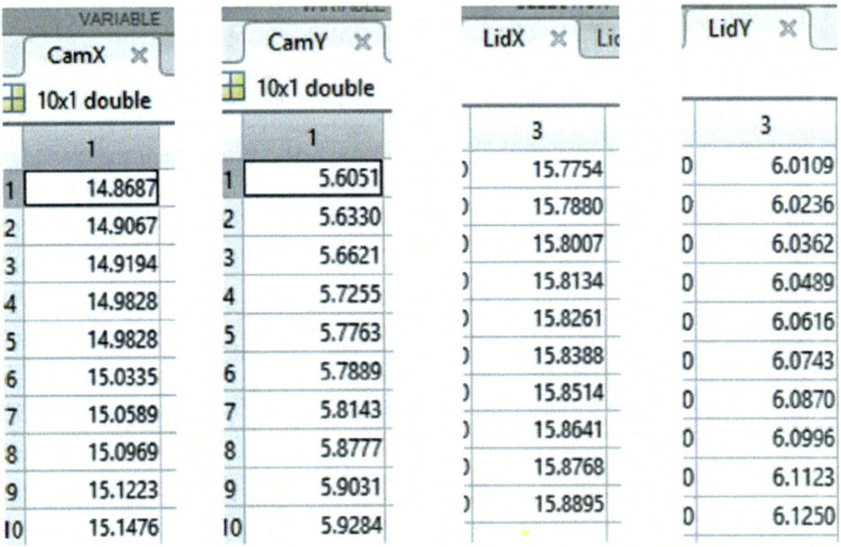

VARIABLE

CamX
10x1 double

	1
1	14.8687
2	14.9067
3	14.9194
4	14.9828
5	14.9828
6	15.0335
7	15.0589
8	15.0969
9	15.1223
10	15.1476

CamY
10x1 double

	1
1	5.6051
2	5.6330
3	5.6621
4	5.7255
5	5.7763
6	5.7889
7	5.8143
8	5.8777
9	5.9031
10	5.9284

LidX

3
15.7754
15.7880
15.8007
15.8134
15.8261
15.8388
15.8514
15.8641
15.8768
15.8895

LidY

3
6.0109
6.0236
6.0362
6.0489
6.0616
6.0743
6.0870
6.0996
6.1123
6.1250

Fig. 8 Centroid coordinates in terms of physical distance

5 Data Association

Now that we have the coordinates of the centroid of the obstacle obtained from different frames of the camera and Lidar in terms of physical distances, we need to make sure that the coordinates procured by camera and Lidar belong to the same obstacle. If not, the next step which is Fusion will yield highly erratic results. In order to do this step, we need to calculate the distance between the centroid coordinates obtained by the 2 sensors. If this calculated distance lies below a certain preset threshold, which is subject to real-world sizes of objects, then we can safely say that the coordinates belong to the same object. Another thing to be noted here is that, when we use the original Lidar point cluster, the centroid of the cluster will not be in the center like the one received from the image because in Lidar, the distribution of point clouds is not homogeneous or uniform. So, to calculate the distance between such a centroid of the cluster, we use Mahalanobis Distance instead of the regular Euclidean Distance. The simplified version of this is given in Eq. 3. Here the denominator stands for the variance of the data. If the same object is detected, "same target located" is printed.

$$d_k(a, b) = \frac{(x_a - x_b)^2}{\sigma_{x_a}^2 + \sigma_{x_b}^2} + \frac{(y_a - y_b)^2}{\sigma_{y_a}^2 + \sigma_{y_b}^2}. \tag{3}$$

6 Bayes Fusion

Once it is confirmed that the centroids belong to the same object, it is now time to fuse these values to get a less noisy result. The sensors are employed to detect the positions of obstacles. Position uncertainty is represented using the standard variation. Therefore, if X is the true position of the detected object, by using the Bayesian fusion, the probability of fused position $P_F\ [x_F\ y_F]^T$ by the two sensors is given in Eq. 4.

$$P_{\text{prob}}\left(\frac{P}{X}\right) = \frac{e^{\frac{-(P-X)^T R^{-1}(P-X)}{2}}}{2\pi R^{0.5}}. \tag{4}$$

where P is the fused position and R is the covariance matrix are given as:

$$P = \frac{\frac{P_1}{R_1} + \frac{P_2}{R_2}}{\frac{1}{R_1} + \frac{1}{R_2}}. \tag{5}$$

$$\frac{1}{R} = \frac{1}{R_1} + \frac{1}{R_2}. \tag{6}$$

where P_1 and R_1 are respectively the position and covariance matrix of sensor 1 and P_2 and R_2 are that of sensor 2. So, the outcome is a combination of the two measurements weighted by their noise covariance matrices. The fused results using Bayesian approach follow the measurements provided by the sensor which has the smallest covariance matrix and gives more trust to it. [6] So finally, the value of the fused measurement vector is given in Eq. 7.

$$\overbrace{y} = \left\{\sum_{k=1}^{m} R_k^{-1}\right\}^{-1} \sum_{k=1}^{m}\{R_k^{-1} y_k\}. \tag{7}$$

In case of 2 sensors, the equation further simplifies to Eq. 8.

$$\overbrace{y} = \{R_1^{-1} + R_2^{-1}\}^{-1}\{R_1^{-1} y_1 + R_2^{-1} y_2\}. \tag{8}$$

R is the covariance matrix. It is an $n \times n$ positive definite symmetric matrix.

$$R_{cam} = \begin{bmatrix} \sigma^2_{\text{camm}_x} & 0 \\ 0 & \sigma^2_{\text{cam}_y} \end{bmatrix}.$$

$$R_{lid} = \begin{bmatrix} \sigma^2_{\text{lid}_x} & 0 \\ 0 & \sigma^2_{\text{lid}_y} \end{bmatrix}.$$

Sensor fusion is a technique that has to be implemented in real time on a moving car and has to take in continuously changing values. In such situations, 2 types of errors are prominent. One is called measurement noise. It is the error inherent in any sensing system. The more important error is the plant or driver noise. When a sensor is calibrated, it is done using the simplest model: a target moving in a straight line at constant speed. But when implemented practically, the sensor is mounted on a moving car. Also, the targets need not be moving in a straight line or at constant speed. So, their paths will be an accelerated curvy flight which causes errors with the Bayesian filter because it doesn't have any mechanism to update the values according to the error value. At each time step, the sensor values are influenced by a number of factors like acceleration, gravity, angular velocity, air friction, streamline effects, etc. Hence the error keeps accumulating. Another technique called the Kalman filter technique is efficient in handling continuously changing real-time data. It has 2 cycles: a prediction and an update cycle. This allows it to update the next values based on the error of the previous stage. Thus, it minimizes error with each iteration of the algorithm. Also, it can handle both position and velocity (state) of the target simultaneously. It also doesn't need to remember previous data which significantly saves time and cost.

7 Kalman Filter

The Kalman filter is the most commonly used algorithm in the tracking and prediction tasks. Its use in the analysis of visual motion can be seen quite frequently. The purpose of filtering is to extract the required information from a signal, ignoring everything else. The Kalman filter has found great importance as the filter is defined in terms of state-space methods which help to simplify the implementation of the filter in the discrete domain. Since this project is based on simulation results and not experimental setup, the initial values of a number of parameters that have to be practically derived are deficient. Hence, we had to stick to the most basic model, i.e., the constant velocity model designed based on kinematics equations. The ego vehicle is assumed to be moving at a constant velocity of 0.37 m/s.

The Kalman filter is a closed form solution to the Bayesian Filter. Equations 9–14 have been taken from Ref. [7] and substituted with required parameters. The filter assumes linear Gaussian motion and measurement models and the prior distribution.

$$X_0 \sim N(x_0, P_0).$$

Where x_0 and P_0 are the initial state and covariance matrices.

$$x_0 = \begin{pmatrix} \text{Initial\_}x\text{\_position} \\ \text{Initial\_}y\text{\_position} \end{pmatrix}.$$

$$P_0 = \begin{pmatrix} \text{Error\_in\_predicted\_position} \\ \text{Error\_in\_predicted\_velocity} \end{pmatrix} = \begin{pmatrix} 0.001 \\ 0.0001 \end{pmatrix}$$

With the previous assumptions, we get the following set of equations.

$$x_{kp} = AX_{(k-1)} + w_k. \tag{9}$$

where x_{kp} is the predicted state matrix. A is the state transition matrix of the motion model. It is a square matrix $\begin{bmatrix} 1 & \Delta t \\ 0 & 1 \end{bmatrix}$. Δt is the time stamp gap. It is taken equal to 0.1. w_k is the error in prediction. We neglect it here.

$$P_{kp} = AP_{(k-1)}A^{\mathrm{T}} + Q_k. \tag{10}$$

where P_{kp} is the Process Covariance Matrix. Q_k is the error in estimation. It is also called the Driver Noise Matrix. These 2 steps are called the prediction step. Next is the update step. It is here we calculate the Kalman Gain matrix.

$$K = \frac{P_{kp}H^{\mathrm{T}}}{HP_{kp}H^{\mathrm{T}} + R}. \tag{11}$$

Here, H is an identity matrix used to correct order. It transforms the given state vector into measurement. R is the random measurement noise. It is also called observed error. Its value is taken as $0.03 * H$. Using the Kalman Gain Matrix, we can estimate the next position or velocity to a high degree of accuracy.

$$X_k = X_{kp} + K\big(Y - HX_{kp}\big). \tag{12}$$

X_k is the estimated state vector, i.e., position and velocity. It is based on the predicted value as well as the actual measurement, Y. As we can see, the Kalman Gain Matrix acts like a weightage. Depending on whichever has the least covariance in the previous step, the Kalman Gain Matrix will be adjusted to support that particular variable in the next round which sort of acts like a self-check.

$$Y_k = HY_k + Z_m. \tag{13}$$

Y_k is the observed value or the input measurement, the data that comes from the sensors. In our case, this will be the coordinates obtained after Bayes fusion. Z_m is the random measurement noise, taken to be zero.

$$P_k = (I - KH)P_{(k-1)}. \tag{14}$$

One more value has to be updated, i.e., P_k is the updated process covariance matrix. I is the identity matrix used for order correction. In the next iteration, the estimated state vector, X_k and updated process covariance matrix, P_k becomes the $X_{(k-1)}$ and $P_{(k-1)}$.

8 Observation and Results

After implementing the Kalman filter on the Bayes fused results, we plotted a graph of the obstacle path on the Cartesian plane. 2 graphs were plotted. The orange line is the path plotted using Bayes fused values from the sensor input while the blue line is plotted using values estimated by the Kalman filter. We know that in order to avoid complications, we have gone for a constant velocity model, which means that the direction of the ego vehicle or the obstacle will not have changed over the entire duration of the journey. As we can clearly see from Fig. 9, the blue line resembles the constant velocity model more than the orange line. Hence, we can conclude that an estimation filter like Kalman filter can predict the path of the tracked objects more accurately and reduce the noise in measurement of the sensors rather than simply employing a fusion algorithm on the sensor data.

It can also be observed that the Kalman filter generates a fused value that is closer to the original value. This in turn proves that we are tracking the object as desired. In order to get these results, only data from online sources have been made use of. Experimental setup and real-world data collection were hampered due to pandemic

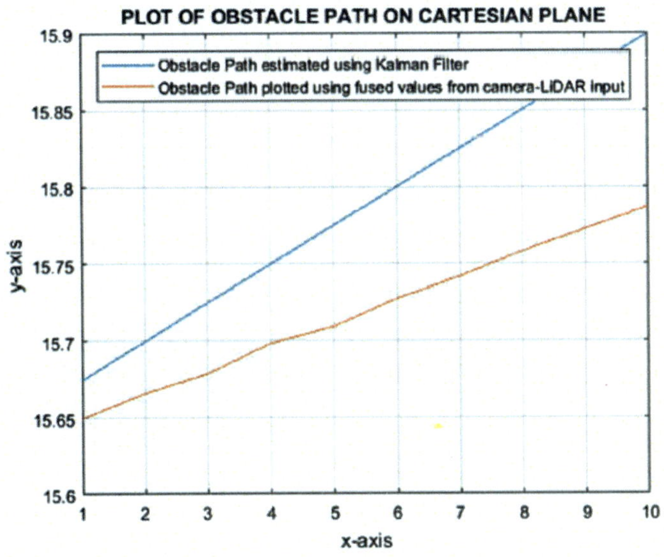

Fig. 9 Plot of obstacle path on Cartesian plane

restrictions. Parameters involved in using Kalman filter require accurate data from experimental setup, many of which were absent from the online database. Approximate values have been substituted wherever possible. But such approximations would affect the performance of the Kalman filter and accuracy of the final result. In a highway simulation setting involving a physical system and more parameters, the Kalman filter can generate the sensor fusion values much faster and accurately that can help the ego vehicle to make proper decisions.

9 Conclusion

The main purpose of this project was to achieve the fusion of data obtained from the Lidar and Camera sensors attached to the ego vehicle. The system was modeled successfully based on the state-space model and an algorithm was implemented to fuse the data. The fusion algorithm devised helped to achieve the prediction and Kalman gain calculations that are important to the operation. It helped to perform specific tasks like data association which in turn resulted in coordinate transformation, data fusion, etc.

It can be observed that even though we have considered a constant velocity model, the final result is very similar to the original path. When other parameters are brought in and more factors are modeled, the accuracy of the system will improve and so will the complexity. For now, under the scope of this project, the constant velocity model serves the purpose of understanding Sensor fusion using Kalman filter for Lidar and Camera sensors.

Thus, the fusion of data from the camera and radar sensor was achieved successfully using the Kalman filter. This research can be extended to an experimental setup using real-world values for the parameters and by considering the effect of many practical factors that affect the system, the accuracy of prediction and updation by Kalman filter can be further improved. Also, based on the generated result, a decision-making system can also be created so as to control the operation of the ego vehicle in real time.

References

1. Ziebinski A, Cupek R, Erdogan H, Waechter S (2016) A survey of ADAS technologies for the future perspective of sensor fusion. In: Nguyen NT et al (eds) ICCCI 2016. Part II. LNAI 9876, pp 135–146
2. SAE J3016 (2016) Taxonomy and definitions for terms related to driving automation systems for on-road motor vehicles
3. Bouain M, Karim MAA, Berdjag D, Fakhfakh N, Atitallah RB (2018) An embedded multi-sensor data fusion design for vehicle perception tasks. J Commun 13(1)
4. Khaleghi B, Khamis A, Karray FO, Razavi SN (2013) Multisensor data fusion: a review of the state-of-the-art. Inf Fusion 14(1):28–44

5. Ogawa T, Takagi K (2016) Lane recognition using on-vehicle lidar. In: Proceedings IEEE intelligent vehicles symposium, pp 540–545
6. Becker J (1999) Fusion of data from the object-detecting sensors of an autonomous vehicle. In: Proceedings of 1999 IEEE international conference on intelligent transportation systems, pp 362–367
7. Kalman R (1960) A new approach to linear filtering and prediction problems. ASME J Basic Eng 82:35–45

Chapter 33
Multi-phase Essential Repair Analysis for Multi-server Queue Under Multiple Working Vacation Using ANFIS

Richa Sharma and Gireesh Kumar

1 Introduction

In the recent era of industrial revolution, MWV models play a very important role for controlling congestion situations. In MWV models, servers work during his vacation period, if the necessity occurs. During the vacation period, the servers lower down their service rate for providing service. MWV queueing systems found an ample of applications such as networking modeling, telecommunications, production systems, and many more. Servi and Finn [1] analyzed a queue with WV and derived various results in his study. Since the server cannot find any waiting clients in the system, it can leave for the first essential vacation [2]. Multiple vacation policy [3] was applied for determining the performance of an unreliable queueing system. Sharma and Kumar [4] studied a machining system with multiple working vacations.

Multi-server queuing models have had a wide range of application in recent past years. Many authentication queue systems deal with unpredicted failures by servers which can be repaired by the repair facility. In this way, it is very important to recognize the effect of failure on the level for the proper execution of the queuing system. Sharma [5] analyzed k-phases hyper-exponential-based systems. Further, the concept of server breakdown is also considered in her study due to which service of ongoing customers can be interrupted. An infinite server system with catastrophes and server failures at the service station was investigated by Sophia and Murali [6]. Also, the transient and steady-state probabilities of an infinite server-queueing system was provided in their study. Dudin and Dudina [7] examined a multi-server retrial queue with unreliable servers. Very recently, a Markovian queue with vacations under break down was considered in the study of [8]. Further, a reduced service rate was

R. Sharma (✉)
Department of Mathematics, JK Lakshmipat University, Jaipur 302026, India

G. Kumar
Department of CSE, JK Lakshmipat University, Jaipur 302026, India

A. Sheth et al. (eds.), *Intelligent Systems*, Algorithms for Intelligent Systems,
https://doi.org/10.1007/978-981-16-2248-9_33

opted by the backup server for serving the customers during his vacation or may be under repair state.

Soft computing approach namely neuro-fuzzy has been employed by many prominent researchers for analyzing many composite systems in their study. Relationships have been developed by Ciaramella et al. [9] for neural networks. Performance prediction of the systems using ANIFS is provided in the study of [10]. Using neuro-fuzzy technique, they obtained various performance indices in their investigation. Jain and Upadhyaya [11] worked on degraded machining systems. The queuing model was investigated by Sharma [12] by applying neuro-fuzzy technique.

In our study, MPER system has been developed wherein the repair has been done in the multi-essential phases. Moreover, the concept of MWV is also taken into account. In literature, the combination of MPER and MWV was not considered together for developing the queueing system. In Sect. 2, we provide assumptions along with notations for developing the MPER system. The proposed model was further investigated in Sect. 3. Governing equation of the MPER system is also given in this section. Next, metrics of MPER are presented in Sect. 4. Section 5 is devoted to numerical results on the basis of transient analysis. Concluding notes are given in Sect. 6.

2 Model Description

In our study, we study the MPER system wherein the repairman provides 'l' phase repair to the failed component. As there are no awaited customers in MPER system, then 'K' servers may go for working vacation. For governing the MPER system model, assumptions are considered which are given as:

- The system's states are symbolized using (m, n); n $(n = 0, 1, 2, \ldots)$ denotes customers present in MPER system, and m $(m = V, B, R_i)$ describes the state of the server as:

$$\varphi(t) = \{V, \text{when the servers are on working vacations} \\ B, \text{when the servers are busy in providing service} \\ R_i, \text{when the servers are breakdown and under} \\ i\text{th } (i = 1, 2, 3, \ldots, l) \text{ phase essential repair}$$

- In the MPER system, there are 'c' $(1 \leq c \leq K)$ servers who serve with service rate μ_b. The 'c' servers may go for vacation as there is no one available in the system. The servers change their service rate and serve with lower rate $\mu_v (\mu_v < \mu_b)$.
- All arriving customers join the system with state-dependent rates λ_v, λ_b and λ_r during WV, busy, and repair state. The service has been provided by the servers using first-in first-out (FIFO) approach.

- When the working vacation ends, servers join the system back with rate η. If servers find there are no awaited customers in the MPER system, they may go for another second optional vacation.
- As servers provide services to arriving customers, the current customer service may be stopped due to a server error known as server breakdown with rate α. As breakdown occurs, the repairman provides the essential repair in phases with rate β_i $(i = 1, 2, \ldots, l)$.

Notation:

$P_{m,n}(t)$ Prob. of customers during 'm' states at time 't' when there are 'n' customers in the system

3 The Analysis

For determining the MPER system, we provide the governing Eqs. (1)–(13) of MPER with the help of Fig. 1 which are given as follows:

A. Governing Equations:

- Exponential Working Vacation Stat

$$P'_{V,0}(t) = -\lambda_v P_{V,0}(t) + \mu_v P_{V,1}(t) + \mu_b P_{B,1}(t) \tag{1}$$

$$\begin{aligned} P'_{V,n}(t) = & -(\lambda_v + \eta + n\mu_v) P_{V,n}(t) + \lambda_v P_{V,n-1}(t) \\ & + (n+1)\mu_v P_{V,n+1}(t), \quad 1 \le n \le c-1 \end{aligned} \tag{2}$$

$$P'_{V,n}(t) = -(\lambda_v + \eta + c\mu_v) P_{V,n}(t) + \lambda_v P_{V,n-1}(t) + c\mu_v P_{V,n+1}(t), \quad n \ge c \tag{3}$$

- Busy State

$$P'_{B,1}(t) = -(\lambda_b + \alpha + \mu_b) P_{B,1}(t) + 2\mu_b P_{B,2}(t) + \eta P_{V,1}(t) + \beta_2 P_{R_1,2}(t) \tag{4}$$

$$\begin{aligned} P'_{B,n}(t) = & -(\lambda_b + \alpha + n\mu_b) P_{B,n}(t) + \lambda_b P_{B,n-1}(t)(n+1)\mu_b P_{B,n+1}(t) \\ & + \eta P_{V,n}(t) + \beta_2 P_{R_n,2}(t), \quad 2 \le n \le c-1 \end{aligned} \tag{5}$$

$$\begin{aligned} P'_{B,n}(t) = & -(\lambda_b + \alpha + c\mu_b) P_{B,n}(t) + \lambda_b P_{B,n-1}(t) + c\mu_b P_{B,n+1}(t) \\ & + \eta P_{V,n}(t) + \beta_2 P_{R_n,2}(t), \quad n \ge c \end{aligned} \tag{6}$$

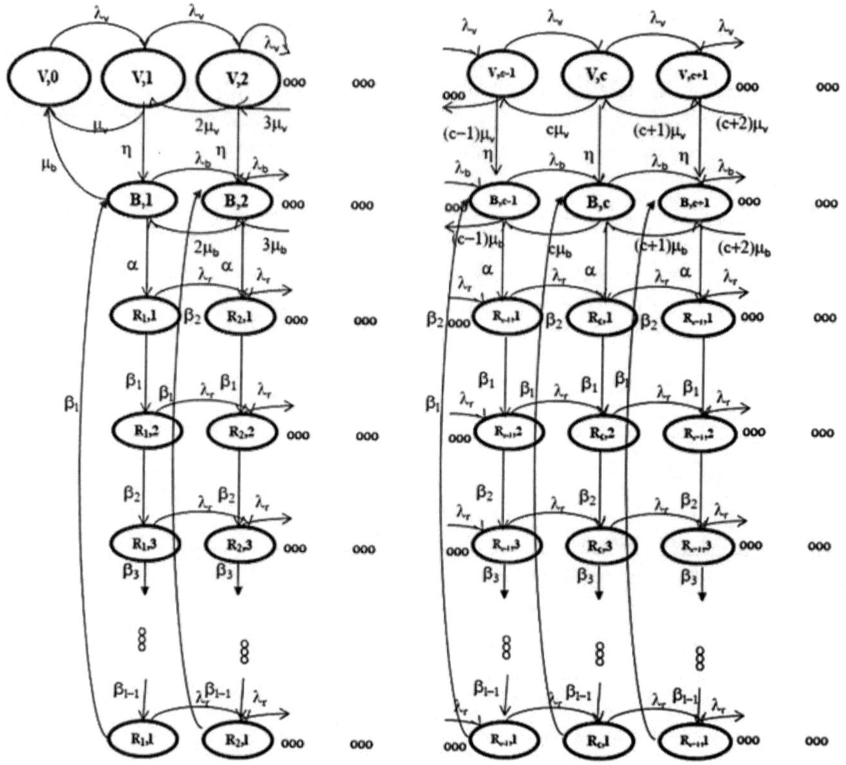

Fig. 1 MPER system with multi-server and multiple working vacations

- MPER State

$$P'_{R_1,1}(t) = -(\lambda_r + \beta_1)P_{R_1,,1}(t) + \alpha P_{B,1}(t) \tag{7}$$

$$P'_{R_1,i}(t) = -(\lambda_r + \beta_i)P_{R_n,i}(t) + \beta_{i-1}P_{R_1,i-1}(t), \quad 2 \leq i \leq l-1 \tag{8}$$

$$P'_{R_1,l}(t) = -(\lambda_r + \beta_l)P_{R_1,,l}(t) + \beta_{l-1}P_{R_1,,l-1}(t) \tag{9}$$

$$P'_{R_n,1}(t) = -(\lambda_r + \beta_1)P_{R_n,1}(t) + \alpha P_{B,n}(t) + \lambda_r P_{R_{n-1},,1}(t), \quad n \geq 2 \tag{10}$$

$$\begin{aligned} P'_{R_n,i}(t) = & -(\lambda_r + \beta_i)P_{R_n,i}(t) + \beta_{i-1}P_{R_n,,i-1}(t) \\ & + \lambda_r P_{R_{n-1},,i}(t), \quad n \geq 2, \ 2 \leq i \leq l-1 \end{aligned} \tag{11}$$

$$P'_{R_n,l}(t) = -(\lambda_r + \beta_l)P_{R_n,2}(t) + \beta_{l-1}P_{R_n,l-1}(t) + \lambda_r P_{R_{n-1},l}(t), \quad n \geq 2 \tag{12}$$

A. Normalization condition:

$$\sum_{n=0}^{K} P_{V,n}(t) + \sum_{n=1}^{K} P_{B,n}(t) + \sum_{i=1}^{l}\sum_{n=1}^{K} P_{R_n,i}(t) = 1 \tag{13}$$

4 Performance Measures

To control MPER system at time 't', we calculate various performance measures given in Eqs. (14)–(17) such as probability of the server during WV, busy, and under repair state. Further, MPER system length at time 't' is given as under:

- Prob. of server during WV state at time 't' is

$$P_{WV}(t) = \sum_{n=0}^{K} P_{V,n}(t) \tag{14}$$

- Prob. of server during busy state at time 't' is

$$P_B(t) = \sum_{n=1}^{K} P_{B,n}(t) \tag{15}$$

- Prob. of server system under ith ($i = (1, 2, 3, \ldots, l)$) phase of essential repair at time 't' is

$$P_{R_i}(t) = \sum_{n=1}^{K} P_{R_n,i}(t), \quad i = (1, 2, 3, \ldots, l) \tag{16}$$

- Average number of the customers in the MPER system at time 't' is

$$AN_S(t) = \sum_{n=0}^{K} nP_{V,n}(t) + \sum_{n=1}^{K} nP_{B,n}(t) + \sum_{i=1}^{l}\sum_{n=1}^{K} nP_{R_n,i}(t) \tag{17}$$

5 Numerical Analysis

Numerical results through tables and graphs are displayed for validation purposes. The results are given by showing the different parameters using the different set values for the probability of the server during working vacation, busy, and repair state and average queue length. The "ode45" function of MATLAB software has been applied for determining outcomes. For computation, we fix default parameter such as $\lambda_v = 0.5$, $\lambda_b = 0.8$, $\lambda_r = 0.2$, $\mu_v = 0.6$, $\mu_b = 0.95$, $\alpha = 0.5$, $\eta = 0.1$, $\beta_1 = 0.1$, $\beta_2 = 0.2$, $\beta_3 = 0.4$, $\beta_4 = 0.5$ and $\beta_5 = 0.9$, $c = 8$, and $K = 10$.

In Table 1, we provide the effect service and failure rates on the MPER system. In Table 1, as the values of service rate (μ_b) increases, the probability of working vacation increases. On the other hand, probability of busy ($P_B(t)$) and repair ($P_{R_i}(t)$, $i = 1, 2, 3$ and 4) states decrease for increasing values of μ_b. The reverse trend has been seen with respect to failure rate (α). From Figs. 2, 3, 4, and 5, the impact of arrival, service, failure, and repair rates with respect to time can be seen on average number of the customers ($AN_S(t)$) in the MPER system. From Figs. 2, 3, 4,

Table 1 Effect of time on performance measures for MPER systems by varying service and failure rate

	$t = 1.5$					
μ_b	PWV (t)	PB (t)	$P_{R_1}(t)$	$P_{R_2}(t)$	$P_{R_3}(t)$	$P_{R_4}(t)$
1	0.10	0.15	0.12	0.11	0.07	0.068
2	0.16	0.12	0.11	0.08	0.06	0.05
3	0.19	0.11	0.09	0.07	0.05	0.04
4	0.20	0.10	0.07	0.06	0.04	0.03
5	0.21	0.09	0.06	0.05	0.03	0.02
α	PWV (t)	PB (t)	$P_{R_1}(t)$	$P_{R_2}(t)$	$P_{R_3}(t)$	$P_{R_4}(t)$
0.1	0.19	0.28	0.17	0.20	0.24	0.28
0.2	0.18	0.30	0.20	0.22	0.26	0.30
0.3	0.11	0.31	0.21	0.23	0.28	0.34
0.4	0.09	0.32	0.24	0.27	0.31	0.38
0.5	0.05	0.35	0.25	0.30	0.32	0.40

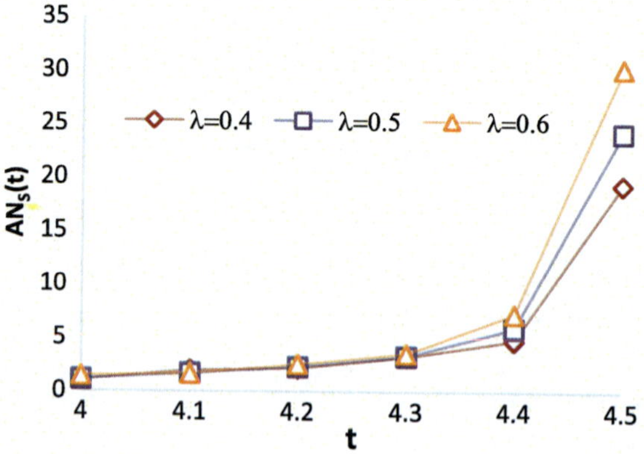

Fig. 2 $AN_S(t)$ versus t by varying λ for MPER system

Fig. 3 $AN_S(t)$ versus t by varying μ_b for the MPER system

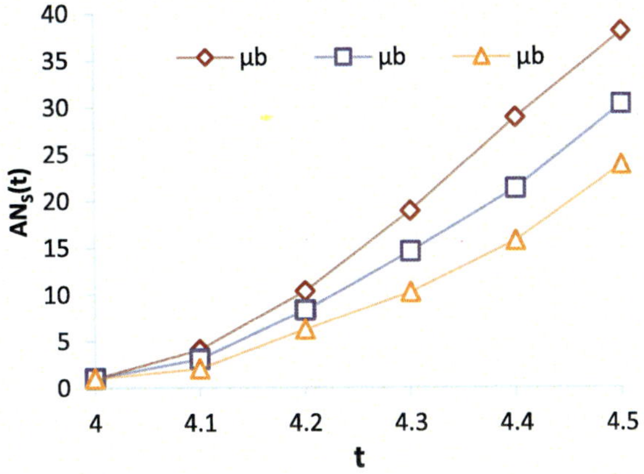

Fig. 4 $AN_S(t)$ versus t by varying α for the MPER system

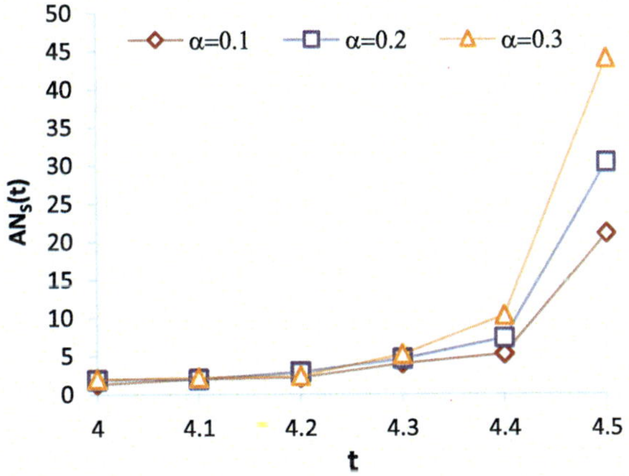

Fig. 5 $AN_S(t)$ versus t by varying β_5 for the MPER system

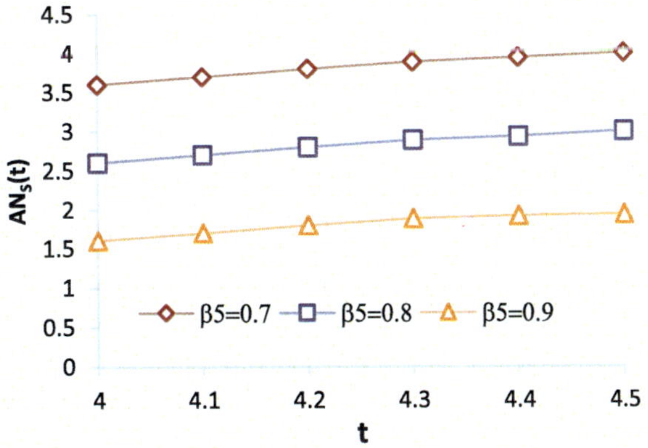

and 5, the $AN_S(t)$ increases as time reaches the higher values. Further, the average number of the customers in the MPER system decreases (increases) as there is an increment of the values in μ_b and β_5 (λ and α). ANFIS network results for $AN_S(t)$ are displayed in Figs. 6 and 7. For this purpose, 15 epochs iterations are used following the toolbox namely fuzzy using MATLAB package. The linguistic variables for input

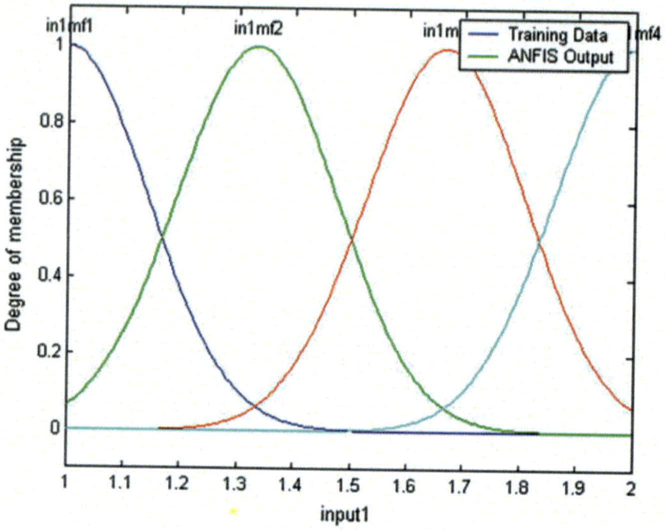

Fig. 6 Membership functions for input parameter λ

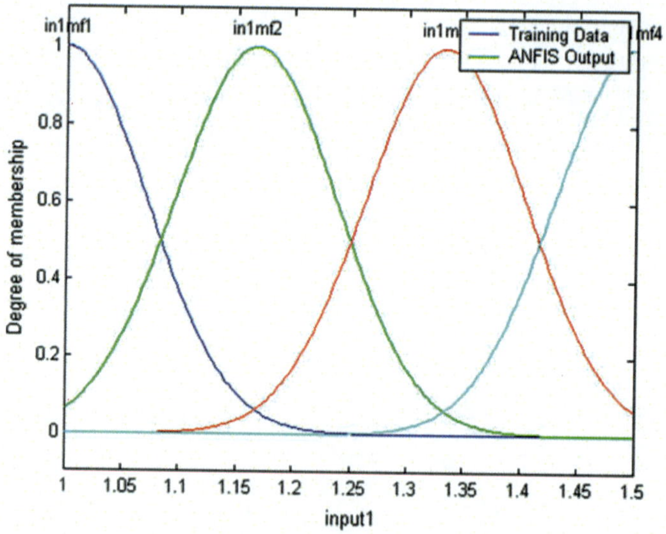

Fig. 7 Membership functions for input parameter μ

variables λ and μ are Low, Average, High, and Very high. Gaussian function was used to show the membership functions (mfs) for linguistic variables. The shapes of mfs for Figs. 2 and 3 are displayed in Figs. 6 and 7.

At last, based on the above results, the authors pointed out that if certain sensitive parameters (i.e., service and failure rate) can be treated well then the average number of customers can be controlled. The results derived in this section matches with many real scenarios. ANFIS results provide a simple, fast solution comparable to analytical results, helping analysts and decision makers manage queuing systems with accurate results.

6 Conclusion and Future Scope

In this manuscript, we have examined the MPER queuing systems with multiple working vacations and unreliable servers that encounter many practical congestion situations related to computer networks, communication systems, production systems, and many others. The concept of unreliable servers provides valuable insight for developing the cost-efficient model for real time systems.

In future, the current investigation can be extended by including the batch arrival along with batch service of the customer.

References

1. Servi LD, Finn SG (2002) M/M/1 queues with working vacations (M/M/1/WV). Perf Eval 50:41–52
2. Jain M, Sharma GC, Sharma R (2013) Unreliable server M/G/1 queue with multi-optional services and multi-optional vacations. Int J Math in Oper Res 5:145–169. https://doi.org/10.1504/ijmor.2013.052458
3. Jain M, Sharma R, Sharma GC (2013) Multiple vacation policy for $M^X/H_k/1$ queue with unreliable server. J Ind Eng Int 9:1–11. https://doi.org/10.1186/2251-712X-9-36
4. Sharma R, Kumar G (2017) Availability improvement for the successive K-out-of-N machining system using standby with multiple working vacations. Int J Rel Saf 11:256–267. https://doi.org/10.1504/IJISE.2015.069918
5. Sharma R (2016) Vacation queue with server breakdown for MX/HK/1 queue under N-Policy. Int J Comp Info Sci 17:33–41
6. Sophia S, Murali TS (2018) Transient analysis of an infinite server queue with catastrophes and server failures". Int J Pure App Math 118:253–262
7. Dudin DO (2019) Retrial multi-server queuing system with PHF service time distribution as a model of a channel with unreliable transmission of information. Appl Math Mode 65:676–695
8. Chakravarthy SR, Shruti KR (2020) A queueing model with server breakdowns, repairs, vacations, and backup server. Oper Rese Pers 7:1–13
9. Ciaramella A, Taglia Ferri R, Pedrycz W, Nola AD (2006) Fuzzy relational neural network. Int J Approx Reas 41(2):146–163
10. Jain M, Upadhyaya S (2009) Threshold N-policy for degraded machining system with multiple type spares and multiple vacations. Qual Tech Quant Manage 6(2):185–203

11. Jain M, Maheshwari S, Baghel KP (2008) Queueing network modelling of flexible manufacturing system using mean value analysis. Appl Math Mode 32(5):700–711
12. Sharma R (2010) Threshold N-Policy for MX/H2/1 queuing system with un-reliable server and vacations. J Int Acad Phys Sci 14(1):41–51

Chapter 34
Performance Comparison of Benchmark Activation Function ReLU, Swish and Mish for Facial Mask Detection Using Convolutional Neural Network

Rupshali Dasgupta, Yuvraj Sinha Chowdhury, and Sarita Nanda

1 Introduction

In the field of image processing and computer vision, face detection has been a compelling nodus. It comprises a diverse range of utilization, from capturing facial motion to recognition of face, which permits the face to be identified with sterling precision at the beginning. Face detection is important to an absolute extent today since it is not just used for images, besides that, it has practical applications that involve videos such as live monitoring and video face detection. With the advances of Convolutional Networks, high-precision image recognition is now possible.

The global pandemic of COVID-19 has seriously affected the world and now more than 35 million individuals [1] are affected by it worldwide, according to data collected by the World Health Organization. Two of the necessary protection precautions must be observed in public places to stop the virus from spreading are wearing face masks [2] and adopting safe social distance. Creating a low-risk ambience that contributes to public safety, a paper proposed [3] suggests a productive real-time instantaneous automated surveillance of people, based on computer vision approach to detect whether a person is wearing a mask or not integrated from pre-installed CCTV cameras, consisting of a convolutional neural network (CNN) architecture. This will help control safety breaches, facilitate the use of face masks, and maintain a safe working atmosphere.

The novelty of this work is training of the proposed model using Mish activation function and Swish activation function, besides ReLU as was proposed by the some of the previous works, in order to find something that works better. This paper describes a comparison between activation functions implemented on the dataset created by Bhandary [4] for detection of face masks against COVID-19 from real-time camera footage captured using a webcam. Each of the activation functions would be used

R. Dasgupta · Y. S. Chowdhury · S. Nanda (✉)
Kalinga Institute of Industrial Technology, Bhubaneswar, Odisha 751024, India
e-mail: snandafet@kiit.ac.in

A. Sheth et al. (eds.), *Intelligent Systems*, Algorithms for Intelligent Systems,
https://doi.org/10.1007/978-981-16-2248-9_34

in the same model. The comparison criteria would be based on training accuracy, testing accuracy and computational time. The face detection is done using the Haar cascade algorithm for the cascade classifier.

1.1 Related Works

Choosing the most suitable activation functions for the convolutional neural networks has a huge influence on the training time and efficiency of the task. The Rectified Linear Unit (ReLU) is the most prevalent activation function (AF) nowadays. Despite its "dying ReLU" problem and several attempts to replace it with something better, it is still considered as a default option for network creation, which was also the first choice for this problem as well. Reference [5] mentions a new promising feature regarding a new activation function. "A combination of exhaustive and reinforcement learning-based search" was used to obtain the proposed function called "Swish". Simply replacing ReLU with Swish units, according to the authors, increases "top 1 classification accuracy on ImageNet for Mobile NASNet-A by 0.9% and Inception-ResNet-v2 by 0.6%". In Reference [6], the author came up with a new activation function "Mish" and experiments that show that "Mish tends to work better than both ReLU and Swish along with other standard activation functions in many deep networks across challenging datasets". The technology of face detection has achieved a triumph with the very popular Viola Jones Detector [7], that significantly enhanced concurrent detection. The Haar [8] features were optimized by the Viola Jones detector, but it failed to solve the problems of the real world and was affected by different parameters such as face brightness and face orientation. These problems have led researchers to strive on creating new models of face detection relying on deep learning in order to obtain better outcomes for various facial features. Toshanlal Meenpal developed a model for face detection that used the Multi-Human Parsing Dataset [9], focused on completely fully convolutional networks, it can for that matter, detect the face in any frontal or non-frontal geometric condition.

2 Methodology

To construct a binary face classifier that is capable of facial recognition in any orientation, we propose this paper intending to compare three different activation functions, namely ReLU, Swish and Mish. The fundamental function of the model is extracting features and predicting class. The model output is a function which is optimized by the technique of gradient descent and the loss function utilized is binary cross-entropy. There are 1376 images, 690 face images with masks and 686 without masks in the initial dataset as shown in Figs. 1 and 2 [4].

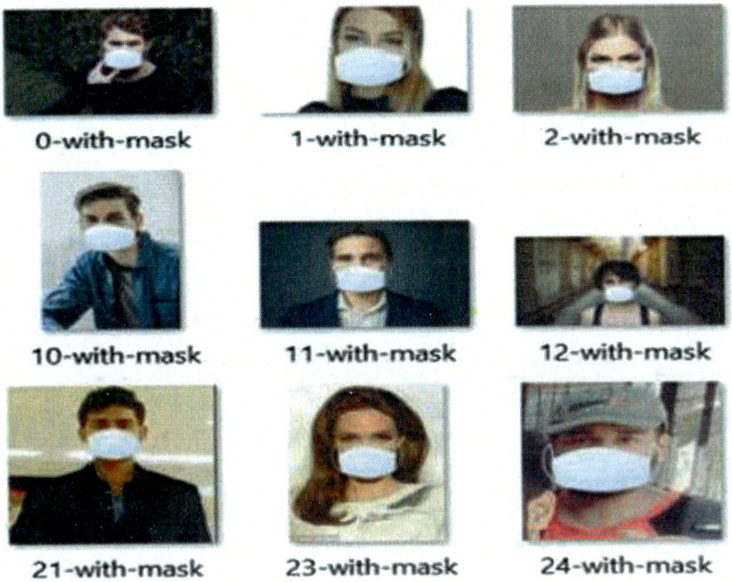

Fig. 1 Example images with masks

Fig. 2 Example of images without masks

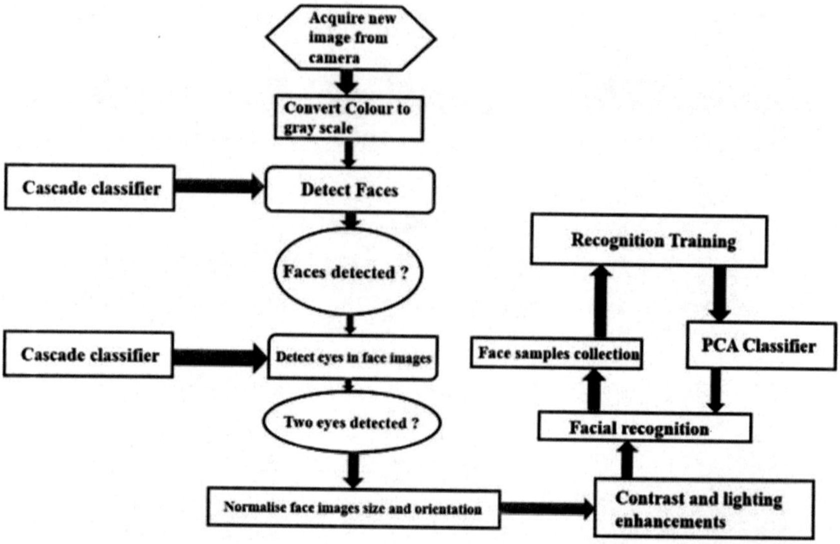

Fig. 3 Flowchart of the Haar cascade algorithm

2.1 *Workflow of the Model*

The input RGB images of any arbitrary size are first converted into grayscale images and then resized to 100 × 100 since we need common sized images for all inputs. The resized grayscale images, which act as our data for the network, are then delivered to the CNN model for extraction of features and class prediction. The output of this model is then put through to post-processing. The face detection is done using the Haar cascade algorithm [10, 11], shown in Fig. 3. The Haar cascade algorithm is used in Machine Learning to detect objects in video or images. The faces are detected using a cascade classifier which is followed with eye detection. After both the eyes are detected successfully, the face images and orientation are normalized. Facial recognition is completed by collecting face samples and training the model to recognize followed by PCA Classifier.

2.2 *Convolutional Neural Network*

The model, as shown in Fig. 4, consists of the first CNN layer with an activation function accompanied by a MaxPooling layer. The second convolution layer with the same activation function is accompanied by another MaxPooling layer. The input data is convoluted by the convolution layer with another window while the MaxPooling layer makes sure that the size of the function vector that is being generated in each layer is then reduced to half to minimize the number of parameters. This is followed

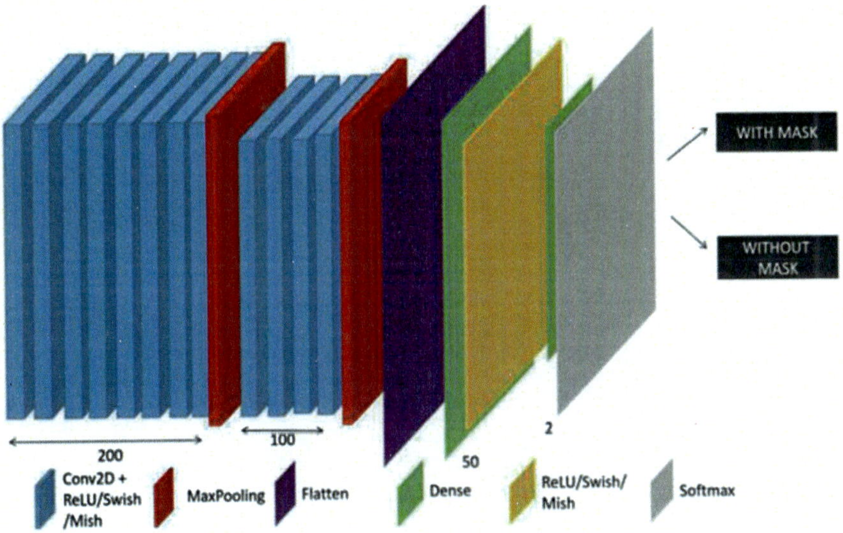

Fig. 4 Architecture of convolutional neural network

by a Flatten layer which converts the 2D matrix of the features from the second convolutional layer into a vector and is accompanied by a Dense layer of 50 nodes, with the same activation function as used in the convolution layers.

We built our network model using Keras and TensorFlow libraries available in Python 3. Our study is based on training the identical template models utilizing separate activation functions. Each node has the same activation function each and every time we assess one of the three activation functions, except the final layer which invariably conforms to softmax as shown in Fig. 5, using Keras and TensorFlow modules.

3 Activation Functions (AF)

The activation function decides whether or not the neuron should be enabled by evaluating the weighted sum and adding additional bias. The activation function (AF) aims to introduce non-linearity into the neuron output. It is known that the neural network has neurons that operate in correlation with the weight, bias and their activation function. Concerning the weights of the neural network, the process, known as backpropagation, measures the gradient of the error function. Activation functions facilitate backpropagation because the gradients are provided together with the error of updating the weights and biases.

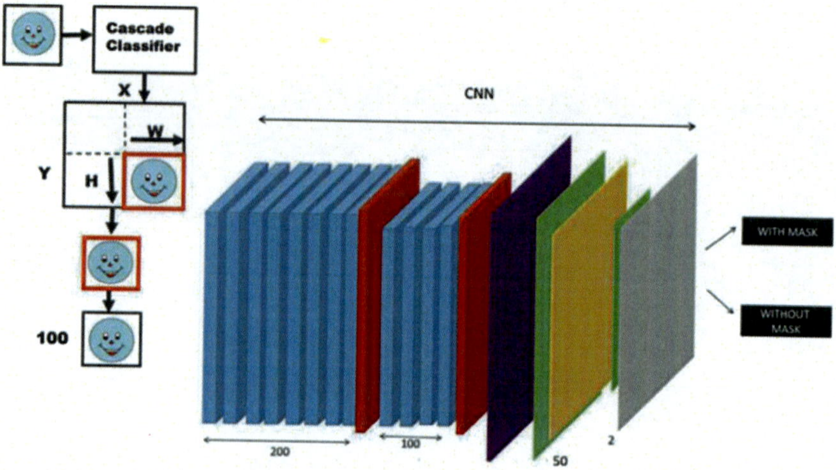

Fig. 5 Workflow of the model

3.1 Rectified Linear Unit (ReLU) Activation Function

The rectified linear unit (ReLU) AF (Fig. 6) was introduced by Nair and Hinton [12], and ever since has been the most prevalent AF for deep learning uses. The ReLU is a faster AF for learning, which has proven to be the most effective and widely used function [4]. It supplies better performance and generalization in deep learning in comparison with the Sigmoid and Tanh AF [13]. The ReLU is an almost total representation of the linear function and therefore retains the properties of linear models that help them in easily getting optimized, with gradient-descent methods. For each input element, the ReLU AF performs a threshold operation where value is zero for negative arguments and the variable itself for non-negative arguments, and

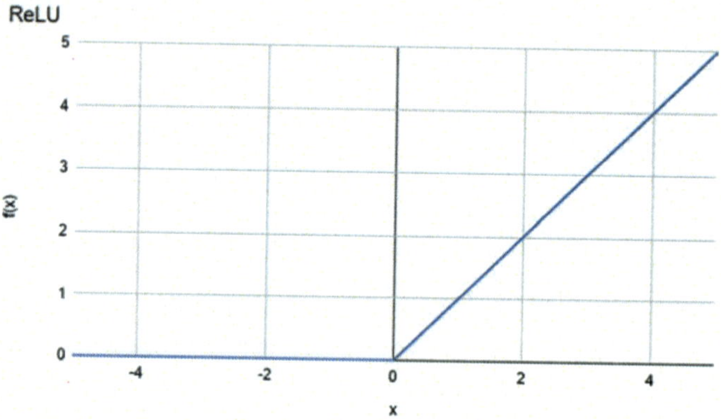

Fig. 6 Graphical representation of ReLU

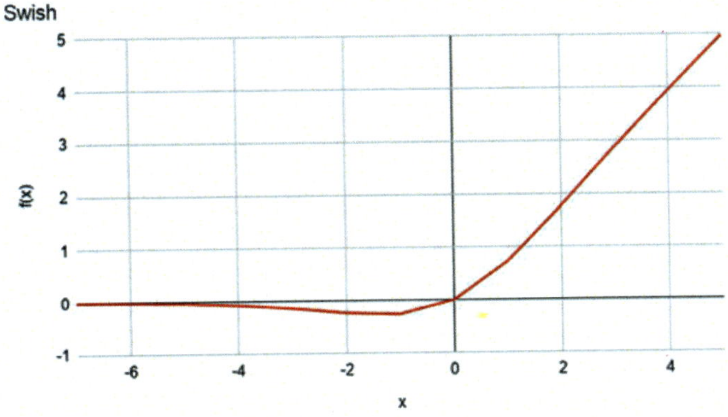

Fig. 7 Graphical representation of Swish

thus, ReLU is represented by

$$f(x) = \max(0, x) \tag{1}$$

3.2 *Swish Activation Function*

The Swish AF (Fig. 7) is among the first AF compounds stated to achieve a hybrid AF by combining the sigmoid AF and the input function. Ramachandran et al., 2017, proposed the Swish activation [4], and it uses the automatic search method that is based on reinforcement learning to obtain the function. Smoothness, non-monotonic, bounded below and unbounded in the upper limits are some of the properties of the Swish function. The property of smoothness makes the Swish function when used in training deep learning architectures, delivering better optimization and generalization performance [14]. The Swish function is represented by

$$f(x) = \text{sigmoid}(x) * x = x/\left(1 + \mathrm{e}^{-x}\right) \tag{2}$$

3.3 *Mish Activation Function*

One of the latest activation networks is the Mish AF (Fig. 8). Most studies show that Mish performs better than ReLU, Sigmoid as well as Swish [5]. For any AF, being unbounded above is a beneficial asset as it prevents saturation, which usually causes training to decline drastically due to near-zero gradients. And it is also beneficial to

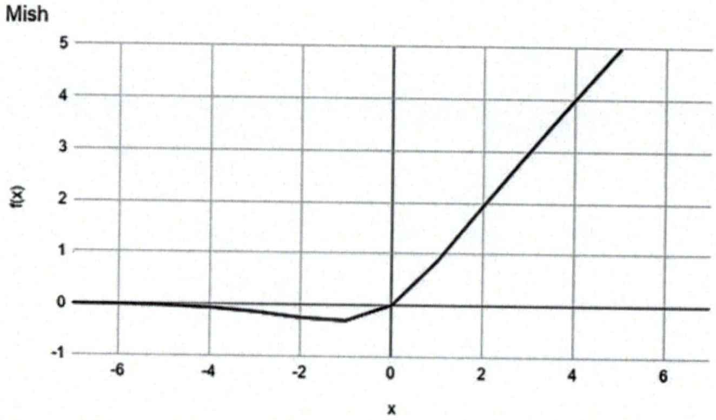

Fig. 8 Graphical representation of Mish

be restricted below because it brings about good regulatory effects and minimizes overfitting. As ReLU is not continuously differentiable, it has an order of continuity as zero and may create some problems in gradient-based optimization which doesn't occur in the case of Mish. Thus, Mish function is represented by

$$f(x) = x * \tan h(\text{softplus}(x)) = x * \tan h\left(\ln\left(1 + e^{x}\right)\right) \tag{3}$$

Comparing ReLU, Swish and Mish graphically (Fig. 9).

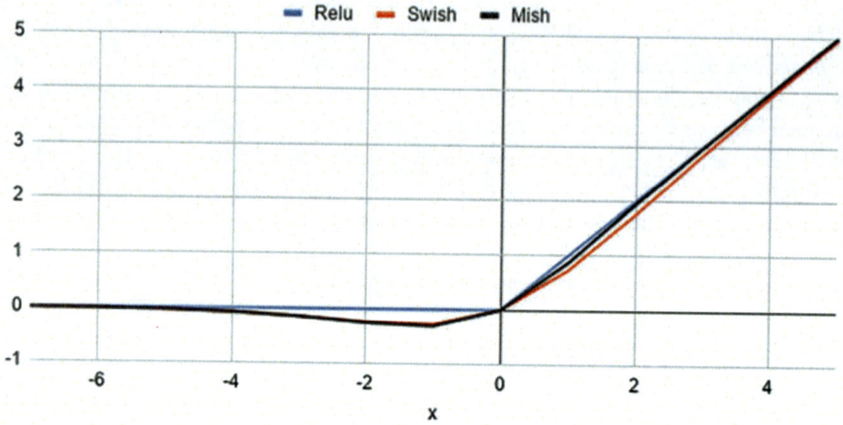

Fig. 9 Graphical representation of ReLU, Swish and Mish

4 Results and Observations

The paper affirms to what extent an activation function is crucial. Choosing an activation function is one of the foremost parameters of a neural network [15]. The performance of various activation functions indicates that there is not much disparity between them. Once a network is successfully trained, with activation functions such as ReLU, Swish and Mish, it has a congruent effect on the network. It is an essential task to choose an activation function for a network or its particular layers. But, as the observations demonstrate, if a network is successfully trained with ReLU activation function it is found that the accuracy is 95.3544%, whereas in case of Mish it is slightly lower with an accuracy of 92.5326% however it is observed that accuracy of Swish is 89.7912% and is lower than both Mish and ReLU (Fig. 12). ReLU takes the least amount of computational time, followed by Swish and then Mish (Fig. 14). The model was trained in 990 samples and validated on 248 samples. Figures 10, 11, 12, 13 and 14 show the results of our experiments.

Table 1 describes the training and testing accuracy results, training and testing losses and the computational time taken by each of the activation functions to train the model.

4.1 Confusion Matrix and Results

The confusion matrices of all the models when used with ReLU, Swish and Mish are given below in Figs. 15, 16 and 17, respectively. Accordingly, the parameters accuracy, precision, recall, sensitivity, specificity and F1 score were calculated and displayed in a tabular format in Table 2.

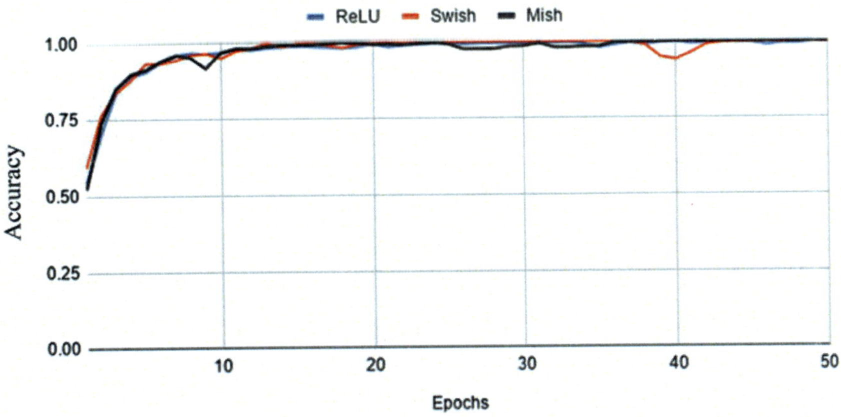

Fig. 10 Pictorial representation of a comparison of training accuracy of 50 epochs for ReLU, Swish and Mish

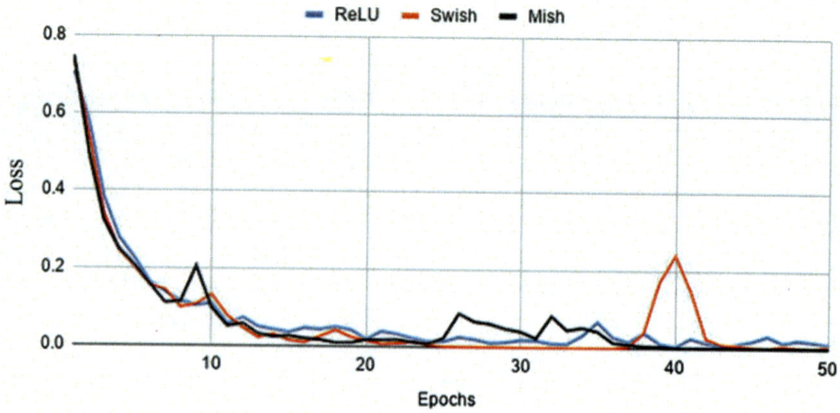

Fig. 11 Pictorial representation of a comparison of training loss of 50 epochs for ReLU, Swish and Mish

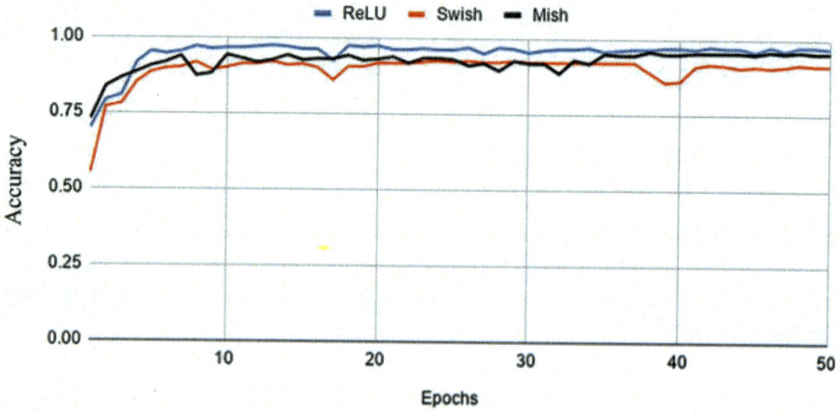

Fig. 12 Pictorial representation of a comparison of testing accuracy of 50 epochs for ReLU, Swish and Mish

5 Conclusion

Gathering from Tables 1 and 2, we can conclude that, unlike our expectations, Swish and Mish couldn't outperform ReLU, however, Mish does outperform Swish. Conclusively it implies that in problems such as facial mask detection, Swish or Mish do not offer more than ReLU. Mish overperforms Swish, but it could not overperform ReLU, which is very evident from Table 2. In addition to that, in the matter of computational time, ReLU shines above both Mish and Swish. Similar results have been obtained by Szandała [16] and Sharmain [17] in their study to draw a contrast between Swish and ReLU. But hereafter comparing ReLU with Mish, we conclude that even Mish has failed to outperform ReLU. Hence it seems the "dying ReLU"

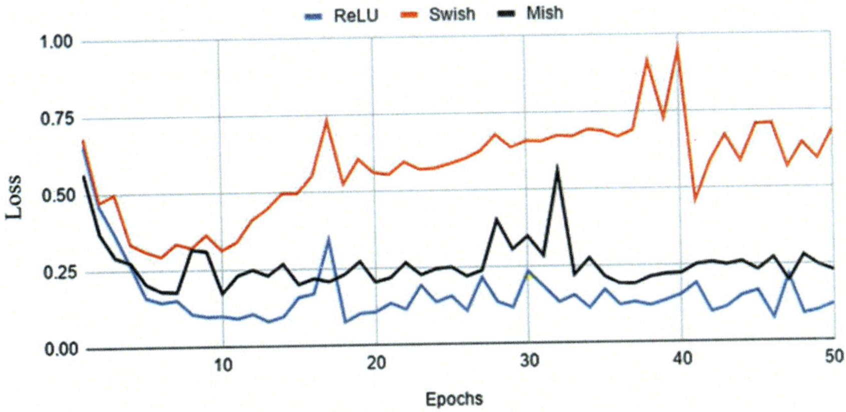

Fig. 13 Pictorial representation of a comparison of testing loss of 50 epochs for ReLU, Swish and Mish

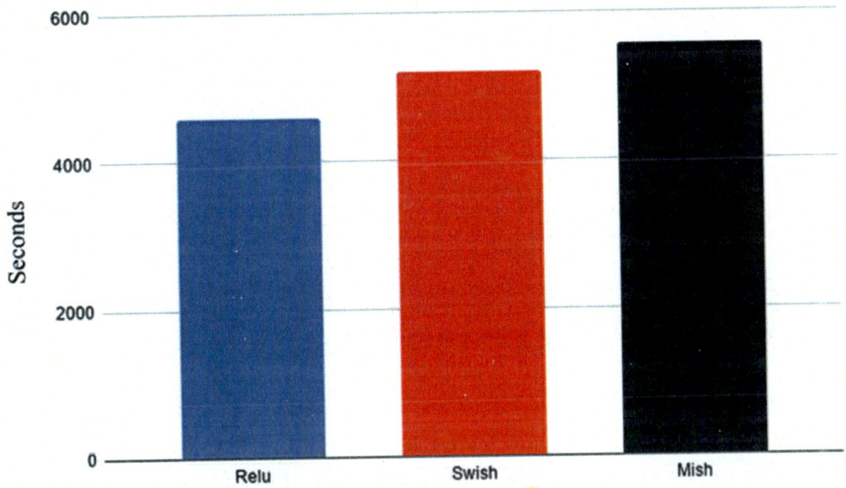

Fig. 14 Histogram of a comparison of computational time of 50 epochs for ReLU, Swish and Mish

Table 1 Training accuracy, training loss, testing accuracy, testing accuracy, testing loss and computational time of ReLU, Swish and Mish Activation Functions

Activation function	Training accuracy	Training loss	Testing accuracy	Testing loss	Computational time in seconds
Rectified Linear Unit (ReLU)	0.966046	0.0775	0.953544	0.161548	4586.671
Swish	0.968872	0.0754	0.897912	0.57208	5187.128
Mish	0.96715	0.072932	0.925326	0.258602	5556.61

Fig. 15 Confusion matrix of ReLU

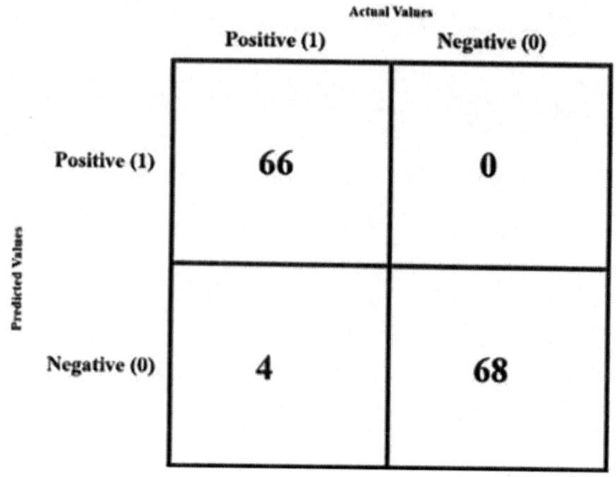

Fig. 16 Confusion matrix of Swish

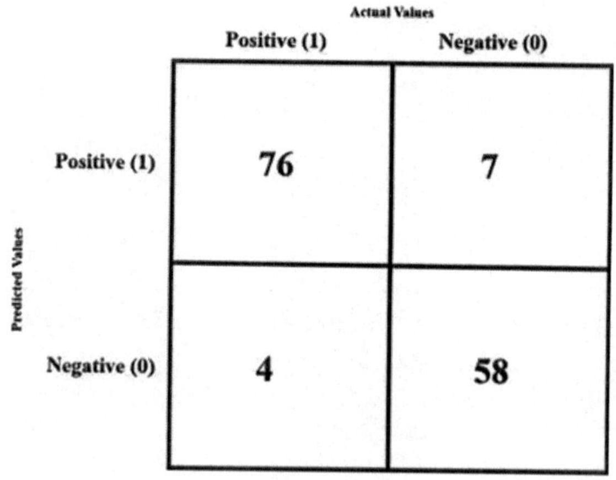

Fig. 17 Confusion matrix of Mish

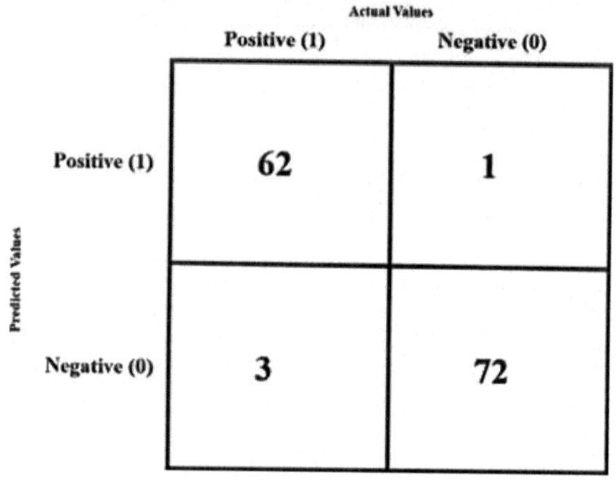

Table 2 Comparison of accuracy, precision, recall, sensitivity, specificity and F1 score of the three confusion matrices

Measure	ReLU	Swish	Mish
Accuracy	0.9710	0.9241	0.9710
Precision	1.0000	0.9157	0.9841
Recall	0.9428	0.9156	0.9841
Sensitivity	0.9429	0.9500	0.9538
Specificity	1.0000	0.8923	0.9863
$F1$ score	0.9706	0.9325	0.9688

problem is not that problematic as there has been no activation function to perform more accurately than ReLU. Hence ReLU remains the first choice for preparing and training network models.

References

1. Roser M et al (2020) Coronavirus pandemic (COVID-19). Our world in data
2. Howard J et al (2020) Face masks against COVID-19: an evidence review. https://doi.org/10.20944/preprints202004.0203.v1
3. Chavda A et al (2020) Multi-stage CNN architecture for face mask detection. arXiv preprint arXiv:2009.07627
4. Bhandary (2020) Mask detector. https://github.com/prajnasb/maskdetector
5. Ramachandran P, Zoph B, Le QV (2017) Searching for activation functions. arXiv preprint arXiv:1710.05941
6. Misra D (2019) Mish: a self-regularized non-monotonic neural activation function. arXiv preprint arXiv:1908.08681
7. Viola P, Jones MJ (2004) Robust real-time face detection. Int J Comput Vision 57(2):137–154. https://doi.org/10.1023/B:VISI.0000013087.49260.fb
8. Viola P, Jones M (2001) Rapid object detection using a boosted cascade of simple features. In: Proceedings of the 2001 IEEE computer society conference on computer vision and pattern recognition. CVPR 2001, vol 1. IEEE. https://doi.org/10.1109/CVPR.2001.990517
9. Meenpal T, Balakrishnan A, Verma A (2019) Facial mask detection using semantic segmentation. In: 2019 4th international conference on computing, communications and security (ICCCS). IEEE. https://doi.org/10.1109/CCCS.2019.8888092
10. Padilla R, Costa Filho CFF, Costa MGF (2012) Evaluation of haar cascade classifiers designed for face detection. World Acad Sci Eng Technol 64:362–365
11. Zhang X, Gonnot T, Saniie J (2017) Real-time face detection and recognition in complex backgrounds. J Signal Inf Process 8(2):99–112
12. Nair V, Hinton GE (2010) Rectified linear units improve restricted boltzmann machines. Icml
13. Nwankpa C et al (2018) Activation functions: comparison of trends in practice and research for deep learning. arXiv preprint arXiv:1811.03378
14. Ramachandran P, Zoph B, Le QV (2017) Swish: a self-gated activation function. arXiv preprint arXiv:1710.0594171
15. Sharma S (2017) Activation functions in neural networks. Towards Data Sci 6
16. Szandała T (2020) Benchmarking comparison of Swish versus other activation functions on CIFAR-10 Imageset. In: International conference on dependability and complex systems. Springer, Cham. https://doi.org/10.1007/978-3-030-19501-4-49
17. Sharma J (2017) Swish in depth: a comparison of Swish and ReLU on CIFAR-10. Medium. Medium, 22 Oct 2017. Web

Chapter 35
Time-Aware Online QoS Prediction Using LSTM and Non-negative Matrix Factorization

Parth Sahu, S. Raghavan, K. Chandrasekaran, and Divakarla Usha

1 Introduction

Cloud services are becoming the backbone of large workforces. On-demand nature, high scalability, comparative costs, and flexibility to user's requirements are some of its key features. With the advancement of cloud services, its demand has increased multifold. More providers bring in more competition which builds up a confusion for the customer in choosing the services. Recommending a cloud service to the customer is a real challenge for cloud service providers. Quality of service (QoS) parameters have a crucial role in this regard. On-demand nature, high scalability, comparative costs, and flexibility to user's requirements are some of its key features of cloud services. With the advancement of cloud services, its demand has also increased multifold. Service with good QoS values tends to be liked by the customer, so if we can predict QoS values for services, surely, we can recommend new services to the user based on it. One of the biggest challenges in such recommendations is that a large number of users generally invoke a very few numbers of services, so the data is highly sparse. Such data mainly consist of QoS parameter values like response time, throughput, latency, etc. These values are experienced by the customer when they invoke a service. A user's experience may highly depend on factors like geographical location and type of service. But the task of collecting QoS experiences on the customers' side is a challenging and costly operation. As there are many such services, most of them are paid to be used, so it is impossible for a single user to invoke them all. Thus, researchers try to build such systems that can take highly sparse data as input and predict those missing values with at most accuracy.

Memory-based collaborative filtering technique [1–5] was one of the initial algorithms in this field, which used to calculate similarity among the users and services,

P. Sahu · S. Raghavan (✉) · K. Chandrasekaran · D. Usha
National Institute of Technology Karnataka, Surathkal, Mangalore 575025, India

K. Chandrasekaran
e-mail: kchnitk@ieee.org

A. Sheth et al. (eds.), *Intelligent Systems*, Algorithms for Intelligent Systems,
https://doi.org/10.1007/978-981-16-2248-9_35

to find top neighbor users and services. They create feature matrices and predict according to these values. The main shortcoming of such methods is the cold start problem, and in high sparsity, it is challenging to find neighbor users and services accurately. Then came model-based methods with variations of matrix factorization [1, 6–8], such methods pretty much alleviated the problem of cold start. These methods need to define a model in which data is fitted, and then, predictions are made. With the advancement of artificial intelligence, methods based on machine learning algorithms [9, 10] and deep learning [11–13] have been proposed. These methods are much more capable of learning complex relations and non-convex pattern recognition in a high sparse environment than previous statistical-based methods.

Prediction in the time-aware environment is a completely different area where methods like memory-based collaborative filtering cannot learn temporal aspects. Methods based on machine learning algorithms are not flexible in incorporating time factors with more recent methods based on tensor factorization, and factor models have shown some potential. But none of the methods was originally drafted for time-series data. The problem is that all such methods consider data values to be independent of each other, which is not the case here. Previous values directly impact future values. So we propose a model based on LSTM, a type of recurrent neural network, specially designed to hold long-term dependencies, combined with matrix factorization to solve the cold start problem and extract local features. Experiments show our prediction accuracy to be second to none.

The remaining paper is organized as follows. Section 2 discusses the work carried out in the field over the last few years, glance over various methods, and present advancements, followed by Sect. 3 where we have explained the working principle of our approach, step-wise learning of features and building of the model. Section 4 has details about the environmental setup, dataset, performance statistics, and comparison and other details of the experimentation carried out. Section 5 concludes the work.

2 Literature Review

QoS prediction in the dynamics of a time-aware environment is the most crucial problem in the field of service selection and recommendation. The restraint of time incorporates real-time challenges and difficulties. QoS recommendation is being researched thoroughly, mainly since the past decade. There are two completely different areas in QoS prediction, first is QoS prediction in a single time frame or static environment, and second is QoS prediction in the dynamic or time-aware environment. For the first case, collaborative filtering (CF)-based methods are used most commonly. This technique depends upon historical QoS information for finding patterns in users' choices with other similar users. Collaborative filtering techniques can be implemented in two ways, memory-based approaches, and model-based approaches. Memory-based methods are also known as neighborhood methods and were among the earliest algorithms developed for CF. These methods are easy to implement; they do not require costly training [14]. Memory-based methods depend upon similar neighbors of the target user for finding personalized factors which may affect user choices [1]. Memory-based approaches can be further divided into three categories, user-based [2], service-based [3], and hybrid approaches [5].

Model-based techniques are completely different from memory-based models. They use machine learning algorithms to make a model that learns latent features of data by training. Model-based approaches are widely implemented through matrix factorization (MF) [7, 8]. For the second area, where time has a crucial role, approaches are different and complex. Zhang et al. proposed WSPred, a model-based approach that uses tensor factorization to learn user-specific, service-specific, and time-specific latent features [4]. Zhang et al. further update their work in [15], and introduced a new approach of non-negative tensor factorization. Li et al. describe another model-based technique, TMF, which is based on the adaptive matrix factorization with temporal smoothing on a three-dimensional user-service-time matrix [6]. Yu and Huang have presented a memory-based approach though rare in a time-aware environment. Xiong et al. [12] elaborate that though matrix factorization is an effective modeling technique, it is still not capable of learning temporal dependencies and unable to make optimum utilization of past observation. Hence, we proposed a unique solution by combining matrix factorization with the LSTM. LSTM is a type of recurrent neural network (RNN), which can effectively learn temporal dependencies.

3 Our Approach

The problem considered in this paper is precisely quality of service prediction for cloud service recommendation in a time-aware environment. Let us consider an original QoS data matrix R. With the dimension $m\ n\ k$, where m is the number of users (Total number of rows), n is the number of services (Total number of columns), and k is the third dimension denoting total number of time frame. Users are denoted as U_x where $x \leq m$ Ex. (u_0, u_1,u_2,u_{m-1},u_m), services are denoted as S_y where $y \leq n$ Ex (s_0, s_1,s_2, …..s_{n-1}, s_n) and time frames are denoted as t_z where $z \leq k$ Ex (t_0, t_1, t_2, t_{k-1}, t_k). So, any particular QoS value in the R can be identified with the help of the following triplet (u_x, s_y, t_z) where $x \in m$, $y \in n$ and $z \in k$. Now the challenge is to predict missing QoS values at time frame t_k with the help of QoS value available in all time frames t_x where $x < k$. Overall matrix M is sparse.

3.1 *Phase 1: Role of NMF*

As we have established the fact that our dataset is very sparse, we can see that method cannot learn proper time-sequential correlation because, for predicting value at time frame t_k, it will need QoS values of previous time frames like t_{k-1}, t_{k-2} … depending upon training window size. Hence if we can approximate the original missing QoS values with the help of some other method, it will enhance the temporal learning capability and forward it in the right direction. Matrix factorization was developed to work around the cold start problem (i.e., availability of very few values to start with) and is highly effective in filling out missing values in highly sparse conditions. One more crucial point to be noted is that QoS values are quality parameters whose values are never negative, so if we do not incorporate this fact, then the slightest of negative values can throw away a model from its right track of learning. NMF is an approach where both the basic elements and the weights are assumed to be component-wise non-negative; hence, we used non-negative matrix factorization. NMF uses Frobenius norm as its objective function to assess the quality of the approximation.

The advantage of using Frobenius norm is that it implicitly assumes noise to be Gaussian, which suits many practical situations, hence ours, and it can efficiently compute the approximation using singular value decomposition. Now another important aspect in NMF is the rank or dimension of the W and H, and it is important because it signifies the number of latent features that belong to users and services. These features can be any attribute or properties related to user or service; an optimum number of features will allow approximating better. These are features that can also be considered as local information, which will later be combined with the temporal aspect. W_z and H_z $\forall z \leq k$ are initialized as non-negative random matrices, scaled with *sqrt*(D_z*.mean()/Number of features*). NMF has been used with a multiplicative update approach for updating weight. The multiplicative approach is simple to implement and has better scalability with the sparse matrices though it may converge slowly. For the stopping mechanism, we can define a maximum number of iterations or the objective function itself stops after converging [16].

3.2 Phase 2: Role of LSTM

After completion of phase 1, we have successfully gathered local information in the form of the features of user and service and created an approximation matrix which is as close to our original matrices with missing values filled. In this phase, we will try to learn temporal correlation by training the LSTM model. The aim of performing this operation is to modify data shape, so that we can work on the time axis. We need to select the window size of the time frame, let us call it T, that is, how many numbers of previous values will the model require to predict the next value. A tiny window will not be able to retain the possible pattern, and a much larger window may introduce incorrect patterns. This window keeps on sliding by a single time frame and creates a set of training data. This is a kind of hyperparameter whose optimum value will be calculated later in the experiment part. Hidden layer output for time frame t works as input for the time frame $t + 1$. Cell state keeps track of learning, same as memory, and the whole process is repeated iteratively for a fixed number of epochs.

4 Experiment

The experiment involves two important components to be discussed; one is the setup used for doing experiments, and the next is the dataset description.

4.1 Environmental Setup

All the experiments are performed on the Intel® Core i7-4790 CPU @ 3.60 GHz 8 with 16 GB of RAM running 64-bit Ubuntu 18.04 LTS operating System. For implementation of various functions and creating models, we have used Keras library functions and TensorFlow as backend and some libraries from Scikit learn also.

4.2 Dataset

For this experiment, we have used a time-aware QoS dataset from WS-Dream originally provided by Zheng et al. [2, 17, 18]. The dataset consists of real-world QoS invocation experience of 4500 services by 142 users worldwide for 64 consecutive time intervals (15 min gap each). The original density of the dataset is around $\approx$67%, which is quite higher, if we compare it to the real-world service invocation scenarios. Hence, we increased the sparsity of the data to make it closer to real-world scenarios and randomly removed values to create a dataset with density = 5%, i.e., only 5% of known values were kept and the rest were made zero. Now, this dataset has been used to carry out all the operations.

4.3 Performance Comparison

From Fig. 1a, b, we can infer that the performance results of our approach are consistently better. Here, metrics used for measurement are absolute error and root mean square error. Lower value means better results. Our approach managed to perform exceedingly well at a 5% density level of data, which is a high sparsity scenario. If we compare with the basic NMF, there is an improvement of more than 50% with both absolute error and root mean squared error. It shows the efficiency of LSTM over NMF data, and results are justified as NMF only learns local relationships, while LSTM is able to learn temporal correlations over it. If we compare it with other time-aware approaches, then we can see there is still a large improvement over WSPred, while there is a slight improvement over PLMF and RTF in RMSE as these approaches are based on the same domain, that is a recurrent neural network,

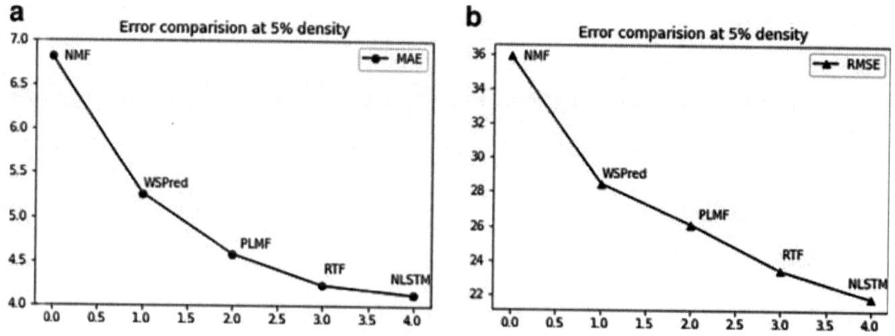

Fig. 1 **a** MAE error representation. **b** RMSE error representation

still better missing values handling and pre-data processing using NMF provide an advantage to out method.

4.4 *Impact of Hyperparameters*

Hyperparameters are crucial tuning elements of a model. The model itself does not learn these parameters and we have to choose their optimum values for the best results. In the case of NMF, a number of features define the rank of smaller matrices, which are used to approximate the original matrix. An optimum number of features can accommodate all the data's properties, so that it can fit better. As shown in Fig. 2a, b, we can see that error is quite high in starting as the number of features selected to represent the data was a low number. But on increasing the number of features, error reduced steeply, until a saturation point which can be taken as optimum value for the number of features. As the sparsity of data changes, it may require a greater number of features to represent data properly.

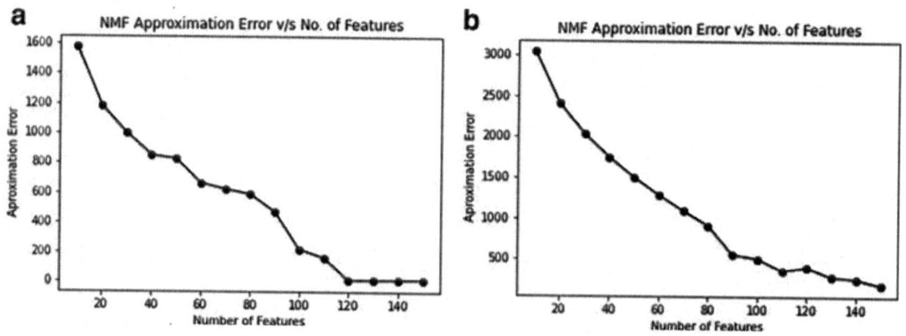

Fig. 2 **a** Error statistics at 5% data density. **b** Error statistics at 40% data density

5 Conclusion

We have taken advantage of the proven effectiveness of recurrent neural network, LSTM, for time sequencing data like QoS values in time-aware situations. We have proposed a method by combining two very popular and proven methods; for the recommender system, it is non-negative matrix factorization and for time-series prediction is LSTM. Dataset is processed (sparse) in advance to make it closer to the real-world scenario. Our method step-wise learns the local aspect first using matrix factorization then temporal aspects using LSTM with the help of past QoS data. Big plus point for our method is that it requires very few (4 in this case) past QoS values to train the model, which can be very fast and memory-efficient in real-life online environments. Instead of making the problem all about the prediction of missing values, we have taken it as a problem for prediction in a time-aware situation. On the limitations side, our approach only uses past QoS invocations, but a QoS value can also be affected by other factors like geographical location, high demand areas, different time intervals, etc. For that to incorporate, we will need more contextual information to add to the model.

References

1. Aggarwal C (2016) Neighborhood-based collaborative filtering, pp 29–70
2. Breese JS, Heckerman D, Kadie C (1998) Empirical analysis of predictive algorithms for collaborative filtering. In: Fourteenth conference on uncertainty in artificial intelligence. Morgan Kaufmann Publishers Inc., pp 43–52
3. Shao L, Zhang J, Wei Y, Zhao J, Xie B, Mei H (2007) Personalized QoS prediction for web services via collaborative filtering. In: IEEE international conference on web services (ICWS 2007), pp 439–446. http://doi.org/10.1109/ICWS.2007.140
4. Zhang Y, Zheng Z, Lyu MR (2011) Wspred: a time-aware personalized QoS prediction framework for web services. In: 2011 IEEE 22nd international symposium on software reliability engineering, pp 210–219. http://doi.org/10.1109/ISSRE.2011.17
5. Zheng Z, Ma H, Lyu MR, King I (2011) QoS-aware web service recommendation by collaborative filtering. IEEE Trans Serv Comput 4(2):140–152. https://doi.org/10.1109/TSC.2010.52
6. Li S, Wen J, Luo F, Ranzi G (2018) Time-aware QoS prediction for cloud service recommendation based on matrix factorization. IEEE Access 6:77716–77724. https://doi.org/10.1109/ACCESS.2018.2883939
7. Mnih A, Salakhutdinov RR (2008) Probabilistic matrix factorization. In: Advances in neural information processing systems, pp 1257–1264
8. Zheng Z, Ma H, Lyu MR, King I (2013) Collaborative web service QoS prediction via neighborhood integrated matrix factorization. IEEE Trans Serv Comput 6(3):289–299. ISSN 2372-0204. http://doi.org/10.1109/TSC.2011.59
9. Luo X, Lv Y, Li R, Chen Y (2015) Web service QoS prediction based on adaptive dynamic programming using fuzzy neural networks for cloud services. IEEE Access 3:2260–2269. https://doi.org/10.1109/ACCESS.2015.2498191
10. Ren L, Wang W (2018) An SVM-based collaborative filtering approach for top-n web services recommendation. Futur Gener Comput Syst 78:531–543. https://doi.org/10.1016/j.future.2017.07.027

11. Wang W, Wang L, Lu W (2016) An intelligent QoS identification for un-trustworthy web services via two-phase neural networks. In: IEEE international conference on web services (ICWS). IEEE, pp 139–146. http://doi.org/10.1109/ICWS.2016.26
12. Xiong R, Wang J, Li Z, Li B, Hung PC (2018) Personalized LSTM based matrix factorization for online QoS prediction. In: IEEE international conference on web services (ICWS). IEEE, pp 34–41. http://doi.org/10.1109/ICWS.2018.00012
13. Yin Y, Zhang W, Xu Y, Zhang H, Mai Z, Yu L (2019) QoS prediction for mobile edge service recommendation with auto-encoder. IEEE Access 7:62312–62324. https://doi.org/10.1109/ACCESS.2019.2914737
14. Desrosiers C, Karypis G (2011) A comprehensive survey of neighborhood-based recommendation methods. Springer US, Boston, pp 107–144. ISBN 978-0-387-85820-3. http://doi.org/10.1007/978-0-387-85820-3_4
15. Zhang W, Sun H, Liu X, Guo X (2014) Temporal QoS-aware web service recommendation via non-negative tensor factorization. In: 23rd International conference on world wide web, pp 585–596. http://doi.org/10.1145/2566486.2568001
16. Gillis N (2014) The why and how of non-negative matrix factorization
17. Zheng Z, Zhang Y, Lyu MR (2014) Investigating QoS of real-world web services. IEEE Trans Serv Comput 7(1):32–39. ISSN 2372-0204. http://doi.org/10.1109/TSC.2012.34
18. Yu C, Huang L (2014) Time-aware collaborative filtering for QoS-based service recommendation. In: IEEE international conference on web services. IEEE, pp 265–272. http://doi.org/10.1109/ICWS.2014.47

Chapter 36
An Efficient Wavelet-Based Image Denoising Technique for Retinal Fundus Images

S. Valarmathi and R. Vijayabhanu

1 Introduction

Diabetic Retinopathy has been found to be the common cause of vision loss in diabetic patients. Proper vision screening is vital in Diabetic Retinopathy detection. Diabetic Retinopathy stages range from normal vision (No DR) to PDR (Proliferative Diabetic Retinopathy). DR is commonly an asymptomatic eye disease, but it shows various characteristics when it progresses, according to the severity levels.

Image denoising is a low-level preliminary process performed before executing successive high-level image processing tasks such as remote sensing, machine vision, object recognition, pattern recognition, and so on. Image noise causes a difference in intensity variation that produces a grain-like appearance in the image. In common, all of the digital images contain some noise, which may arise due to many reasons such as scattering, inaccurate capturing devices, sensor heat, misaligned lenses, and adverse atmospheric conditions, and so on. The Digital image gets distorted due to several types of noise. Some of the noises are Gaussian, Poisson, Speckle, and Salt and Pepper.

Wavelet-based transform proves to be powerful in image noise removal [1]. The original input signal is decomposed into different scales that represent distinct signal components, unlike the input image. The thresholding [2–4] and other operations are performed at each scale to reduce the image noise. The output denoised image can be obtained by reconstructing or transforming back the wavelet coefficients. In this proposed DWT_K-SVD method, the wavelet coefficients can be denoised by a redundant dictionary learned over the extricated patches from the sparse representation of the distorted image. The test result proves that the proposed DWT_K-SVD method surpasses the other denoising techniques, such as median filtering, wiener filtering, and DWT.

S. Valarmathi (✉) · R. Vijayabhanu
Department of Computer Science, Avinashilingam Institute for Home Science and Higher Education for Women, Coimbatore, Tamil Nadu 641043, India

A. Sheth et al. (eds.), *Intelligent Systems*, Algorithms for Intelligent Systems,
https://doi.org/10.1007/978-981-16-2248-9_36

This paper is structured as follows: Sect. 2 discusses some of the related works on image denoising. Section 3 examines a few denoising approaches considered in this work. The median filter, wiener filter, discrete wavelet transform (DWT), and a wavelet-based proposed technique DWT_K-SVD are tested on two Diabetic Retinopathy (DR) datasets. Section 4 discusses and analyzes the performance results on two different DR Datasets. Section 5 concludes our work with future recommendations.

2 Related Work

Beck and Teboulle [5] developed a general framework that covers a few non-smooth regularizers for constrained TV using a gradient-based technique. Even though the proposed method increases the PSNR value, it only focuses on local image characteristics. Buades et al. [6] proposed a non-local means (NLM) approach that used the Non-local Self-Similarity (NSS) weighted filtering before denoising images. It is one of the significant improvisations in solving an image denoising problem. The main aim is to develop a point-wise image estimation. Here, each pixel is acquired as a weighted average of region-centered pixels, which resembles the estimated pixels.

Sutour et al. [7] developed an adaptive method that integrates both NLM and TV regularization methods. The combinational approach provides better image noise removal results, but structural information is not maintained well, which lowers the image quality. Dong et al. [8] developed an SVD-based low-rank method to build the sparse representation of image patches. For image noise removal, the Bayes Shrink framework was used in this method. The major drawback in this method is too low or high rank values lead to information loss or noise preservation, respectively. Gu et al. [9, 10] developed a WNNM model that set weights to single values of various sizes adaptively. The soft thresholding methods are used to denoise the image.

3 Approach

This section examines a few denoising approaches considered in this work. The median filter, wiener filter, discrete wavelet transform (DWT), and a wavelet-based proposed technique DWT_K-SVD are tested on two Diabetic Retinopathy (DR) datasets.

3.1 Median Filter

The median filter is a type of non-linear filtering method that is mainly used to eliminate image noise. It is one of the commonly applied preprocessing methods which improves the image quality, highlights significant details before doing intense image operations. It effectively removes image noise and preserves edges simultaneously. The main aim of this median filter technique is to replace each entry with the adjacent entries median value. If the entries are odd, it is simple to process, and when it is even, it requires little processing. The median filters are robust and used as smoothers in image and signal processing [11].

3.2 Wiener Filter

In 1940, Norbert Wiener proposed the concept called Wiener filter (WF), which was used to reduce or remove noises in the signal. Inverse filtering helps in recovering the image when it is blurred by a low-pass filter. However, the inverse filtering is sensitized to additive noise. Due to which wiener filtering makes a compromise between inverse filtering and noise-smoothening. Simultaneously WF eliminates the additive noise and inverts the image blur. The wiener filtering is a statistical-based approach that minimizes the error between the evaluated noise and the preferred noiseless signal. The WF is commonly used due to its processing speed and simplicity [12]. This filtering technique is found to be very effective in Gaussian noises.

3.3 Discrete Wavelet Transform (DWT)

The discrete wavelet transform (DWT) proves to be a prominent method in wavelet analysis. The DWT is based on a multi-resolution analysis, which was introduced by Mallat [13]. The non-redundant representation of an image produced by the DWT image signal results in more accurate spectral and spatial image localization. When compared with other multi-scale methods, DWT has been a well-known technique for image denoising, which possesses various benefits such as multi-scale approximation, sparse representation, image quality retention, and so on. The DWT technique can be stated as a decomposition of the image signal in a set of independent, spatial-based frequency channels. The image signal is fed through 2 filters, which outputs 2 different coefficients, approximation and details. This process is known as analysis or decomposition. The image components are reassembled into the original signal by retaining the image quality. This is called synthesis or reconstruction. The process of manipulating decomposition and reconstruction is called DWT and Inverse DWT [14]. The wavelet techniques were found to be very effective in Gaussian noise. The DWT-based image denoising methodology has been presented in Fig. 1.

The DWT-based image denoising technique consists of three steps in common,

Step 1: The original input noisy image is converted into an orthogonal domain using DWT. The image signals are decomposed to produce wavelet coefficients.

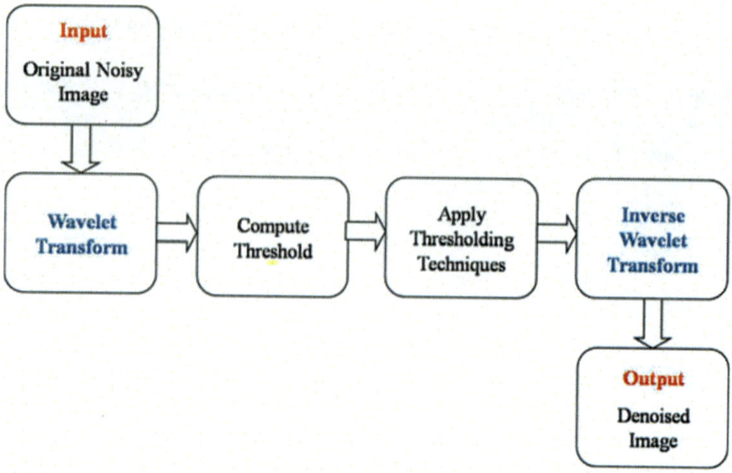

Fig. 1 DWT-based image denoising framework

Step 2: The wavelet thresholding techniques are applied to denoise the wavelet coefficients.

Step 3: The inverse DWT is performed to acquire the denoised image.

3.4 The Proposed DWT_K-SVD Method

The DWT-based denoising method seems to be effective when compared to the other methods, median filter and wiener filter. In order to further improve the performance measures such as PSNR, MSE and SSIM and also to minimize the cost function, the K-SVD method is used in a wavelet domain (DWT). The proposed DWT_K-SVD method produces good performance on both EyePACS and Messidor-2 datasets.

The K-SVD [15] denoising has its origin from k-means clustering and singular value decomposition. A dictionary-based learning method can obtain a redundant dictionary that produces sparse representations. The prior learning strategy can be used when both the signal and the initial dictionary are provided. According to the GOMP (Generalized Orthogonal Matching Pursuit) method, the sparse coding step is altered, and a dictionary is updated based on SVD.

Consider x is the original noisy image, y is the noiseless image corrupted by additive noise and Gaussian noise z with σ (standard deviation). The input image x is given in Eq. 1,

$$x = y + z \tag{1}$$

In this proposed technique, we tend to eliminate the image noise z from x to produce a denoised image $y1$ which seems closer to the original image y. The steps

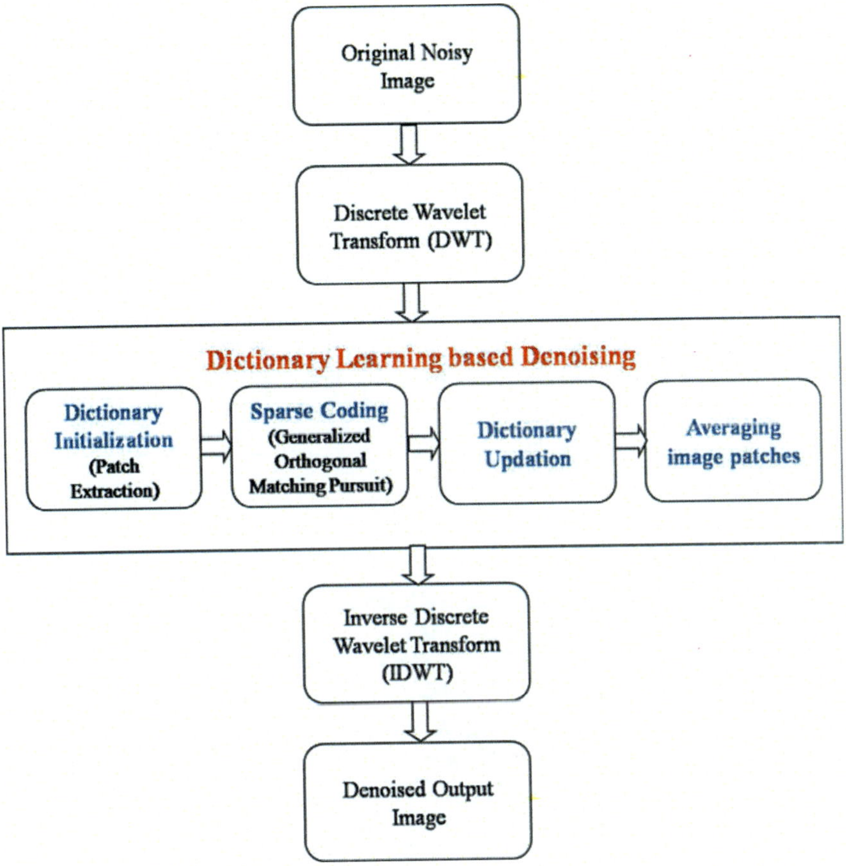

Fig. 2 The proposed DWT_K-SVD framework for image denoising

involved in this proposed DWT_K-SVD method are discussed below, and Fig. 2 shows the DWT_K-SVD image denoising framework.

Step 1: The DWT is applied on the original image x, which produces approximation and wavelet coefficients.
Step 2: The wavelet coefficients can be denoised by a dictionary learning-based algorithm called K-SVD. Here, a dictionary is learned from the wavelet representation of the extricated patches adaptively.
Step 3: The inverse DWT is applied for image reconstruction, which produces the denoised output image $y1$.

From Eq. 1, we take a noiseless image y of size $\sqrt{M} \times \sqrt{M}$ pixels, and DWT is applied on the original image x. When using the wavelet property on the Eq. 1, we get Eq. 2.

$$W_x = W_y + W_z \tag{2}$$

The wavelet transforms of x, y and z are denoted as W_x, W_y and W_z respectively in Eq. 2. The K-SVD algorithm is used where the dictionary is learned on the small image patches which were extracted, $W_{yij} = K_{ij} W_y$ of size $\sqrt{m} \times \sqrt{m}$ in each location (i,j) of W_y. The block size of the image W_y is extracted using the matrix K_{ij}. The dictionary (redundant) D is provided and the sparse representations ($\widehat{\alpha}$) of every extracted patch (W_{yij}) with a bounded error is given in Eq. 3.

$$\hat{\alpha}_{ij} = \arg\arg \|\alpha_{ij}\|_0 \quad \text{such that} \; \| W y_{ij} - D\alpha_{ij} \|_2^2 \leq (C\sigma)^2 \tag{3}$$

The Lagrange form of the above Eq. 3 is given in Eq. 4.

$$\hat{\alpha}_{ij} = \arg\arg \| W y_{ij} - D\alpha_{ij} \|_2^2 + \mu_{ij} \|\alpha_{ij}\|_0 \tag{4}$$

The GOMP (Generalized Orthogonal Matching Pursuit) [16] is a greedy technique which is used to discover the sparse representation of the signal with a given dictionary. In each iteration, it attempts to choose more than one atom (S, $S \geq 1$). The GOMP technique is one of the improvised OMP (Orthogonal Matching Pursuit) techniques which boosts the computational efficiency and signal recovery performance. The K-SVD algorithm helps in minimizing the cost function and finds the coding (sparse) using the GOMP method and the dictionary is initialized. The resulting code attempts to minimize the error iteratively. After sparse coding, the atoms of the dictionary are updated. Then the averaging can be done on the image patches. The process can be repeated until the image gets denoised well. Finally, the inverse DWT can be performed on the wavelet coefficients to get the denoised output image.

4 Results and Analysis

In the results and analysis section, we discuss the test results of the image denoising techniques for two DR datasets such as EyePACS and Messidor-2. The proposed DWT_K-SVD method is analyzed and compared with three different image denoising techniques such as median filter, wiener filter and discrete wavelet transform (DWT). The EyePACS and Messidor-2 datasets are considered for our work which contains 88,702 images and 1748 images, respectively. Each of the images in the dataset was graded according to the severity levels. The original noisy image and the denoised image by DWT_K-SVD on both the datasets are shown in Figs. 3 and 4.

In this work, we have used three different performance metrics to assess our proposed DWT_K-SVD method. The peak signal-to-noise ratio (PSNR), mean square error (MSE), and structural similarity index measure (SSIM) are the metrics considered to assess the performance. The PSNR values are computed using MSE

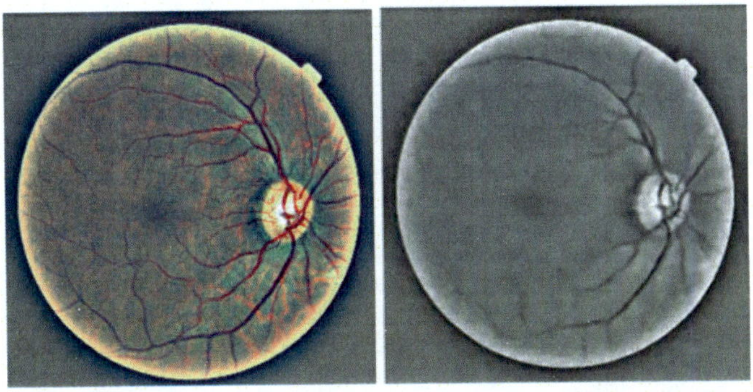

(a) (b)

Fig. 3 **a** Original noisy image. **b** Denoised image by DWT_K-SVD (proposed method) for EyePACS dataset

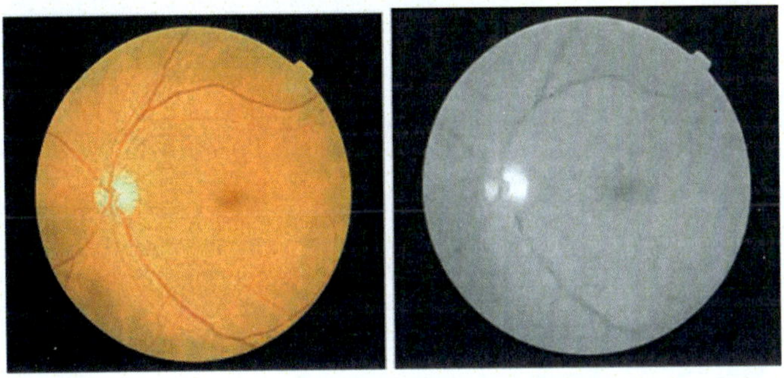

(a) (b)

Fig. 4 **a** Original noisy image. **b** Denoised image by DWT_K-Svd (proposed method) for Messidor-2 dataset

values, and it can be calculated using Eqs. 5 and 6.

$$\text{MSE} = \frac{1}{M_1 \times M_2} \sum_{i=1}^{M_2} \sum_{j=1}^{M_1} \left[I(i, j) - I'(i, j) \right]^2 \tag{5}$$

$$\text{PSNR} = 10 \times \log\left(\frac{255^2}{\text{MSE}}\right) \tag{6}$$

where I and I' are the 2 images of size $M_1 \times M_2$. $I(i,j)$ and $I'(i,j)$ represent the value (pixel) on ith row and jth column of I and I'. The SSIM values can be calculated using Eq. 7.

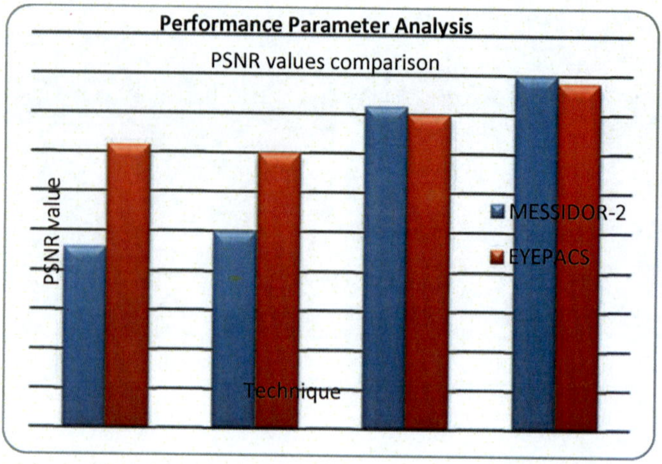

Fig. 5 The PSNR results comparison of denoising methods on Messidor-2 and EyePACS dataset

$$\mathrm{SSIM}\left(I, I'\right) = \frac{(2\mu_I \mu_{I'} + c_1)(2\sigma_{II'} + c_2)}{\left(\mu_I^2 + \mu_{I'}^2 + c_1\right)\left(\sigma_I^2 + \sigma_{I'}^2 + c_2\right)} \tag{7}$$

where μ_I and σ_I are the gray values in average and I variance. $\sigma_{II'}$ are the covariance between the images I and I', C_1 and C_2 are the 2 constants.

The greater PSNR and SSIM values result in more accurate denoising performance, which seems closer to the original image. The statistical comparison of PSNR, MSE and SSIM for four denoising methods on two different datasets is discussed further in this section.

The PSNR values are compared on Messidor-2 and EyePACS datasets in Fig. 5 which shows that our proposed DWT_K-SVD method outperforms the other methods on both datasets. The PSNR values achieved on Messidor-2 and EyePACS datasets are 45 dB and 44 dB, respectively.

The MSE values are compared on Messidor-2 and EyePACS datasets in Fig. 6, which shows that our proposed DWT_K-SVD method outperforms the other methods on both datasets. The MSE values should be lesser, unlike PSNR and SSIM. It has achieved 0.17 MSE value on both Messidor-2 and EyePACS datasets.

The SSIM values are compared on Messidor-2 and EyePACS datasets in Fig. 7, which shows that our proposed DWT_K-SVD method outperforms the other methods on both datasets. The SSIM values achieved on Messidor-2 and EyePACS datasets are 0.98 and 0.9, respectively. Our proposed method DWT_K-SVD performs better than the other techniques considered in our work. The dictionary learning-based approach in a wavelet domain proved to be so effective for denoising retinal fundus images.

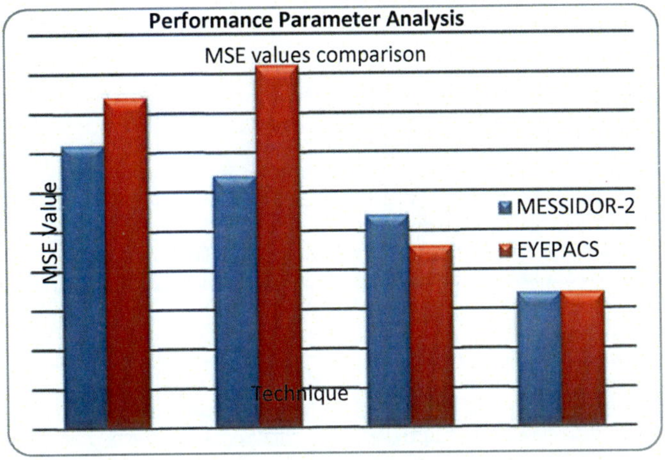

Fig. 6 The MSE results comparison of denoising methods on Messidor-2 and EyePACS dataset

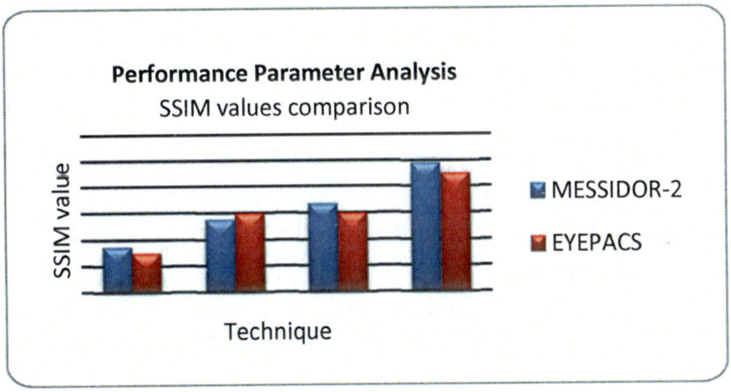

Fig. 7 The SSIM results comparison of denoising methods on Messidor-2 and EyePACS dataset

5 Conclusion

In this work, some of the image denoising techniques are analyzed for retinal fundus images on the EyePACS and Messidor-2 dataset. The proposed DWT_K-SVD method provides better results than the median filter, wiener filter, and DWT. The K-SVD was used in a wavelet domain, which further reduces the cost function and improves the performance measures such as PSNR, MSE, and SSIM. The computational cost and the processing time can further be improved in the future by using additional filters to retain the image quality.

References

1. Mallat S (2008) A wavelet tour of signal processing. In: The sparse way, 3rd edn. Academic Press, Elsevier, Burlington
2. Grace Chang S, Yu B, Vattereli M (2000) Adaptive wavelet thresholding for image denoising and compression. IEEE Trans Image Process 9:1532–1546
3. Donoho DL, Johnstone IM (1995) Adapting to unknown smoothness via wavelet shrinkage. J Am Stat Assoc 90(432):24–1200
4. Donoho DL (1995) De-noising by soft-thresholding. IEEE Trans Inf Theory 41(3):613–627
5. Beck A, Teboulle M (2009) Fast gradient-based algorithms for constrained total variation image denoising and deblurring problems. IEEE Trans Image Process 18(11):2419–2434
6. Buades A, Coll B, Morel JM (2005) A non-local algorithm for image denoising. In: Abstracts of 2005 IEEE computer society conference on computer vision and pattern recognition. IEEE, San Diego, pp 60–65
7. Sutour C, Deledalle CA, Aujol JF (2014) Adaptive regularization of the nl-means: application to image and video denoising. IEEE Trans Image Process 23(8):3506–3521
8. Dong WS, Shi GM, Li X (2013) Nonlocal image restoration with bilateral variance estimation: a low-rank approach. IEEE Trans Image Process 22(2):700–711
9. Gu SH, Xie Q, Meng DY, Zuo WM, Feng XC, Zhang L (2017) Weighted nuclear norm minimization and its applications to low level vision. Int J Comput Vis 121(2):183–208
10. Gu SH, Zhang L, Zuo WM, Feng XC (2014) Weighted nuclear norm minimization with application to image denoising. In: Abstracts of 2014 IEEE conference on computer vision and pattern recognition. IEEE, Columbus, pp 2862–2869
11. Church JC, Chen Y, Stephen V (2008) A spatial median filter for noise removal in digital images. IEEE, pp 618–623
12. Anilet B, Chiranjeeb H, Punith C (2014) Image denoising method using curvelet transform and wiener filter. Int J Adv Res Electr Electron Instrum Eng 3(1)
13. Mallat S (1980) A theory for multiresolution signal decomposition: the wavelet representation. IEEE Trans Pattern Anal Mach Intell 12(7):674–693
14. Kother Mohideen S, Arumuga Perumal S, Mohamed Sathik M (2008) Image de-noising using discrete wavelet transform. IJCSNS Int J Comput Sci Netw Secur 8(1):213–216
15. Aharon M, Elad M, Bruckstein AM (2006) The K-SVD: an algorithm for designing of over-complete dictionaries for sparse representation. IEEE Trans Signal Process 54(11):4311–4322
16. Zhao L, Liu Y (2017) A new generalized orthogonal matching pursuit method. J Electr Comput Eng 2017:Article ID 3458054

Chapter 37
An Intelligent System for Spam Message Detection

Sahil Sartaj and Ayatullah Faruk Mollah

1 Introduction

Email or other types of messages are known as advanced, secure, cheap, reliable, and fast technology for information exchange in this era. In recent times, the number of email users is rapidly increasing. In parallel, the size of unsolicited emails called spam is also increasing, causing trouble over the Internet. Senders of spam messages are traditionally known as spammers. Spammers usually collect email addresses from different sources [1]. The huge volume of spam messages wastes the storage and memory space of servers, network bandwidth, computational power, and usage time [2]. Email spam could be one of the most prevalent crimes that emerged over recent years that almost every Internet user probably has faced at some point of time. According to Kaspersky 2013 report [3], globally, more than 77% of the emails are spam mails.

Many email providers such as Gmail, Hotmail, and Yahoo have developed their own spam classifier to detect unsolicited commercial emails in email inboxes. However, if the classier mistakenly identifies an important message as spam, it could create a great chaos to the users. Most of the spam filter software is static. Spammers apply several procedures to surpass static filtering approaches [4]. To effectively combat these spam tactics, adaptive new techniques are needed that may lie in machine learning methods. Many machine learning algorithms have been widely studied for spam classification [5] such as Naive Bayes [6], support vector machine [7, 8], k-nearest neighbor [5, 9], artificial neural networks [10], and random forest [11]. Several other works on spam classification by various approaches have been published by Dada and Joseph [11].

S. Sartaj (✉) · A. F. Mollah
Department of Computer Science and Engineering, Aliah University, IIA/27 New Town, Kolkata 700160, India

A. F. Mollah
e-mail: afmollah@aliah.ac.in

A. Sheth et al. (eds.), *Intelligent Systems*, Algorithms for Intelligent Systems,
https://doi.org/10.1007/978-981-16-2248-9_37

However, such methods are not satisfactory to meet the changing nature of spam. In this paper, we report the development of an adaptive spam classification system based on lexical preprocessing, stemming, lemmatization, feature extraction, and classification with random forest. In Sect. 2, we discuss related works for spam prediction. Working methodology is presented in Sect. 3 and experimental results are reported in Sect. 4. Finally, the conclusion with future scope of work is presented in Sect. 5.

2 Related Works and Motivation

Spam filtering is a matter of contention among researchers. In this section, we present recent spam trends and the works carried out so far to address them. Word obfuscation refers to representation of words in different ways to escape filtering. For instance, the word ‘money’ may be represented as ‘m-o-n-e-y’ or ‘mo <! - -> n <! - -> ey’. There are many other techniques for obscuring words such as misplacing spaces, embedding special characters, purposeful misspellings, etc. Thus, spammers could escape early days heuristic filters. Spammers also mislead the tokenization process by modifying words, e.g., ‘URGENT’ as ‘U-R-G-E-N-T’. In phishing [12], someone pretends to be part of a legitimate organization [13, 14] and communicates to an individual with a sense of urgency to obtain important data. Most often, phishing messages contain image content, attachment, or misleading links such as www.l00 tales.com (‘1’ instead of small L). Besides obfuscation and phishing, other techniques are being gradually uncovered such as image spam, botnet, and spam.

Obied [15] proposed a spam filtering method based on Bayesian analysis to classify spam or ham messages. In [16], Wang et al. experimented over different techniques to find out the unwanted spam emails deprived of using machine learning methods or spam architectures. Bhowmick and Hazarika [17] presented a high-level review on some reported email-spam filtering approaches with the focus on spam filtering based on machine learning approaches. Karthika and Visalakshi [18] applied ant colony optimization for feature selection and support vector machine for classification, which assists to be more efficient.

Awad and ELseuofi [5] applied pattern classifiers, artificial immune system, and rough sets to classify spam emails. Deshpande et al. [19] applied Bayes classifier for blocking spam emails. Mishra and Thakur [20] played a significant role in eliminating unwanted commercial emails, worms, trojans, viruses, and electronic frauds by applying a classification approach. Term frequency-inverse document frequency (TF-IDF)-based spam filter methods have been reviewed in [21], and a random forest classifier with this feature yields an accuracy of 97.50%.

A common issue with many of these techniques is unsatisfactory performance in spam detection. In addition, such methods are not adequately robust and they often find it difficult to deal with the evolutionary character of spams [11]. Adaptive and more sophisticated methods need to be designed to increase prediction accuracy and most importantly to meet the changing characteristics of spam evolution.

3 Methodology

At first, messages are processed to make them suitable for feature extraction as discussed in Sect. 3.1. Then, features are extracted using the method presented in Sect. 3.2. Finally, in Sect. 3.3, we discuss training of an ensemble of decision trees and prediction of recall samples. A block diagram having these operations is shown in Fig. 1.

3.1 Preprocessing

The dataset contains a set of messages and their class labels indicating whether they are spam or ham. As every information in a message may not be necessary, less informative as well as noisy terms may be removed to lower the dimensionality of feature space and boost the classification performance with machine learning. Hence, these messages need to undergo a preprocessing stage as a preparation for feature extraction. So, to eliminate all these unwanted characters, we follow these standard procedures.

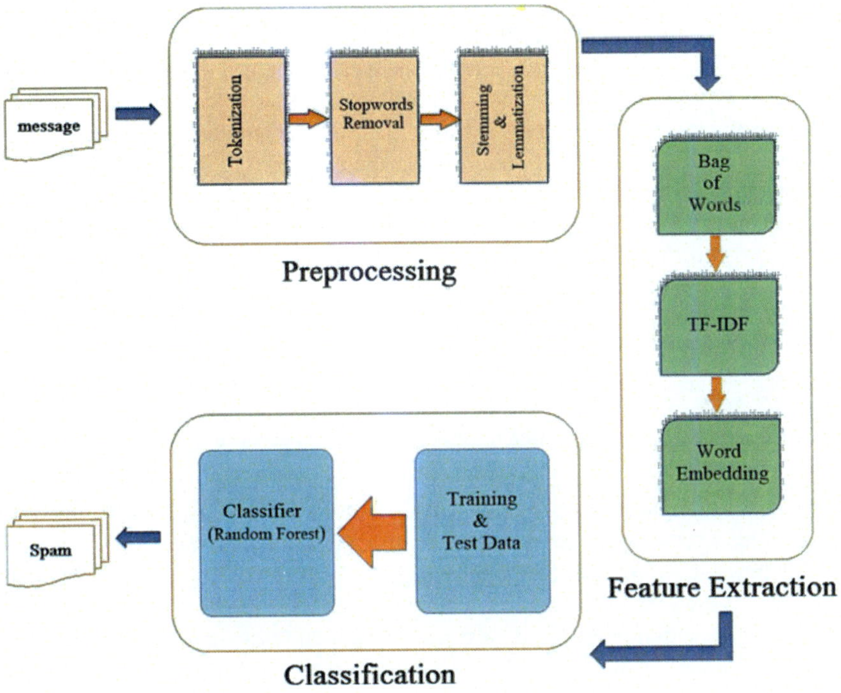

Fig. 1 Block diagram showing the major steps of the spam prediction system

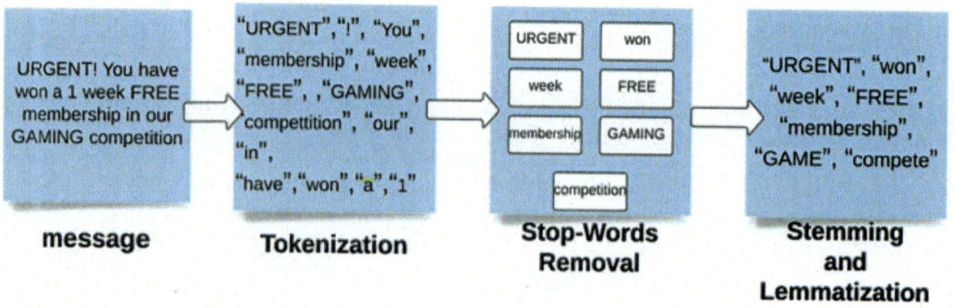

Fig. 2 Step-by-step preprocessing operations for a sample text message

Lexical Analysis (Tokenization). Each sentence of a message is tokenized, i.e., the whole message is broken into small raw chucks or words so as to substantiate the candidate words to be received as appropriate spam or ham terms.

Stop-words Removal. Non-informative words that are frequently, e.g., 'a', 'an', 'the', 'he', 'she', 'it', 'is', 'about', etc., are referred to as stop-words. These words are just taking up the dataset size and valuable processing time. For this, we remove all those words which are irrelevant to this experiment.

Stemming and Lemmatization. Word stemming is a process of converting words to its morphological base forms. It mainly omits the end or the beginning of the words, plurals, tenses, prefixes, and suffixes. Lemmatization, followed by stemming, utilizes dictionaries and contributes in converting a word back to its base. The main privilege of utilizing word stemming and lemmatization are dimensionality reduction and improved classification performance. For example, the word 'consolidate' consists of the base word 'consolid' and the suffix 'ate'. Stemming converts 'consolidate' to 'consolid'. Given the word 'ran', which is the past form of the base word 'run', lemmatization will convert it to 'run'. Another example may be taken: the word 'studies' turns to 'studi' in stemming and 'study' in lemmatization. It is, however, slower than stemming, but it retains the context of the word in the given sentence(s) as this is a lexicon-based approach. For a single message, all the steps involved in preprocessing are illustrated in Fig. 2.

3.2 Extraction of Features

A classification engine requires features of the same dimension for all object samples. So, stemmed words need to be represented with a fixed number of numeric attributes. For this purpose, we exploit the popular bag of words (BOW) approach, in which the words of a message are used as features. This BOW method throws away all the order information in the words and just focuses on frequencies of words in the message by assigning numbers (i.e., frequency) to each word. There are many ways to clarify a word and encode it into a vector such as CountVectorizer, TdfVectorizer, and Word Embedding. Every message may thus be transformed to a feature vector of 1-6000 attributes approximately. The CountVectorizer keeps a dictionary of every word and its respective ID. Here, if a word is present, 1 is considered, else 0 is considered. TdfVectorizer is an alternative method. It utilizes word counts across multiple messages to downscale them.

In case of word embedding, words with similar meaning are represented to signify similar distributed representation, which eventually represents a word as a vector or a point in higher dimensional similarity space. Therefore, the feature extraction module extracts a feature dictionary and provides feature vectors that are passed to a classifier as input. In the current work, the number of features obtained from each message is 6292.

3.3 Training and Prediction

Random forest is, nowadays, a popular pattern classifier which is basically an ensemble of decision tree-based learning algorithms. It is usually trained with the bagging method which means this classifier is a group of decision trees from a randomly selected subset of the training set. The set of decision tree classifiers is referred to as the forest. Decision trees are constructed using attribute fitness indicators such as entropy and gini index. The classifier aggregates the votes of the individual trees to determine the final class of a test object. Thus, recall messages are classified as spam or ham based on their computed feature vector.

4 Experimental Results and Discussion

In this section, we report the details of experiments carried out in the present work and results obtained followed by an insightful discussion. At first, we present a brief outline of the dataset containing spam and messages in Sect. 4.1. Next, in Sect. 4.2, standard evaluation metrics that have been used to quantify the performance of this work are discussed. Prediction performance in terms of these standard metrics is reported in Sect. 4.3. Then, overall discussion along with merits and demerits of the current work is included in Sect. 4.4.

4.1 Brief Description of Dataset

In the current experiment, UCI spam collection dataset [22] has been used. It contains a total of 5572 non-blank messages with a spam rate of 13.40%. Sample messages of both spam and ham categories are shown in Table 1. We split this dataset into 4:1 ratio (for a single fold in a fivefold cross-validation experiment). Thus, 80% of samples are considered for training and the rest 20% samples are considered for

Table 1 Sample spam and ham messages from UCI SMS spam collection dataset [22]

S. No.	Class	Message
1	Spam	WINNER!! As a valued network customer, you have been selected to receive a £900 prize reward! To claim call 09061701461. Claim code KL341. Valid 12 h only
2	Spam	Had your mobile 11 months or more? U R entitled to Update to the latest colour mobiles with camera for Free! Call The Mobile Update Co FREE on 08002986030
3	Ham	Just forced myself to eat a slice. I'm really not hungry tho. This sucks. Mark is getting worried. He knows I'm sick when I turn down pizza. Lol
4	Ham	Yup... Ok i go home look at the timings then i msg¨u again... Xuhui going to learn on 2nd may too but her lesson is at 8 am

Table 2 Distribution of message samples of different classes, intro training, and test sets

Category of message	Training set	Testing set	Training + Test set
Spam messages	598	149	747
Ham messages	3859	966	4825
Total	4457	1115	5572

testing. Distribution of samples for various sets and message types is shown in Table 2.

4.2 Evaluation Metrics

As spam filtering is basically a binary classification problem, its performance may be quantified with true positive (TP), false positive (FP), true negative (TN), and false negative (FN) counts. It may be noted that when a spam is predicted as spam, it is considered as TP, else it is considered as FN. Similarly, if a ham is classified as ham, it is considered as TN, else it is considered as FP. Consequently, for a standalone experiment, spam precision (SP), spam recall (SR), and accuracy (ACC) may be computed as $\mathrm{SP} = \frac{\mathrm{TP}}{\mathrm{TP+FP}}$, $\mathrm{SR} = \frac{\mathrm{TP}}{\mathrm{TP+FN}}$, $\mathrm{ACC} = \frac{\mathrm{TP+TN}}{\mathrm{TP+FP+FN+TN}}$.

Spam precision signifies how accurate are the predicted spam messages and spam recall signifies how accurately actual spam messages are predicted, whereas accuracy signifies overall prediction performance. Besides these metrics, another metric called F-measure or $f1$-score is also popularly used to represent quantitative performance to avoid class-biased performance quantification. It is basically the harmonic mean of SP and SR. This $f1$-score is computed as $f1\ \mathrm{score} = \frac{2\times\mathrm{SP}\times\mathrm{SR}}{\mathrm{SP+SR}}$.

Table 3 Summary of spam prediction performance of different experiments with varying number of folds for cross-validation

Cross-validation	Spam precision	Spam recall	$F1$-score	Accuracy
Threefold	1.00	0.84	0.91	97.84
Fivefold	1.00	0.83	0.90	97.66
Tenfold	1.00	0.83	0.91	97.75
20-fold	1.00	0.84	0.91	98.02

4.3 Prediction Performance

Multiple sets of experiments have been carried out and obtained results have been reported with multiple standard metrics discussed in Sect. 4.2. Four different cross-validations, viz., threefold, fivefold, tenfold, and 20-fold, have been performed and the summary of prediction performance is presented in Table 3. It may be mentioned that for this spam/ham classification experiment, a random forest classifier of the Python scikit-learn package has been employed. Values of the tuning parameters are as follows:

n_estimator = 100, min_samples_split = 2, criterion = gini, min_samples_leaf = 1, max_features = auto, max_depth = None, max_leaf_nodes = None.

It is worth mentioning that in case of n-fold cross-validation, samples of the dataset are divided into n equal (or nearly equal, if splitting in equal folds is not possible) folds. Each of these folds is used as a testset and the remaining $(n-1)$ folds are used in the training set. Thus, when the value of n increases, the number of training samples also increases while the number of test samples decreases. Due to the increase in the number of training samples, one may state that training of a classifier may be relatively better or around. It may be noted from Table 3 that prediction accuracy for 20-fold cross-validation is higher (i.e., 98.02%) than all other cross-validations.

4.4 Discussion

As evident from Table 3, prediction accuracy is 98.02% which is reasonably high and suitable for practical applications. Moreover, it may be observed that the spam precision rate is 1.00 in all experiments, which reflects that all the messages predicted as spam are correct and ham messages are not classified as spam at all. This reveals the desirable fact that the chance of predicting a spam message as ham is nil. However, few spam messages may get predicted as ham, which is nonetheless acceptable. In a nutshell, it may be stated that from a practical point of view, the system filters 98.02% spam messages, which may be highly useful.

5 Conclusion

In this paper, we present a method for detecting spam from incoming messages based on appropriate lexical analysis, stop-words removal, stemming, lemmatization, bag of words feature extraction, and classification with random forest classifier. Multiple experiments on UCI spam collection dataset reveal high performance (such as 98.02% for 20-fold cross-validation). Moreover, the precision rates of the developed system are 1.0 in all experiments. It signifies that the developed system is very accurate in terms of predicted spam messages. In the spam or ham classification problem, such a situation is desirable since a spam predicted as ham is acceptable, but the reverse is not. The developed system is also hosted at https://spam-detection-app-sartaj.herokuapp.com for wider usage. In future, we plan to work on more complex messages involving images, links, etc.

References

1. Awad M, Foqaha M (2016) Email spam classification using hybrid approach of RBF neural network and particle swarm optimization. Int J Netw Secur Appl 8(4):17–28. http://doi.org/10.5121/ijnsa.2016.8402
2. Fonseca O, Fazzion E, Cunha I, Las-Casas PHB, Guedes D, Meira W, Hoepers C, Steding-Jessen K, Chaves MHP (2016) Measuring, characterizing, and avoiding spam traffic costs. IEEE Int Comp 20(4):16–24. http://doi.org/10.1109/MIC.2016.53
3. Kaspersky Lab Report. https://www.kaspersky.com/about/press-releases/2013_kaspersky-lab-report-37-3-million-users-experienced-phishing-attacks-in-the-last-year. Accessed on 1 Feb 2021
4. Cormack GV, Smucker MD, Clarke CL (2011) Efficient and effective spam filtering and re-ranking for large web datasets. Inf Retrieval 14(5):441–465. arXiv:1004.5168v1
5. Awad WA, ELseuofi SM (2011) Machine learning methods for spam e-mail classification. Int J Comput Sci Inf Technol 3(1):173–184. http://doi.org/10.5121/ijcsit.2011.3112.173
6. Marsono MN, El-Kharashi MW, Gebali F (2008) Binary LNS-based naive Bayes inference engine for spam control: noise analysis and FPGA synthesis. IET Comput Digit Tech 2(1):56–62. http://doi.org/10.1049/iet-cdt:20050180
7. Amayri O (2009) On email spam filtering using a support vector machine. Doctoral dissertation, Concordia University. https://spectrum.library.concordia.ca/976212/
8. Torabi ZS, Nadimi-Shahraki MH, Nabiollahi A (2015) Efficient support vector machines for spam detection: a survey. Int J Comput Sci Inf Secur 13(1):11–28
9. Chawla G, Saini R (2016) Implementation of improved KNN algorithm for email spam detection. Int J Trends Res Dev 3(5):479–483
10. Cao Y, Liao X, Li Y (2004) An e-mail filtering approach using neural network. In: International symposium on neural networks, pp 688–694. http://doi.org/10.1007/978-3-540-28648-6_110
11. Dada EG, Joseph SB (2018) Random forests machine learning technique for email spam filtering. Semin Ser 9(1):29–36
12. Sheng S, Holbrook M, Kumaraguru P, Cranor LF, Downs J (2010) Who falls for phish? A demographic analysis of phishing susceptibility and effectiveness of interventions. In: Proceedings of the SIGCHI conference on human factors in computing systems, pp 373–382. http://doi.org/10.1145/1753326.1753383
13. Akinyelu AA, Adewumi AO (2014) Classification of phishing email using random forest machine learning technique. J Appl Math 2014(425731):1–6. https://doi.org/10.1155/2014/425731

14. Khonji M, Iraqi Y, Jones A (2013) Phishing detection: a literature survey. IEEE Commun Surv Tutorials 15(4):2091–2121. https://doi.org/10.1109/SURV.2013.032213.00009
15. Obied A (2007) Bayesian spam filtering. Department of Computer Science, University of Calgary
16. Wang XL (2005) Learning to classify email: a survey. In: International conference on machine learning and cybernetics. IEEE, pp 5716–5719. http://doi.org/10.1109/ICMLC.2005.1527956
17. Bhowmick A, Hazarika SM (2018) E-mail spam filtering: a review of techniques and trends. In: Advances in electronics, communication and computing, pp 583–590. http://doi.org/10.1007/978-981-10-4765-7
18. Karthika R, Visalakshi P (2015) A hybrid ACO based feature selection method for email spam classification. WSEAS Trans Comput 14:171–177
19. Deshpande VP, Erbacher RF, Harris C (2007) An evaluation of naive bayesian anti-spam filtering techniques. In: IEEE SMC information assurance and security workshop, pp 333–340. http://doi.org/10.1109/IAW.2007.381951
20. Mishra R, Thakur RS (2013) Analysis of random forest and Naive Bayes for spam mail using feature selection categorization. Int J Comput Appl 80(3):42–47. http://doi.org/10.5120/13844-1670
21. Sjarif NNA, Azmi NFM, Chuprat S, Sarkan HM, Yahya Y, Sam SM (2019) SMS spam message detection using term frequency-inverse document frequency and random forest algorithm. Procedia Comput Sci 161:509–515. http://doi.org/10.1016/j.procs.2019.11.150
22. UCI SMS Spam Collection Dataset. https://www.kaggle.com/uciml/sms-spam-collection-dataset. Accessed 17 Sept 2020

Chapter 38
Energy Minimization in a Cloud Computing Environment

Sanna Mehraj Kak, Parul Agarwal, and M. Afshar Alam

1 Introduction

Cloud computing is the availability and accessibility of storage of data and computing power among other resources of the computer system without being directly managed by the user where the user only pays as per he uses the services.

Cloud computing services cover the maximum needs of a user, varying from basic needs of storage to complex office applications. Lately, the rising technology of cloud computing offers new quality models whose assets, for example, online utilization, figuring force, and system framework can be shared as administrations through the Internet [1, 2]. The most common cloud computing model used by few known CSPs (e.g., Amazon EC2) is a motivating aspect for buyers whose requests on virtual or online assets fluctuate as the time changes. Energy utilization is the key issue in content dissemination framework and most appropriated frameworks. This utilization is the main concern in cloud systems. The energy utilization of data centers checked all over the world is calculated at around 26 GW compared to about 1.4% of the world's electrical energy utilization with a development pace of 12% every year [3]. The Barcelona medium-size supercomputing center which is the basic data center that pays a yearly bill of about 1 million pounds just for the energy utilization of 1.2 MW, which can be comparable to the intensity of 1200 houses [4]. Hence, limiting the total energy utilization has a significant effect on the overall reliability, the output and accessibility of the framework. To manage such type of issues and ensure the future development of cloud computing, data centers is viable in an energy

S. M. Kak · P. Agarwal (✉) · M. Afshar Alam
Department of Computer Science and Engineering, Jamia Hamdard, New Delhi 110062, India
e-mail: pagarwal@jamiahamdard.ac.in

S. M. Kak
e-mail: sanah.mehraj@gmail.com

M. Afshar Alam
e-mail: aalam@jamiahamdard.ac.in

A. Sheth et al. (eds.), *Intelligent Systems*, Algorithms for Intelligent Systems,
https://doi.org/10.1007/978-981-16-2248-9_38

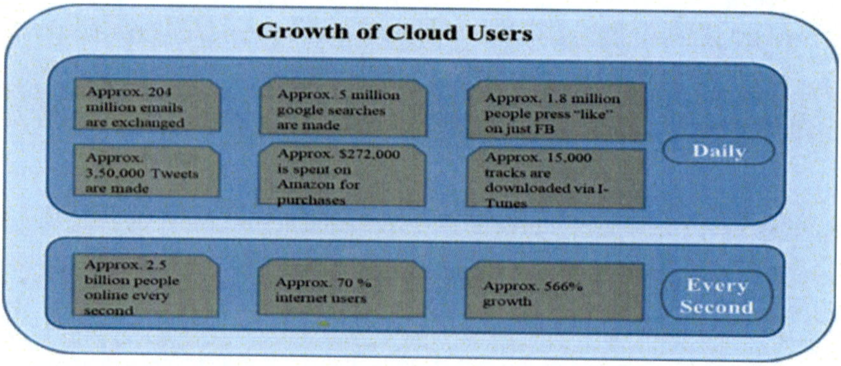

Fig. 1 Growth of cloud users

proficient way, especially with cloud assets to fulfill QoS prerequisites indicated by customers through SLAs, consequent decrease in the energy utilization is essential. Data centers have always been the strength to deal with all the online workloads. Figure 1 suggests the approximate growth of cloud users in today's struggle.

2 Literature Survey

A few issues about reducing energy in cloud and green computing are acquiring enormous attention in the research field. Several authors have been forming energy-efficient models reducing the cost and managing the work distribution [5].

There are energy consumption models that use energy at a higher scale. Firstly, some equipment includes the processor, display system, etc., which uses energy but on the customer side. Secondly, the energy is used by the data centers in providing 24 × 7 services to all the customers. To save energy, the servers not in use are to be turned off which leads to a distributed load to just a few servers resulting in degraded performance, which is one of the important issues that need to be sorted. In [6], the authors stated that the key to decreasing the usage of energy in a data center was by VM powering meter. In their paper, they investigated the issues related to this VM powering meter. They also analyzed their efficiency and performance. They discussed the details of implementation of the VM powering meter including the tools for the collection of data, the modeling methods used and the estimations made. The authors mainly concentrated on assessing the VM power at the software level. On analyzing the efficiency and performance, they deduced that the black box using PMC data for modeling is chosen over any other method as it maintains integrity. They also analyzed that machine learning methods can be used to improve the accuracy of the current method used in their VM power meter. They suggested that VM service billing, power budgeting, and power aware scheduling would be the future for making data centers green.

Microsoft, Google, Amazon, and Facebook are among the major cloud providers that readily depend on the cloud data centers to support the computational and application services. The carbon emission and financial price nullify the influence of sustainability in cloud services. In [7], the author proposes a theoretical model for holistic management of resources which is planned to decrease the carbon footprint and increase the energy efficiency. In the paper, the author also suggests some of the areas in making data centers sustainable [8] such as:

(i) Use of renewable energy as a substitute for grid energy created from fossil fuels (ii) Use of waste heat produced from the heat dissipated from servers (iii) Adoption of the free cooling mechanism (iv) Adoption of energy-efficient mechanism.

The conceptual model proposed by the author decreases the carbon footprint and makes the system efficient and more sustainable. The author also successfully mentioned that using the holistic approach, the energy efficiency of power structures and cooling devices was widely improved.

In [9], the author discusses the various applications of cloud computing. Several cloud providers offer pay per use services that require the management of QoS to check the services through IoT which needs to follow the SLA. Though providing the customer's dynamic QoS requirements without interrupting, the SLA violation is one of the challenges that the cloud computing environment is facing. They presented STAR, SLA-aware autonomic resource management technique which targeted on decreasing the violation rate of SLA for a proficient delivery of cloud services. The author suggests some future work where STAR can adopt the properties of self-configuration and self-protection from cyber-attacks. STAR can also adapt some energy efficiency parameters such as resource utilization, scalability, and attack detection rate which can be enhanced to work with STAR. They also suggested some future work where STAR can adopt the properties of self-configuration and self-protection from cyber-attacks. STAR can also adapt some energy efficiency parameters such as resource utilization, scalability, and attack detection rate which can be enhanced to work with STAR.

In [10], the author discusses the advantages of using green cloud computing thus by reducing the energy usage and carbon footprint. Even though the energy is saved in a system, performance degradation can occur which can result in the violation of SLA. Challenges faced by the author include:

(i) The necessity of a new optimization technique. (ii) Minimization of architectural complexity. (iii) The necessity of efficient data centers. (iv) Cooling of data centers. (v) Reducing the carbon footprint. (vi) Degradation of server performance increases the power consumption and energy output. (vii) To build cost-efficient techniques. The author finally concludes his paper with the implication that a lot of issues are yet to be dealt with in the future.

In [11], the author presented solutions and formulas for green computing environment (GCE) to decrease the influence on the environment and energy usage by introducing new models taking static and dynamic shares of the cloud component into consideration. The proposed model presents the generic model for them. The authors investigate the energy usage configurations and prove that by using the proper and suitable optimization policies focused via their energy consumption models, and it is even imaginable to save about 20% of energy usage in a cloud data center. The results from their research can be assimilated into cloud computing systems to keep a check on energy usage and maintain static and dynamic system level optimization. The authors depicted energy usage formulas for computing energy usage in the cloud environment. They also stated the tools and experimental analysis for energy usage including a broad energy usage model for the servers that are in idle or active state which resulted in the managed mechanisms for decreasing energy usage in a cloud computing environment.

In [12], the authors analyzed a number of methods that have been used and implemented to decrease the use of power at a high scale in a cloud computing environment. The authors concluded the paper by mentioning the necessity to optimize the data centers, so that they can be energy efficient. Thus, many techniques have been proposed that could decrease the usage of power. These methods revolve around virtualization of the computer system. They suggested a table comparing various techniques that have been used in making cloud computing more energy efficient and deduced that the main solution is virtualization.

In [13], the authors discuss how growth in cloud computing resulted in the inefficient use of energy for the processing of data, communication, or for storage. This inefficiency results in a carbon footprint. Thus, the main aim of involvement in an eco-friendly cloud computing environment is to decrease the carbon footprint thus by decreasing the use of energy. SDN is the latest prototype that can be used to centralize the network control plane and program the configuration of separate network elements. An SDN-based operation known as S-Core has been proposed which is an accessible, topology aware mitigation algorithm that decreases the cost of communication in a pairwise VM traffic flow. Their VM management systems have increasingly decreased the cost of communication specifically for high cost and overcrowded combinations and the core levels of the data centers. Their results showed an increase in network wide throughput by six times. Also, a reduction in the communication cost of about 60% by mitigating less than 50% of VM was also found.

In [14], the main focus of the authors was on the design and use of an energy-efficient basis for green cloud data centers that improves the efficiency, decreases the cost and fulfills the required QoS. A dynamic migration algorithm was proposed which could minimize the cost of energy taking SLAs into consideration. This approach uses one of the effective technologies in the field of server virtualization area of research called the SDN using OpenFlow technology. The results thus obtained indicate that the efficiency of resource usage and decreased consumption of power by cloud infrastructure can work together with SLAs while keeping low cost of penalties and minimum power consumption. The increased development and demand for the power used for computation by various sectors like scientific, business, and web application which has increased the formation of data centers that use huge amounts of power. The main focus of this paper was to design and use an energy-efficient basis for green cloud data centers that improves the efficiency, decreases the cost, and fulfils the required QoS.

In [15], the authors have emphasized various issues related to increasing the QoS while decreasing the usage of energy by data centers. A concept of energy-efficient controllers for data centers has been introduced. This controller can manage the resources of a data center with the least consequence on QoS. The authors also discuss the hardware and software-based techniques for the data center including the architectures. The high usage of data centers calls for the optimization of energy and all the resources used. Advancement in the energy-efficient hardware devices like the SSDs is not widely used in the cloud environment due to their expensive cost. The authors analyzed various mechanisms that can be used to control and organize resources of the data center for an efficient energy setup. They also analyzed various mechanisms that can be used to control and organize resources of the data center for an efficient energy setup. The authors suggested that if the workload on the data centers is low, it can be switched to a less power state or could even be switched off.

It is usually useful and easy to start working on a new cloud environment but moving the existing data or software and their application on a cloud maybe dangerous and more expensive at the same time as there can be a huge loss of data while transferring, and there is no assurance of the safe data transfer which can also be followed by a third-party attack which is still one of the threats present in cloud structures. The amount of power and energy used by the data centers in a cloud environment has been increasing at a high scale which calls for the need of the application of energy aware management of resources. Frequent access to the data has highly increased the need of web-based cloud storage services. Some of the famous CSPs that are vastly popular are Amazon S3 and Google Drive. Various algorithms have been proposed where the data that has been uploaded on a CSP is encrypted and can be decrypted once being downloaded using encryption and decryption keys which help in keeping the data of a user secure [16]. Figure 2 suggests various research issues that sustainable cloud computing has been facing.

The main purpose of our research is to find ways to decrease the carbon footprint within a cloud data center where the temperature is decreased with a decreased cost which can only be achieved with the implementation of sustainable cloud computing.

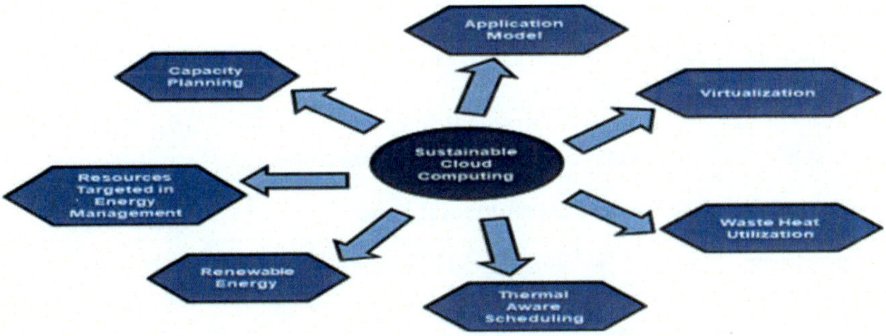

Fig. 2 Research challenges in sustainable cloud computing

3 Major Issues with Cloud Computing w.r.t Energy

Energy consumption is among the main issues in employing cloud computing. The usage of power in a cloud data center is very high which results in huge carbon emissions. There are various issues and challenges in using cloud computing, and some are discussed here.

3.1 *Minimizing the Energy Consumption*

In cloud computing, the main issue is decreasing the total amount of energy consumed. These energy saving algorithms have to be implemented from the very beginning thus saving the energy in the idle state of the cloud. While achieving a similar outcome, the amount of energy used should be decreased; this suggests the success of the algorithm implemented.

3.2 *Challenges Faced in Using Cloud Computing as Green Technology*

Green cloud computing faces various challenges in the current situations. One of the main challenges is to minimize the use of energy thus providing QoS. Therefore, the count of issues that occur when we use cloud computing as a green application is very high. Some of them are:

Energy aware dynamic resource allocation

In a cloud computing environment, every single physical machine congregates a virtual machine on which the requests are executed. The reliability is negatively affected by the increased cycling of power within servers in a cloud computing environment. Thus, in a dynamic cloud computing environment, any kind of disruption of energy can highly affect the services provided by the CSP. Also, any virtual machine would not be able to record the exact timing behavior of a physical machine which usually leads to issues of time keeping and incorrect measurement of time in a virtual machine which results in inaccurate implementation of SLA [17].

Enhancing awareness of environmental issues

The growth and expansion of IT industries have improved the software quality into an increased research area. Therefore, a number of institutes and organizations need to work together and take initiatives to make it more efficient. For achieving the power and cooling requirement of data centers, renewable resources such as wind, solar, or hydro energy should be used for electricity generation which will save energy and decrease the emission of CO_2. Apple, Google, and Facebook have started to use the renewable sources of energy for power generation. However, Apple is leading in terms of saving energy as it is said to cover 87% of the total energy used by Apple all over the globe.

Cloud data management

Cloud setups are categorized by the increase in large volumes of data. The CSPs depend on the cloud infrastructure providers for maintaining the security of data from any sort of breach. Decreasing performances of a server finally result in an increased usage of energy and power throughput which results in performance deprivation. The resources on a cloud data center are handled by a virtualization layer that is present over physical resources. For providing an advanced level interface for users and applications, the hardware layer is extracted by the virtualization layer.

Interoperability

Among various public cloud computing systems, a few are closed public cloud systems that are not built to relate with each other. Thus, to let the CSPs to build an interoperable cloud platform, industry standards have to be generated such as the open grid forum which is an industry group that works with open cloud interfacing to provide an API for handling and organizing different platforms of cloud [18]. The epitome of interoperability is that the interfaces are uniform in such a way that they are all interoperable so that the customer who has an ongoing request can shift from one CSP to another with the least influence on the customer availing the services.

Security

The identification, security, and authentication of data are very important and crucial especially for the data of the government. Since the whole government system is complex, the use of cloud computing comprises policy changes, implementation of dynamic applications, and providing security to the dynamic environment [19]. Cloud computing is done at two levels: the provider level and the user level. Security provided by a CSP can be tested whenever an attacker breaches a system by posing as a legitimate user thus infecting the entire cloud. This will further affect the users that are using the shared cloud which is now infected. There are certain issues that arise including the privacy issue, data issue, and infected applications [20]. Security can be heavily increased by using encryption techniques. A coordinated symmetric and asymmetric algorithm has been proposed to increase the security of data where the maximum concentration is put on the validation of the servers that improve the main agenda of security and secure the sensitive data of the clients from exploitation [21].

4 Conclusion

Among the most viable aspects in the current green cloud environment, change and instability from energy inefficiency to its sustainability stay the top most priority. The suggested work of various authors, a few mentioned in this paper solidifies the fact that utilization of the cloud resources can be reduced to an extensive measure thus by decreasing the energy utilized by the resources. We reviewed the possible impact of energy saving approaches for managing unified systems that comprise computer systems and networks. Due to the high employment of ICT and its proximity, it has become the fundamental requirement to make the system energy efficient. When attention is given to energy efficiency at various levels of IoT, it results in green IoT (GIoT) which finally decreases the energy cost keeping the human health threat into consideration. If the combination of grid computing with renewable energy is used to run a data center, the usage of energy can be minimized to a great extent. Currently, vast research has been going on in this direction but there is a need for more progressive and developed research to guarantee the energy efficiency and sustainability of cloud services.

References

1. Armbrust M (2009) Above the clouds: a Berkeley view of cloud computing. Technical report UCB/EECS-2009-28
2. BONE project (2009) WP 21 tropical project green optical networks: report on year 1 and undated plan for activities. No. FP7-ICT-2007-1216863 BONE project
3. Kogge P (2011) The tops in flops. IEEE Spectrum, pp 49–54
4. Feng WC, Feng X, Rong C (2008) Green supercomputing comes of age. IT Professional 10(1):17–23
5. Li X, Li Y, Liu T, Qiu J, Wang F (2009) The method and tool of cost analysis for cloud computing. In: IEEE international conference on cloud computing (CLOUD 2009), Bangalore, India, pp 93–100
6. Gu C, Huang H, Jia X (2014) Power metering for virtual machines in cloud computing-challenges and opportunities. IEEE Access 2:1106–1116
7. Buyya R, Gill SS (2018) Sustainable cloud computing: foundations and future directions. In: Business technology & digital transformation strategies, vol 21, no 6. Cutter Consortium, pp 1–9
8. Li X, Garraghan P, Jiang X, Wu Z, Xu J (2017) Holistic virtual machine scheduling in cloud datacenters towards minimizing total energy. IEEE Trans Parallel Distrib Syst
9. Singh S, Chana I, Buyya R (2017) STAR: SLA-aware autonomic management of cloud resources. IEEE Trans Cloud Comput
10. Rawat S, Kumar P, Singh S, Singh I, Garg K (2017) An analytical evaluation of challenges in green cloud computing. In: International conference on infocom technologies and unmanned systems (ICTUS'2017)
11. Uchechukwu A, Li K, Shen Y (2014) Energy consumption in cloud computing data centres. Int J Cloud Comput Serv Sci (IJ-CLOSER) 3(3). ISSN 2089-3337
12. Shakeel F, Sharma S (2017) Green cloud computing: a review on efficiency of data centers and virtualization of servers. In: International conference on computing and automation (ICCCA2017). IEEE. ISBN: 978-1-5090-6471-7/17

13. Cziva R, Jouet S, Stapleton D, Tso FP, Pezaro DP (2016) SDN based virtual machine management for cloud data centers. IEEE Trans Netw Serv Manag 13(2)
14. Anan M, Naseer N (2015) SLA-based optimization of energy efficiency for green cloud computing. IEEE
15. Shuja J, Bilal K, Madani SA, Othman M, Ranjan R, Balaji P, Khan SU (2016) Survey of techniques and architectures for designing energy-efficient datacenters. IEEE Syst J 10(2):507–519
16. Aggarwal P (2017) Cryptography based security for cloud computing system. Int J Adv Res Comput Sci 8(5)
17. Kalange Pooja R (2013) Applications of green cloud computing in energy efficiency and environmental sustainability. J Comput Eng 1:25–33
18. Padhy RP, Patra MR, Satapathy SC (2011) Cloud computing: security issues and research challenges. Int J Comput Sci Inf Technol Secur 1(2):136–146
19. Paquette S, Jaeger PT, Wilson SC (2010) Identifying the security risks associated with governmental use of cloud computing. Gov Inf Q 27:245–253
20. Mohammad A, Kak SM, Alam MA (2017) Cloud computing: issues and security challenges. Int J Adv Res Comput Sci 8(2)
21. Tariq H, Agarwal P (2018) Secure keyword search using dual encryption in cloud computing. Int J Inf Technol

Chapter 39
Automated Classification of Sleep Stages Based on Electroencephalogram Signal Using Machine Learning Techniques

Santosh Kumar Satapathy, D. Loganathan, M. V. Sangameswar, and Deepika Vodnala

1 Introduction

Sleep is an integral section of the human physiological and directly connected with the different physiological activities of the body. The proper sleep patterns decide the mental and physical stableness of the human body. It has been seen that some of the basic functionality is directly associated with sleep such as learning ability, memory concentration, and cognitive behaviour [1]. It has been also found that deprivation of sleep may induce the various critical diseases like increase of hypertension [2], obesity [3], sleep apnoea [4], Parkinson's diseases [5], Alzheimer's diseases [6], heart disease [7, 8], epilepsy detection [9], cardiovascular disease [10], and finally it also affects the immunity system of the body. To analyse and diagnose the above-mentioned sleep-related diseases, the first most important step is an accurate classification of the sleep stages and analysis of the sleep patterns. Polysomnography test is one of the best evaluation methods for different types of sleep diseases. Generally, PSG recordings are combinations of electroencephalogram (EEG), electrooculogram (EOG), electrocardiogram (ECG), electromyogram (EMG), pulse oximetry, and respiratory signals. The entire sleep staging process followed the Rechtschaffean and Kales (R&K) rules [11], based on the R&K, the whole sleep duration is divided into six sleep stages: wake, non-rapid eye movement stage (NREM), and rapid eye movement stage (REM) wake (W), non-rapid eye movement (NREM)

S. K. Satapathy (✉) · D. Loganathan
Pondicherry Engineering College, Puducherry 605014, India
e-mail: santosh.satapathy@pec.edu

D. Loganathan
e-mail: drloganathan@pec.edu

M. V. Sangameswar
Godavari Institute of Engineering and Technology, Rajahmundry, Andhra Pradesh 533296, India

D. Vodnala
Vignana Bharathi Institute of Technology, Medchal, Telangana 501301, India

A. Sheth et al. (eds.), *Intelligent Systems*, Algorithms for Intelligent Systems,
https://doi.org/10.1007/978-981-16-2248-9_39

N1 stage, N2 stage, N3 stage, and N4 stage, REM (R) stage. According to a new sleep standard edited by the AASM in the year 2007 (Updated 2017) [12] the whole sleep cycle segmented into five-sleep states, the only changes reflected in the NREM stage, combined (N3 + N4) as a single stage named as N3 or SWS. Each sleep stage is associated with the different neuronal features, which helps to discriminate the sleep stages of a person. But it has been seen that the PSG test is one of the time-consuming, labour-intensive, and tedious tasks. It also produced variations on sleep staging results depending upon the variations upon the different sleep experts. Additionally, the PSG test also created some uncomfortable situations for the patients because more connectivity of the electrodes and wires are fixed in the body during the recordings of the sleep behaviour. Therefore, developing an automated sleep staging system could be more helpful in this context. Among all the physiological signals, EEG signals are more effective with subject to sleep staging because it provides direct brain behaviour of the subjects, which is quite more helpful during the analysis of sleep behaviour. To provide a comfortable and reliable sleep scoring system, consider a single-channel EEG signal for data acquisition. It has been seen that sleep staging using machine learning (ML) techniques widely used so far by different researchers. The performance of the automated sleep scoring system completely depends on the power of feature properties, which helps to discriminate the sleep behaviour with respect to the individual sleep classes. It has been found that, most of the existing sleep studies used linear and nonlinear features [13–17]. Similarly, the feature reduction step is also equally important for sleep staging method because the selection of the suitable feature are also important challenges because it directly affects the classification performance of the model. It has been noticed that many of the features may not be suitable for all the cases, so different feature selection algorithms obtained by the different authors such as ReliefF [18], fast correlation-based filter (FCBF) [19], minimum redundancy maximal relevance [20], information gain [21], and recursive feature elimination [22]. The final important phase of sleep staging is selection of classification model. The mostly widely obtained classifier by the researcher are: Support vector machine (SVM) [14], decision tree (DT) [23], K-nearest neighbour [24], linear discriminant analysis (LDA) [25], Naïve Bayes [24], random forest (RF) [24], artificial neural network (ANN)[13], AdaBoost [24], and multi-layer perceptron [26]. Some of the existing contributed sleep studies are discussed below here. In [15], the author used Tunable-Q factor wavelets transform (TQWT) and obtained the RF classifier for classifying the sleep stages. The model achieved the classification accuracy for six-to-two sleep stages classification is 90.38, 91.50, 92.11, 94.80, and 97.50% using the sleep-EDF dataset. In [27], the author considered different structural properties and spectral features from the recorded signals, and the final selected features are forwarded into the least-squares support vector machine (LS-SVM). The proposed classification model result is compared with the K-nearest neighbour, Naïve Bayes, and multi-class SVM. The entire sleep staging is worked upon by the AASM and R&K rules. The model achieved 96.74% and 96%, respectively. In [28] presented the new sleep staging system based on the statistical features and weighted brain networks using multiple-channel of EEG signals under both R&K and AASM sleep scoring guidelines. The proposed model has been performed on the two most popular public datasets namely ISRUC-Sleep and S-EDF dataset. The model reported an average accuracy of 96.74% with C3-A2 channel under the AASM scoring standards and 96% with Pz-Oz channel under the R&K standards. In [29] used minimum redundancy and maximal relevance feature selection algorithm obtained for selecting the suitable features from the pool of extracted features, and the selected features are forwarded into the HMM and RF classification model. The model performed an overall accuracy of ranges 79.4–87.4% and 77.6–80.4% based on the R&K and AASM sleeping rules, respectively. In [30], the author obtained fuzzy entropy and log energy features from the recorded signal. This study achieved an overall accuracy of 91.5 and 88.5% using the small and long dataset for six sleep classes. Hassan [30] used ensemble empirical mode of decomposition techniques for the decomposition of the EEG signals in to different frequency sub-bands, and extracted the statistical features from the processed signals and the model achieved the accuracy of 93.55, 92.95, 93.11, 96.86, and 98.10% for six-two sleep states classification on the sleep-EDF dataset. In [14], the author extracted the visibility graph properties from the EEG signals, and the overall accuracy of 87.5% was reported using SVM classifier. In [31], the proposed sleep study based on the single-channel EEG signals, extracted the entropy and autoregressive features and the selected features are classified through LDA classifier. The model resulted in 87.5% average classification accuracy for five-class sleep states classification. In [24], the author used bagged decision trees for the classification of five-class sleep stages. The author segments the EEG signal using intrinsic mode functions and after that static behaviour are extracted from these segmented signals. The model achieved 86.8% for six-class sleep stages classification.

It has been noticed from the above-mentioned literature that many of the studies did not perform well for five-class sleep stages classification. Therefore, in the present study, we proposed an efficient automated sleep stages system using a single-channel of EEG signals. The proposed methods extracted different categories of features like time and frequency domain features from the signals. Afterwards, this proposed study obtains the feature screening techniques as ReliefF weight, which improves towards selecting the suitable features, which ultimately helps to improve the sleep staging performance. Finally, we obtained the RF classifier for classifying the five-sleep state's classification. This proposed sleep study evaluated on the ISRUC-Sleep subgroup-I dataset. The reported results signify that our proposed methodology better than existing published sleep staging studies. Further, this research work is presented as accordingly: Sect. 2 presents the methodology which includes the experimental data preparation, data preprocessing, feature extraction, feature screening, classification, and performance evaluation metrics. In Sect. 3, we briefly discuss our proposed methodology results and make a result analysis with the state-of-the-art method. Section 4 ends with concluding remarks with future work description.

2 Methodology

Figure 1 briefly presents the proposed layout of the automated sleep staging system. First of all, the 30s epochs of the EEG signal (6000) sample points input into the model. The entire research workflows in the four basic steps such as preprocessing, feature extraction, feature selection, and classification. The main importance behind obtaining the preprocessing stage is to remove the irrelevant signal compositions included during the sleep recordings. After that, the signal features were extracted from the pre-processed signals, which alternatively help to discriminate the sleep stages with regards to the time and frequency domains. Next to the feature extraction, it is important to screen the features. Finally, we obtained the shallow learning classifier, which took the final selected features as input for the classification model.

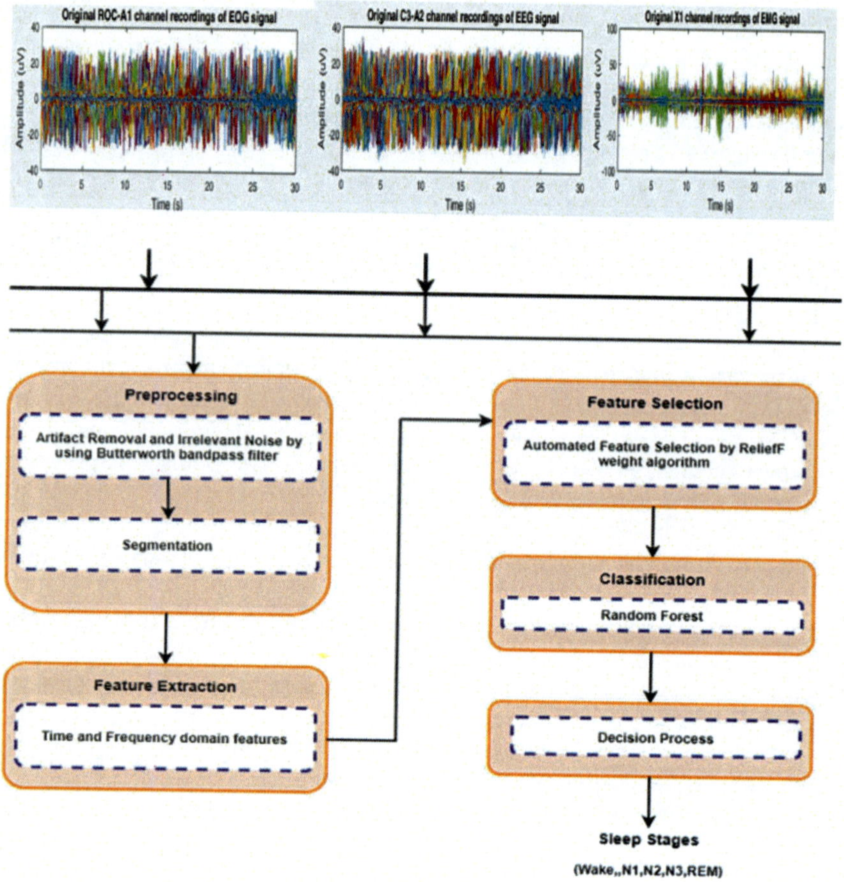

Fig. 1 Proposed research work framework

Table 1 Sleep epoch information with respect to the individual sleep stage

Dataset	Sleep states					Total epoch
	W	N1	N2	N3	R	
ISRUC-sleep subgroup-I	1003	516	1211	591	429	3750

A brief description of the individual stages of the proposed sleep study is described below here.

2.1 *Experimental Data*

In this research work, we used the ISRUC-Sleep public database, which was prepared by the groups of sleep experts in the sleep medicine centre, Hospital of Coimbra University [32]. This dataset contained the different subgroups of data. The first subgroup section contained 100 subjects with one session sleep recordings and having symptoms of mild-sleep problems. Similarly, the second subgroup section contained 20 subjects with two-session sleep recordings. All these 20 subjects were affected by sleep-related disorders. Finally, the third subgroup section contains 10 healthy controlled subjects' sleep recordings. All were completely health controlled and before not having any types of medications. All these sections of sleep recordings are labelled according to the AASM sleep standards. Each epoch of length is the 30 s. In this study, we considered C3-A2 channel of EEG signal and the retrieved sleep epoch information are described in Table 1.

2.2 *Preprocessing*

During this section, we removed the irrelevant signal compositions which are added during recordings of the sleep behaviour of the subjects. Sometimes, different types of the artefacts such as muscle movements, eye blinks, and irregular noises. We obtained a 10th order Butterworth band-pass filter for reducing the effects of those artefacts and contaminates.

2.3 *Features Extraction*

It is very difficult to analyse the sleep behaviour directly from the recorded signals during the sleep time because the EEG signals are highly random and their behaviour is continuously changing with time and frequency ranges. So, it is important to analyse the signal behaviour which gives important information on the characteristics of the change of the sleep during different sleep stages. In this sleep study, we obtained both linear (time and frequency domain) and nonlinear features for discriminating the sleep characteristics with the changes in amplitude and frequency levels. A total of 29 features are extracted from the processed signals, out of that 12 time-domain features, 15 frequency domain features, and 2 nonlinear features.

2.4 *Feature Selection*

Next to feature extraction, it is a very important task to select the suitable features. It has been found that most of the time all the extracted features may not be suitable for each and individual subject cases. So, it is highly necessary to screen the feature before proceeding into the classification model. In this paper, we considered the ReliefF algorithm [33] to find out the most relevant features, which prevents producing biased results. ReliefF feature selection algorithm screened the best weightage feature by assigning the weight value against each feature, and finally, the feature considered more relevant, which feature having higher the weight value.

2.5 *Classification*

It has been found that the RF classifier performed well in the sleep stages classification system because of its well in adaptability and robust nature [34]. This algorithm is introduced by Breimant. Generally, this algorithm uses multiple tree structures for multiple variables, and each tree structure used to train with the different training samples, and finally, the decision will be taken based on the maximum vote received by the decision tree.

3 Results and Discussion

The proposed sleep staging is based on the single-channel (C3-A2) of the EEG signal. We have obtained sleep recordings of the subjects who were already having the symptoms of the sleep problems. The whole experiment was executed under the AASM sleep scoring rules. For further removing the artefacts information from the recorded signals, we used Butterworth band-pass filtering techniques of the order 10th. For analysing the sleep behaviour, we extracted the linear and nonlinear properties of the signal, which ultimately helpful for studying the changes in sleep behaviour in the different sleep stages. Finally, we obtained the RF classification model for five-sleep stages classification. The recorded dataset split into the training and testing dataset into the ratio 70:30, respectively. To evaluate the performance, we used a performance index as accuracy [35], sensitivity [36], precision [37], and F1-score [38–42].

Table 2 Confusion matrix

Training samples						Testing sample					
	W	N1	N2	N3	R		W	N1	N2	N3	R
W	691	4	2	1	1	W	123	1	2	1	0
N1	6	303	3	2	3	N1	1	188	5	3	2
N2	5	4	790	2	2	N2	2	3	489	2	2
N3	1	2	5	404	2	N3	3	2	3	189	1
R	2	2	10	7	370	R	2	2	1	2	96

Table 3 Performance evaluation results using training and testing data

CT-5	Performance metrics	REM (%)	N1 (%)	N2 (%)	N3 (%)	Wake (%)	Overall (%)
Training data	Accuracy	98.88	98.99	99.03	99.15	99.15	99.04
	Precision	97.88	96.19	98.50	97.12	98.01	99.43
	Sensitivity	94.63	95.58	98.38	97.58	98.86	97.54
	Specificity	99.64	99.47	99.33	99.45	99.26	97.01
	F1-score	96.23	95.89	98.44	97.35	98.43	97.27
Testing data	Accuracy	98.91	98.28	98.46	98.46	98.91	98.60
	Precision	95.05	95.92	98.39	95.94	93.89	99.12
	Sensitivity	93.20	94.47	98.19	95.45	96.85	95.84
	Specificity	99.50	99.12	98.68	99.12	99.18	95.63
	F1-score	94.12	95.19	98.29	95.70	95.35	95.73

The entire experiment executed and coded using MATLAB (2017b) software with windows 10 operating system and 8-GB REM. The reported confusion matrix is presented in Table 2 and the result of the performance metrics is shown in Table 3. Table 4 presents comparison analysis in between the performance of the proposed model and the existing state-of-the-art works.

From Table 3, it has found that, the overall performance reported of 99.04, 99.43, 97.54, 97.01, and 97.27% with respect to classification accuracy, precision, sensitivity, specificity, and F1-score using training dataset. Similarly, the same model achieved 98.60, 99.12, 95.84, 95.63, and 95.73% with the testing data.

4 Conclusion

This proposed study considers developing an automated sleep staging classification system using single-channel signals. To improve the sleep staging performance, the proposed method presents an automated sleep staging system. The proposed model was evaluated on the one public dataset such as ISRUC-Sleep subgroup-I data under

Table 4 Performance comparisons of the existing published contributions with proposed model performance

Published study	Year	Input signal	Classification model	Overall accuracy (%)
Ref. [39]	2018	EEG signal	DT	80.07
Ref. [36]	2018		DNN	73.28
Ref. [40]	2017		SVM	92.09
Ref. [41]	2017		CNN	91.22
Ref. [42]	2019		Ensemble learning stacking model	96.67
Proposed study	**Present**		**RF**	**98.60**

the AASM sleep scoring rules. The most important part of this research work is discriminating the sleep characteristics using both linear and nonlinear features of the subject as one of the features, which ultimately provides better improvements results for the obtained dataset. The reported results from the proposed model show that the proposed model provides higher sleep staging accuracy incomparable to the existing state-of-the-art works. The proposed methodology in this paper significantly improves the automated sleep staging performance, which supports the clinical practices for the various types of sleep-related diseases.

References

1. Ellenbogen JM, Payne JD, Stickgold R (2006) The role of sleep-in declarative memory consolidation: passive, permissive, active or none? Curr Opin Neurobiol 16(6):716–722. https://doi.org/10.1016/j.conb.2006.10.006
2. Lu K, Chen J, Wu S, Chen J, Hu D (2015) Interaction of sleep duration and sleep quality on hypertension prevalence in adult Chinese males. J Epidemiol 25(6):415–422 https://doi.org/10.2188/jea.JE20140139
3. Liu J, Hay J, Faught BE (2013) The association of sleep disorder, obesity status, and diabetes mellitus among US adults The NHANES 20092010 survey results. Int J Endocrinol 2013(12):234129. https://doi.org/10.1155/2013/234129
4. Rodríguez-Sotelo JL, Osorio-Forero A, Jiménez-Rodríguez A, Cuesta-Frau D, Cirugeda-Roldán E, Peluffo D (2014) Automatic sleep stages classification using EEG entropy features and unsupervised pattern analysis techniques. Entropy. https://doi.org/10.3390/e16126573
5. Wulff K, Gatti S, Wettstein JG, Foster RG (2010) Sleep and circadian rhythm disruption in psychiatric and neurodegenerative disease. Nat Rev Neurosci 11(8):589–599. https://doi.org/10.1038/nrn2868
6. Olsson M, Ärlig J, Hedner J, Blennow K, Zetterberg H (2018) Sleep deprivation and CSF biomarkers for Alzheimer disease. Sleep 41(1):18. https://doi.org/10.1093/sleep/zsy025
7. Gottlieb DJ, Redline S, Nieto FJ, Baldwin CM, Newman AB, Resnick HE, Punjabi NM (2006) Association of usual sleep duration with hypertension: the sleep heart health study. Sleep 29(8):1009–1014. https://doi.org/10.1093/sleep/29.8.1009
8. Arruda-Olson AM, Olson LJ, Nehra A, Somers VK (2003) Sleep apnea and cardiovascular disease: implications for understanding erectile dysfunction. Herz 28(4):298–303. https://doi.org/10.1007/s00059-003-2482-z

9. Ullah I, Hussain M, Qazi E-H, Aboalsamh H (2018) An automated system for epilepsy detection using EEG brain signals based on a deep learning approach. Expert Syst Appl 107:61–71. https://doi.org/10.1016/j.eswa.2018.04.021
10. Bertisch SM, Pollock BD, Mittleman MA, Buysse DJ, Bazzano LA, Gottlieb DJ, Redline S (2018) Insomnia with objective short sleep duration and risk of incident cardiovascular disease and all-cause mortality: sleep heart health study. Sleep 41(6). https://doi.org/10.1093/sleep/zsy047
11. Wolpert EA (1969) A manual of standardized terminology, techniques and scoring system for sleep stages of human subjects. Arch Gen Psychiatry 20(2):246. https://doi.org/10.1046/j.1440-1819.2001.00810.x
12. Berry RB, Brooks R, Gamaldo C, Harding SM, Lloyd RM, Quan SF, Troester MT, Vaughn BV (2017) AASM scoring manual updates for 2017 (version 2.4). J Clin Sleep Med 13(05):665–666. https://doi.org/10.5664/jcsm.6576
13. Ronzhina M (2012) Sleep scoring using artificial neural networks. Sleep Med Rev 16(3):251–263. https://doi.org/10.1007/978-3-319-67934-13
14. Zhu G, Li Y, Wen P (2014) Analysis and classification of sleep stages based on difference visibility graphs from a single-channel EEG signal. IEEE J Biomed Health Inform 18(6):1813–1821 https://doi.org/10.1007/978-3-319-67934-13
15. Hassan, AR, Bhuiyan, MIH (2016a) A decision support system for automatic sleep staging from EEG signals using tunable Q-Factor wavelet transform and spectral features. J Neurosci Methods 271:107–118. https://doi.org/10.1016/j.jneumeth.2016.07.012
16. Sharma R, Pachori RB, Upadhyay A (2017) Automatic sleep stage classification based on iterative filtering of electroencephalogram signals. Neural Comput Appl 28(10):2959–2978. https://doi.org/10.1007/s00521-017-2919-6
17. Seifpour S, Niknazar H, Mikaeili M, Nasrabadi AM (2018) A new automatic sleep staging system based on statistical behavior of local extrema using single channel EEG signal. Expert Syst Appl 104:277–293. https://doi.org/10.1016/j.eswa.2018.03.020
18. Kira K, Rendell LA (1992) The feature selection problem: traditional methods and a new algorithm. In: Proceeding of 10th National Conference on Artificial Intelligent, pp 129–134
19. Yu L, Liu H (2003) Feature selection for high-dimensional data: a fast correlation-based filter solution. In: Proceedings of the 20th international conference on machine learning (ICML-03), pp 856–863
20. Peng H, Long F, Ding C (2005) Feature selection based on mutual information criteria of max-dependency, max-relevance, and min-redundancy. IEEE Trans Pattern Anal Mach Intell 27(8):1226–1238. https://doi.org/10.1109/TPAMI.2005.159
21. Quinlan JR (1992) C4.5: programs for machine learning, 1st edn. Morgan Kaufmann, San Mateo, CA, USA, pp 313–320
22. Guyon I, Weston J, Barnhill S, Vapnik V (2002) Gene selection for cancer classification using support vector machines. Mach Learn 46(1–3):389–422. https://doi.org/10.1023/A:1012487302797
23. Imtiaz SA, Rodriguez-Villegas E (2015) Automatic sleep staging using state machine-controlled decision trees. In: 2015 37th Annual International Conference of the IEEE Engineering in Medicine and Biology Society (EMBC), pp 378–381. https://doi.org/10.1109/EMBC.2015.7318378
24. Hassan AR, Bhuiyan MIH (2016b) Automatic sleep scoring using statistical features in the EMD domain and ensemble methods. Biocybern Biomed Eng 36(1):248–255. https://doi.org/10.1016/J.BBE.2015.11.001
25. Liang S-F et al (2012) Automatic stage scoring of single-channel sleep EEG by using multiscale entropy and autoregressive models. IEEE Trans Instrum Measur 61(6):1649–1657. https://doi.org/10.1109/TIM.2012.2187242
26. Huang W, Guo B, Shen Y, Tang X, ZhangT, Li D, Jiang Z (2019) Sleep staging algorithm based on multichannel data adding and multi feature screening. Comput Methods Programs Biomed 105253 https://doi.org/10.1016/j.cmpb.2019.105253

27. Sharma R, Pachori RB, Upadhyay A (2017) Automatic sleep stages classification based on iterative filtering of electroencephalogram signals. Neural Comput Appl 28(10):2959–2978 https://doi.org/10.1007/s00521-017-2919-6
28. Diykh M, Li Y, Abdulla S (2019) EEG sleep stages identification based on weighted undirected complex networks. Comput Methods Programs Biomed. https://doi.org/10.1016/j.cmpb.2019.105116
29. Abdulla S, Diykh M, Laft RL, Saleh K, Deo RC (2019) Sleep EEG signal analysis based on correlation graph similarity coupled with an ensemble extreme machine learning algorithm. Expert Syst Appl 138:112790–112804. https://doi.org/10.3390/s20174677
30. Ghimatgar H, Kazemi K, Helfroush MS, Aarabi A (2019) An automatic single-channel EEG-based sleep stage scoring method based on a hidden Markov model. J Neurosci Methods 324:180320–180336. https://doi.org/10.1016/j.jneumeth.2019.108320
31. Hassan AR, Bhuiyan MIH (2017) Automated identification of sleep states from EEG signals by means of ensemble empirical mode decomposition and random under sampling boosting. Comput Methods Programs Biomed 140:201–210. https://doi.org/10.1016/j.cmpb.2016.12.015
32. Liang S-F, Kuo C-E, Hu Y-H, Pan Y-H, Wang Y-H (2012) Automatic stage scoring of single-channel sleep EEG by using multiscale entropy and autoregressive models. IEEE Trans Instrum Meas 61(6):1649–1657. https://doi.org/10.1109/TIM.2012.2187242
33. Khalighi S, Sousa T, Santos JM, Nunes U (2016) ISRUC sleep: a comprehensive public dataset for sleep researchers. Comput Methods Programs Biomed 124:180–192. https://doi.org/10.1016/j.cmpb.2015.10.013.
34. Robnik-Šikonja M, Kononenko I (2003) Theoretical and empirical analysis of ReliefF and RReliefF. Mach Learn 2003(53):23–69. https://doi.org/10.1023/A:1025667309714
35. Shabani F, Kumar L, Solhjou-Fard S (2017) Variances in the projections, resulting from CLIMEX, boosted regression trees and random forests techniques. Theor Appl Climatol1–14. https://doi.org/10.1007/s00704-016-1812-z.
36. Sanders TH, McCurry M, Clements MA (2014) Sleep stage classification with cross frequency coupling. In: 2014 36th annual international conference of the IEEE engineering in medicine and biology society (EMBC), pp 4579–4582. https://doi.org/10.1109/EMBC.2014.6944643
37. Bajaj V, Pachori RB (2013) Automatic classification of sleep stages based on the time frequency image of EEG signals. Comput Methods Programs Biomed 112(3):320–328. https://doi.org/10.1016/j.cmpb.2013.07.006
38. Yildiz A, Akin M, Poyraz M, Kirbas G (2009) Application of adaptive neuro-fuzzy inference system for vigilance level estimation by using wavelet-entropy feature extraction. Expert Syst Appl 36(4):7390–7399. https://doi.org/10.1016/j.eswa.2008.09.003
39. Powers, David and Ailab: Evaluation: from precision, recall and F-measure to ROC, informedness, markedness and correlation. J Mach Learn Technol 2:2229–3981. https://doi.org/10.9735/2229-3981
40. Gunnarsdottir KM, Gamaldo CE, Salas RME, Ewen JB, Allen RP. Sarma SV (2018) A novel sleep stage scoring system: combining expert-based rules with a decision tree classifier. In: 2018 40th annual international conference of the IEEE engineering in medicine and biology society (EMBC). https://doi.org/10.1109/EMBC.2018.8513039
41. Nakamura T, Adjei T, Alqurashi Y, Looney D, Morrell MJ, Mandic DP (2017) Complexity science for sleep stage classification from EEG. In: Proceedings of the international joint conference on neural networks, Anchorage, AK, USA. https://doi.org/10.1109/IJCNN.2017.7966411
42. Hassan AR, Subasi A (2017) A decision support system for automated identification of sleep stages from single-channel EEG signals. Knowl-Based Syst 128:115–124. https://doi.org/10.1016/j.knosys.2017.05.005

Chapter 40
B-Tree Versus Buffer Tree: A Review of I/O Efficient Algorithms

Soni Harshit and Verma Santosh

1 Introduction

Data structures are evolving everyday as with the change of requirements that is being faced by the industries. Research has been made on the structures that can hold more data in lesser nodes, which is the root behind formation of B-tree. Various researchers have shown their interest to optimize the I/O data structures. The author has made an attempt in [2] to highlight the progress made in evolution of data structures by B-tree techniques as well as its operations. In [5], the author has depicted the various cases of operations and detailing of complexities on further operations, whereas in [1, 4, 5], authors have explained the analysis of its further evolution in the Buffer tree, its implementation, the logical possibilities, and their complexity analysis. Through the study of references, a scope is identified to study and review the comparison of both the data structures.

S. Harshit · V. Santosh (✉)
Institute of Engineering and Technology, JK Lakshmipat University, Jaipur, Rajasthan 302026, India
e-mail: santosh.verma@jklu.edu.in

S. Harshit
e-mail: harshitsoni@jklu.edu.in

A. Sheth et al. (eds.), *Intelligent Systems*, Algorithms for Intelligent Systems,
https://doi.org/10.1007/978-981-16-2248-9_40

1.1 B-tree

B-tree is a special modified version of m-way [1–7] search tree where every node can have m children with m-1 keys. The problem with m-way search tree was that there was no rule to insert nodes which could eventually make the tree 'wild' in terms of its height and biased structure. As with the growing data structures, it has been a requirement that the formed structure needs to be compact and concentrated. With such a requirement, it is necessary to work in self-balancing trees that balance themselves to maintain their height and balance of weights on both sides. B-tree is an even advanced version of self-balancing trees with more than one key on a node. The overall idea is to keep its height as small as possible. An m-way search tree, to be called as a B-tree, needs to follow some of the rules, A. Node should have either m/2 children or no children except the root node. B. There can be two or no children of root node. C. Leaf nodes are to be of the same level.

Efficiency in indexing of databases indexing of tuples in a table of a database has been a very crucial process as searching in huge amounts of data needs to be efficient in terms of time complexity. With the concept of B-tree, a better version of it was introduced, i.e., B+ tree with just a small modification in which the leaf nodes had the copy in the pre-order of the value of its parent. This forms a linked list of all the elements in a sorted order. This data structure facilitates with a property of binary trees as well as linked lists. With the property of these, the complexity for sorting, insertion, and deletion is reduced to O(log(n)). This way, it helps in a faster processing of data. Now the indices of tuples are stored in a B+ tree and can be searched, inserted, and deleted in a faster manner and are found to be already sorted in the linked list.

Pseudocode through the following algorithm, B-tree can be implemented.

```
Pseudo code:
Operations () :
        Repeat
                Val=input('filename')
                Index=search(tree,val)
                  Insert(Tree,Index,val)
                  Delete(Tree,Index,val)
        While the operations are left
End

Insert(Tree,Index,val):
        Tree[Index]=val
        Check(Tree,Index)
End
```

```
Check(Tree,Index):
        If tree has more than M elements
        Split(Tree,Index)
        Check(Tree,Parent_Index)
End

Split(Tree,Index):
        New_node1=Left_of_middle_element
        New_node2=Right_of_middle_element
        Append New_node1&New_node2 to tree
End
```

B-tree evolution with the growth in capacity of internal memory, there has been scope for the efficient use of it. This has led to the development of algorithms that can use as much of internal memory so to reduce the interaction with the external memory. In this situation, the algorithms should be ready to use memory accordingly. This way, B-tree is supposed to be modified in such a way that it uses an amount of internal memory according to the requirement of operations on it. It is to be redefined that it is capacity to hold a number of operations change with the size of the tree. This has given rise to the idea of the Buffer tree. With the inclusion of B-tree, furthermore evolution is supposed to be done in the same field to make it furthermost faster and efficient in terms of all the situations that it can face. The better and modified version came into existence, i.e., Buffer tree.

1.2 The Buffer Tree

As the B-tree has grown well and has given better results, its modification is needed to be suitable for the processing of very massive dataset. As in the former section, it has been pointed out that due to slow external disks and fast internal memory and due to their speed not being in sync has led the processor to be idle for a span of time. The solution is to reduce the I/Os as every input or output process costs a lot of time, and the optimal use of the processor as well as the internal memory is skipped. The number of I/O operations are very high due to communication between the internal and external memories. This can be done if the operations are transferred in batches. Batches are groups of elements stored in a large packet and act as a single unit in an I/O.

The Buffer method like B-tree, rules, and restrictions are the same for insertions and deletions in buffer trees. There is a buffer included at every node. With the splitting of nodes, buffers keep increasing with new nodes. The buffer is a kind of priority queue where the order or the pattern in which the operations are done is maintained, and implementation is done in the same way only. So, operations coming in batches to the internal memory are filled firstly to the buffer of the root node. When the root node is filled completely and gets full, the elements of the buffer are pushed to the buffer of its child nodes according to their respective key child. This process keeps going on until the buffer of any leaf node gets full. As the buffer of any leaf node gets full, all the operations inside it gets implemented until that buffer is empty. So, this method of pushing into the buffer and cleaning a buffer keeps going on until all the I/Os get completed [6, 7]. This way, operations are not implemented at a time but are implemented in a group with a 'lazy' manner. Through this particular manner of executing the process, it helps in reducing the total I/Os and overall comparisons in respect to the B-tree. In this process, the buffer size is decided depending on the size of internal memory, the batch size, and total number of elements in a problem. If a node is supposed to split, then its buffer is distributed to the two new nodes accordingly.

Pseudocode through the following algorithm, Buffer trees can be implemented in the same way.

```
Pseudo code:
Operations() :
        Repeat
               Val[batch]=input('filename')
                Repeat
                        Append item to Root_buffer
                                Clearbuffer(Tree,Root)
                While the items in batch are left
        While the operations are left
End
```

```
Clearbuffer(Tree,Index):
        If buffer is full
                If Index is a leaf node
                        Repeat
                                Call operation
                        While the buffer is empty
                Else
                        Append elements to the buffer of child node
                        Delete element from buffer
                        Clearbuffer(Tree,child_node)
End

Insert(Tree,Index,val):
        Tree[Index]=val
        Check(Tree,Index)
End

Split(Tree,Index):
        New_node1=Left_of_middle_element
        New_node2=Right_of_middle_element
        Append New_node1&New_node2 to tree
End
```

1.3 Comparative Analysis

A comparison on the practical analysis made between B-tree and Buffer tree through both of their programming implementations has shown following cases where Buffer tree proves to be better than B-tree.

As already pointed out, the overall I/O interaction of the B-tree has been much higher in comparison with the Buffer tree. This has made a big difference between the time taken between both of their working. The I/O on every operation has created this gap.

Another benefit of Buffer tree over B-tree is that in Buffer tree, the program is already aware of the coming payload that is going to be executed in near future. As every operation is executed after reaching the exact leaf node where it is meant to be executed, the cost to search it again in case of B-tree is saved here, and hence, this can save a small interval of time in many cases as the overall comparisons made would be lesser than the B-tree.

Following Fig. 2 can be used to refer to the example.

As per this example, it can be observed that 99 is in the buffer of the (72,99) node. After further insertions and buffer clearing process, the tree would look exactly same with (Fig. 1), but if checked into that figure, the height has been increased from 2

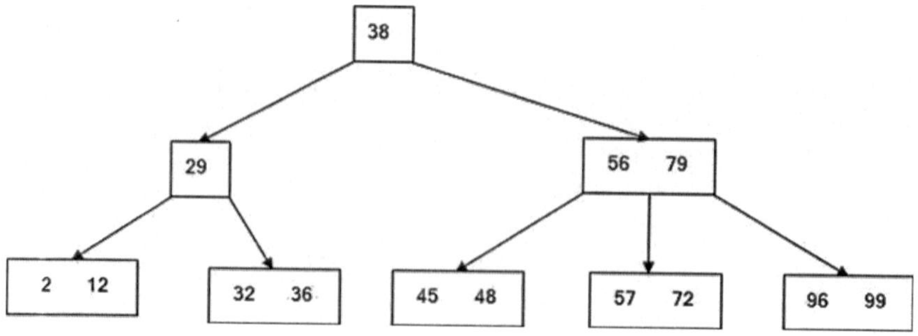

Fig. 1 B-tree sample image

to 3 which means that if the compiler had to search for suitable position of 99 in a B-tree, it would cost it total 2 comparisons, but now in this case of Buffer tree, it eventually cost it only a single comparison as when it is pushed into leaf buffer, the height of the tree is only 2. This way, in many of the cases, comparisons would be avoided, and it would save the total time taken. There would be a few cases when there are operations which are exactly opposite to each other. Like in Fig. 2, there is an operation for inserting 59, and then, there is an opposite operation of deleting 59. If observed, B-tree would implement both the operations, and after the execution, the tree remains the same as it was which means that there was no need for both of the operations. In case of Buffer tree, both the operations are cancelled out in the buffer only and this again saves the time in comparison with B-tree. Example of canceling of opposite operations (Fig. 3):

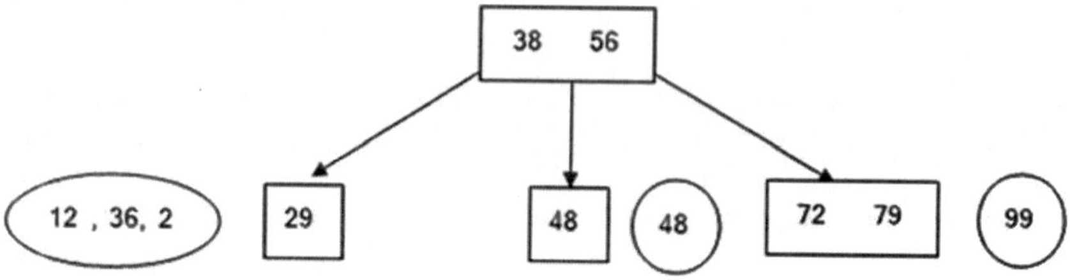

Fig. 2 Buffer tree sample image

Insert 43 | ~~Insert 59~~ | Search 37 | ~~Delete 59~~ | Insert 97

Fig. 3 Example of cancelling of opposite operations in Buffer tree

2 Result

The logical difference between B-tree and Buffer tree has been explained, and real-time programs has been designed for practical analysis and has been made on the time taken for further operations. Recorded time taken after keeping the I/O constant, i.e., running only internal memory for both the trees, the results are as follows:

1. For B-tree: 0.0046297 s
2. For Buffer tree: 0.0031973 s.

2.1 Analysis with I/Os from External Memory

The same programs were modified, so that they could take I/Os from external memory. Further variation with change in batch size and block size was observed, and results are given in Tables 1, 2 and 3.

Iteration 1 for 100 datasets:

1. For B-tree: 0.0789511 s
2. For Buffer tree: Considering different batches.

Iteration 2 for 500 datasets:

1. For B-tree: 0.2078714 s
2. For Buffer tree: Considering different batches.

Further analyzed on number of I/O operations of Buffer tree and B-tree, it is observed that the exact count of operations in B-tree is actually that much count of I/Os, while in Buffer tree, I/Os count vary according to the batch size of processing the operations and buffer size. A sample I/O count on both trees are given below.

Table 1 Running time of buffer tree

Batch size	Time taken (s)
5	0.0156254
10	0.0134953
15	0.0072689
20	0.0059959

Table 2 Running time of Buffer tree

Batch size	Time taken (s)
5	0.0899431
10	0.0589635
15	0.0359771
20	0.0312488

Table 3 Count of I/Os of buffer tree versus B-tree per 600 operations

Batch size	I/O count of buffer tree	I/O count of B-tree
5	120	600
10	60	600
15	40	600
20	30	600

Through this comparative analysis, Buffer tree is found to be an I/O efficient data structure in comparison with B-tree in terms of running time complexity and the reduced I/O operations.

3 Conclusion and Future Work

As per the analysis, it can be clearly observed that Buffer tree is a proven development of B-tree with positive results with respect to the running time complexity. Growth of internal memory has made the I/O efficient and external algorithms helpful to solve problems with massive datasets. With our experimental results, Buffer tree is found to be an I/O efficient data structure in comparison with B-tree in terms of running time complexity and the reduced I/O operations. With the results achieved so far, Buffer tree works slightly better for data in internal memory but shows big differences when the dataset is external. With the development of B-tree to Buffer tree, algorithms have proven to be working efficiently, so further efforts can be made to improve it more in any such possible ways.

As it is a developing algorithm, it is further improved, and another version is persistent data structure which allows the user to view and use the previous version of the structure. We will review the persistence buffer tree and its variant in our further study.

References

1. Arge L (2003) The buffer tree: a technique for designing batched external data structures. Algorithmica 37(1):1–24
2. Dominic SJ, Sajith G (2004) The persistent buffer tree: an I/O-efficient index for temporal data. arXiv preprintcs/0404033
3. Arge L, Danner A, Teh S-M (2003) I/O-efficient point location using persistent B-trees. J Exp Algorithmics (JEA) 8:1–2
4. Arge L (1995) The buffer tree: a new technique for optimal I/O-algorithms. In; Workshop on algorithms and data structures. Springer, Berlin, Heidelberg, pp 334–345
5. Comer D (1979) Ubiquitous B-tree. ACM Comput Surv (CSUR) 11(2):121–137

6. Grigore R, Kiefer S (2015) Tree buffers. In: International conference on computer aided verification. Springer, Cham, pp 290–306
7. Sitchinava N, Zeh N (2012) A parallel buffer tree. In: Proceedings of the twenty-fourth annual ACM symposium on Parallelism in algorithms and architectures, pp 214–223

Chapter 41
A Heterogeneous Ensemble-Based Approach for Psychological Stress Prediction During Pandemic

Shruti Jain, Sakshi, and Jaskaranpreet Kaur

1 Introduction

Coronavirus disease (COVID-19) has brought a drastic alteration to the way of life. Implementation of nationwide lockdown has left millions of people without work. People have faced a lack of routine and difficulties of isolation. Abnormal sleeping time, loss/excess of appetite, and disagreements at home has been reported by many.

Existing studies show that lifestyle has a significant impact on mental health [1]. Stress is a physiological reaction to the mental, emotional, and physical challenges faced by humans in their daily-life activities. Continuous exposure to stress can lead to serious health problems, such as physical illness, behavioral changes, and social isolation issues [2–4].

Coronavirus is a notable threat to mental health globally through increased anxiety, depression, stress disorder, and negative social behavior [5]. Recent Lanzet Psychiatry paper calls for urgent improvement in mental health services, requiring important and immediate mental well-being evaluations based on previous pandemic experiences [6]. As the virus spreads, strategies for early detection and neutralization of mental health problems are required.

In the past few decades, the computational intelligence and machine learning community has attracted increasing attention in multi-level classification systems, also called stacking models. Ensemble systems have proved extremely effective and versatile in a wide range of problems.

This paper introduces a psychological stress meter (PSD). It is an ensemble learning approach for early detection of stress based on the 18 behavioral markers observed during the lockdown. These markers include—arguments at home, time spent at performing physical activities, loss of appetite, sleep anomalies, financial pressure, etc. The markers are obtained on a four-point scale from the participant. The method used in this study involves capturing current lifestyle status through a

S. Jain (✉) · Sakshi · J. Kaur
Bharat Electronics Limited, Ghaziabad, Uttar Pradesh 201010, India

A. Sheth et al. (eds.), *Intelligent Systems*, Algorithms for Intelligent Systems,
https://doi.org/10.1007/978-981-16-2248-9_41

mental-state survey. Survey is an efficient choice for this study because questionnaires are a practical way to gather data. They offer actionable information and are scalable. The collected non-obtrusive data is used to train the models. Different ML based techniques are applied to analyze the patterns and to narrow down the factors that determine stress levels. Eventually after model comparison and evaluation, the best performance model is selected. PSD can then predict the level of stress a person has based on the value of input markers.

2 Related Work

It is well-known that the pandemic poses a consequential threat to mental health around the world due to extensive national lockdown and economic reasons.

A WHO technical note (2020), stated that "the main psychological impact today is elevated rates of stress or anxiety," with a warning that "as new measures and impacts are introduced—especially quarantine and its effects on many people's usual activities, routines, or livelihoods—levels of loneliness, depression, harmful alcohol and drug use, and self-harm or suicidal behavior are also expected to rise." Number of scientific papers analyzing effects on mental health of past pandemics and recently published articles specifically on mental health and the outbreak of COVID-19 show that prompt action needs to be taken to minimize damage to mental health [7].

During the 2003 SARS outbreak, out of 129 quarantined people, 28.9% polled that they experienced post-traumatic stress disorder (PTSD) and 31.2% showed symptoms of depression [8]. Similarly, in COVID-19, 21.5% of 5000 Chinese citizens have registered PTSD symptoms, illustrating a similar effect on mental health as the other pandemic [9].

In March 2020, a coronavirus poll in the USA reported that about one-third of the total adults (32%) felt worried and stressed, including 14% who said they had a "significant" impact on their mental health [10]. Authors in [11] have elaborated on the direct and the indirect relationship between physical and mental health. Their results imply that the current mental health of an individual is impacted by past physical health and past social interaction. The global pandemic has led to health problems like insomnia, anger, restlessness, anxiety, etc. The emerging problems with mental health can develop into prolonged health issues, isolation, and stigma [12].

The key focus of this research is to use digital technology and develop an intelligent system for stress prediction and mitigation. Here, stacking generalization [13] which is an ensemble learning framework is used. Experiments are performed using various baseline machine learning algorithms like Support Vector Machine (SVM) [14], Decision Trees (DT), Logistic Regression (LR), k-Nearest Neighbors (KNN), and Naive Bayes (NB). Other than these, two homogeneous tree-based ensemble algorithms—Random Forest (RF) and XGBoost (XB)—are also used as baseline models. Selection of baseline algorithms as base-learner for stacking has been discussed by the authors in [15].

3 Proposed Methodology

This section describes the experimental design and setup.

Section 3.1 explains the data collection process and describes the dataset in brief. Steps used for preprocessing of the collected data are described in Sect. 3.2. Further, in Sect. 3.3, experimental setup has been discussed. The block diagram in Fig. 1 depicts the overview of the proposed methodology.

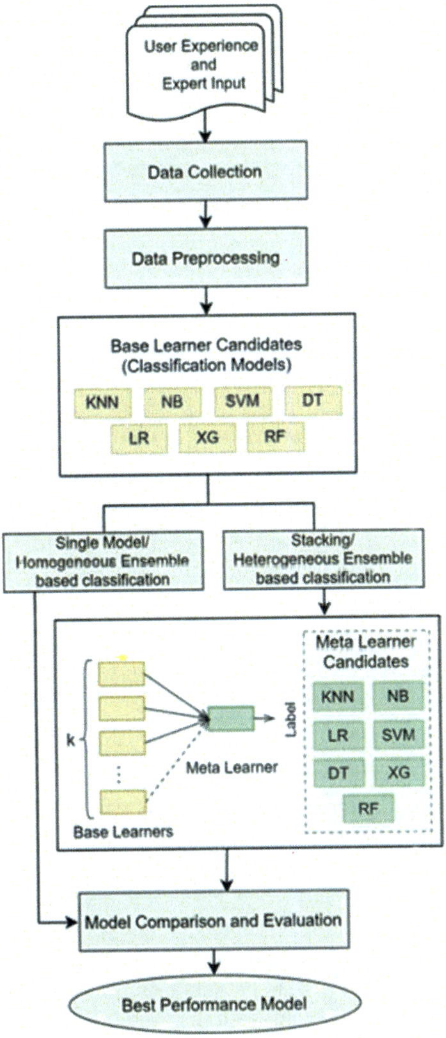

Fig. 1 Overview of proposed comparative study-based model

3.1 Dataset Description

The dataset is obtained from responses to an online questionnaire survey. The questionnaire was distributed to around 700 people, out of which 392 responded. The collected data involves observations on the psychological and mental state experienced by the respondents during the lockdown phase of COVID-19. Survey form included sixteen multiple choice and one open-ended question. General details like participant name (optional), gender, age, and occupation were also asked. This study involves strict respondent anonymity. Following is the detailed questionnaire:

Name
Age
Gender
Occupation
During the period of lockdown, how often have you:
- Experienced arguments at home? (1 - 4)
- Had disagreement between you and your loved ones? (1 - 4)
- Craved for family members, relatives or friends? (1 - 4)
- Felt anxious about future uncertainty? (1 - 4)
- Felt financial pressure/need? (1 - 4)
- Felt overburdened by studying/working from home? (1 - 4)
- Felt the pressure to meet your future goals? (1 - 4)
- Found yourself with insufficient time to do things you really enjoy? (1 - 4)
- Lost your appetite? (1 - 4)
- Found yourself constantly nibbling at snack food? (1 - 4)
- Been restless? (1 - 4)
- Found yourself getting angry or upset? (1 - 4)
- Suffered from headache? (1 - 4)
- Performed physical exercise to stay fit? (1 - 4)
- Found it difficult to fall asleep at night? (1 - 4)
- Were you stuck away from your family during lockdown? (Yes/No)

Any other comment or suggestion (Open ended)

Individuals were asked to rate each question (from Q.5 to Q.19) on the scale of 1–4 as—(1) Almost Never (2) Seldom (3) Often (4) Almost Always.

Based on the responses to the behavioral markers in the questionnaire, dataset was created. Instances of the dataset were classified into five classes, namely "Severe Stress", "High Stress", "Moderate Stress", "Low Stress", and "No Stress". PSM shall predict one of these classes as the stress level.

3.2 Data Preprocessing

Each question corresponds to a feature/behavioral marker as shown in Fig. 2. To evaluate and compare the importance of markers, expert advice was taken from three psychiatrists (field experts). They were asked to rate each feature on a scale of 1–5, where 1 meant no impact on causing stress and 5 meant severe impact. Based on the input from experts, weights were associated with each question, marking its importance in predicting stress. Weights allotted by three different experts were

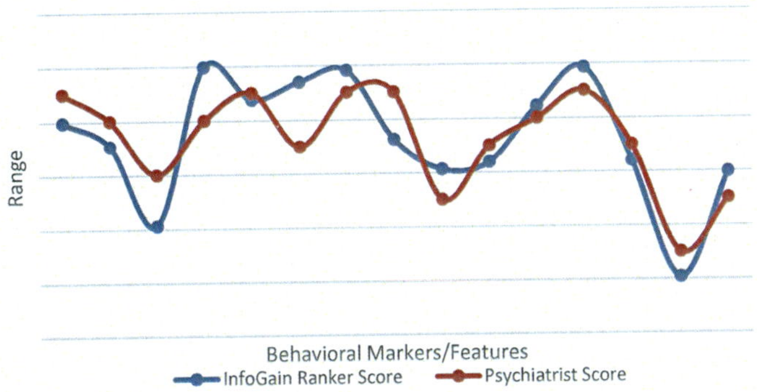

Fig. 2 InfoGain ranker and psychiatrist score for different behavioral markers

averaged and used to calculate the psychological stress score (PSS). Higher values of PSS implied greater influence. PSS was used to label the dataset.

3.3 *Experimental Setup*

The models used in this paper have been implemented in Python using Scikit Learn. Accuracy is used as a performance metric. It depicts the percentage of instances correctly classified.

Baseline Model

In the first experiment, each of the five single classifiers (KNN, SVM, DT, LR, and NB) and two homogeneous classifiers (RF and XG) were used with fivefold cross-validation. The accuracy was calculated both with and without hyper-parameter tuning. GridSearchCV was employed for auto-tuning of hyper-parameters.

Heterogeneous Ensemble Model

Second part of the experiment deals with stacking classifiers. Stacked generalization is a usual ensemble learning approach for constructing classifier ensembles. Here, the capabilities of the best performing baseline models are further explored. By using stacking, we combine the predictive powers of baseline learners (level 0) using a meta-learner (level 1). The motivation is that higher prediction precision can be obtained by amalgamating the best performing base-learners.

In this experiment, two stacking variants were considered which differ in the number of base-learners, $k (k = 2 \text{ or } k = 3)$. All possible combinations have been explored with all selected classifiers as meta-learners.

4 Results and Discussion

Table 1 depicts the performance of baseline models. It can be observed that SVM has outperformed all other baseline models with an accuracy of 90.39%. On tuning the hyper-parameters, accuracy increased to 91.05%. Best results were achieved with $C = 1$, gamma = 01, and radial basis function as the kernel. Accuracy of other models is 89.74% (RF), 88.43% (KNN), 86.69% (XG), 85.37% (LR), 84.73% (NB), and 80.34% (DT). Top-five learners are SVC, RF, KNN, XG, and LR.

Tables 2 and 3 show the outcome on heterogeneous ensemble methods with $k = 2$ and $k = 3$, respectively. Top-five baseline classifiers are grouped in pairs of two and used as base-learners in stacking as shown in Table 2. As expected, stacking increased the accuracy of the model when two best performers (SVC and RF) were used as base-learners. Ensemble has increased the prediction precision to 92.37% when LR is used as the meta-classifier.

In Table 3, a blend of three learners is used. Combination of SVC, KNN, and RF (best baseline models) gives the highest accuracy of 93.23% which is significantly higher than the initial prediction values. This is obtained when SVC is used as a meta-learner. Similarly, a combination of SVC, KNN, and XG also produced 93.23% accuracy.

Experiments were conducted by taking groups of four base-learners, $k = 4$, at a time, but no notable accuracy improvement was observed.

Table 1 Performance comparison of baseline learners (Accuracy in %age)

Base-learner	Accuracy	Accuracy with hyper-parameter tuning	Best parameters
SVC	90.39	91.05	‘C’: 1, ‘gamma’: 0.1, ‘kernel’: ‘rbf’g
LR	80.35	85.37	‘C’: 1438.44988828766, ‘penalty’: ‘l2’, ‘solver’: ‘newton-cg’
KNN	85.58	88.43	‘algorithm’: ‘brute’, ‘n neighbors’: 1, ‘weights’: ‘uniform’
XG	84.94	86.69	‘learning rate’: 0.3, ‘n estimators’: 100, ‘objective’: ‘multi:logistic’
DT	74.24	80.34	‘criterion’: ‘gini’, ‘max depth’: 20, ‘min samples split’: 3
NB	83.41	84.73	Priors = None, var smoothing = 1e–09
RF	88.21	89.74	‘criterion’: ‘gini’, ‘max features’: ‘log2’, ‘min samples split’: 3, ‘n estimators’: 130 g

Table 2 Performance comparison of base-learners ($k = 2$) with meta-learners

	Meta-learners						
Base-learners ($k = 2$)	SVC	LR	KNN	XG	DT	NB	RF
SVC + RF	91.94	92.37	91.93	91.28	91.05	90.39	90.62
SVC + KNN	91.92	91.05	91.48	90.62	90.4	91.7	91.49
SVC + LR	90.82	90.39	90.62	89.3	89.3	90.39	89.3
SVC + XG	91.49	91.27	91.05	89.75	91.49	90.17	90.18
KNN + RF	90.39	90.83	89.96	89.52	89.74	90.61	89.52
KNN + LR	88.87	88.65	89.08	88.65	88.87	89.95	88.65
KNN + XG	89.53	89.09	89.53	89.31	89.53	88.66	88.65
RF + LR	88.21	88.65	88	87.99	87.78	87.34	88.21
RF + XG	87.35	87.35	87.35	87.35	88.43	87.79	87.35
LR + XG	88.22	88	88	88	87.34	89.09	88

Table 3 Performance comparison of base-learners ($k = 3$) with meta-learners

	Meta-learners						
Base-learners ($k = 3$)	SVC	LR	KNN	XG	DT	NB	RF
SVC + LR + KNN	91.27	90.17	91.05	91.05	90.83	91.49	90.83
SVC + LR + XG	90.83	91.05	90.62	90.18	89.31	91.05	90.61
SVC + LR + RF	91.26	91.48	91.04	90.83	90.39	89.96	90.39
SVC + KNN + XG	93.23	92.58	92.36	91.49	91.7	91.27	91.49
SVC + KNN + NB	92.14	90.84	91.92	90.83	91.05	91.7	91.7
SVC + KNN + RF	93.23	92.58	92.57	92.14	91.93	91.7	91.71
SVC + XG + RF	91.49	91.49	88.42	90.19	90.63	89.53	89.96
LR + KNN + XG	90.83	91.05	90.18	90.82	91.27	91.05	91.05
LR + KNN + RF	91.92	91.04	90.61	90.83	91.26	90.62	91.05
LR + XG + RF	88.87	88.87	89.09	89.09	88.87	88.88	88.87
KNN + XG + NB	91.06	90.4	90.39	90.4	91.05	88.66	91.05
KNN + XG + RF	91.04	90.83	90.83	89.96	90.84	89.73	90.18
KNN + NB + RF	90.39	91.26	91.05	91.26	91.27	88.22	91.04
XG + NB + RF	89.75	89.1	89.54	89.1	89.75	86.92	89.32

5 Conclusion

In this paper, several ML techniques were investigated for early detection of psychological stress during COVID-19 pandemic.

It can be concluded from the research that all 15 behavioral markers play a crucial role in causing stress-related disorder. Selecting a few out of 15 leads to information loss. Heterogeneous stacking model outperformed the result of baseline models. Best performers of the baseline models proved to be suitable as base-learners in stacking. Psychological stress meter can compute the level of psychological stress with an accuracy of 93.23%. Early prediction of level of stress on a five-point scale can aid the individual to react and adapt better to the situation.

References

1. Dale H, Brassington L, King K (2014) The impact of healthy lifestyle interventions on mental health and wellbeing: a systematic review. Ment Health Rev J. https://doi.org/10.1108/MHRJ-05-2013-0016
2. Glanz K, Rimer BK, Viswanath K (eds) (2008) Health behavior and health education: theory, research, and practice. Wiley, pp 23–38
3. Korabik K, McDonald LM, Rosin HM (1993) Stress, coping, and social support among women managers. Women, Work Coping 133–153
4. Maslach C, Schaufeli WB, Leiter MP (2001) Job burnout. Annu Rev Psychol 52(1):397–422. https://doi.org/10.1146/annurev.psych.52.1.397
5. Shigemura J, Ursano RJ, Morganstein JC, Kurosawa M, Benedek DM (2020) Public responses to the novel 2019 coronavirus (2019-nCoV) in Japan: mental health consequences and target populations. Psychiatry Clin Neurosci 74(4):281. https://doi.org/10.1111/pcn.12988
6. Xiang YT, Yang Y, Li W, Zhang L, Zhang Q, Cheung T, Ng CH (2020) Timely mental health care for the 2019 novel coronavirus outbreak is urgently needed. Lancet Psychiatry 7(3):228–229. https://doi.org/10.1111/pcn.12988
7. World Health Organization (2020) Coronavirus disease (COVID-19) outbreak-technical guidance-EUROPE: mental health and COVID-19. http://www.euro.who.int/en/healthtopics/health-emergencies/coronavirus-covid-19/novel-coronavirus-2019-ncov-technicalguidance/coronavirus-disease-covid-19-outbreak-technical-guidance-europe/mental-healthand-covid-19
8. Hawryluck L, Gold WL, Robinson S, Pogorski S, Galea S, Styra R (2004) SARS control and psychological effects of quarantine, Toronto Canada. Emerg Infect Dis 10(7):1206. https://doi.org/10.3201/eid1007.030703
9. Kirton D (2020) Chinese public dial in for support as corona-virus takes mental toll. Reuters, 13 Feb 2020. https://www.reuters.com/article/us-china-health-mental/chinese-public-dial-infor-support-as-coronavirus-takes-mental-tollidUSKBN2070H2
10. Hamel L, Lopes L, Munana C, Kates J, Michaud J, Brodie M (2020) KFF Coronavirus poll, Mar 2020. https://www.kff.org/global-health-policy/poll-finding/kffcoronavirus-poll-march-2020/
11. Ohrnberger J, Fichera E, Sutton M (2017) The relationship between physical and mental health: a mediation analysis. Soc Sci Med 195:42–49. https://doi.org/10.1016/j.socscimed.2017.11.008
12. Torales J, O'Higgins M, Castaldelli-Maia JM, Ventriglio A (2020) The outbreak of COVID-19 coronavirus and its impact on global mental health. Int J Soc Psychiatry. https://doi.org/10.1177/0020764020915212
13. Domingos P, Pazzani M (1997) On the optimality of the simple Bayesian classifier under zero-one loss. Mach Learn 29(2–3):103–130. https://doi.org/10.1023/A:1007413511361
14. Li Y, Gao J, Li Q, Fan W (2014) Ensemble learning. In: Data classification, pp 511–538. Chapman and Hall/CRC
15. Kavzoglu T, Colkesen I (2009) A kernel functions analysis for support vector machines for land cover classification. Int J Appl Earth Obs Geoinf 11(5):352–359. https://doi.org/10.1016/j.jag.2009.06.002

Chapter 42
A Survey on Missing Values Handling Methods for Time Series Data

Siddharth Thakur, Jaytrilok Choudhary, and Dhirendra Pratap Singh

1 Introduction

A time series is a sequence of data points each indexed with a timestamp. The usage of time series data includes: fitting a model to forecast, monitor, and provide feedback and feedforward control and obtaining an understanding of the underlying trends and structure that produced the data. Time series prediction is used heavily in the decision-making process in various domains such as financial analysis, control engineering, weather prediction, and industrial monitoring. The various reasons affecting the values of a data point in a time series are the components of a time series [1] as shown in Fig. 1.

Trend also called as long-term movement represents the variations of low frequency. It is the component with the longest time period. Seasonal variations are the underlying forces that operate in a periodic and regular manner over a span of less than a year. These variations are a result of some natural forces. Cyclic variations have time periods usually longer than a year. One oscillation of these movements is called business cycle. Random or irregular movements are fluctuations that are purely random, unpredictable, uncontrollable, and are erratic.

Data collection involves measuring and gathering relevant information on targeted variables in a well-organized manner. It is an important step because the quality of data directly affects the end results. One of the major problems encountered in data collection and preprocessing is the problem of missing values. Missing values are unobserved values that would be useful for analysis if observed [2]. Improper handling of missing values may lead to inaccurate inferences about the data. Due to

S. Thakur (✉)
Artificial Intelligence Center, Maulana Azad National Institute of Technology, Bhopal 462003, India

J. Choudhary · D. P. Singh
Department of Computer Science and Engineering, Maulana Azad National Institute of Technology, Bhopal 462003, India

A. Sheth et al. (eds.), *Intelligent Systems*, Algorithms for Intelligent Systems,
https://doi.org/10.1007/978-981-16-2248-9_42

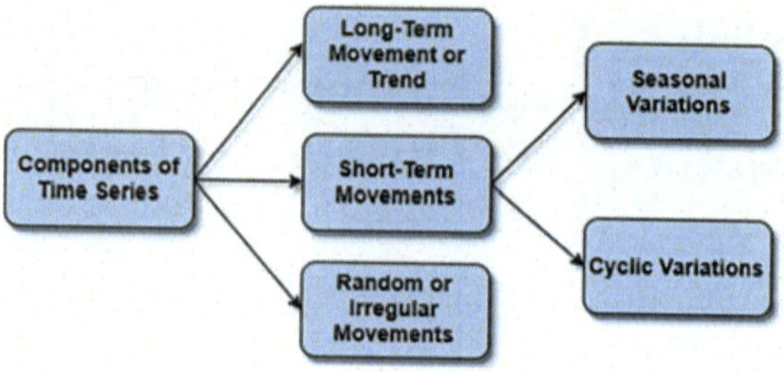

Fig. 1 Components of time series [1]

the serious effect of missing values on end results, it becomes important to understand the reasons for missing values. Data with missing values is usually categorized into three different mechanisms, also known as the missingness mechanism [2]. Consider X as a set of independent variables, Y as the dependent variable, and P is the probability of a data point missing a value. Thus, these mechanisms can be categorized as:

i. Missing Completely At Random (MCAR): P neither depends on X and Y. In other words, missing values are randomly located across all observations. Missing data is completely unrelated to the other variables in the data. For example, accidental deletion of some responses. A formal test for MCAR is to use Little's MCAR test [3].
ii. Missing At Random (MAR): P is dependent on X, but it is independent of Y. In MAR, the probability of missingness depends on the available information. For example, old people in the survey are less likely to respond to their political opinion.
iii. Missing Not At Random (NMAR): P is dependent on X, and it is possible that there are other additional influences of X on P. For example, participants with higher salary report their salary less often.

Figure 2 shows an illustration of different types of response mechanisms. The three plots show a correlation between variables X and Y. For MCAR, the plot shows no significant difference between the distributions of missing and observed values. For MAR, the distribution of red points is only determined by X, and not by Y. Y

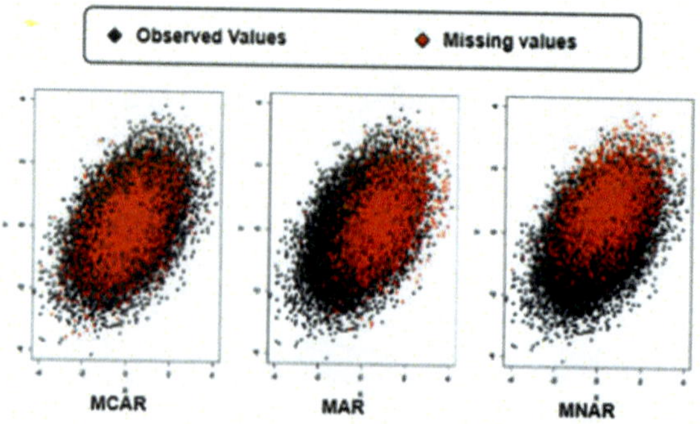

Fig. 2 Response mechanism [2]

has no direct influence on the distribution of missing values. For MNAR, the plot is exactly opposite to the plot for MAR indicating the influence of Y on P. However, this example does not show any influence of X on P.

The aim of this paper is to provide a review of several techniques used to handle missing values in time series data. This paper consists of four sections. In the first section, we will discuss time series data and things related to missing values. The second section will contain some explanation on various handling methods for missing values, from conventional methods to more complex and advanced methods. The third section contains a brief discussion about the methods listed in section second. The last section ends with a conclusion with plausible options of estimation methods.

2 Materials and Techniques

Once the missing pattern is observed, the next step is to handle these missing values. Missing values handling is an important part of data preprocessing. It is important to identify the correct approach for handling our missing values. Missing values can be handled in two ways: data deletion and missing value imputation. In the first approach, missing values are ignored completely by either deleting an entire variable with missing values or deleting the observations with missing values. The model is built based only on the observed data which may produce biased estimates and results in the loss of efficiency depending on the missing rate. In other words, this approach is suitable only when the missing rates are small. The second method involves finding suitable values for the missing data. Furthermore, missing value imputation methods can be further divided into two classes, namely conventional methods and advanced methods, described in the next sections.

2.1 Conventional Methods

Conventional methods make use of simple statistical techniques for imputing values in incomplete data. Some of these methods are briefly described below:

2.1.1 Mean/Mode Imputation

In this imputation method, the missing values are replaced by attribute mean in case of continuous variables or by mode in case of nominal values [4, 5]. The method is easy to implement but has a lot of drawbacks. Firstly, it does not preserve the relationship among variables. Secondly, this method leads to an underestimate of standard errors.

2.1.2 Hot and Cold Deck Imputation

These methods are effective when there is a large incomplete dataset [6]. In hot deck imputation, missing values are imputed by replacing them with values from different observed data with similar values for observed variables. In cold deck imputation, missing values are replaced by values from an external source from the same domain. These methods are somewhat efficient, but these methods often assume that the data is MCAR, which is not always true for real-life datasets. This method may also not perform better when there is no correlation between variables [7].

2.1.3 Multiple Imputation

This method produces N complete datasets from the incomplete dataset by imputing the missing values N times by some method. Each of these N completed dataset is analyzed, and the methods are combined to achieve inference [8]. Multiple imputation is an iterative form of stochastic imputation where the distribution of observed data is used to approximate multiple values that denote uncertainty around the true value. However, there are some conditions that must be satisfied before performing multiple imputation. These conditions require that the data should be MAR, and the model used for imputation must be appropriate.

2.2 *Advanced Methods*

Advanced methods for data imputation require a learning phase in which the methods are trying to extract some hidden information from the observed incomplete dataset. Some of these methods are described below:

2.2.1 XGBoost Imputation Using SMILES

Zhang et al. proposed a method called SMILES (xgbooSt MIssing vaLues in timE Series) which is used to impute missing values in a time series data [9]. XGBoost is a distributed gradient boosting library which is highly optimized that allows fast and accurate predictions. XGBoost is well known for mitigating the effects of over-fitting, which is a common problem in any machine learning task. It is insensitive to outliers. The algorithm is able to combine the contextual information across related variables. SMILES is able to focus on both cross-sectional and longitudinal features simultaneously.

The workflow of SMILES framework includes following steps:

i. The first step is to prefill all the missing values in a time series data. A different prefilling strategy is used for each variable which has missing values. An appropriate strategy is chosen for each variable by evaluating different prefilling strategies. Some of the prefilling strategies are: (1) global mean, (2) local mean, (3) iterative SVD [10], and (4) soft impute [11].
ii. A distinct model is trained on prefilled data for each variable. The input features are first normalized to train models.
iii. Use the trained model to impute missing values for each variable.

2.2.2 Genetic Programming and Lagrange Interpolation

Resende et al. [12] proposed an imputation method which uses Lagrange interpolation and genetic programming. The method builds an interpretable regression model which explores the statistical features such as mean, variance, and auto-correlation. The method uses a genetic programming algorithm called LGPImpute. This algorithm uses Lagrange interpolation for estimating the missing values in a multivariate time series data. Lagrange interpolation is used as a pre-imputation method that regresses the missing values. This allows the algorithm to use all the instances of the data to build regression functions for each attribute. The method uses a multi-criteria fitness function that considers three metrics: mean, variance, and auto-correlation [13].

The mean, given by Eq. (1), is the ratio of the sum of all data points divided by the number of data points.

$$\mu_x = \frac{1}{n}\sum_{i-1}^{n} x_i. \tag{1}$$

Variance is given by Eq. (2) indicating the spread of the data.

$$v_x = \frac{1}{n-1}\sum_{t=1}^{n}(x_t - \mu_x)^2. \tag{2}$$

The auto-covariance expressed by Eq. (3) is represented by the distance of a data point x_t to another data point x_{t+h} where h is the lag.

$$\gamma(h) = \frac{1}{n}\sum_{t=1}^{n-|h|}\left(x_{t+|h|} - \mu_x\right)(x_t - \mu_x), -n < h < n. \tag{3}$$

The auto-correlation expressed by Eq. (4) is defined as the ratio of the auto-covariance of two data points with lag h and with lag zero. Its value lies between −1 and 1.

$$\rho(h) = \frac{\gamma(h)}{\gamma(0)}, -n < h < n. \tag{4}$$

The dataset is pre-imputed using Lagrange's interpolation before the computation of auto-covariance. Lagrange interpolation polynomial given by Eq. (5) estimates new points based on observed data points.

$$L_j(p) = \frac{\pi_{k=0,k\neq j}^{n}(p - p_k)}{\pi_{k=0,k\neq j}^{n}(p_j - p_k)} \tag{5}$$

Equation (6) describes the multi-criteria fitness function where each of its parameters is responsible for maintaining the original characteristics of time series data.

$$F = |u_x - u_{x'}| + |v_x - v_{x'}| + \sum_{h=1}^{H} \frac{\rho(h) - \rho'(h)}{H}. \tag{6}$$

The algorithm aims at minimizing the difference between the statistical values of two successive temporary datasets. The algorithm captures the missing pattern efficiently. However, to guarantee an optimal solution, the algorithm must be run multiple times.

2.2.3 Data Imputation Using Dynamic Bayesian Network (DBN)

Dehghanpour et al. [14] proposed a method used to maintain the correlation between the variables. The Bayesian network is very efficient in modeling multivariate data in the form of a graph that can show the relationship between variables. DBN is used for modeling time series data. It is a combination of hidden Markov model (HMM) and Bayesian network. These models are usually categorized as:

i. Constraint based: The probabilistic relations are first derived from the Markov properties on Bayesian network and then analyzed. The d-separation principle [15] is then used to construct the graph.
ii. Score-based: Each structure is given a score, and the structure with the highest score is selected. Some of the score-based algorithms are Bayesian—Dirichlet equivalent [16], Akaike information criterion, Bayesian information criterion, etc.

Before the data can be fed into DBN, data needs to fulfill some characteristics such as that all the values of all variables must be an integer, non-negative, and must be on the same scale. Once the data satisfies these conditions, it goes through some preprocessing steps in the sequence: (1) Shift all the lowest points in data, (2) Normalize the data, (3) Rounded off the values, and (4) Conversion of all categorical attributes into some integer encoding. Once the data passes through these preprocessing steps, a DBN is built in which the data is converted into graph form. Once the DBN is constructed between variables, support vector regression (SVR) is used for predicting the values of the variables from the set of variables connected in the DBN to the target value. DBN results in a smaller error rate [17].

2.2.4 Deep Learning-Based Time Series Imputation

Recently, deep Learning approaches have been used for the task of time series imputation, and they have shown remarkable results. Deep learning approaches for time series imputation mainly use two of the most common deep learning architectures—generative adversarial network (GAN) and recurrent neural network (RNN). RNN and its variations like LSTM and GRU are very powerful models which are very useful in handling sequential data. GAN are very successful in generation and imputation tasks. Most of these deep learning approaches make use of these architectures in combination with minor variations. Some of these hybrid deep learning methods that are proposed for data imputation include E^2GAN [18], BRITS [19], GRU-D [20], GRUI-GAN [21], and NAOMI [22]. Most of these methods mainly focus on the structure of RNN along with the use of bidirectional RNN, GAN, and auto-encoder structure to enhance the performance of the model. These methods are very efficient in achieving their objective, but they are often inaccurate in cases where the time series data has missing time stamps [23].

3 Discussion

Data imputation techniques are required to handle incomplete datasets. The problem of missing value is inevitable in any data collection and preprocessing process. Several techniques have been used to solve this problem, but very few techniques are able to provide an optimal solution to this problem. This is mainly because most of the techniques make certain assumptions about the incomplete data which is rarely true in the real-world scenario. We have discussed some of the conventional methods which are accurate only in a small percentage of the cases where the missing percentage is very low. These conventional methods are easier to implement but often result in biased estimates as these methods are not able to capture the correlation across the variables and the inherent characteristics of the time series data are not considered under these methods. We have also discussed some of the advanced methods which are relatively more complex than the conventional methods but are able to capture the inherent characteristics of the observed incomplete data. These methods also make certain assumptions about the data, but they are still able to outperform the conventional methods in terms of accuracy.

4 Conclusion

Handling of missing values can be done in a number of ways depending upon the missingness pattern and problems caused by these missing values. An efficient and accurate imputation method is very crucial to the data mining process as it can directly or indirectly affect the results of the process. Rather than simply ignoring the missing values, estimation techniques are probably the best option for handling the missing values as it is often the case that most of the machine learning algorithms are able to perform better when there is enough data to learn. This paper widens the separation line between the various imputation methods from the conventional one to the most recent ones. This overview may provide some of the missing values handling methods with their advantages and disadvantages that should be taken into consideration before arriving at the decision of using a certain method for data imputation.

References

1. Chatfield C (2003) The analysis of time series: an introduction, 6th edn. Chapman and Hall/CRC
2. Roderick J, Little A, Rubin DB (2002) Statistical analysis with missing data, 2nd edn. Wiley
3. Roderick J, Little A (1988) A test of missing completely at random for multivariate data with missing values. J Am Stat Assoc 83
4. Allison PD (2001) Missing data. Sage publication
5. Bishop C (2002) Pattern recognition and machine learning. Wiley
6. Schafer JL (1997) Analysis of incomplete multivariate data. Chapman & Hall, London, U.K.
7. Hermsen J, Koblitz H, Klerk GJ (2004) Book Review. Edinb Med J 18:256–262
8. Little RJA, Rubin DB (2002) Statistical analysis with missing data, 2nd edn. Wiley
9. Zhang X et al (2019) XGBoost imputation for time series data. In: IEEE international conference on healthcare informatics (ICHI), pp 1–3
10. Hastie T et al (2015) Matrix completion and low-rank SVD via fast alternating least squares. J Mach Learn Res JMLR 16:3367–3402
11. Mazumder R, Hastie T, Tibshirani R (2010) Spectral regularization algorithms for learning large incomplete matrices. J Mach Learn Res 2287–2322
12. Resende D, Santana A, Lobato F (2016) Time series imputation using genetic programming and Lagrange interpolation. In: 2016 5th Brazilian conference on intelligent systems (BRACIS), pp 169–174
13. Garcıa JF, Kalenatic D, Amilcar LopezBello C (2010) In: An evolutionary approach for imputing missing data in time series. J Circ Syst Comput 19:107–121
14. Dehghanpour K et al (2016) Agent-based modeling in electrical energy markets using dynamic Bayesian networks. IEEE Trans Power Syst 4744–4754
15. Geiger D, Verma T, Pearl J (1990) Mach Intell Pattern Recogn 139–148
16. Koski T, Noble J (2012) A review of Bayesian networks and structure learning. Math Applicanda 40:51–103
17. Susanti SP, Azizah F (2017) Imputation of missing value using dynamic Bayesian network for multivariate time series data. In: International conference on data and software engineering, pp 1–5
18. Luo Y, Zhang Y, Cai X, Yuan X (2019) End to end generative adversarial network for multivariate time series imputation. In: Twenty eighth international joint conference on artificial intelligence, IJCAI 2019, Macao, China, pp 3094–3100

19. Cao W, Wang D, Li J, Zhou H, Li L, Li Y (2018) BRITS: bidirectional recurrent imputation for time series. In: Advances in neural information processing systems 31: annual conference on neural information processing systems 2018, Canada, pp 6776–6786
20. Che Z, Purushotham S, Cho K, SontagD, Liu Y (2018) Recurrent neural networks for multivariate time series with missing values. Sci Rep 8(1):6085
21. Luo Y, Cai X, Zhang Y, Xu J, Yuan X (2018) Multivariate time series imputation with generative adversarial networks. In: Advances in neural information processing systems 31, NeurIPS 2018, Canada, pp 1603–1614
22. Liu Y, Yu R, Zheng S, Zhan E, Yue Y (2019) NAOMI: non-autoregressive multiresolution sequence imputation. In: Advances in neural information processing systems 32: annual conference on neural information processing systems 2019, NeurIPS 2019, Vancouver, BC, Canada, pp 11236–11246
23. SongS, Cao Y, Wang J (2016) Cleaning timestamps with temporal constraints. PVLDB

Chapter 43
Unsupervised Land Cover Classification on SAR Images by Clustering Backscatter Coefficients

Emily Jenifer and Natarajan Sudha

1 Introduction

Remote sensing is the process of acquisition of the earth images via satellites and unmanned aerial vehicles without having direct contact with the surface. From the acquired images, one can obtain the terrain details associated with land use, land cover, tree species identification, oil spill in ocean, minerals mapping, management of disasters due to natural calamities and many more.

Many conventional image processing techniques have been employed in the past decade to extract the maximum needed information from the remotely sensed images [1] and [2]. In recent years, machine learning has become an effective replacement for the conventional approaches [3]. The intensive tasks of the remote sensing applications are efficiently performed by the machine learning algorithms [4].

1.1 Synthetic Aperture Radar

SAR images come under the classification of active remote sensing. Active sensor has its own energy to sense the scene of surveillance. The advantages of SAR are its all-weather usage, penetration of clouds and other atmospheric dust, day and night use and sensitivity to rough and smooth surfaces. SAR images do have its disadvantage in terms of speckle noise which leads to difficulty in interpretation thus, few preprocessing needs to be performed to eliminate such noise [5]. The backscattered values from the SAR sensors are recorded for giving the needed information. The

E. Jenifer · N. Sudha (✉)
School of Computing, SASTRA Deemed University, Thanjavur, Tamilnadu 613401, India
e-mail: sudha@cse.sastra.edu

E. Jenifer
e-mail: emilyjenifer@sastra.ac.in

A. Sheth et al. (eds.), *Intelligent Systems*, Algorithms for Intelligent Systems,
https://doi.org/10.1007/978-981-16-2248-9_43

backscattering coefficient σ represents the average value of the radar reflection per unit area as given in Eq. 1. To normalize the backscattering value, it is converted to,

$$\sigma = n \frac{\text{Received}}{\text{Emitted}} \tag{1}$$

where n is a calibration coefficient. The pixels in the SAR images are the result of the physical process of the measurements. Thus, the SAR images can be interpreted by understanding the physical process.

2 Related Works

Considerable work is available in the literature on using the unsupervised clustering-based segmentation for the analysis of the remotely sensed data. In [6], the authors proposed a new clustering algorithm with K-means to produce two different clusters for a time series data, and the difference between them was used to generate the change map. The change map obtained was smooth and effective when compared to other traditional methods like principal component analysis. In the literature [7], the authors proposed a clustering based on EM algorithm to classify the terrain area. The result was efficient for classifying two classes (wet soil and water) out of four. More recent work [8] proposes a segmentation based on the log estimations of the SAR image. The unsupervised image segmentation was helpful in segmenting the coast, land and sea areas effectively in terms of accuracy and cross region fitting when compared to the existing segmentation algorithm.

In this paper, the unsupervised land cover classification is done using two popular clustering algorithms. The results are compared and analyzed.

3 Data Collection and Preprocessing

This section discusses data collection and processing techniques.

3.1 Data Collection

The sentinel-1 SAR data can be downloaded from Alaska satellite facility (ASF) [9]. It operates with two polar orbiting satellites, namely sentinel-1A and 2A, which provides the images of the entire earth for every 6 days. These satellites have the sensor for C-Band and an operating frequency of 5.405 GHz. The data is downloaded with interferometric wide (IW) swath, ground range detected (GRD), and it is in full resolution. The IW swath has a 250 km swath width and 5 m × 20 m ground resolution. The acquired image is a dual (VV/VH) polarized image. GRD data can be used for processing by applying certain basic preprocessing [10].

The preprocessing was done in SNAP, which is a common Sentinel-1 toolbox. The basic preprocessing steps are listed below.

3.2 Data Preprocessing

Subset the Data. The acquired data covers a larger scene; thus, by specifying the latitudes and longitudes, the region of interest can be obtained for processing. The data used in this paper is of N-10.415, W-79.328, S-10.35 and E-79.397 with the acquisition date of November 8, 2018. It covers the region around Pudukkottai district of Tamilnadu which is a southern state of India.

Add Orbit File. Adding the orbit file to the image is necessary because updating the orbit metadata to the file helps in acquiring the detailed position and speed information of the satellite.

Radiometric Calibration. The pixel values are converted into normalized backscattered values σ. These calibrated values are very much helpful in the process of interpreting the data.

Geometric Correction. The SAR image downloaded has multiple view angles, thus distortion will be caused. This distortion can be removed by the geometric correction. Range Doppler terrain correction is applied to remove the shadows and elevations in pixels. It will result in images with dimensions that are closer to scaled down real-world dimensions.

Speckle Noise Removal. SAR images have granular noise due to the scattering of the received and emitted waves from the sensor [11]. Thus, the noise should be removed for better visualization. Gamma map single product speckle filter operator of window size 7 × 7 was used to produce noise-free data.

Conversion to dB and Exporting. The final and important step is to convert the backscattering sigma values to dB (decibels) using a logarithmic conversion as given in Eq. 2. For easier calculations and representations, the σ values are converted to logarithmic scale,

$$\sigma = 10.\log_{10}(\sigma)\,\text{dB}. \tag{2}$$

Finally, the processed and converted image can be exported as GeoTIFF files, and it has all the geographic data lodged within it. The preprocessed images are given below (see Fig. 1). Further algorithms for analysis are applied to the preprocessed image.

(a) Radiometric corrected (b) Geometric corrected (c) Noise removed

Fig. 1 Preprocessed images

4 Unsupervised Clustering

Clustering is an unsupervised process of separating data into clusters which are similar in nature. It splits into clusters based on the similarity calculated by a distance metric [12]. Of all the clustering algorithms, the K-means and EM clustering are popular. Both the K-means and EM algorithm perform the iterative process to generate the clusters. In K-means, the clusters are formed by calculating the distance between the featured data; while in EM clustering, it analyses the statistical inferences underlying the data [13] and [14].

4.1 K-Means Clustering

The K-means algorithm uses the probabilistic approach, and it calculates the sum of intercluster variance and minimizes it [15] and [16], since the SAR data that we processed do not have any ground truth. K-means can be used for its clustering without any prior assumptions of the data present in it.

K-means for k clusters and n data points are expressed in the Eq. 3,

$$K = \sum_{k=1}^{j} \sum_{i=1}^{n} Z_j^i \mid x_i - \mu_j \mid^2 . \tag{3}$$

where Z_j^i takes value 0 or 1. If ith point belongs to kth cluster, then the value becomes 1 else 0. μ_j is the mean of jth cluster, and x_i is the ith data point in d dimension.

4.2 EM Clustering

It is a statistical approach which maximizes the likelihood of the data arrival from the original set. EM clustering finds the distribution underlying the dataset and uses it for clustering. It has E-steps for updating and M-steps for maximization.

EM clustering is given by the Eq. 4,

$$f\left(\frac{k}{\theta}\right) = \sum_{j=1}^{i} \tau_i \aleph\left(\frac{k}{\theta_j}\right). \tag{4}$$

where $\aleph\left(\frac{k}{\theta_j}\right)$ is individual Gaussian density function and τ_i is the weight for Gaussian distribution.

5 Experimental Results and Analysis

Experiments were carried out to classify the land cover using the two clustering algorithms. Initially, RGB composite and Google earth images are taken as reference to classify water, urban areas, vegetation and tall trees.

5.1 RGB Colour Composite

SAR images are difficult to interpret without any processing, thus adding colours (RGB) can make the image more meaningful. The backscattering values from the vegetative area depending upon its moisture content and height represent the variation of colour from green to cyan [17]. Water (wet) areas render blue colour and barren dry land will exhibit Persian blue range in RGB composite due to low backscatter [18]. The formula used for RGB colour composite is:

R—Sigma0_VV; G—Sigma0_VH; B—Sigma0_VV/Sigma0_VH.

The RGB colour composite and the Google earth image are displayed in Fig. 2 and can be compared visibly.

5.2 K-Means Clustering

K-means clustering was performed on the acquired SAR image of size 768 × 723. With 60 iterations, the K-means algorithm was able to generate output with 2 classes and 4 classes. The clustering results when $K = 2$ and $K = 4$ are shown in Fig. 3. The frequency and the colours of the clusters ($K = 4$) are listed in Table 1.

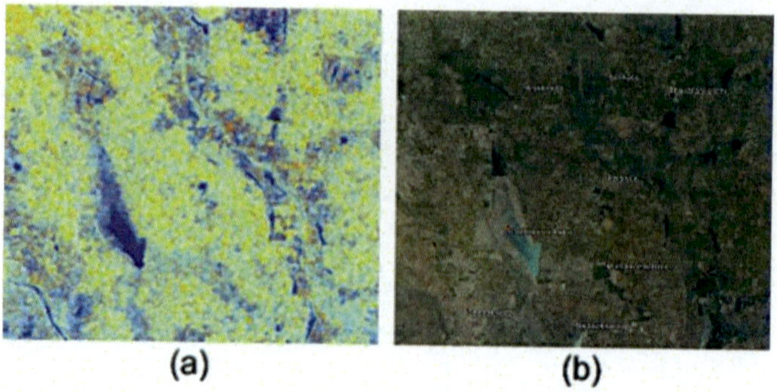

Fig. 2 **a** RGB colour composite. **b** Google earth image of the target area

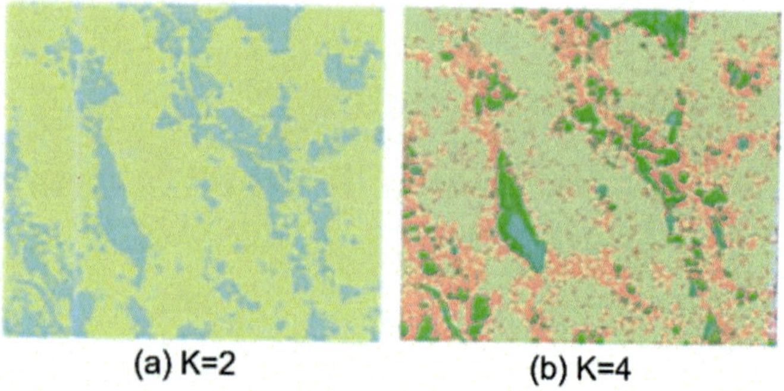

Fig. 3 Results of K-means clustering

Table 1 Frequency of the cluster data points in K-means ($K = 4$)

Clusters	Colour	Frequency (in %)
Urban area	Yellow	45.489
Bare land	Peach	35.171
Vegetation	Green	13.404
Water	Blue	7.936

5.3 *EM Clustering*

EM clustering was also performed on the same SAR for the cluster number to be kept as 2 and 4 and performed for the same 60 iterations. A comparison of the clustering when 2 clusters and 4 clusters is shown in Fig. 4. The frequency and the colours of the 4 clusters are listed in Table 2.

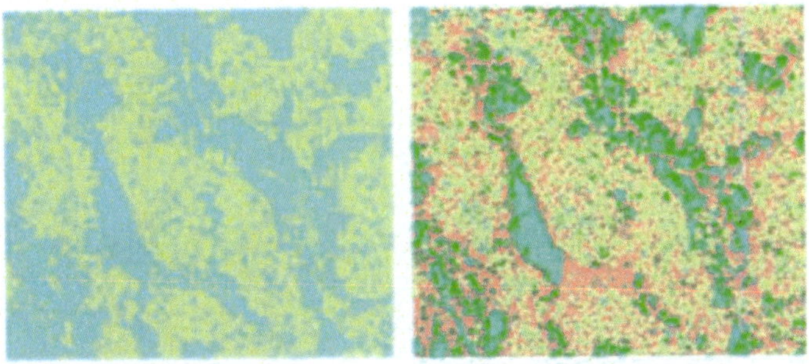

Fig. 4 Results of EM clustering

Table 2 Frequency of the cluster data points in EM algorithm (4 classes)

Clusters	Colour	Frequency (in %)
Urban area	Yellow	25.152
Bare land	Peach	39.834
Vegetation	Green	15.010
Water	Blue	22.004

5.4 *Comparison Study*

A comparison study on the K-means and EM estimation is done in the target area. In visual comparison, K-means pictured exactly the lake Chelli Kuruchi and the nearby urban area much better than the EM algorithm.

The study area's σ_0 value was fed to the two algorithms and the resulting unsupervised clusters were plotted in scatter-plot diagrams (Figs. 5 and 6). The predicted cluster distance was taken into account for cluster validation purpose, and outputs are shown in Table 3.

The cluster validation performed by Davies-Bouldin (DB) index [19] is given in Eqs. 5 and 6. It is performed by calculating the average similarity between clusters. It is an internal cluster validation with no ground truth. It evaluates the clusters by calculating the average of all cluster similarities. The smaller the index, the performance of the algorithm can be validated as a better one.

DB index is calculated as the average similarity index of clusters C_i. The number of clusters varies from 1 to Z. Similarly,

$$R_{ij} = \frac{d_i + d_j}{S_{ij}} \quad (5)$$

d_i—average distance between each point of cluster i and the cluster centroid; d_j—average distance between each point of cluster j and the cluster centroid; S_{ij}—distance between the cluster centroids i and$j$. Therefore,

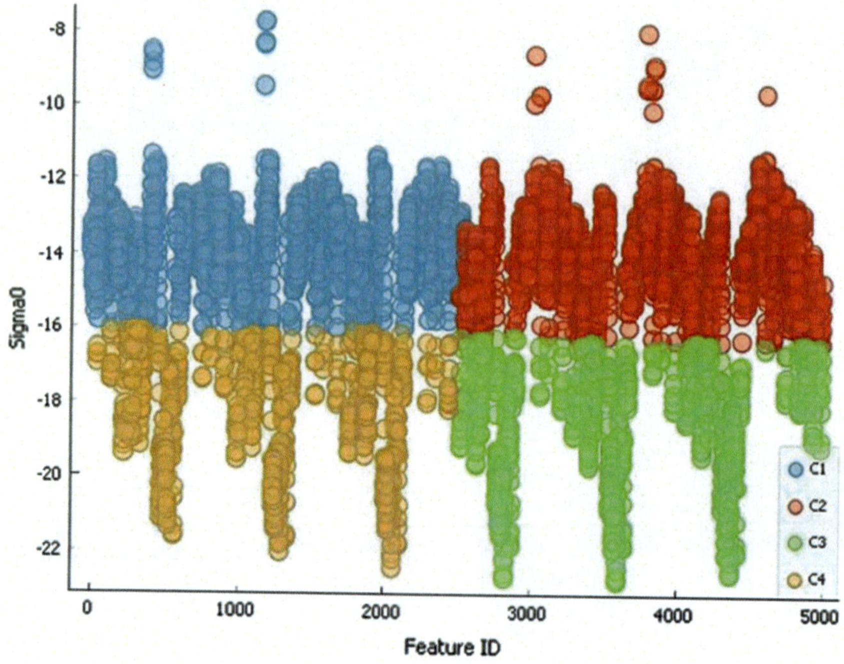

Fig. 5 Four clusters formed by EM algorithm and plotted as scatter plot

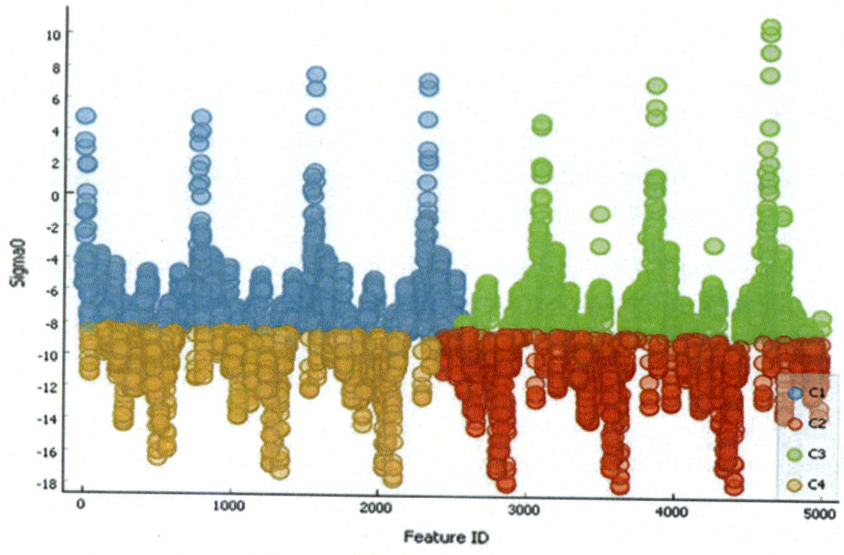

Fig. 6 Four clusters formed by K-means and plotted as scatter plots

Table 3 Comparative results of K-means and EM expectation for 4 clusters

Algorithm	DB index	Computational time (s)
K-Means	0.985	49
EM	1.379	57

$$DB = \frac{1}{Z} \sum_{i=1}^{Z} \max_{(i \neq j)} R_{ij} \tag{6}$$

6 Conclusion

The land cover classification was performed successfully for the target study location of the SAR image. The cluster-based analysis was carried out, and two different clustering algorithms were analysed. As a result, K-means clustering turns out to be the better performing algorithm than the EM estimation on the basis of DB index, computational time and visual comparison. Since the SAR data is high dimensional in nature, deep neural networks are expected to perform better for effective SAR data interpretation in various applications such as agriculture monitoring, land use land cover classification and many others. Future work will focus on deep learning-based approaches.

References

1. Blaschke T, Burnett C, Pekkarinen A (2004) Image segmentation methods for object-based analysis and classification. In: Remote sensing image analysis: including the spatial domain. Springer, Dordrecht, pp 211–236
2. Du P, Liu P, Xia J, Feng L, Liu S, Tan K, Cheng L (2014) Remote sensing image interpretation for urban environment analysis: methods, system and examples. Remote Sens 6:9458–9474
3. Lary DJ, Alavi AH, Gandomi AH, Walker AL (2016) Machine learning in geosciences and remote sensing. Geosci Front 7:3–10
4. Hansch R, Schulz K, Sorgel U (2018) Machine learning methods for remote sensing applications: an overview. In: Earth resources and environmental remote sensing/GIS applications IX. International Society for Optics and Photonics, 10790
5. Remote sensing and SAR radar images processing. ESA Earth online. https://earth.esa.int/c/document_library. Accessed 25 Sept 2020
6. Zhang X, Li Z, Hou B, Jiao L (2011) Spectral clustering based unsupervised change detection in SAR images. In: 2011 IEEE international geoscience and remote sensing symposium. IEEE, pp 712–715
7. Kayabol K, Krylov VA, Zerubia J (2012) Unsupervised classification of SAR images using hierarchical agglomeration and EM. In: International workshop on computational intelligence for multimedia understanding. Springer, Berlin, 7252 LNCS, pp 54–65
8. Nogueira FEA, Marques RCP, Medeiros FNS (2020) SAR image segmentation based on unsupervised classification of log-cumulants estimates. IEEE Geosci Remote Sens Lett 17:1287–1289

9. Alaska Satellite Facility. https://search.asf.alaska.edu/. Accessed 5 May 2020
10. Filipponi F (2019) Sentinel-1 GRD preprocessing workflow. In: Multidisciplinary digital publishing institute proceedings, p 18(11)
11. Lee JS, Jurkevich L, Dewaele P, Wambacq P, Oosterlinck A (1994) Speckle filtering of synthetic aperture radar images: a review. Remote Sens Rev 8:313–340
12. Jung YG, Kang MS, Heo J (2014) Clustering performance comparison using K-means and expectation maximization algorithms. Biotechnol Biotechnol Equip 28:S44–S48
13. Abbas OA (2008) Comparisons between data clustering algorithms. Int Arab J Inf Technol 5(3)
14. Alldrin N, Smith A, Turnbull D (2003) Clustering with EM and K-means. University of San Diego, California, Technical report, pp 261–295
15. Hamada MA, Kanat Y, Abiche AE (2019) Multi-spectral image segmentation based on the K-means clustering. Int J Innov Technol Explor Eng 9:1016–1019
16. Xie X, Zhao J, Li H, Zhang W, Yuan L (2012) A SPA-based K-means clustering algorithm for the remote sensing information extraction. In: 2012 IEEE international geoscience and remote sensing symposium. IEEE, pp 6111–6114
17. Fung A (1979) Scattering from a vegetation layer. IEEE Trans Geosci Electron 17:1–6
18. Amitrano D, Di Martino G, Iodice A, Riccio D, Ruello G (2016) RGB SAR products: methods and applications. Eur J Remote Sens 49:777–793
19. Liu Y, Li Z, Xiong H, Gao X, Wu J (2010) Understanding of internal clustering validation measures. In: 2010 IEEE international conference on data mining. IEEE, pp 911–916

Chapter 44
Cloud Algorithms: A Computational Paradigm for Managing Big Data Analytics on Clouds

Syed Owais Bukhari

1 Introduction

There is not even a slightest doubt in the fact that cloud computing has induced a revolutionary transformation in the process of big data analytics. This applies to the domains of business intelligence, business analytics, and business process outsourcing. The services like storage, infrastructure, servers, intelligence, and analytics can now be delivered on demand. This is definitely what we like to call the next level of digital service delivery that has sparked off this revolutionary transformation. Let us first examine the broad contours of this computing revolution [1, 2].

When we delve into a specific cloud environment, we observe that a voluminous quantity of processes and activities are being executed simultaneously. To maintain this data flow uninterrupted, resource allocation comes first on the priority list. This is where Johnson's algorithm steps in. It enables us to monitor, synchronize, and coordinate activities and processes between working cloud environs. The effectiveness of the algorithm lies in the fact that it paves the way for a switching mechanism for activity execution between clouds. It even enables the hypothetical creation of dummy clouds so that the time lag for processes between clouds is reduced significantly. For the provision of efficient allocation and consistency, the data and services are stored in numerous sites or locations. These resources at specific locations are channelized using a given network-public or private or hybrid [3, 4].

This manuscript is developed in four sections. The starting section probes the limitations, challenges, and shortcomings of the cloud computing ecosystem. The second section provides the literature review of the current approaches, processes, and mechanisms to handle cloud data and services. The third section provides the details of the proposed cloud algorithm, that is Johnson's algorithm, including its architectural framework and working mechanism. The fourth section provides some

S. O. Bukhari (✉)
School of Engineering Science and Technology, Jamia Hamdard, New Delhi 110062, India

A. Sheth et al. (eds.), *Intelligent Systems*, Algorithms for Intelligent Systems,
https://doi.org/10.1007/978-981-16-2248-9_44

quantitative conclusions and argumentative discussions and identifies a genre of future trends in cloud computing.

2 A Brief About the Limitations and Challenges of the Cloud Computing Ecosystem

The arrival of the robust ecosystem of cloud computing has paved the way for a new architecture in which organizations can offer their solutions over the Internet and can be accessed through an on-demand mechanism of service delivery. While it is difficult to negate the multitude of benefits, especially for micro, small and medium enterprises, such as the lower costs (of installing a complete infrastructure locally) and faster deployment, there arise several shortcomings to this emerging ecosystem that are briefly described here. The special concerns of privacy and security of data and adjoining structures are two of the notable reasons that need to be countered while navigating across the platform of cloud computing. Given the existing distribution of multiple resources by many customers across various platforms, the domains of security and privacy are further challenged. The clauses of security rest with the agent providing cloud services and these lie beyond the reach of the company, thereby increasing susceptibility to a possible digital invasion. Critical dependency on service providers becomes an unavoidable reality as each one uses their specific infrastructure, hardware, and software which make migration between providers a grey digital area. The unavailability of service on demand (as the case may be) is the sole responsibility of the provider. This is because those who are responsible for providing the solutions in the cloud environment can land in a compromising situation with clients who want a 24/7 functional service as and when required [5]. The relative absence of a documented service agreement is another limitation of a client dealing with cloud services. This becomes the bone of contention and is responsible for the uncertainty relating to the features that the provider has to make available across a simulated environment within the stipulated time. It is thus clear that this is what prevents easier migration between different providers in the cloud environment. Apart from this, the efficiency of various components of cloud architecture is drastically impacted at times of high load as the limited resources are shared among numerous clients at the same time. The growing need for data storage space amplification may not be a possibility at times as the cloud service vendors have their database connectivity solutions and additional space provisioning may not become possible in a quick, flexible, and easy way. This is what may become a limiting factor for cloud service providers. Moreover, given the simultaneous deployment of resources by active clients, one small error may have sharp consequences for all online clients, thereby making them devoid of all cloud services. At times, a digital intrusion by a worm may occur, affecting all clients who are onboard over a particular cloud environment [6]. All this is sufficient to lay the initial framework for cloud algorithms which may prove to be a nouveau digital remedy in the effective and secure deployment of various resources in the ecosystem of clouds.

3 Motivation

There are a variety of computational tasks where computing service delivery from many clouds are required by the clients at the same time. In such a scenario, it becomes extremely vital to have coordinating mechanisms for providing a perfect synchronization between numerous participating clouds. Whenever a cloud takes a while to respond due to overloading or becomes unavailable temporarily, a similar kind of cloud known as the dummy cloud can be created or cloned, and the needs of the client can be serviced using that dummy cloud. In many situations, a user may require the services and requests for execution of a specific task and the need to devolve all the resources to this client may not be a good idea. All we require to do in such a case is to choose only the required services and resources of the clouds under operation. This can become feasible once we can identify and allocate the resources from multiple clouds and devolve the necessary ones toward the execution of a particular operation. Motivated by these factors, the current manuscript pops up the concept of cloud algorithms (CA) in general and Johnson's algorithm in particular which can prove to be the real panacea to manage big data analytics using cloud services.

The primary concern when it comes to cloud computing is both the security and synchronization of voluminous databases stored at numerous locations. This is because many copies of identical data sets are available at easily accessible but remote locations. What becomes necessary is that if any sort of update is carried out on one copy, it should be reflected in all other concurrent and remote copies. This means any time lag between the operating resources may affect the entire cloud environment. This is where cloud algorithms would come to our rescue under such circumstances. Just like the concept of advanced and computational algorithms to query and process the relational databases, cloud algorithms can be used to process, query, and manage cloud databases as well as other cloud-based services and resources. The proposed cloud algorithms in general and Johnson's algorithm, in particular, will not only work for cloud databases but will also work for cloud resources, cloud management, cloud operations, etc.

4 Related Works

Cloud computing has been one of the most sought-after areas of research for data scientists worldwide. Cloud technology has emerged as a savior of modern-day computing technology as it has provided a sustainable computing environment with the help of advanced algorithms. This section presents some of the significant works done in the areas of advanced algorithms as well as cloud computing domains covering the aspects like privacy, functionality, security, safety, reliability, rationality, scalability, efficiency, effectiveness, interoperability, and ease of use. The authors in [1] presented an overview of cloud computing and other related technologies. They described "cloud computing" as on-demand dispatch of software and hardware as a provision of digital service over the Internet to the users. This ensures a highly cost-effective yet lucrative business model for start-ups and enterprises. The concept of public and private cloud vis-a-vis hybrid cloud was explained highlighting the advantages and limitations of both types of clouds. Furthermore, the paper also discussed opportunities with cloud computing technology with a special focus on divergent features. The authors in [2] discussed the architecture along with the neural schema of the cloud computing paradigm and the open research challenges associated with its wide implementation and usage. The paper starts by highlighting the notable features and characteristics of cloud computing. In addition to this, the paper describes several commercial cloud service providers and the various types of services offered by them. The authors in [3] give an industrial cloud architecture and the corresponding hardware schema. Moreover, the paper discusses the models for delivering IT-as-a-service in its first place. In the end, a broad comparison of different commercial cloud service domains is laid down in the paper.

The authors in [4] pictured a comprehensive architecture for managing data stored in the clouds. Their paper featured a "three schema architecture" and "three-levels object-oriented database" model. The authors discussed three major perspectives relating to various types of data processing centers. Moreover, the authors in [5] proposed a concept that could act as a working paradigm for managing data stored on the clouds. The authors stressed on four key operations on cloud databases, namely "addition," "deletion," "distribution," and "management" of databases. The paper also featured the description for mining the data items from the clouds with an illustration. In addition to this, the authors described the handling of unstructured data which may even be in its raw form using an illustration from graph mining and performing a series of operations on an imaginary database. The author in [6] laid down the concept of "Big Data" for performing analytics in terms of data science and computational science. The presentation of operations for "data modeling, analysis, and management" for effectively handling big data structures deserves a notable mention. The authors in [7, 8] discussed the various aspects of data management in a cloud ecosystem.

The authors stressed on working out a novel management system involving databases for workstations operating via the cloud ecosystems. In [9, 10], numerous other approaches for data-intensive applications across the cloud ecosystem are discussed in deeper detail. The urgency to carve out a kit of tools to handle the data flow in a cloud environment is elaborately described. The work presented in [11, 12] shows the various architectural design choices for framing an efficient data management system that is quite reliable to provide services in a cloud ecosystem. The plethora of applications of cloud computing has attracted a large number of industries to opt for cloud-based services. Given this voluminous amount of data available in the cloud ecosystem, there arises a need for efficient data management and processing techniques that can prove to be a one-stop solution for businesses opting for this model. In the literature survey presented in this section, we have attempted to pen down a holistic survey of the researches in the area of cloud computing and advanced algorithms and the relation between the two. We have also tried to analyze similar approaches for data management on the cloud. Various kinds of approaches and models have been presented in the past to handle the increasing flow of data in the cloud [13, 14]. Having said that, there has been no significant work related to the drafting of an algorithmic paradigm for effective working of the cloud ecosystem. This calls for an effort to bridge the research gaps in the previous works and apply Johnson's algorithm for handling data in the cloud environs. Let us take this up in deeper detail by first presenting the general algorithmic procedure and then applying the same via a working model.

5 Cloud Algorithms: The Proposal

In this module, we discuss the problem of determining the sequence (order) in which several processes should be performed on different clouds to make effective use of available facilities and achieve greater output. If there are n processes that are to be performed on m different clouds, then the problem on hand is to frame a sequential algorithm for the execution of processes, that lowers the duration of relay, which means, the duration from the beginning of the starting process to the end of the final process. Suppose that 2 or 3 processes are to be processed on 2 or 3 clouds. Then, the process sequencing can be done by the method of enumeration. If, however, the number of processes and/or clouds increases, then the problem becomes complicated, and the method of enumeration is not suitable. Let us suppose that we work on n clouds and m processes. This means that we are actually executing $(n!)\ m$ types of various processes. When $n = 9$ and $m = 9$, we may actually be running $(9!)9 =$ 81,000,000,000 types of different processes. This seems to be a herculean task to obtain various types of execution formats and choosing the required one from all types. Hence, this calls for a dire need to come up with a sequencing algorithm for clouds.

5.1 *Principal Assumptions*

- The processing time on each cloud is known.
- The time for execution of a process does not depend on the format of processes executed.
- No cloud can process more than one process simultaneously.
- The relay time in changing from one cloud to the next may be ignored.
- Every process that begins on a cloud needs execution up to the end on that very cloud.

5.2 *Various Procedures for Determining Process Sequencing Tasks*

Different ways of process sequencing tasks are encountered in the cloud computing ecosystem. All sequencing problems cannot be solved using the same model. Here, we consider the following four types of process sequencing problem: [15–19].

I. n processes need execution on two clouds, suppose, cloud A and cloud B according to format AB. We comprehend that each process needs execution first on cloud A and afterward on cloud B.

II. n processes need execution on m clouds in the assigned format.

5.3 *The Working Model*

The working model is described in the following stages.

5.3.1 Processing of *n* Processes Through 2 Clouds

Suppose that n processes need execution on two clouds, suppose, A and B. That very process needs execution via the parallel series of commands in the similar way. When a process is executed on cloud A, it is sent to cloud B. When cloud B is engaged, in that case, the process gets into the transition phase. Every process from transition phase is sent to cloud B [20–23].

Consider

Ai = duration of computational flow for jth process on cloud A.
Bi = duration of computational flow for jth process on cloud B.
Tn = total duration of relay.

The task on hand is to find the way in which m processes can be executed via clouds A and B, so that the total duration of relay is least.

Johnson's Algorithm

Step 1. Obtain the least duration of computational flow in all Ai's and Bi's. When we have Ar, we can execute the Rth process in the beginning. When we have Bs, we can execute the Sth process at the end [24–26].
Step 2. Whenever we get a draw or tie while choosing the lowest of all the computations, we proceed in such a situation with the following procedure:

(i) When the least of the duration of computational flow is Ar, that may be approximated to Bs, in that case, least of $(Ai, Bi) = Ar = Bs$, we strive to execute the rth process in the beginning and sth process at the end.

(ii) When the least of $(Ai, Bi) = Ar$, however, $Ar = Ak$, which means, we get a tie for lowest entry among Ai's, we may choose the one at will [27–29].

(iii) When the least of $(Ai, Bi) = Bs$, however, $Bs = Bt$, which means, we obtain a tie for the least entry in Bi's, we may choose the one at will [30–32].

Step 3. Thereafter, we do away with the process that we initially computed and execute steps first and second till we get the required solution [33–35].

Let us frame a pseudo code to obtain the computational sequencing procedure using Johnson's algorithm.

```
# processing from a shown array
# a function to place the processes taking second
# placing arguments array and scheduling the processes
  printprocessScheduling:
        # sequence which is given
        m = seq(array)
        # Sorting given tasks in
        # top down order
        for j in (m):
              for k in (m - 1 - i):
                    if array[k][2] < array[k + 1][2]:
array[k], array[k + 1] = array[1 + 1], array[k]
        # Keeping a vigil on open slots
        solution = [wrong] * u
        # To embed solution (placing of task)
        task = ['-1'] * u
        # Repeating across various tasks
        for j in order(length(array)):
              # Obtaining an open slot for given task
              # starting from last entry
for k in range(lowest(u - 1, array[j][1] - 1), -
1, -1):
                    # obtaining entry
                    if solution[k] is wrong:
                          solution[k] = correct
                          task[k] = array[j][0]
        write(task)
```

Suppose that there are five processes, and all of these need to be executed on two clouds A and B in the manner AB. The duration of computational flow is briefed in the next table (Table 1).

Table 1 Processing time of clouds *A* and *B*

Process	Cloud *A*	Cloud *B*
1	6	3
2	2	7
3	10	8
4	4	9
5	11	5

Table 2 First reduced set of processing times

Process	Cloud *A*	Cloud *B*
1	6	3
3	10	8
4	4	9
5	11	5

Table 3 Second reduced set of processing times

Process	Cloud *A*	Cloud *Bs*
3	10	8
4	4	9
5	11	5

Demonstration 1: Let us determine a sequence where such processes need to be executed to lower the total duration of computational flow.

Procedure: The lowered duration of computational flow shown in table is 2, that refers to process 2 on cloud A. Henceforth, the allocation of processes will start as 2. Thereafter, we do away with process 2. The next tally of duration of computational flow is shown as follows; (Table 2).

Now, the least duration of computational flow is 2.999 for process 1 on cloud B. As such, this process may be executed at the final stage. The sequence of processes till this point would be 2 1. After removal of process 1, the condensed tally of duration of computational flow is as follows: (Table 3).

Likewise, the duration of computational flow is obtained as 2 4 3 5 1. Based on this optimal sequence, the minimum elapsed time is obtained from the following table (Table 4) as 36 ms.

Further, time for which no task is processed on cloud $A =$ total duration of relay—duration when the final task is sent to cloud $A = 35.999 - 33.999 = 2.000$ microseconds. Duration for which no task is executed on cloud $B =$ duration during which the starting task is executed on cloud $A +$ (duration for the ith task to start on cloud B) − (duration for the $(i - 1)$th process to get executed on cloud B). Therefore, duration for which no process is executed on cloud $B = 1.9999 + (8.999 - 8.9999) + (17.999 - 17.9999) + (26.999 - 25.999) + (32.9999 - 31.9999) = 3.9999$ ms [36–40].

Table 4 Table for obtaining minimum elapsed time

Process	Cloud *A*	Cloud *A*	Cloud *B*	Cloud *B*
	Time in	Time out	Time in	Time out
B or 2	0.001	1.99	1.9	8.99
D or 4	1.99	5.99	8.99	17.99
C or 3	5.999	15.99	17.99	25.99
E or 5	15.99	26.99	26.99	31.99
A or 1	26.99	32.99	32.99	35.99

Table 5 Processing time of clouds *A* and *B*

Process	Cloud *A*	Cloud *B*
1	3	8
2	12	10
3	15	10
4	6	6
5	10	12
6	11	1
7	9	3

Table 6 First reduced set of processing times

Process	Cloud *A*	Cloud *B*
1	3	8
2	12	10
3	15	10
4	6	6
5	10	12
7	9	13

Demonstration 2: There are seven processes, all of them require to be transitioned across clouds *A* and *B* in the way *AB*. The duration of computational flow is given in the adjoining table: (Table 5).

Let us arrive at the order of these processes that would lower the duration of the relay period. We strive to find the total elapsed duration and idle clouds *A* and *B*.

Procedure: The smallest duration of computation is 1 microsecond for process 6 on cloud *B*. Thus, process 6 will be processed last on cloud *A* as shown below: ------ 6.

The reduced set of processing times becomes: (Table 6).

There are two equal minimal values: processing time of 3 ms for process 1 on cloud *A* and processing time of 3 microseconds for process 7 on cloud *B*. According to Johnson's rules, process 1 is scheduled first and process 7 next to 6 as shown below:

Table 7 Second reduced set of processing times

Process	Cloud *A*	Cloud *B*
2	12	10
3	15	10
4	6	6
5	10	12

Table 8 Third reduced set of processing times

Process	Cloud *A*	Cloud B
2	12	10
3	15	10
5	10	12

1 ---- 7 6.

The reduced set of processing times becomes; (Table 7).

Again, there are two equal minimal values; processing time of 6 microseconds for process 4 on cloud A as well as on cloud B. We may choose arbitrarily to process 4 next to 1 or next to process 7 as shown below:

1 4--- 7 6 or 1--- 4 7 6.

The reduced set of processing times becomes; (Table 8).

There are three equal minimal values: processing time of 10 ms for process 5 on cloud A and for processes 2 and 3 on cloud *B*. According to rules, process 5 is scheduled next to process 4 in the first or next to process 1 in the second schedule. Process 2 then is scheduled next to process 7 in the first schedule or next to process 4 in the second schedule. The optimal sequences are shown below:

1 4 5 3 2 7 6 or 1 5 3 2 4 7 6.

The calculation of both sequencing for total elapsed time for clouds *A* and *B* are shown in the following tables: (Table 9).

From the above tables, we see that the total elapsed time in both sequencing is 67 ms, and idle time for cloud *A* is 1 ms; idle time for cloud *B* is 17 ms.

5.3.2 Processing *n* Tasks Through *m* Clouds

In the following paragraph, we elaborate on the application of Johnson's procedure for scheduling *n* processes on 3 primary clouds *A*, *B*, and *C* corresponding to the sequence *ABC*. This list of processes with their duration of processing on three clouds *A*, *B*, and *C* are given below; (Table 10).

The approximated value to the corresponding sequence may be computed if one or both of the corresponding situations are met: [41–43].

Table 9 Tables for obtaining minimum elapsed time

Process	Cloud *A*	Cloud *A*	Cloud *B*	Cloud *B*
	Time in	Time out	Time in	Time out
A or 1	0.99	2.9	2.9	10.99
D or 4	2.99	8.99	10.999	16.99
E or 5	8.9	18.999	10.989	30.889
C or 3	18.9	33.9	33.999	43.000
B or 2	33.99	45.889	45.99	55.0079
G or 7	45.999	54.8879	55.9	58.999
F or 6	54.9	65.9	65.9	66.001
Process	Cloud *A*	Cloud *A*	Cloud *B*	Cloud *B*
	Time in	Time out	Time in	Time out
1	0.00	2.9	2.9	10.999
5	2.9	12.99	13.0001	24.999
3	12.999	27.99	27.999	37.9
2	27.8999	39.999	40.001	49.0001
4	39.99	45.899	49.0001	55.5599
7	45.9	54.99	55.9999	59.99
6	54.99999	65.8999	65.9998	66.99999

Table 10 Tables for obtaining minimum elapsed time

Processing time on cloud	Process alpha 1	Process beta 2	Process gamma 3	…	Process *abg- n*
Cloud *A*	$t11$	$t12$	$t13$	$t14, t15$…	$t1n$
Cloud *B*	$t21$	$t22$	$t23$	$t24, t25$…	$t2n$
Cloud *C*	$t31$	$t32$	$t33$	$t34, t35$…	$t3n$

1. When the least duration of computational flow on cloud *A* is approximately as high as the top most or peak duration on cloud *B*, which means, min $1j \geq$ max $t2j$, for $j = 1, 2, 3, 4, 5, 6, 7, 8, 9, \ldots, n$.
2. When the least duration of computational flow on cloud *C* is approximately as high as the top most or peak duration on cloud *B*, which means, min $t3j \geq$ max $t2j$, for $j = 1, 2, 3, 4, 5, 6, 7, 8, 9, \ldots, n$.

Whenever one or both of the corresponding situations are met, then the Johnson's algorithm can be briefed in the given way:

Firstly: Evaluate the duration of computational flow of the processes on given 3 clouds. Whenever one or both of the corresponding situations are met, we may proceed to the next step of the process; lest, the algorithm cannot be applied.

Secondly: We may like to go ahead with two dummy clouds, say Geta or G and Heta or H, with the duration of computational flow briefed by;

(i) time Gj = time $1j$ + time $2j$, $j = 1, 2, 3, 4, 5, 6, 7, 8, 9, \ldots, n$, i.e., the duration of computational flow on cloud G is the total sum or addition of the duration of computational flow on clouds A and B.
(ii) time Hj = time $2j$ + time $3j$, $j = 1, 2, 3, 4, 5, 6, 7, 8, 9, \ldots, n$, i.e., the duration of computational flow on cloud H is the total sum or addition of the duration of computational flow on clouds B and C.

Thirdly: We strive to get the computational sequence for n processes and 2 cloud equivalent duration of computational flow with the approximate sequence GH in a very similar manner as briefed above.

6 Experimental Results

To test the model experimentally, we made use of a machine viz. Alpha P5498, Intel® Xeon® Compiler F6-3581 v3 with 9 cores of 4.11 GHz each, 19 GB RAM, 121 GB HD, Microsoft Ultimate 32 bits, V1 C++ 999, and appropriate domains to compute results.

The Hamming distance is achieved by analyzing the hierarchy of computational processes on various clouds. When the computational approximation (say Cx) obtained at the end is less than 1 or zero as far as Cbest (say) is concerned, then the value of Cx approaches to Cbest, and we obtain a parallel solution. As a corollary, when the value of Cx is approximated to be very large, e.g., 300 (which can be treated as the maxima of the curve of the process in consideration), then, in such a case, Cbest is to be treated as an entity which has no similarity with Cx. After the computation of the approximate values as shown in the graph below, we obtain a solution set $S = \{CH1\_c, CH2\_c, CH3\_c, \ldots, CHn\_c\}$ in consonance with the Cbest value. We also obtain a distributed variety in various values approximated for S, where n gives the values of the total number of computational processes. Figure 1, shows that without the deployment of Johnson's algorithm, more diversity exists than in the cloud processes. As the process steers along a streamlined path via Cbest, we witness relatively lower levels of diversity. In other words, after the deployment of Johnson's algorithm, we witness a relatively less solution diversity. Figure 1a, b showing the computational range of a diverse category of processes. A total of thirty processes.

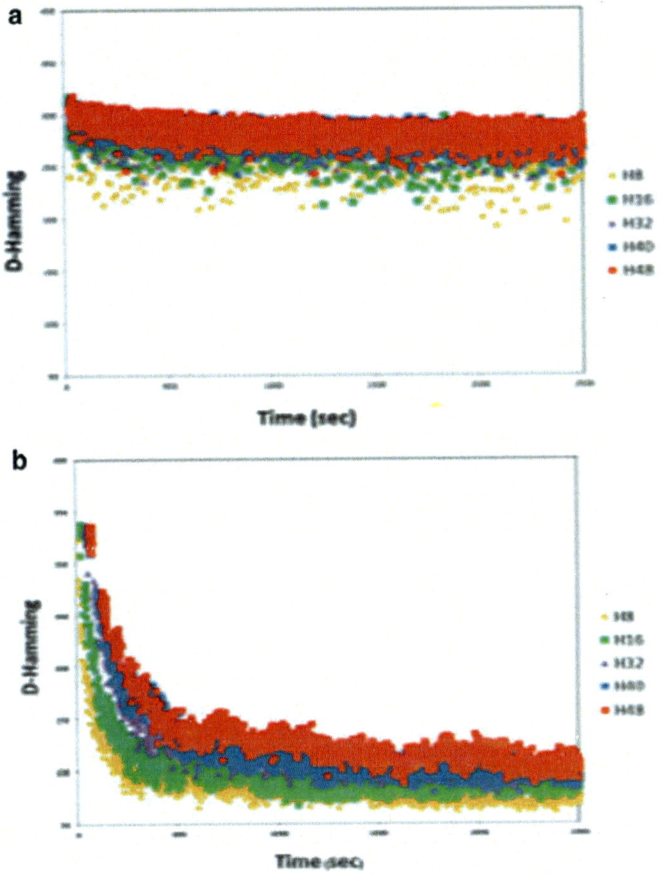

Fig. 1 **a** Excluding Johnson's algorithm. **b** Including Johnson's algorithm

7 Conclusion and Future Scope

7.1 Future Scope of the Proposed Research

In the erstwhile cloud computing systems, the cloud manager played an unavoidable role. All the processes needed to be operated systematically through the cloud manager which functioned as a workstation similar to the application programming interface but not the same. Hence, the dependency on the cloud manager remained an area of concern leaving security and privacy gaps in the entire cloud architecture. However, in case of the proposed cloud algorithms approach, the role of the cloud manager has been done away with. The working cloud algorithmic architecture also takes care of the conflict of interest between the users, authentication issues, or access privilege conflicts. Moreover, it also leads to a large-scale reduction in latency.

Having said that, the algorithmic cloud architecture provides a befitting framework to deal with data management and services. In the end, we cannot limit the scope of this approach to the briefing described above. In fact, the applications of this proposed procedure extend to the entire domain of big data analytics.

7.2 Concluding Remarks

The roadmap that this paper lays down is testimony to the fact that advanced algorithms find overwhelming applications in cloud ecosystems for the management of data. That said, the proposed cloud algorithmic approach can prove to be a great leap in attaining efficient, fast, and reliable computing services. There is no doubt in the fact that cloud service providers offer a diverse genre of purposeful services and products to users. The proposed cloud algorithmic architecture could work as an interface between cloud service providers and clients worldwide. This is because such a proposal is devoid of privacy and security threats and is also able to manage, compute, and coordinate voluminous amounts of data that flow across the digital systems every day. The need of the hour is to spark off quick research in this genre so that the rapid flow of data is captured by the combination of cloud computing and advanced algorithms giving rise to an ecosystem titled, "Cloud Algorithms".

References

1. Armbrust M, Fox A, Griffith R, Joseph AD, Katz R, Konwinski A, Lee G, Patterson D, Rabkin A, Stoica I, Zaharia, M (2010) A view of cloud computing. Commun ACM 53(4):50–58
2. Zhang Q, Cheng L, Boutaba R (2010) Cloud computing: state-of-the-art and research challenges. J Int Serv Appl 1(1):7–18
3. Buyya R, Yeo CS, Venugopal S (2008) Market-oriented cloud computing: vision, hype, and reality for delivering its services as computing utilities. In: 10th IEEE international conference on high performance computing and communications, 2008. HPCC'08. IEEE, pp 5–13
4. Alam M, Shakil KA (2013) Cloud database management system architecture. UACEE Int J Comput Sci Appl 3(1):27–31
5. Alam M (2012) Cloud algebra for handling unstructured data in cloud database management system. Int J Cloud Comput Serv Archit 2(6)
6. Wang Y (2015) Big data algebra: a rigorous approach to big data analytics and engineering. In: The 17th international conference on mathematical and computational methods in science and engineering
7. Biswas R (2015) Atrain distributed system (ADS): an infinitely scalable architecture for processing big data of any 4Vs. In: Computational intelligence for big data analysis. Springer, Cham, pp 3–54
8. Zhu M, Risch T (2011) Querying combined cloud-based and relational databases. In: 2011 international conference on cloud and service computing, pp 330–335
9. Chen D, He Y (2010) A study on secure data storage strategy in cloud computing. J Converg Inf Technol 5(7):175–179
10. Olmsted A (2019) Relational algebra for heterogeneous cloud data sources. Cloud Comput 2019:130

11. Ghrist R (2018) Homological algebra and data. Math Data 25:273
12. Abadi DJ (2009) Data management in the cloud: limitations and opportunities. IEEE Data Eng Bull 32(1):3–12
13. Sakr S, Liu A, Batista DM, Alomari M (2011) A survey of large-scale data management approaches in cloud environments. IEEE Commun Surv Tutor 13(3):311–336
14. Agrawal D, El Abbadi A, Antony S, Das S (2010) Data management challenges in cloud computing infrastructures. In: International workshop on databases in networked information systems. Springer, Berlin, pp 1–10
15. Wu W, Tsai WT, Jin C, Qi G, Luo J (2014) Test-algebra execution in a cloud environment. In: 2014 IEEE 8th international symposium on service-oriented system engineering. IEEE, pp 59–69
16. Islam MM, Morshed S, Goswami P (2013) Cloud computing: a survey on its limitations and potential solutions. Int J Comput Sci 10:159–163
17. Wang P, Gao R, Fan Z (2015) Cloud computing for cloud manufacturing: benefits and limitations. J Manufact Sci Eng 137. https://doi.org/10.1115/1.4030209
18. Ahad MA, Biswas R (2018) PPS-ADS: a framework for privacy-preserved and secured distributed system architecture for handling big data. Int J Adv Sci Eng Inf Technol 8(4):1333–1342
19. Ahad MA, Biswas R (2019) Request-based, secured and energy-efficient (RBSEE) architecture for handling IoT big data. J Inf Sci 45(2):227–238
20. Ahad MA, Biswas R (2018) Dynamic merging based small file storage (DM-SFS) architecture for efficiently storing small size files in Hadoop. Proc Comput Sci 132:1626–1635
21. Demirkol E, Mehta S, Uzsoy R (1997) A computational study of shifting bottleneck procedures for shop scheduling problems. J Heuristics 3:111–137
22. Watson J-P, Beck J, Howe AE, Whitley L (2003) Problem difficulty for tabu search in job-shop scheduling. Artif Intell 143:189–217
23. Goncalves JF, Resende MGC (2014) An extended Akers graphical method with a biased random-key genetic algorithm for job-shop scheduling. Int Trans Oper Res 21:215–246
24. Kolonko M (1999) Some new results on simulated annealing applied to the job shop scheduling problem. Eur J Oper Res 113:123–136
25. Kuo-Ling H, Ching-Jong L (2008) Ant colony optimization combined with taboo search for the job shop scheduling problem. Comput Oper Res 35:1030–1046
26. Sha D, Hsu C-Y (2006) A hybrid particle swarm optimization for job shop scheduling problem. Comput Ind Eng 51:791–808
27. Kurdi M (2015) A new hybrid island model genetic algorithm for job shop scheduling problem. Comput Ind Eng 88:273–283
28. Kurdi M (2016) An effective new island model genetic algorithm for job shop scheduling problem. Comput Oper Res 67:132–142
29. Kurdi M (2017) An improved island model memetic algorithm with a new cooperation phase for multi-objective job shop scheduling problem. Comput Ind Eng 111:183–201
30. Hernández-Ramírez L, Frausto-Solis J, Castilla-Valdez G, González-Barbosa JJ, Terán-Villanueva D, Morales-Rodríguez ML (2019) A hybrid simulated annealing for job shop scheduling problem. Int J Comb Optim Probl Inform 10:6–15
31. Amirghasemi M, Zamani R, Amirghasemi M (2015) An effective asexual genetic algorithm for solving the job shop scheduling problem. Comput Ind Eng 83:123–138
32. Nagata Y, Ono I (2018) A guided local search with iterative ejections of bottleneck operations for the job shop scheduling problem. Comput Oper Res 90:60–71
33. Peng B, Lu Z, Cheng T (2015) A tabu search/path relinking algorithm to solve the job shop scheduling problem. Comput Oper Res 53:154–164
34. Cruz-Chávez MA (2015) Neighborhood generation mechanism applied in simulated annealing to job shop scheduling problems. Int J Syst Sci 46:2673–2685
35. Aksenov V (2018) Synchronization costs in parallel programs and concurrent data structures. In: Distributed, parallel, and cluster computing [cs.DC]. ITMO University. Paris Diderot University

36. Tanenbaum AS, Bos H (2016) Modern operating systems, 4th edn. Pearson Education, Amsterdam, p 1136
37. Zhang CY, Li P, Rao Y, Guan Z (2008) A very fast TS/SA algorithm for the job shop scheduling problem. Comput Oper Res 35:282–294
38. Pongchairerks P (2019) A two-level metaheuristic algorithm for the job-shop scheduling problem. Complexity 2019:8683472
39. Asadzadeh L (2015) A local search genetic algorithm for the job shop scheduling problem with intelligent agents. Comput Ind Eng 85:376–383
40. Kurdi M (2019) An effective genetic algorithm with a critical-path-guided Giffler and Thompson crossover operator for a job shop scheduling problem. Int J Intell Syst Appl Eng 7:13–18
41. Asadzadeh L (2016) A parallel artificial bee colony algorithm for the job shop scheduling problem with a dynamic migration strategy. Comput Ind Eng 102:359–367
42. Cruz-Chávez MA, Cruz-Rosales MH, Zavala-Díaz JC, Hernández-Aguilar JA, Rodríguez-León A, Prince-Avelino JC, Luna ME, Salina OH (2019) Hybrid micro genetic multi-population algorithm with collective communication for the job shop scheduling problem. IEEE Access 7:82358–82376
43. Bryson K (2019) Global HPC leaders join to support new platform. NVIDIA, Santa Clara, CA, USA

Chapter 45
Analysis of Machine Learning and Deep Learning Classifiers to Detect and Classify Breast Cancer

Alarsh Tiwari, Ambuje Gupta, Harsh Kataria, and Gaurav Singal

1 Introduction

Cancer is a chronic disease. It can exist in various forms: leukemia, lung, prostate, breast, or skin cancer. It involves body or organ cells to behave abnormally where they divide uncontrollably and damage the surrounding area [1]. They restrict the regular processing of the body and increase in size and numbers at rapid rates. This paper talks about breast cancer, a form of cancer that affects the cells of breasts, typically in women [2]. Julienne E. Bower in [3] illustrates various symptoms in patients and survivors of breast cancer, including lump formations, fatigue, insomnia, and cognitive disturbances. Breast cancer may sometimes even show delayed symptoms as well, as discussed by C. Burgess et al. in [4], making the need for a predictive algorithm for them is even more prominent. Several studies like [5] have concluded that the severe effects of cancer are not restricted just to the physical health of an individual but also their mental well-being. The medical severity and unpredictability of cancer call for a solution that can help patients prone to it.

Machine Learning or generally referred to as ML is a system or set of algorithms capable of automatically consuming and reproducing information by experience. A system can learn knowledge, information, and data [6]. There are different ML algorithms: unsupervised, supervised, semi-supervised, etc. [7]. Unsupervised algorithms work in a restrictive and predefined manner. In contrast, supervised algorithms use existing data to train a system and learn from the labeled samples and make predictions based on the gained experience. Prediction is an output from a machine learning algorithm that essentially represents the features of a future scenario. Prediction can be in the form of a classification of certain data points of observation into a certain set of classes. Classifications have now found applications in several domains affecting

A. Tiwari (✉) · A. Gupta · H. Kataria · G. Singal
Department of Computer Science Engineering, Bennett University, Greater Noida 201310, India

G. Singal
e-mail: gaurav.singal@bennett.edu.in

A. Sheth et al. (eds.), *Intelligent Systems*, Algorithms for Intelligent Systems,
https://doi.org/10.1007/978-981-16-2248-9_45

regular lives [8]. In this paper, the authors have attempted to study and present the performances of some supervised machine learning algorithms, namely SVC, KNN, GaussianNB, Random Forest Classifier, Logistic Regression, and Decision Tree classifier.

This research aims to find the foremost classification technique on the Breast Cancer dataset on various evaluation metrics.

2 Classification Algorithm

This section discusses the various Deep Learning and Machine Learning algorithms which were tried on either dataset, the Wisconsin Diagnostic Breast Cancer (WDBC) dataset and the Wisconsin Breast Cancer Dataset (WBCD).

2.1 Convolution Neural Network

The datasets used by authors are referred to as structured data, consisting of 30 and 9 features, respectively. The authors changed the convolution layer to work in one-dimension space as the dataset used is sequential.

$$y = \text{conv1}d(z, d, b). \tag{1}$$

Equation (1) represents the 1D convolution function. In Eq. (1) z is the input to the convolution function, d stands for the filter, and the output is represented by y. The input data (z) has $L \times N$ dimensions, where features are denoted by L and N represents the number of channels. Dimensions of d are $L_F \times N \times N'$ where L_F is the filter size. It works on vector z with N channels, producing y along with N feature vectors as shown in Eq. (2).

$$y_{i'N'} = \sum_{iN} d_{iNN'} z_{i+i',N} + b_{N'}. \tag{2}$$

2.2 GRU-SVM

The authors used a neural network approach to solve the binary classification problem where they have used recurrent neural networks, more commonly known as RNN. They have amalgamated a form of RNN that is GRU (Gated recurrent unit) and Support vector machine, also known as SVM [9].

$$\sigma(X_a * [g_t - 1, y_t]). \tag{3}$$

$$a = \sigma(X_a * [g_t - 1, y_t]). \tag{4}$$

$$\widetilde{g_t} = \tanh \tanh(X * [b_t * g_t - 1, y_t]). \tag{5}$$

$$g_t = (1 - a_t) * g_t - 1 + a_t * \widetilde{g_t}. \tag{6}$$

In Eqs. (3) and (4), a is the update gate of a gated recurrent unit in a recurrent neural network. In Eqs. (5) and (6), $\widetilde{g_t}$ is the candidate value, the next RNN cell state value is represented by g_t and b_t is its reset gate of the RNN. y and X are predictor and learning parameters respectively. The loss, hence calculated, is passed through the Adam optimizer.

2.3 *Artificial Neural Network*

Artificial neural networks are a very famous tool for intelligent systems when it comes to classification and recognition tasks. It is also widely used for other cancer-related diagnostic and classification problems. We used ANN for this binary classification due to its flexibility toward complex to non-linear classification tasks. ANNs consist of layers of neurons much similar to the biological neural networks of the brain. It works in a feed-forward mechanism where it consists of an input layer, hidden layers, and the output layer. To learn from data, it uses backpropagation to adjust the individual weights of the neurons by error and cost calculation.

2.4 *Logistic Regression*

Logistic Regression [10] is generally used when the target value (dependant value) is categorical and dichotomous. However, lately, logistic regression is used for multi-class data as well. Logistic regression uses the function of the logistic sigmoid which returns a probability that can be then mapped to individual classes. Based on categories, logistic regression has three sub-types: binary regression, multinomial regression, and ordinal regression. We can explain the logistic regression by describing the standard logistic function, which is simply a sigmoid function. A sigmoid function is one that takes real values as its input and gives a value between zero and one as its output. The standard logistic function ($f(t)$) is described in Eq. (7).

$$f(t) = \frac{e^t}{e^t + 1} = \frac{1}{1 + e^t}. \tag{7}$$

Logistic Regression, unlike linear regression, does not compulsorily require continuous data. For the prediction of group membership, the logistic regression uses the log odds ratio and not the least-squares for fitting the model. This gives the researcher much more freedom when he makes use of LR over Linear Regression. Hence this approach is better for data that is not normally distributed.

2.5 K-Nearest Neighbor Classifiers

The K in KNN stands for the number of neighbors closest to a particular data point. K is the core determining factor in the functioning of a KNN algorithm. The data point is assigned to the class depending on its K closest neighbors. Each of its neighbors has an assigned class and the class with the most votes is taken as the prediction for the data point. For the sake of finding the closest points that are similar, one can make use of various measures such as the Manhattan distance and the Euclidean distance.

2.6 Gaussian Naive Bayes Classifier (GNBC)

Gaussian Naive Bayes [11] is a variation of the commonly known Naive Bayes algorithm but with a Gaussian distribution (also known as the normal distribution). This distribution makes it easy to understand the data since the estimation of mean and standard deviation becomes relatively simpler. These values majorly contribute to a better understanding of the used dataset.

Training Gaussian Naive Bayes Model involves finding the standard deviation and mean of each data point for each class. Mean $(\underline{X})$ is given by the sum of all input variable values of training data divided by the total number of instances encountered, as described in Eq. (8).

$$\text{Mean}(\underline{X}) = \frac{1}{n} * \sum_{i=1}^{N} (x_i). \tag{8}$$

The standard deviation (σ) is described in Eq. (9):

$$\text{standard deviation } (\sigma) = \sqrt{\frac{1}{n} * \sum_{i=1}^{N} (x_i - \text{mean}\{(x)\})^2}. \tag{9}$$

For prediction, the probabilities of a new data point are calculated using the Gaussian Probability Density Function (PDF) described in equation with the required parameters:

$$\mathrm{PDF}(x, \underline{X}, \sigma) = \frac{1}{\left(\sqrt{2 * \pi}\right) * \sigma} * \mathrm{e}^{-\frac{(x-\underline{X})^2}{2 * \sigma^2}}. \tag{10}$$

Here, PDF represents the Gaussian Probability Density Function, π is the numerical constant and e is Euler's number.

3 Dataset and Pre-processing

Two datasets have been studied and experimented upon in this paper. Their respective pre-processing has been explained in Sects. 3.1 and 3.2.

3.1 Wisconsin Diagnostic Breast Cancer (WDBC)

The original dataset contains 569 instances with 32 features. Among those 32 features, one of them was a unique ID, which was dropped from the dataset. Among the remaining 31 features, one was the dependent variable (diagnosis) and the rest were independent. After analyzing the dataset, it was concluded that there were very similar features; for example, radius mean, and area mean were very similar to each other because both were dependent on a single variable. Similar features would not help the model to learn effectively, instead of training with more similar features would make the model computationally heavy. To overcome this problem, correlation analysis was performed on the dataset consisting of 30 features. Correlation analysis is considered as an important tool in feature engineering. It is used to analyze the dependency of one variable over another. Its value varies from −1 to 1. Each feature is compared against all the features and the correlation score is calculated. While training a machine learning model, one would want the data to stay uncorrelated. This paper has chosen 0.9 as its threshold value, i.e., if any variable's correlation score comes out to be more than 0.9, it would be discarded. If the chosen correlation value is below 0.75, then there stands a good chance that important features would be eliminated which would ultimately result in lower accuracies. Figure 1a shows the heatmap after implementing correlation analysis for the WDBC dataset.

Figure 1a illustrates which features are closely related. The correlation of a feature with itself will always be 1. If Matrix[i, j] (Heatmap) has a light shade, then the value will be close to 1, and if the color is toward the darker side, then features are uncorrelated with each other. After implementing correlation analysis, only 20 features were left, and the algorithms were applied on them.

Table 1 shows the 20 features that were selected after performing the correlation analysis.

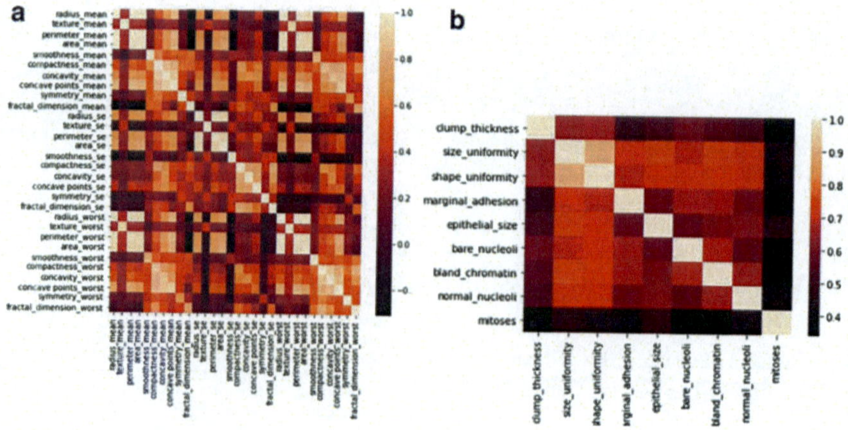

Fig. 1 Heatmaps for **a** Wisconsin Diagnostic Breast Cancer and **b** Wisconsin Breast Cancer Dataset

Table 1 Features selected after correlation analysis—Wisconsin Diagnostic Breast Cancer (WDBC) dataset

Radius mean	Texture mean	Smoothness mean	Compactness mean
Concavity mean	Symmetry mean	Fractal dimension mean	Radius Se
Texture Se	Smoothness Se	Compactness Se	Concavity Se
Concave points Se	Symmetry Se	Fractal dimension Se	Smoothness worst
Compactness worst	Concavity worst	Symmetry worst	Fractal dimension worst

The dependent variable (Y) is a binary classification, i.e., benign (not dangerous to health) and malignant (can be dangerous). The two classes' value count in the data is 357 and 212, respectively.

3.2 *Wisconsin Breast Cancer Dataset (WBCD)*

This dataset consisted of 699 instances and 10 features. One feature was the dependent variable, and the other 9 were independent. Similar pre-processing techniques were performed on this dataset as in WDBC. After correlation analysis, 8 features were left remaining on which analysis of different algorithms was performed. Figure 1b depicts a heatmap for the dataset. Table 2 shows the features that were considered after implementing correlation analysis on the WBCD dataset.

Table 2 Features selected after correlation analysis-Wisconsin Breast Cancer Dataset (WBCD)

Clump thickness	Size uniformity	Marginal adhesion	Epithelial size
Bare nucleoli	Bland chromatin	Normal nucleoli	Mitoses

There were some missing values, and after removing them, there were a total of 683 instances. The dependent variable (Y) is a binary classification, i.e., benign (not dangerous to health) and malignant (can be dangerous). The value count of the two classes in the data is as-Benign (444) Malignant (239).

4 Results

For training and testing purposes, the dataset was split in the ratio of 83:17. The analysis was performed with the help of the confusion matrix that gave us a deeper insight into the obtained results. Accuracy is one of the metrics that could be evaluated after computing the confusion matrix. It is the percentage of classes that were right predicted. Classes that get correctly classified are referred to as the true-positive rate (TPR) or Sensitivity. The True-Negative Rate (TNR) or Specificity helps us to analyze how many negative classes are classified correctly. The Receiver Operating Characteristic (ROC) curve has been used to analyze the result, which is a trade-off between the true-positive rate and the false-positive rate (FPR). The false-positive rate is calculated as, (1 − Specificity). Equations (11), (12), (13) and (14) describe accuracy, sensitivity, specificity, and the false-positive rate, respectively. In these equations, T.P, T.N, F.P, F.N refers to true positive, true negative, false positive, and false negative, respectively.

$$\text{Accuracy} = \frac{\text{T.P} + \text{T.N}}{\text{T.P} + \text{T.N} + \text{F.P} + \text{F.N}}. \tag{11}$$

$$\text{Sensitivity} = \frac{\text{T.P}}{\text{T.P} + \text{F.N}}. \tag{12}$$

$$\text{Specificity} = \frac{\text{T.N}}{\text{T.N} + \text{F.P}}. \tag{13}$$

$$\text{False Positive Rate} = \frac{\text{F.P}}{\text{T.N} + \text{F.P}}. \tag{14}$$

Usually, for binary classification problems, ROC curve analysis is performed as it tests the potential of the classifier to differentiate between the classes. Let us look at the results of various machine learning, and deep learning algorithms applied to the datasets. Figure 2a, b represent the comparison of various machine learning algorithms on "Wisconsin Diagnostic Breast Cancer" and "Wisconsin Breast Cancer

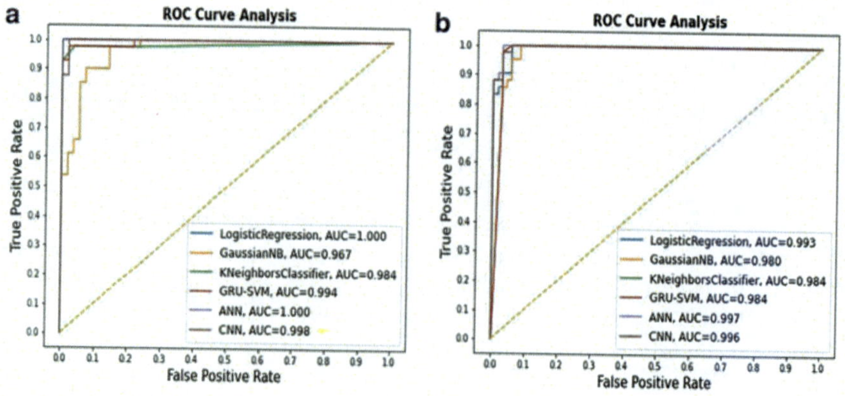

Fig. 2 Comparison of different Machine Learning and Deep Learning models in terms of ROC-AUC on **a** Wisconsin Diagnostic Breast Cancer (WDBC) dataset and **b** Wisconsin Breast Cancer Dataset (WBCD)

Dataset" dataset, respectively, and the evaluation metrics considered is the ROC curve.

From Fig. 2a, we can analyze and draw an inference that ANN and Logistic Regression has AUC = 1.0, which means that these algorithms are able to separate the negative and the positive classes perfectly. From Fig. 2b, we can observe that ANN has an AUC score of 0.997. Figure 3 shows a trade-off between validation loss and epochs in the analysis. The authors have also considered different parameters like accuracies, false-positive rate (FPR), and specificity of the considered models on either dataset as their evaluation metrics. Different accuracies for different models were observed for either dataset. Table 3 represents hyperparameters used for training and experimental results performed on the WDBC dataset. Artificial Neural Network (ANN) got the

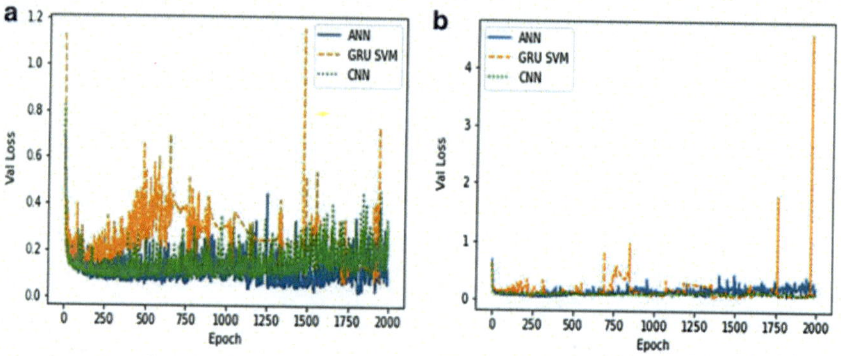

Fig. 3 Val Loss versus Epochs **a** Wisconsin Diagnostic Breast Cancer (WDBC) dataset and **b** Wisconsin Breast Cancer Dataset (WBCD)

Table 3 Hyperparameters and experimental results of WDBC dataset

Parameters	CNN	GRU-SVM	ANN	LR	KNN	GNBC
Batch size	20	20	20	NA	NA	NA
Epochs	2000	2000	2000	NA	NA	NA
Normalization	NA	L2	NA	L2	L2	NA
FPR	1.018	1.01	0	0.017	0	0.071
Specificity	0.982	0.982	1	0.982	1	0.928
Accuracy	0.9896	0.973	1	0.989	0.969	0.917

Table 4 Hyperparameters and experimental results of WBCD

Parameters	CNN	GRU-SVM	ANN	LR	KNN	GNBC
Batch size	16	16	16	NA	NA	NA
Epochs	2000	2000	2000	NA	NA	NA
Normalization	NA	L2	NA	L2	L2	NA
FPR	0.053	0.026	0.026	0.053	0.026	0.08
Specificity	0.9466	0.9733	0.973	0.9466	0.9733	0.92
Accuracy	0.9743	0.9829	0.9829	0.940	0.974	0.948

highest accuracy of one. WBCD dataset hyperparameters and experimental results are shown in Table 4. Here also, ANN has the best accuracy.

5 Conclusion

In the presented work, we compared 6 widely used ML and DL methods for classification, on two separate datasets. Their respective performances were evaluated on several evaluation metrics; we wanted a classifier that could carry out the required binary classification with consistency over these metrics on either dataset. We concluded that ANN performed best across all the considered evaluation metrics for the WDBC dataset and was ranked a close second for the WBCD. Such consistency was not observed in the respective performances of other algorithms.

References

1. Ruoslahti E (1996) How cancer spreads. Sci Am 275(3):72–77
2. Kelsey JL, Gammon MD, John EM (1993) Reproductive factors and breast cancer. Epidemiol Rev 15(1):36

3. Bower JE (2008) Behavioral symptoms in breast cancer patients and survivors: fatigue, insomnia, depression, and cognitive disturbance. J Clin Oncol Off J Am Soc Clin Oncol 26(5):768
4. Burgess C, Hunter MS, Ramirez AJ (2001) A qualitative study of delay among women reporting symptoms of breast cancer. Br J Gen Pract 51(473):967–971
5. Rasic DT, Belik S-L, Bolton JM, Chochinov HM, Sareen J (2008) Cancer, mental disorders, suicidal ideation and attempts in a large community sample. Psychol Oncol J Phychol Soc Behav Dim Cancer 17(7):660–667
6. Alpaydin E (2020) Introduction to machine learning. MIT Press
7. Ayodele TO (2010) Types of machine learning algorithms. New Adv Mach Learn 3:19–48
8. Kotsiantis SB, Zaharakis I, Pintelas P (2007) Supervised machine learning: a review of classification techniques. Emerg Artif Intell Appl Comput Eng 160(1):3–24
9. Agarap AFM (2018) A neural network architecture combining gated recurrent unit (GRU) and support vector machine (SVM) for intrusion detection in network traffic data. In: Proceedings of the 2018 10th international conference on machine learning and computing, 2018, pp 26–30
10. Hosmer DW Jr, Lemeshow S, Sturdivant RX (2013) Applied logistic regression, vol 398. Wiley
11. Gayathri B, Sumathi C (2016) An automated technique using gaussian naive bayes classifier to classify breast cancer. Int J Comput Appl 148(6):16–21

Chapter 46
Clinical Decision Support for Primary Health Centers to Combat COVID-19 Pandemic

Vinu Sherimon, Sherimon Puliprathu Cherian, Renchi Mathew, Sandeep M. Kumar, Rahul V. Nair, Khalid Shaikh, Hilal Khalid Al Ghafri, and Huda Salim Al Shuaily

1 Introduction

Severe acute respiratory syndrome coronavirus 2 (SARS-CoV-2) is increasing anomalously around the universe. The virus was first reported in December 2019 from Wuhan, the capital of the Hubei Province in China, and on March 11, 2020, the World Health Organization (WHO 2020) announced this threat as a pandemic [1]. As of 13 November 2020, according to WHO (2020), the worldwide number of live cases is 14,618,008 and the number of deaths is 1,301,818 [2]. Countries have issued travel bans, and 14–21 days of full suspension have been implemented by several nations. The sad reality is that the majority of the people are afraid to come out of

V. Sherimon (✉) · H. S. Al Shuaily
Department of IT, University of Technology and Applied Sciences, P.O. 74, Muscat, Sultanate of Oman

H. S. Al Shuaily
e-mail: huda.alshuaily@hct.edu.om

S. Puliprathu Cherian
Faculty of Computer Studies, Arab Open University, P.O. 1596, Muscat, Sultanate of Oman
e-mail: sherimon@aou.edu.om

R. Mathew · H. K. Al Ghafri
Internal Medicine, Royal Oman Police Hospital, P.O. 116, Muscat, Sultanate of Oman

S. M. Kumar
Family Medicine and Tropical Medicine, Royal Oman Police Hospital, P.O. 116, Muscat, Sultanate of Oman

R. V. Nair
Emergency Department, Royal Oman Police Hospital, P.O. 116, Muscat, Sultanate of Oman

K. Shaikh
Internal Medicine and Diabetes, Royal Oman Police Hospital, P.O. 116, Muscat, Sultanate of Oman

A. Sheth et al. (eds.), *Intelligent Systems*, Algorithms for Intelligent Systems,
https://doi.org/10.1007/978-981-16-2248-9_46

the houses and have depression, concern, and discomfort. The required support and direction are provided by the appropriate authorities of almost all countries to its people.

The initial case of COVID-19 was registered on February 24, 2020, by the Ministry of Health (MOH) in the Sultanate of Oman [3]. There are now 8066 active patients, and 1326 deaths happened in Oman as of 13 November 2020 [3]. As part of the properly designed interventions to stop the spread of the virus, the authorities imposed several lockdowns. The health industry in the Sultanate of Oman is administered by MOH. Royal Oman Police (ROP) hospital is always at the frontline of serving the citizens of Oman. ROP manages multiple satellite clinics in the remote areas and villages of the Sultanate. To make a diagnosis in critical situations, these centers do not have specialized physicians. Often, some of these centers may not have the proper infrastructure to provide suitable treatments. Because of fear, the majority of the people are hesitant to move to clinics in the capital city to access proficient support and treatment.

In recent decades, the healthcare sector has been involved in the implementation of decision support systems, including routine clinical practices. Clinical decision support systems (CDSS) depict knowledge of the field of application and provide some reasoning strategies to derive new knowledge from current knowledge [4, 5]. The knowledge that a CDSS utilizes resides in a knowledge base [4]. One study [6] relates to published solutions, methods, and insights on artificial intelligence and machine learning, aimed at advancing solutions for the extraction, collection, management, and analysis of clinical data to improve decisions and pave the way for healthcare discovery. Another study [7] outlines the clinical decision support system (CDSS) as well as the obstacles, implementation, and efficacy of improving clinical practice in Saudi Arabia's healthcare sector, using real examples.

In this context, we suggest a clinical decision support system (CDSS) with the adoption of teleconference/video conference facilities to render simple, efficient diagnosis, and treatment at primary health centers/satellite clinics of the Royal Oman Police. Our CDSS will be based on artificial intelligence techniques. We use ontology to represent domain knowledge and clinical guidelines [8]. Ontology is a formal representation of domain knowledge. It includes concepts (related to medical knowledge), attributes, and semantic relationships between these concepts. Ontologies help to represent the standard medical terms accurately, allow efficient knowledge sharing and reuse, and support automatic reasoning [8]. Healthcare workers at the Royal Oman Police primary health centers/satellite clinics will be able to provide care with expert assistance through the advice of the proposed intelligent CDSS.

The remainder of the paper is outlined as follows: Sect. 2 includes materials and methods. The data collection, ontology development, and implementation of CDSS are presented in this section. Section 3 presents the results and discussion. In Sect. 4, the conclusion and future is presented followed by references.

2 Materials and Methods

The section includes the details of data collection, the creation of ontological classes, and the representation of clinical guidelines in the form of rules. The implementation of ontology is also presented in this section.

2.1 *Data Collection*

Data related to signs, causes, hazards, test data, related health issues, potential problems, seriousness, etc., of coronavirus has been collected from primary and secondary sources. The main data was gathered from the ROP hospital medical team and the secondary data was gathered from articles published in reputed journals. A questionnaire was prepared based on the collected information. Several online meetings were held with the medical team to test the questionnaire.

At the outset, several variants of the clinical guidelines, such as the National Clinical Management Protocol for Hospitalized Patients with COVID-19 and the ICU protocol for the management of COVID-19 [9, 10] were issued by the Ministry of Health (MOH), Oman. We included updated guidelines, titled COVID-19 Infection Guideline-10 for Hospitals, Primary and Private Health Care Institutes (updated in April 2020) to build our framework.

2.2 *Ontology Development*

The top concept is defined initially when ontologies are used in the process of domain modeling. Owl: Thing represents the top concept of a domain. $C = \{C_1, C_2, \ldots, C_n\}$ defines the classes related to the different domain concepts. Let $C_n = \{s_1, s_2, \ldots, s_m\}$, where s_i $(i = 1 \text{ to } m)$ is used to represent the sub-classes in every concept class C_n [11].

The domain concepts in COVID-19 domain are given in Fig. 1. The taxonomy includes all the related concepts.

The *Patient* class includes all the cases of the patient. Various symptoms of SARS-CoV-2 are included in the *Symptom* class. As per the guidelines of the Sultanate of Oman and WHO [9, 10], all the possible symptoms are characterized as sub-classes of *Symptom* class. The *Background_history* class comprises several sub-classes to characterize the epidemiological relation of patients. *RiskFactor* class contains sub-classes that reflect significant predictors for risky cases (for example, patients with comorbidities). Initially, a patient will be classified into one of the groups such as *Not suspected, Suspected, Probable, or Confirmed*. These categories are included as the sub-classes of the *Diagnosis* class. Ontology reasoner will classify a patient into one of the categories automatically. Later, after the diagnosis by the doctor

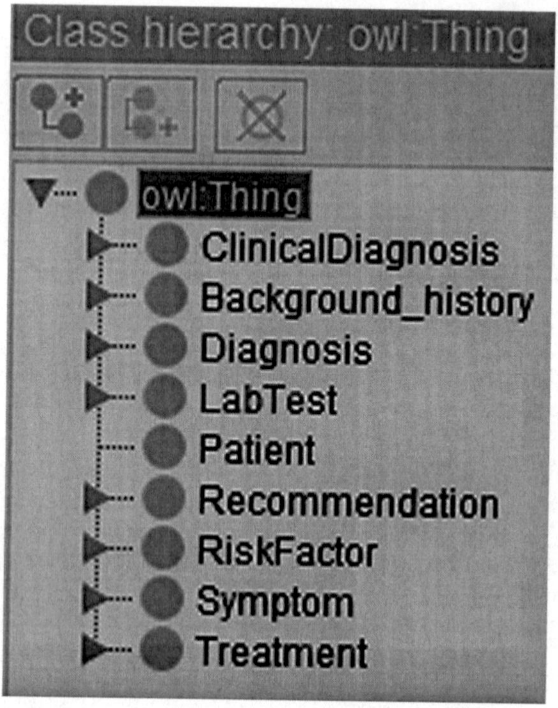

Fig. 1 Taxonomy of main domain concepts [12]

and based on the physical conditions of the patient, the patient will be put into one of the categories such as: mild, moderate, severe, or critical. They are included as sub-classes of *ClinicalDiagnosis* class.

An ontology language, Web Ontology Language (OWL) is used to encode the ontology [13]. The ontology rules are created using Semantic Web Rule Language (SWRL) [14]. As part of the DARPA Agent Markup Language Program (DAML), SWRL was released in 2003 [15]. OWL cannot be used to represent rules such as *If.. Then.. Else..* With the help of SWRL, such weaknesses of OWL can be eliminated [13].

Rule #1

```
Confirmed_case(?x) ^ has_symptom(?x, Breathing_Difficulty) ^
hasRespiratoryRateValue(?x, ?rp) ^ swrlb:lessThanOrEqual(?rp,
30) ^ swrlb:greaterThanOrEqual(?rp, 24) ^ hasCXRResult(?x,
Bilateral_Infiltrates) -> Severe(?x)
```

Rule #1 is used to further diagnose a confirmed case with breathing difficulty. The respiratory rate value and the chest X-ray result are checked to classify the patient into the *Severe* category. There are 4 atoms in the antecedent part of Rule 1. All COVID-19 confirmed cases from the ontology are retrieved using the predicate *Confirmed_case*(?x) and stored into the variable 'x'. To check whether the concerned patient has the symptom of breathing difficulty, the predicate

$has\_symptom(?p, Breathing\_Difficulty)$ is used. $Breathing\_Difficulty$ is an individual of the *Symptom* class. $hasRespiratoryValue(?x, ?rp)$ and the SWRL built-ins `swrlb:lessThanOrEqual` and `swrlb:greaterThanOrEqual` checks whether the respiratory rate is between 24 and 30. The predicate $hasCXRResult(?x, Bilateral\_Infiltrates)$ is used to check the outcome of the chest X-ray. If the antecedents of the rule are satisfied, then the predicate $Severe(?x)$ will automatically classify the patient as the instance of that class.

2.3 Implementation of CDSS

An interface is designed exclusively for the public for self-assessment in case of suspicion of COVID-19. The user must enter the symptoms and if there is a risk of infection suspected, and the system will recommend the next course of action to be taken. Figure 2 represents Part A—symptoms of the self-assessment form. For

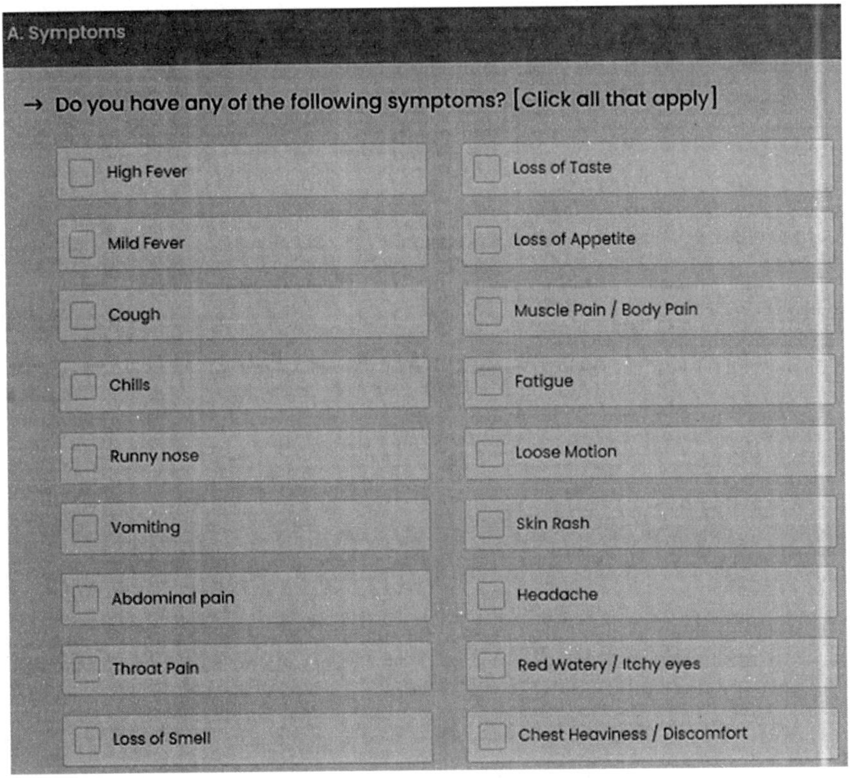

Fig. 2 Part A—symptoms section of patient self-assessment page

example, if the probability is high, the suspected person will be guided to the nearest COVID care center/hospital/health clinic.

We designed separate interfaces for different categories of medical staff. Once the suspected patient visits the satellite clinic, the triage staff will enter the vital details and other readings. Then, the patient will be directed to the doctor for further diagnosis. Based on the current medical condition of the patient, the system provides relevant suggestions to the doctors, suggests laboratory tests/X-ray/ECG, etc., to be done. Also, the system provides alerts to doctors. If the patient's condition is critical, the peripheral clinic doctor will be able to contact the expert doctors in the hospitals in the cities through teleconferencing. We have used WebRTC to implement the teleconferencing facility [16]. It allows the doctors to discuss the condition of the patient and exchange necessary patient documents for further diagnosis.

3 Results and Discussion

Once a user inputs the symptoms as given in Fig. 3, the system provides suitable recommendations as given in Fig. 4. As the symptoms chosen are mild, the recommendation is to call and make an appointment at the nearest health COVID care center.

When the patient arrives at the health center, the nurse examines the patient and does the primary assessment (Fig. 5). Based on the values such as body temperature,

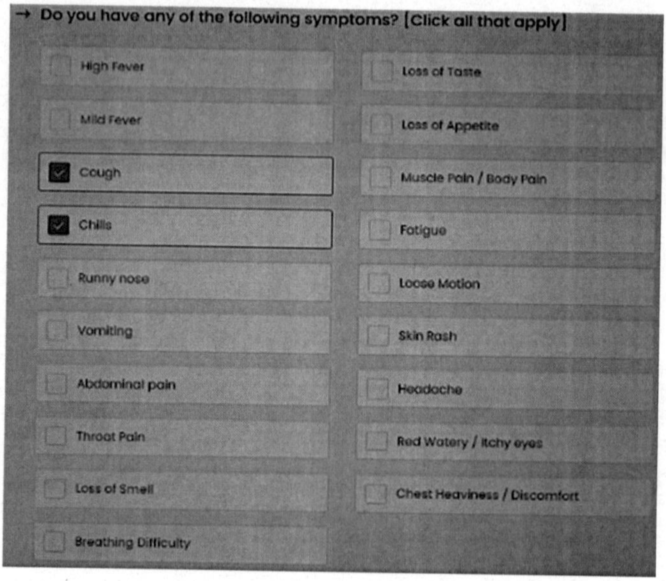

Fig. 3 Self-assessment page

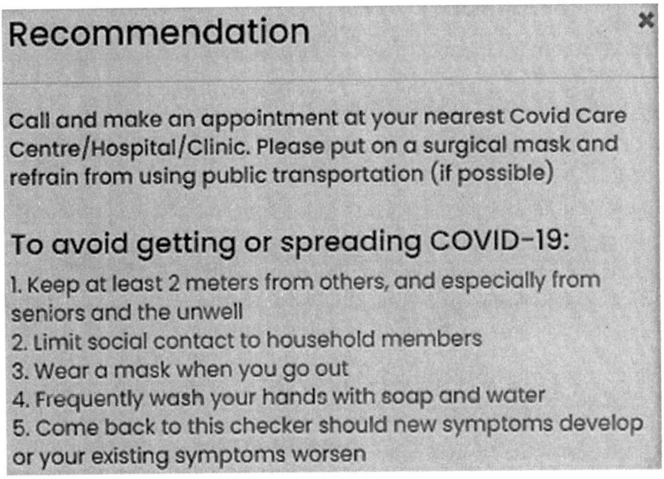

Fig. 4 Mild category recommendations

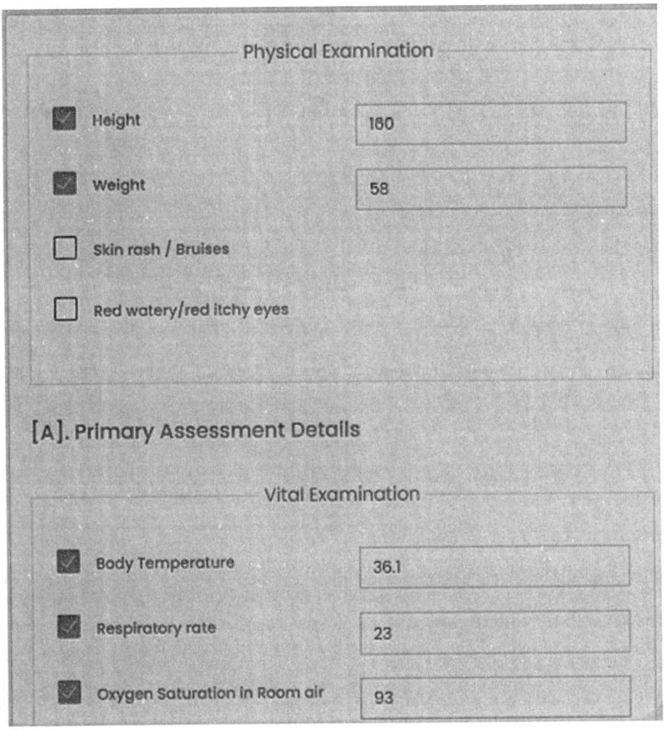

Fig. 5 Nurse examination at the triage

respiratory rate, and oxygen saturation, the system will show recommendations to the nurse too. Once the patient moves to the Doctor, the system shows the recommendations regarding symptomatic treatment (Fig. 6). The condition of the patient is also given as an alert by the system as shown in Fig. 7. He/she can view the pre-assessment done in the triage and patient history. Based on all the data, the doctor may suggest laboratory tests or to take X-ray, ECG, etc. Later, based on the test results, further treatment suggestions and the condition of the patient will be shown automatically by the system.

If the patient is in a critical condition, the doctor in the primary health center has the facility to call expert doctors from other main hospitals to get an opinion about the patient's treatments. WebRTC-based teleconferencing interface is also provided as part of our CDSS (Fig. 8).

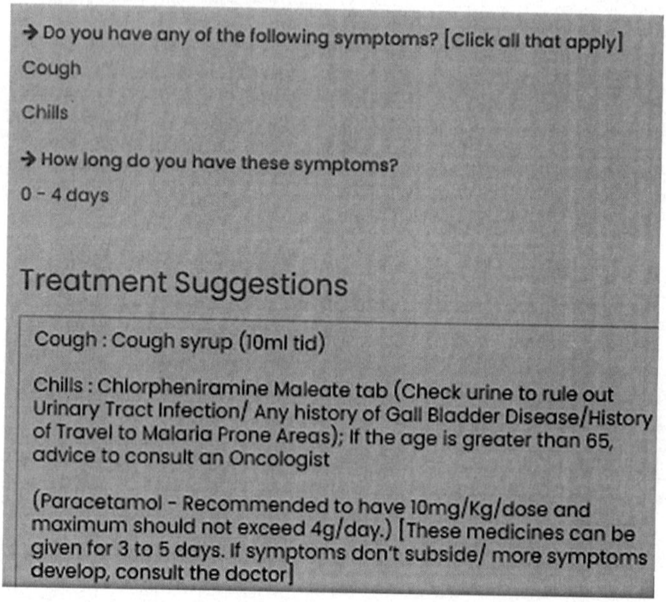

Fig. 6 System recommendation for doctor

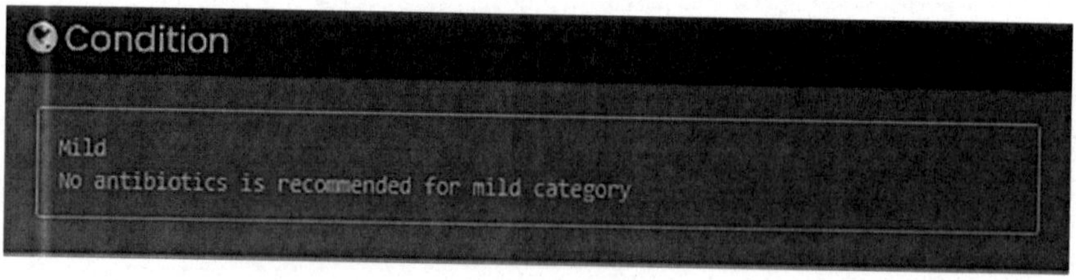

Fig. 7 Automatic categorization of patient into mild category

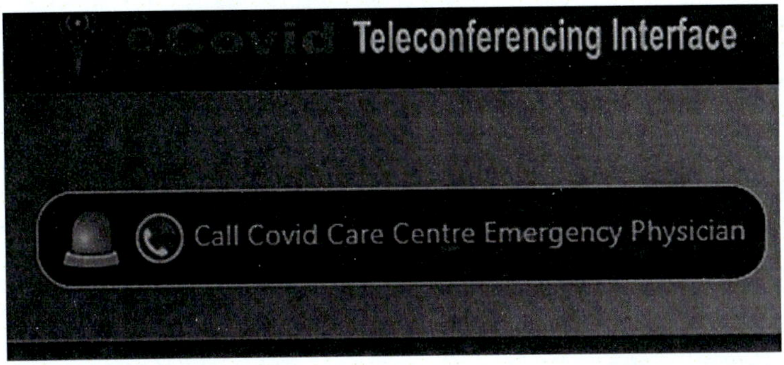

Fig. 8 Interface of teleconferencing

4 Conclusion and Future

This article presents an intelligent CDSS which can assist the primary health centers/satellite clinics of ROP hospital in Oman in diagnosing coronavirus without the aid of expert medical practitioners. Our CDSS has two modules—one module is for the public to do a self-assessment regarding COVID-19. The other module can be used by the primary health centers/satellite clinics of ROP in Oman. The teleconferencing facility is built into the proposed CDSS, where it is possible to contact specialist physicians from a referral hospital or any part of the world. For people in Oman to have a better standard of treatment, the suggested CDSS can be effectively used. To suit other sudden pandemics, it can also be personalized. In the primary health centers/satellite clinics in Oman, prompt follow-up of such critical cases can be performed. We plan to enhance our CDSS by including machine learning-based X-ray prediction.

Funding The research leading to these results has received Research Project Funding from The Research Council (TRC) of the Sultanate of Oman, under Commissioned Research Program, Contract No. TRC/CRP/AOU/COVID-19/20/13.

References

1. World Health Organization (2020) WHO director-general's opening remarks at the media briefing on COVID-19, 11 Mar 2020. Available online: https://www.who.int/dg/speeches/detail/who-director-general-s-opening-remarks-at-the-media-briefing-on-covid-19---11-march-2020. Accessed 27 Aug 2020
2. WHO Coronavirus Disease (COVID-19) Dashboard. https://covid19.who.int/. Accessed 27 Aug 2020
3. Ministry of Health eHealth Portal. MOH registers first two novel coronavirus (COVID-2019) in Oman. Available online: https://www.moh.gov.om/en/-/--1226. Accessed 27 Aug 2020

4. Sherimon PC et al (2016) An ontology-based clinical decision support system for diabetic patients. Arab J Sci Eng 41(3):1145–1160. https://doi.org/10.1007/s13369-015-1959-4
5. Sherimon PC et al (2014) Adaptive questionnaire ontology in gathering patient medical history in diabetes domain. In: Lecture notes in electrical engineering, vol 285. Springer, Singapore, pp 453–460
6. Ahmed Z et al (2020) Artificial intelligence with multi-functional machine learning platform development for better healthcare and precision medicine. Database
7. Sutton RT et al (2020) An overview of clinical decision support systems: benefits, risks, and strategies for success. NPJ Dig Med 3(1):1–10
8. Akerkar R (2009) *Foundations of the semantic web: XML, RDF and ontology*. Alpha Science International, Ltd
9. National Clinical Management Protocol for Hospitalized Patients with Covid-19. Available online: http://ghc.sa/ar-sa/Documents/Oman.pdf. Accessed 27 Aug 2020
10. World Health Organization (2020) Clinical management of severe acute respiratory infection (SARI) when COVID-19 disease is suspected: interim guidance, 13 Mar 2020 (No. WHO/2019-nCoV/clinical/2020.4). Accessed 27 Aug 2020
11. Jin Z (2018) Environment modeling-based requirements engineering for software intensive systems. Morgan Kaufmann
12. Sherimon V et al (2020) Covid-19 ontology engineering-knowledge modeling of severe acute respiratory syndrome coronavirus 2 (SARS-CoV-2). Int J Adv Comput Sci Appl 117–123
13. Zhai Z, Martínez Ortega JF, Lucas Martínez N, Castillejo P (2018) A rule-based reasoner for underwater robots using OWL and SWRL. Sensors (Basel) 18(10):3481. https://doi.org/10.3390/s18103481(2018)
14. SWRL (2021) A semantic web rule language combining OWL and RuleML. Available online: https://www.w3.org/Submission/SWRL/. Accessed 17 Jan 2021
15. DAML Program (2021). Available online: www.daml.org. Accessed 17 Jan 2021
16. WebRTC (2020) Real-time communication for the web. Available online: https://webrtc.org/. Accessed 27 Aug 2020

Author Index

A. Sheth et al. (eds.), *Intelligent Systems*, Algorithms for Intelligent Systems, https://doi.org/10.1007/978-981-16-2248-9